HARNESSING PLANT-STIMULATING MOLECULES

HARNESSING PLANT-STIMULATING MOLECULES

Innovations in Stress Biology and Crop Resilience

Edited by

M. Iqbal R. Khan

Department of Botany, Jamia Hamdard, New Delhi, Delhi, India

Sudhakar Reddy Palakolanu

International Crops Research Institute for the Semi-Arid Tropics (ICRISAT), Hyderabad, Telangana, India

ELSEVIER

ACADEMIC PRESS

An imprint of Elsevier

Academic Press is an imprint of Elsevier
125 London Wall, London EC2Y 5AS, United Kingdom
50 Hampshire Street, 5th Floor, Cambridge, MA 02139, United States

For accessibility purposes, images in electronic versions of this book are accompanied by alt text descriptions provided by Elsevier. For more information, see https://www.elsevier.com/about/accessibility.

Books and Journals published by Elsevier comply with applicable product safety requirements. For any product safety concerns or queries, please contact our authorised representative, Elsevier B.V., at productsafety@elsevier.com.

Publisher's note: Elsevier takes a neutral position with respect to territorial disputes or jurisdictional claims in its published content, including in maps and institutional affiliations.

Notices

Knowledge and best practice in this field are constantly changing. As new research and experience broaden our understanding, changes in research methods, professional practices, or medical treatment may become necessary.

Practitioners and researchers must always rely on their own experience and knowledge in evaluating and using any information, methods, compounds, or experiments described herein. In using such information or methods they should be mindful of their own safety and the safety of others, including parties for whom they have a professional responsibility.

To the fullest extent of the law, neither the Publisher nor the authors, contributors, or editors, assume any liability for any injury and/or damage to persons or property as a matter of products liability, negligence or otherwise, or from any use or operation of any methods, products, instructions, or ideas contained in the material herein.

ISBN: 978-0-443-44994-9

Publisher: Andre Wolff
Acquisitions Editor: Chiara Giglio
Editorial Project Manager: Shreya Jain
Production Project Manager: M. Balakrishnan
Cover Designer: Mark Rogers

Typeset by MPS Limited, Chennai, India

Contents

6. Endophyte-driven strategies for abiotic stress tolerance in plants

Nagarathnamma Yammanuri, Vishnu Babu Vallepu, Krishna Kumar Guduru, Siva Bolupalli, Shajidabi Shaik, Chandra Sekhar Akila and Chandra Obul Reddy Puli

7. Strigolactone-mediated signaling pathways in sorghum: molecular insights into *Striga* resistance

Piyush Panchabhai, Sirisha Kaniganti, Hithaishy Aduru and Sudhakar Reddy Palakolanu

8. Role of reactive carbonyl compounds in plant stress signaling

Prajjal Dey, Venkata H. Limmada, Preethi Vijayaraghavareddy, Sapalika Dogra, Rajneesh Kumar, Ratnakumar Pasala, Chita Sudhakar and Amaranatha R. Vennapusa

9. Integrating gene editing and plant stimulants to develop virus-resilient crops

Ashwini Talakayala, Are Harini, Mohana Durga Prabhakaran, Sandhya Jagtap, Lali Lingfa, Mallikarjuna Garladinne and Srinivas Ankanagari

10. Biotechnological innovations for sustainable agricultur

Dubbala Prshanthi Divya Vani, Narkhede Gopal Wasudeo and Shabir H. Wani

11. Mobile signals in systemic stress response: RNA, peptides, and hormonal interactions

Vaishnavi Tripathi, Gargi Bhardwaj, Sunil Kumar Jaiswal and Sunil Kumar Yadav

12. Heterotrimeric G-protein subunits in plant signal transduction: roles in plant development and stress responses

C. Vismaya, Jismon Jose and Swarup Roy Choudhury

13. CRISPR–Cas-enabled accelerated breeding pipelines for engineering climate-resilient crops under abiotic stress conditions

Aditi Dutt, Arnika Keithellakpam, Adity Laskar, Anita Kachari, Channakeshavaiah Chikkaputtaiah and Banashree Saikia

List of contributors

Hithaishy Aduru International Crops Research Institute for the Semi-Arid Tropics (ICRISAT), Hyderabad, Telangana, India

Parinita Agarwal Halophyte Biology and Biotechnology Division, CSIR-Central Salt and Marine Chemicals Research Institute (CSIR-CSMCRI), Council of Scientific and Industrial Research (CSIR), Gijubhai Badheka Marg, Bhavnagar, Gujarat, India

Pradeep K. Agarwal Halophyte Biology and Biotechnology Division, CSIR-Central Salt and Marine Chemicals Research Institute (CSIR-CSMCRI), Council of Scientific and Industrial Research (CSIR), Gijubhai Badheka Marg, Bhavnagar, Gujarat, India; Academy of Scientific and Innovative Research (AcSIR), Ghaziabad, Uttar Pradesh, India

Chandra Sekhar Akila Department of Botany, Yogi Vemana University, Kadapa, Andhra Pradesh, India

Iqra Aleem Department of Botany, Jamia Hamdard, New Delhi, Delhi, India

Srinivas Ankanagari Department of Genetics, Osmania University, Hyderabad, Telangana, India

Gargi Bhardwaj School of Bioscience and Technology, Galgotias University, Greater Noida, Uttar Pradesh, India

Siva Bolupalli Department of Botany, Yogi Vemana University, Kadapa, Andhra Pradesh, India

Channakeshavaiah Chikkaputtaiah Center for Biotechnology, Biological Sciences and Technology Division, CSIR-North East Institute of Science and Technology (CSIR-NEIST), Jorhat, Assam, India; Academy of Scientific and Innovative Research (AcSIR), Ghaziabad, Uttar Pradesh, India

Anjali Chittora Halophyte Biology and Biotechnology Division, CSIR-Central Salt and Marine Chemicals Research Institute (CSIR-CSMCRI), Council of Scientific and Industrial Research (CSIR), Gijubhai Badheka Marg, Bhavnagar, Gujarat, India; Academy of Scientific and Innovative Research (AcSIR), Ghaziabad, Uttar Pradesh, India

Km Sandhya Devi Department of Botany, University of Allahabad, Prayagraj, Uttar Pradesh, India

Prajjal Dey Division of Basic Sciences and Humanities, Faculty of Agriculture, Sher-e-Kashmir University of Agricultural Sciences and Technology, Wadura, Jammu and Kashmir, India

Sapalika Dogra Division of Basic Sciences and Humanities, Faculty of Agriculture, Sher-e-Kashmir University of Agricultural Sciences and Technology, Wadura, Jammu and Kashmir, India

Aditi Dutt Center for Biotechnology, Biological Sciences and Technology Division, CSIR-North East Institute of Science and Technology (CSIR-NEIST), Jorhat, Assam, India

Farheen Fatima Department of Botany, University of Allahabad, Prayagraj, Uttar Pradesh, India

Mallikarjuna Garladinne Plant Molecular Biology Laboratory, Agri Biotech Foundation, Hyderabad, Telangana, India

Krishna Kumar Guduru Department of Botany, Yogi Vemana University, Kadapa, Andhra Pradesh, India

Aayushi Gupta Halophyte Biology and Biotechnology Division, CSIR-Central Salt and Marine Chemicals Research Institute (CSIR-CSMCRI), Council of Scientific and Industrial Research (CSIR), Gijubhai Badheka Marg, Bhavnagar, Gujarat, India

Are Harini Department of Genetics, Osmania University, Hyderabad, Telangana, India

Noushina Iqbal Department of Botany, School of Chemical and Life Sciences, Jamia Hamdard, New Delhi, Delhi, India

Sandhya Jagtap Department of Genetics, Osmania University, Hyderabad, Telangana, India

Sunil Kumar Jaiswal School of Bioscience and Technology, Galgotias University, Greater Noida, Uttar Pradesh, India

Tabassum Jehan Department of Botany, Zakir Husain Delhi College, University of Delhi, New Delhi, Delhi, India

Anshu Jha Department of Botany, School of Chemical and Life Sciences, Jamia Hamdard, New Delhi, Delhi, India

Jismon Jose Department of Biology, Indian Institute of Science Education and Research (IISER), Tirupati, Andhra Pradesh, India

Anita Kachari Center for Biotechnology, Biological Sciences and Technology Division, CSIR-North East Institute of Science and Technology (CSIR-NEIST), Jorhat, Assam, India

Sirisha Kaniganti International Institute of Tropical Agriculture (IITA), Nairobi, Kenya

Gurleen Kaur Department of Botany, Jamia Hamdard, New Delhi, Delhi, India

Harmanjit Kaur Department of Botany, University of Allahabad, Prayagraj, Uttar Pradesh, India

Arnika Keithellakpam Center for Biotechnology, Biological Sciences and Technology Division, CSIR-North East Institute of Science and Technology (CSIR-NEIST), Jorhat, Assam, India

M. Iqbal R. Khan Department of Botany, Jamia Hamdard, New Delhi, Delhi, India

Amresh Kumar ICAR-National Institute for Plant Biotechnology, New Delhi, Delhi, India

Rajneesh Kumar Division of Genetics and Plant Breeding, Faculty of Agriculture, Sher-e-Kashmir University of Agricultural Sciences and Technology, Wadura, Jammu and Kashmir, India

Sweta Kumari ICAR-Indian Agricultural Research Institute, New Delhi, Delhi, India

Adity Laskar Center for Biotechnology, Biological Sciences and Technology Division, CSIR-North East Institute of Science and Technology (CSIR-NEIST), Jorhat, Assam, India

Venkata H. Limmada Department of Agronomy, Purdue University, IN, United States

Lali Lingfa Department of Genetics, Osmania University, Hyderabad, Telangana, India

Moksh Mahajan Department of Botany, Jamia Hamdard, New Delhi, Delhi, India

Chirag Maheshwari ICAR-Indian Agricultural Research Institute, New Delhi, Delhi, India

Nand Lal Meena ICAR-National Bureau of Plant Genetic Resources, New Delhi, Delhi, India

Sudhakar Reddy Palakolanu International Crops Research Institute for the Semi-Arid Tropics (ICRISAT), Hyderabad, Telangana, India

Piyush Panchabhai International Crops Research Institute for the Semi-Arid Tropics (ICRISAT), Hyderabad, Telangana, India

Ratnakumar Pasala ICAR-Indian Institute of Oilseeds Research, Hyderabad, Telangana, India

Mohana Durga Prabhakaran School of Life Sciences, Bharatidasan University, Tiruchirapalli, Tamilnadu, India

V. Prathap ICAR-Indian Institute of Oil Palm Research, Pedavegi, Andhra Pradesh, India

Chandra Obul Reddy Puli Department of Botany, Yogi Vemana University, Kadapa, Andhra Pradesh, India

Swarup Roy Choudhury Department of Biology, Indian Institute of Science Education and Research (IISER), Tirupati, Andhra Pradesh, India

Banashree Saikia Center for Biotechnology, Biological Sciences and Technology Division, CSIR-North East Institute of Science and Technology (CSIR-NEIST), Jorhat, Assam, India; Academy of Scientific and Innovative Research (AcSIR), Ghaziabad, Uttar Pradesh, India

Shajidabi Shaik Department of Botany, Yogi Vemana University, Kadapa, Andhra Pradesh, India

Ajeet Singh ICAR-Indian Agricultural Research Institute, New Delhi, Delhi, India

Preeti Singh Department of Botany, School of Chemical and Life Sciences, Jamia Hamdard, New Delhi, Delhi, India

Ragini Singh Department of Botany, School of Chemical and Life Sciences, Jamia Hamdard, New Delhi, Delhi, India

Chita Sudhakar Mangalayatan University, Jabalpur, Madhya Pradesh, India

Ashwini Talakayala International Crops Research Institute for the Semi-Arid Tropics (ICRISAT), Patancheru, Telangana, India; Plant Molecular Biology Laboratory, Agri Biotech Foundation, Hyderabad, Telangana, India; Department of Genetics, Osmania University, Hyderabad, Telangana, India

Vaishnavi Tripathi School of Bioscience and Technology, Galgotias University, Greater Noida, Uttar Pradesh, India

Prachi Tyagi ICAR-Indian Agricultural Research Institute, New Delhi, Delhi, India

Shahid Umar Department of Botany, School of Chemical and Life Sciences, Jamia Hamdard, New Delhi, Delhi, India

Vishnu Babu Vallepu Department of Botany, Yogi Vemana University, Kadapa, Andhra Pradesh, India

Dubbala Prshanthi Divya Vani Department of Genetics and Plant Breeding, School of Agriculture, SR University, Warangal, Telangana, India

Amaranatha R. Vennapusa Department of Agriculture and Natural Resources, Delaware State University, Dover, DE, United States

Preethi Vijayaraghavareddy School of Biosciences and Centre for Sustainability, Chanakya University, Devanahalli, Karnataka, India

C. Vismaya Department of Biology, Indian Institute of Science Education and Research (IISER), Tirupati, Andhra Pradesh, India

Nancy Vohra Sher-e-Kasmir University of Agricultural Sciences and Technology, Jammu, Jammu and Kashmir, India

Shabir H. Wani National Seed Project Laboratory, Sher-e-Kashmir University of Agricultural Sciences and Technology of Kashmir, Srinagar, Jammu and Kashmir, India

Narkhede Gopal Wasudeo Department of Genetics and Plant Breeding, School of Agriculture, SR University, Warangal, Telangana, India

Anshika Yadav Department of Botany, University of Allahabad, Prayagraj, Uttar Pradesh, India

Sunil Kumar Yadav School of Bioscience and Technology, Galgotias University, Greater Noida, Uttar Pradesh, India; Department of Bioscience and Biotechnology, Banasthali Vidyapith, Niwai, Rajasthan, India

Nagarathnamma Yammanuri Department of Botany, Yogi Vemana University, Kadapa, Andhra Pradesh, India

About the editors

M. Iqbal R. Khan is an Assistant Professor in the Department of Botany and serves as Deputy Director of the Centre for Environment and Sustainable Development (CESD) at Jamia Hamdard, New Delhi, India. He also holds adjunct appointments at Korea University, Seoul, South Korea, and at The University of Western Australia, Perth, Australia.

Dr. Khan's research focuses on major staple crops, particularly wheat, with the objective of reducing their susceptibility to environmental stress under projected future climatic conditions and contributing to the achievement of the Sustainable Development Goals. His work addresses the challenges posed by environmental stressors to agricultural productivity, which remain a critical concern for global food and nutritional security.

He has published over 100 research articles, has edited several books, and has authored multiple book chapters with leading publishers. He has an established citation record, as reflected in widely used bibliometric indicators.

Dr. Khan has been recognized as a Clarivate Highly Cited Researcher in the Animal and Plant Science category and has been listed among the top 2% of scientists globally by Elsevier and Stanford University (2021–25). He is a member of the National Academy of Sciences, India (NASI), in Plant and Agricultural Sciences, and a recipient of the Indian National Science Academy (INSA) Young Associates award (2024).

His accolades include the R.H. Dastur Gold Medal Award—ISPP (2023), the Young Scientist Platinum Jubilee Award—NASI (2021), and the Young Scientist Award—ISPP (2018). He serves on the editorial boards of many renowned journals, including Plant and Soil, Journal of Plant Growth Regulation, 3-Biotech, and Plant Physiology Reports. He has also served as a guest editor for Agriculture, Frontiers in Plant Science, Sustainability, and Plant Physiology and Biochemistry.

Sudhakar Reddy Palakolanu has more than 14 years of expertise in plant molecular biology, functional genomics, and gene editing. His research focuses on identifying and analyzing genes associated with complex traits using advanced molecular and biotechnological approaches. His work aims to enhance key agronomic traits in dryland crops, particularly millets, through the application of cutting-edge techniques. By integrating modern biotechnological tools, he seeks to validate candidate genes identified through trait dissection, supporting sustainable crop development.

Dr. Reddy's accomplishments have been recognized through several prestigious national awards, including the Young Scientist Award and the INSPIRE Faculty Award from the Department of Science and Technology, Government of India (DST-GOI). He also received a postdoctoral fellowship from the German Academic Exchange Service (DAAD), which supported his research in Germany. In addition, he has served as a guest scientist at the International Rice Research Institute (IRRI) in the Philippines, where he worked on CRISPR–Cas9-mediated genome editing to improve rice grain quality. He has an extensive research portfolio that includes numerous peer-reviewed articles, book chapters, books, and patents. He also serves as a reviewer for leading international journals, including Trends in Plant Science, Genome Biology, Frontiers in Plant Science, Physiologia Plantarum, Plant Cell Reports, Plant Biotechnology Journal, and BMC Plant Biology, among others.

CHAPTER

1

Biostimulants: a sustainable approach to enhancing plant resilience under environmental stress

Nancy Vohra[1], *Amresh Kumar*[2], *Prachi Tyagi*[3], *V. Prathap*[4], *Ajeet Singh*[3], *Nand Lal Meena*[5], *Sweta Kumari*[3], *and Chirag Maheshwari*[3]

[1]Sher-e-Kasmir University of Agricultural Sciences and Technology, Jammu, Jammu and Kashmir, India [2]ICAR-National Institute for Plant Biotechnology, New Delhi, Delhi, India [3]ICAR-Indian Agricultural Research Institute, New Delhi, Delhi, India [4]ICAR-Indian Institute of Oil Palm Research, Pedavegi, Andhra Pradesh, India [5]ICAR-National Bureau of Plant Genetic Resources, New Delhi, Delhi, India

1.1 Introduction

Agriculture constitutes an integral part of human sustenance for providing food, fiber, fuel, and raw materials for industrial development. However, the sector faces unprecedented challenges as the global population is growing rapidly and climate change intensifies. By 2050, the global population is projected to surpass 9.7 billion, demanding a 50%–70% increase in agricultural productivity to ensure food and nutritional security (FAO, 2017). Achieving this target is becoming increasingly difficult due to the widespread occurrence and severity of abiotic stresses such as drought, salinity, extreme temperatures, and nutrient imbalances (Mahajan et al., 2025b). Collectively, these stressors account for approximately 50%–70% of global crop yield losses (Lesk et al., 2016). These stressors frequently occur in combinations that exacerbate their effects rather than acting individually. For instance, heat waves and drought frequently occur together, which causes plants to experience increased physiological stress and drastically lowers the yield (Nazir et al., 2023; Mittler, 2017). Consequently, enhancing resilience and sustainability in agricultural systems, while maintaining environmental integrity, is as imperative as achieving higher productivity.

Harnessing Plant-Stimulating Molecules
DOI: https://doi.org/10.1016/B978-0-443-44994-9.00001-8

Abiotic factors affect plants on multiple levels, including overall plant physiology and cellular homeostasis (Mahajan et al., 2025c). Salinity stress affects photosynthesis and development by limiting stomatal conductance and consumption of carbon (C), while it disturbs the ion equilibrium, resulting in a significant accumulation of sodium (Na^+) and chloride (Cl^-) ions that limit potassium (K) and calcium (Ca) intake and impede enzyme activity (Khatoon et al., 2025; Munns and Tester, 2008). Similarly, low-temperature stress diminishes membrane fluidity, restricts enzymatic activity, and disturbs metabolic balance. In contrast, high-temperature or heat stress accelerates protein denaturation, destabilizes cellular membranes, and promotes the excessive generation of reactive oxygen species (ROS). Collectively, these physiological and biochemical disruptions significantly impair plant growth, development, and productivity under abiotic stress conditions (Hasanuzzaman et al., 2013). These challenges are further aggravated by nutrient deficiencies, which compromise plant defense mechanisms and metabolic functions. Under natural conditions, plants are often exposed to multiple abiotic stresses simultaneously, and the cumulative impact of these concurrent stresses is frequently greater than the sum of their individual effects (Mahajan et al., 2023, 2025d). The unpredictable nature of climatic variations adds another layer of complexity, making it increasingly difficult to predict or manage stress responses effectively. This scenario underscores the urgent need for innovative and sustainable strategies aimed at enhancing plant resilience and productivity without causing adverse environmental consequences.

Conventional strategies for resolving these challenges were reliant primarily on exogenous chemical applications, plant breeding, and agronomic measures (Mahajan et al., 2025a). Fertilizers, insecticides, and growth regulators have undoubtedly contributed to production advances, but they have also resulted in unanticipated consequences such as degradation of soil, water contamination, loss of biodiversity, and risks to human health from residual chemicals (Savci, 2012). Breeding and genetic engineering produced stress-tolerant cultivars. However, these techniques have been limited by narrow genetic diversity, high expenses, prolonged breeding cycles, and consumer aversion to genetically modified organisms (Tester and Langridge, 2010). Improved irrigation systems, crop rotation, and precision agronomy have also contributed to stress mitigation, but these practices are constrained by resource scarcity, rising input costs, and climate uncertainty. Altogether, the limitations of conventional approaches underscore the need for complementary strategies that are sustainable, eco-friendly, and capable of enhancing resilience across diverse cropping systems. Under these circumstances, biostimulants have emerged over the last two decades as promising substances to increase plant tolerance to environmental challenges. Biostimulants, unlike fertilizers and pesticides, act by regulating innate physiological, biochemical, and molecular processes in plants to increase overall performance (Zulfiqar et al., 2024). The European Biostimulant Industry Council defines them as "materials, other than fertilizers, that, when applied to plants or the rhizosphere, stimulate natural processes to enhance nutrient uptake, nutrient efficiency, tolerance to abiotic stress, and crop quality" (European Biostimulant Industry Council (EBIC), 2012). This broad definition encompasses a diverse array of substances and organisms, including humic and fulvic acids derived from soil organic matter, protein hydrolysates rich in amino acids and peptides, seaweed extracts (SWEs) containing polysaccharides and phytohormones, microbial inoculants such as mycorrhizal fungi and plant growth–promoting rhizobacteria (PGPR), and plant-derived bioactive molecules such as flavonoids, phenolics, and melatonin (MT) (Calvo et al., 2014; du Jardin, 2015). Each category exerts unique effects on plant metabolism, but collectively, they improve resilience by enhancing water-use efficiency (WUE), nutrient acquisition, antioxidant defense, osmotic regulation, and stress-responsive gene expression. Ayed et al. (2022), in their research, found that various factors can influence the efficacy of biostimulants, including the type of biostimulant (e.g., microbial inoculants, SWEs, and amino acids), the specific crop variety, application methods, dosage, timing (Maignan et al., 2021), and prevailing environmental conditions (drought, salinity, temperature fluctuations, etc.) (Ma et al., 2022).

The mechanisms underlying the action of biostimulants are intricate and operate across multiple biological levels. At the physiological level, biostimulants enhance chlorophyll stability, improve photosynthetic efficiency, and promote root system development, thereby enabling plants to utilize soil nutrients and water resources more effectively (Rouphael and Colla, 2020). At the biochemical level, they stimulate the activity of antioxidant enzymes such as superoxide dismutase (SOD), catalase (CAT), and peroxidase (POD), which play critical roles in mitigating oxidative damage induced by ROS under stress conditions (Khatoon et al., 2024). Biostimulants also facilitate the accumulation of osmoprotectants, including proline and glycine betaine, which help maintain cellular osmotic balance during drought and salinity stress. Furthermore, they enhance the biosynthesis of secondary metabolites such as flavonoids and phenolic compounds that contribute to plant defense responses. At the molecular level, biostimulants modulate gene expression by activating stress-responsive transcription factors and influencing hormonal signaling pathways associated with abscisic acid (ABA), auxins, and cytokinins (CKs) (Sharma et al., 2019). This comprehensive modulation across physiological, biochemical, and molecular domains renders biostimulants highly effective in mitigating the complex and multifaceted impacts of environmental stresses.

The global relevance of biostimulants is reflected in their rapidly expanding market. The global biostimulant industry, currently valued at over USD 3 billion, is projected to surpass USD 6 billion by 2027, with a compound annual growth rate of around 12% (Markets and Markets, 2022). This growth is driven by the rising demand for organic and residue-free produce, growing awareness of sustainable agricultural practices, and regulatory incentives aimed at reducing the reliance on chemical inputs. European countries such as Italy, Spain, and France have taken the lead in the commercialization and application of biostimulants, supported by favurable policies and strong research infrastructure. In emerging economies, such as India, China, and Brazil, adoption is accelerating as biostimulants align well with government initiatives promoting natural and organic farming (Bulgari et al., 2019). In India, in particular, the use of SWEs, humic substances, and microbial formulations is being encouraged under sustainable intensification programs aimed at improving soil health, enhancing productivity, and doubling farmers' incomes. Despite their immense potential, biostimulants are not devoid of limitations. Diversity in raw material sources, a lack of standardized formulations, and variable efficacy under varying agroclimatic situations have all prevented wider adoption. Research verification and precise guidelines for regulation are still being developed in several areas. Nonetheless, ongoing research is gradually deciphering the mechanisms of action of biostimulants using sophisticated omics technologies, while favorable regulations and extension services are helping farmers acquire confidence. Their potential to supplement existing practices by increasing efficiency, lowering input dependency, and boosting resilience makes them essential instruments in the transition to climate-smart agriculture.

In this context, the present chapter aims to provide a comprehensive overview of biostimulants and their potential role in mitigating the impacts of environmental stresses. It reviews their classification and sources, discusses mechanisms of action at physiological, biochemical, and molecular levels; examines their role in stress-specific tolerance; highlights their importance in rhizosphere interactions and sustainable farming; and evaluates the challenges and prospects for their integration into mainstream agricultural systems. By synthesizing current knowledge, this chapter underscores the potential of biostimulants to not only safeguard yields under adverse conditions but also to serve as a cornerstone of sustainable agricultural intensification in an era of climate uncertainty.

1.2 Classification and sources of biostimulants

Biostimulants encompass a diverse range of naturally derived and synthetically produced substances that promote plant growth, enhance nutrient uptake, and improve stress tolerance without serving as direct sources of nutrients. Owing to their wide array of bioactive compounds and multifaceted modes of action, the classification of biostimulants remains complex and is generally based on their origin and chemical composition. Broadly, biostimulants can be categorized into several major groups. The first category comprises humic substances, which influence soil structure, nutrient availability, and microbial activity. The second includes seaweed and plant extracts, rich in polysaccharides, phytohormones, and antioxidants that contribute to improved growth and stress resilience. The third category consists of protein hydrolysates and amino acids, known to stimulate plant metabolism and enhance tolerance to abiotic stresses. The fourth group includes microbial inoculants, such as beneficial bacteria and fungi, which facilitate nutrient mobilization and promote root health. The fifth category comprises chitosan and other biopolymers, which strengthen plant defense mechanisms against environmental and biological stressors. Additionally, plant-derived molecules and synthetic analogs have been developed to replicate or amplify the natural activity of biostimulants (Chen et al., 2024). All categories contribute distinct mechanisms that improve nutrient use efficiency, physiological responses, and plant resilience against a wide range of stresses (Van Oosten et al., 2017) (Table 1.1).

1.2.1 Humic and fulvic substances

Humic substances, which encompass humic acids, fulvic acids, and humins, are complex organic compounds formed through the decomposition of plant and microbial residues in soil. They play a vital role in improving soil structure, nutrient retention, and water holding capacity (Calvo et al., 2014). Among them, humic acids are particularly important for enhancing nutrient mobilization, photosynthetic efficiency, respiration, water balance, and pigment synthesis (Godara and Bakshi, 2021). Structurally, humic acids contain amino acids, polysaccharide residues, and a mixture of aromatic and aliphatic compounds that are central to nutrient complexation and chelation processes (García et al., 2019). In addition, humic acids exhibit hormone-like activity, primarily through auxin or auxin-like molecules, and can modulate plant hormone signaling pathways (Zandonadi et al., 2013).

TABLE 1.1 Classification of biostimulants, their sources, and key functions.

Category	Primary sources	Key bioactive components	Major functions in plants
Humic and fulvic substances	Soil organic matter, compost, peat	Humic acids, fulvic acids	Enhance nutrient uptake, improve soil structure, stimulate root growth, increase stress tolerance
Seaweed and plant extracts	*Ascophyllum nodosum*, *Ecklonia maxima*, moringa, aloe	Cytokinins, auxins, betaines, polysaccharides	Promote seed germination, improve photosynthesis, enhance osmotic adjustment, stimulate antioxidant defense
Protein hydrolysates and amino acids	Enzymatic/ chemical hydrolysis of plant/animal proteins	Peptides, amino acids (proline, glycine betaine)	Improve nutrient assimilation, act as osmoprotectants, promote root and shoot development, modulate hormonal signaling
Microbial inoculants	PGPR, AMF, N-fixing bacteria, endophytes	Microbial metabolites, phytohormones	Enhance nutrient availability, improve water uptake, induce systemic resistance, enrich soil biodiversity
Chitosan and other biopolymers	Crustacean shells, fungal cell walls, algae	Chitosan, alginates, carrageenans	Induce plant defense responses, improve seed germination, regulate stress-related genes, enhance water-use efficiency
Emerging sources (novel molecules)	Melatonin, brassinosteroids, synthetic analogs	Hormone-like molecules, secondary metabolites	Improve photosynthesis, enhance root development, regulate stress-responsive pathways, offer targeted and consistent field responses

AMF, Arbuscular mycorrhizal fungi; *N*, nitrogen; *PGPR*, plant growth promoting rhizobacteria.

Drought stress severely affects plant growth and productivity by disrupting key metabolic and enzymatic processes (Selmar and Kleinwächter, 2013). The application of humic acids has been shown to alleviate such adverse effects by maintaining relative water content (RWC), enhancing pigment synthesis, and improving photosynthetic performance under water-deficit conditions. Moreover, humic acids improve photosynthetic efficiency by stimulating electron transport activity (Lotfi et al., 2018). At the molecular level, they regulate the expression of stress-responsive genes involved in nutrient uptake, water transport (aquaporins), nitrogen (N) acquisition and distribution (nitrate transporters NRT1 and NRT2), and protein protection mechanisms such as heat shock proteins (Jacob et al., 2017; Sinha et al., 2020; Kumar and Dubey, 2020; Kumar et al., 2022a). These regulatory effects collectively contribute to enhanced drought tolerance in plants; however, the precise molecular mechanisms underlying these responses remain to be fully elucidated.

Similar to humic acids, fulvic acids consist predominantly of phenolic compounds and contain relatively fewer aromatic groups (Canellas et al., 2015). Due to their lower molecular weight and greater solubility, fulvic acids can readily penetrate plant tissues, where they activate enzymatic functions, stimulate chlorophyll biosynthesis, and modulate stress-related metabolic pathways. Fulvic acids may act independently or synergistically with humic acids and SWEs to promote plant growth and stress adaptation. Empirical studies have demonstrated that fulvic acids enhance germination, shoot and root development, and yield in several crops, including *Triticum aestivum* (wheat), *Hordeum vulgare* (barley), and *Beta vulgaris* (sugar beet) (Braziene et al., 2021). At the molecular level, fulvic acids influence transcriptional activity by upregulating genes associated with N metabolism, nutrient transport, hydrolase activity, and lipid biosynthesis (Capstaff et al., 2020).

Beyond their growth-promoting properties, fulvic acids also contribute to abiotic stress tolerance. Under drought stress conditions, they strengthen the antioxidant defense system by stimulating ascorbate, glutathione, and flavonoid metabolism, thereby minimizing oxidative damage caused by ROS (Sun et al., 2020). More broadly, humic substances, including both humic and fulvic acids, have been reported to enhance photosynthetic pigments, carotenoids, phenolics, flavonoids, and overall nutrient content, ultimately improving plant vitality and productivity (Bayat et al., 2021).

1.2.2 Seaweed and plant extracts

SWEs are among the most widely used biostimulants in agriculture. Derived from species such as *Ascophyllum nodosum*, *Ecklonia maxima*, and *Sargassum* spp., these extracts contain bioactive compounds including phytohormones (auxins, CKs, and Gibberellins), polysaccharides, polyphenols, and betaines (Jennings, 1968; Stirk et al., 2009). These bioactive compounds have benefits on plant growth, yield, nutritional quality, stress tolerance, and photosynthetic efficiency and regulate osmotic balance during drought and salinity stress (Battacharyya et al., 2015). Recent evidence reported nonhormonal components, particularly polysaccharides, as key bioactive drivers (Craigie, 2011). For example, *Ascophyllum nodosum* extracts rich in polysaccharides have been shown to modulate transcriptomic, metabolic, and lipid pathways, enhancing growth while improving tolerance to oxidative and abiotic stresses through reductions in ROS and electrolyte leakage (Omidbakhshfard et al., 2020). Beyond seaweed, extracts from plants such as *Moringa oleifera* (drumstick tree) and *Aloe vera* (Aloe) are emerging as promising biostimulants due to their rich profile of bioactive compounds, including phytohormones, amino acids, vitamins, and antioxidants. Extracts from drumstick tree leaves have been reported to enhance seed germination, plant height, and yield in crops such as *Zea mays* (maize) and *Triticum aestivum* (wheat), effects largely attributed to their high CK and antioxidant content (Rouphael et al., 2018). Similarly, *Aloe vera* extracts are gaining attention for their ability to improve nutrient uptake, stimulate root development, and confer tolerance to abiotic stresses, particularly drought and salinity, by enhancing antioxidant defense systems and osmolyte accumulation (Taskos et al., 2019). These terrestrial plant-based extracts represent renewable, eco-friendly resources with broad applicability across cropping systems, offering complementary or alternative strategies to SWEs in sustainable agriculture.

1.2.3 Protein hydrolysates and amino acid–based biostimulants

Food and agricultural industries generate large amounts of mixtures of peptides and free amino acids enriched with secondary metabolites from protein hydrolysates through chemical or enzymatic hydrolysis of plant or animal raw materials (Xu and Geelen, 2018). Recycling these by-products into protein hydrolysates as biostimulants supports both economic and ecological sustainability. Protein hydrolysates promote plant metabolism by providing small peptides and amino acids that modulate hormonal crosstalk, particularly auxin- and ethylene (ET)-mediated pathways; enhancing nutrient uptake, root growth, and enzymatic activity; and improving resistance to oxidative stress and overall plant productivity (Ertani et al., 2009). Similarly, amino acid–based formulations are rich in proline, glutamic acid, and glycine betaine and act as osmoprotectants, stabilizing cellular membranes and proteins during dehydration stress and mitigating oxidative damage (Ashraf and Foolad, 2007). Collectively, these biostimulants enhance plant resilience, productivity, and overall tolerance to environmental stresses, underscoring their potential in promoting sustainable and resource-efficient crop management. Among these, protein hydrolysates have gained particular attention for their multifaceted physiological and molecular effects. Foliar application of protein hydrolysates has been demonstrated to stimulate epiphytic bacterial communities, improve photosynthetic efficiency, and enhance plant biomass and nutrient content, especially when used in conjunction with beneficial microorganisms such as *Trichoderma harzianum* T22 (Almadi et al., 2020).

Protein hydrolysates exert a significant influence on gene expression and metabolic regulation under both optimal and stress conditions. Studies on alfalfa-derived protein hydrolysates have revealed the upregulation of genes associated with photosynthesis, nutrient acquisition, and primary metabolism (Ertani et al., 2009), as well as the positive regulation of nitrate and ammonium transporters and nitrate reductase (NR) activity (Sestili et al., 2018). Under abiotic and biotic stress conditions, the foliar application of protein hydrolysates has been shown to activate defense-related genes, such as catalase 3 (CAT3), oxidative signal–inducible 11 (OXI1), pyruvate dehydrogenase (PDH), glutathione S-transferase F7 (GSTF7), and pathogenesis-related protein 1 (PR1), thereby strengthening tolerance to oxidative stress and pathogen attack (Rentel et al., 2004; Seo et al., 2008).

Furthermore, protein hydrolysates have been reported to modulate the plant metabolome and phytohormonal profiles, influencing the levels of auxins, CKs, and GAs. They also regulate the expression of genes involved in hormone biosynthesis, transport, and signal transduction, particularly under water-deficit conditions. In tomato plants subjected to drought stress, protein hydrolysate treatment resulted in improved physiological performance and altered hormone metabolism, contributing to enhanced stress adaptation (Casadesús et al., 2019). Legume-derived protein hydrolysates also promote root development and increase secondary metabolites (Santi et al., 2017). These molecular effects demonstrate the role of protein hydrolysates in enhancing plant growth and stress tolerance, although responses vary by species to species.

1.2.4 Microbial inoculants

Microbial biostimulants, also known as bioeffectors, are beneficial microorganisms (bacteria and fungi) that interact with plants and play a role in growth, nutrient acquisition, and stress tolerance (Rouphael and Colla, 2020). Among these, the most studied groups are PGPR, arbuscular mycorrhizal fungi (AMF), and N-fixing bacteria. PGPR uses a variety of direct and indirect mechanisms to promote plant growth, including biological N fixation, phosphate solubilization, and siderophore production, which collectively improve nutrient bioavailability (Bhattacharyya and Jha, 2012). Additionally, PGPR are known to synthesize phytohormones such as indole-3-acetic acid (IAA), CKs, and GAs, hence regulating root architecture and shoot development (Glick, 2012). Importantly, several PGPR strains induce systemic resistance (induced systemic resistance [ISR]), activating plant defense signaling pathways and conferring broad-spectrum resistance against pathogens (Pieterse et al., 2014). Similarly, AMF form symbiotic associations with plant roots, and their extensive hyphal networks enhance nutrient (e.g., zinc [Zn], copper [Cu], and N) and water uptake (Smith and Smith, 2011). AMF have been shown to be resistant to abiotic stress and heavy metal toxicity while simultaneously promoting soil aggregation and carbon cycling (Begum et al., 2019). In addition to PGPR and AMF, microbes and fungi that do not cause disease in plant tissues can operate as biostimulants. Santoyo et al. (2016) found that these organisms boost plant performance by influencing host metabolism, increasing stress tolerance, and producing various bioactive chemicals. Combining PGPR, AMF, and other beneficial strains can improve nutrient cycling, stress adaptation, and plant productivity in complex field environments (Backer et al., 2018). Importantly, microbial inoculants not only enhance crop performance but also contribute to soil health, biodiversity, and long-term agroecosystem sustainability, aligning with the principles of environmentally friendly and climate-resilient agriculture (Kumar and Dubey, 2020).

1.2.5 Chitosan and other biopolymers

Chitosan is a natural biopolymer derived from the deacetylation of chitin, which is abundantly present in crustacean shells and fungal cell walls. It stimulates and protects plants' innate immunological responses. According to Pichyangkura and Chadchawan (2015), chitosan can cause systemic resistance by activating PR proteins, stimulating antioxidant enzyme activity, and upregulating genes involved in defense signaling pathways like salicylic acid (SA), jasmonic acid (JA), and ethylene (Et). Chitosan improves seed germination, chlorophyll content, and WUE, leading to increased plant development and resistance to abiotic stresses such as drought and salinity (Iriti and Faoro, 2014; Xing et al., 2015). Marine-derived biopolymers, such as alginates from brown algae and carrageenans from red algae, have strong biostimulant potential. These polysaccharides affect cellular osmotic equilibrium, ion transport, and stress-responsive gene expression, resulting in increased nutritional absorption and resistance to abiotic stressors (Khan et al., 2009; Shukla et al., 2016). Alginates chelate minerals and improve soil structure, promoting root growth and microbial activity in the rhizosphere (Craigie, 2011). Carrageenans activate plant defensive responses, causing the formation of secondary metabolites, including phenolics and flavonoids, which contribute to both growth promotion and disease resistance (El Modafar et al., 2012). Together, these natural biopolymers act as eco-friendly biostimulants, improving plant growth and resistance while lowering the need for synthetic agrochemicals.

1.2.6 Emerging sources: plant-derived molecules and synthetic analogs

In addition to conventional biostimulants, novel plant-derived molecules such as MT, brassinosteroids (BRs), and various categories of secondary metabolites are considered as potential biostimulants. These molecules control stress-responsive pathways, root development, and photosynthetic efficiency in response to abiotic stimuli such as drought, salinity, and heat (Tyagi et al., 2023). MT has received special interest because of its dual role as a growth regulator and a powerful antioxidant. MT applied exogenously has been found to improve drought resistance by modifying root system architecture and increasing the expression of genes that encode antioxidant enzymes (Arnao and Hernández-Ruiz, 2019; Zhang et al., 2013). Similarly, BRs, a type of steroidal phytohormone, enhance photosynthetic performance, membrane stability, and nitrogen assimilation under stress, hence increasing crop yield (Nolan et al., 2017; Vriet et al., 2013). Secondary metabolites, including phenolic chemicals, flavonoids, and terpenoids, are being investigated for their biostimulant potential, particularly in terms of stress tolerance via signaling and oxidative damage regulation (Garg et al., 2024). Synthetic analogs are intended to imitate the actions of natural signaling molecules. These synthetic chemicals are designed to provide higher stability, tailored distribution, and more consistent field-level responses than naturally generated formulations (Yakhin et al., 2017). Although still at an early stage of development, next-generation biostimulants represent a promising factor, combining the ecological safety of natural compounds with the precision of modern chemical and biotechnological tools.

1.3 Mechanisms of action of biostimulants

Biostimulants have emerged as an innovative class of agricultural inputs that operate through mechanisms different from conventional fertilizers and pesticides. While fertilizers primarily supply essential nutrients to plants and pesticides directly suppress pests and pathogens, biostimulants act by stimulating the plants' physiological, biochemical, and molecular processes to optimize growth, development, and stress resilience under both favorable and stress conditions (du Jardin, 2015). Their actions are frequently complex, with direct influences on plant metabolism and growth as well as indirect effects mediated by the soil and rhizosphere microbiota (Calvo et al., 2014). Biostimulants modify molecular signaling pathways that control stress-responsive gene expression and enzyme activity, allowing plants to maintain homeostasis in the face of abiotic challenges like drought, salinity, or nutrient constraint (Battacharyya et al., 2015). They also improve nutrient digestion and translocation by upregulating transporter genes, increasing nutrient solubility, and promoting root system architecture, all of which contribute to greater nutrient utilization efficiency (Sestili et al., 2018). Another important purpose is to increase WUE through improved root shape, aquaporin control, and osmoprotectant accumulation (Qin and Leskovar, 2018). At the molecular level, biostimulants improve the antioxidant defense system by increasing the activity of enzymes such as SOD, CAT, and ascorbate peroxidase (APX), thereby minimizing oxidative damage caused by ROS (Sharma et al., 2019). Furthermore, they modulate hormones, either directly delivering phytohormone-like molecules (e.g., auxins, CKs, GAs, and BRs) or affecting the endogenous balance of plant hormones that govern growth and stress responses (Ertani et al., 2009). Furthermore, biostimulants have a major impact on the rhizosphere environment, improving beneficial microbial communities, nutrient cycling, and soil health, all of which indirectly boost plant growth (Berg et al., 2014).

1.3.1 Modulation of nutrient uptake and assimilation

Biostimulant enhances nutrient uptake, translocation, and assimilation through physiological and molecular mechanisms. Humic acids and fulvic acids are the organic biostimulants that modify root physiology by increasing membrane permeability and root H^+-ATPase activity, thereby enhancing nutrient absorption (Nardi et al., 2021; Nazir et al., 2023). The activation of plasma membrane H^+-ATPases leads to the uptake of essential macronutrients such as nitrate (NO_3^-), phosphate ($H_2PO_4^-$), and potassium ions (K^+) via active transport systems (Zandonadi et al., 2013). Humic substances increase root hair and lateral root proliferation, as a result increasing the roots' surface area and more efficient nutrient uptake (Nardi et al., 2021; Nazir et al., 2023). The other one is protein hydrolysates and amino acid–based biostimulants act as chelating agents that increase the solubility and bioavailability of micronutrients such as iron (Fe), Zn, Cu, and manganese (Mn), particularly in calcareous or alkaline soils where their availability is limited (Colla et al., 2015; du Jardin, 2015). These compounds improve ion mobility and act as signaling molecules that can change the expression of genes associated with nutrient transport and assimilation (Santi et al., 2017). For example, amino acids such as glutamate and glycine betaine act as signaling molecules that influence nitrate sensing and uptake efficiency by activating NO_3^- transporter genes (Paul et al., 2019). Seaweed extracts (SWEs), which contain polysaccharides, phenolics, and phytohormones (auxins, CKs, etc.), upregulate genes involved in N uptake and assimilation, including NR and glutamine synthetase (GS) (Khan et al., 2009). These gene level changes increased the activity of key enzymes in nitrogen and carbon metabolism, resulting in increased photosynthetic efficiency and crop productivity (Colla et al., 2015).

1.3.2 Enhancement of photosynthesis and carbon metabolism

Biostimulants generally showed positive effects on photosynthesis and carbon metabolism. Protein hydrolysates contain bioactive peptides and amino acids that act as signaling molecules, promoting chlorophyll production and shielding the photosystem from abiotic stress (Almadi et al., 2020). Under water- or nutrient-limited conditions, these bioactive components also boost the expression of photosystems I and II genes as well as the Calvin–Benson cycle (Cristofano et al., 2021). SWEs are high in osmoprotective chemicals (betaines, polyamines, and polysaccharides), which protect chloroplast membranes and photochemical stability during drought, salinity, and temperature stresses (Khan et al., 2009; Battacharyya et al., 2015). These chemicals also contribute to improved solar energy use, electron transport, delayed leaf senescence, and chlorophyll retention (Craigie, 2011; Sharma et al., 2019). Microbial biostimulants such as PGPR and AMF improve photosynthetic efficiency, root system architecture, water and nutrient uptake, and soil-plant signaling, resulting in higher C assimilation rates (Ruiz-Lozano et al., 2016; Bulgari et al., 2019). AMF symbiosis improves the delivery of phosphorus (P) and micronutrients to leaves, which aids in ATP generation and carbon metabolism. PGPR stimulates the synthesis of IAA, which promotes leaf expansion and chloroplast growth (Rouphael et al., 2017).

1.3.3 Regulation of plant hormonal balance

Biostimulants are involved in regulating phytohormones, which are central regulators of plant growth, development, and stress responses. Biostimulants either include hormones or enhance the biosynthesis and signaling of essential plant hormones such as auxins, CKs, GAs, BRs, SA, and JAs (Rouphael and Colla, 2020). SWEs contain IAA, CKs, and betaines, all of which promote root initiation, elongation, and lateral branching. Protein hydrolysates increase CK activity, postpone leaf senescence, maintain chlorophyll content, and maintain photosynthetic efficiency under adverse conditions (Colla et al., 2015). Some protein hydrolysates also affect GA and BR signaling, which promotes cell elongation and biomass accumulation (Sestili et al., 2018). Microbial biostimulants influence plant hormonal homeostasis by direct synthesis and indirect regulation of phytohormones. PGPR species such as *Azospirillum*, *Pseudomonas*, and *Bacillus* generate auxins, GAs, and CKs that promote root proliferation, nutrient uptake, and shoot development (Spaepen and Vanderleyden, 2011; Nadeem et al., 2014). AMF modulates ABA signaling in response to drought or salinity stress, encouraging stomatal control, osmotic adjustment, and increased WUE (Ruiz-Lozano et al., 2012; Aroca et al., 2013). Other microbial biostimulants affect the JA and SA pathways, increasing disease resistance (Pieterse et al., 2014). Biostimulants do not just function as external hormone supplements, they also engage with plant hormonal networks to reprogram growth and stress responses.

1.3.4 Induction of antioxidant defense systems

Abiotic stresses in plants lead to the overproduction of ROS like superoxide anions ($O_2\cdot^-$), hydrogen peroxide (H_2O_2), and hydroxyl radicals ($OH\cdot$). Plants accumulate ROS, which causes oxidative damage to lipids, proteins, nucleic acids, and cellular membranes. The use of biostimulants contributes to the prevention of oxidative stress in plants. Protein hydrolysates, SWEs, and microbial inoculants have been shown to enhance enzymatic and nonenzymatic antioxidant systems (Ertani et al., 2009; Bulgari et al., 2019). Protein hydrolysates activate SOD, CAT, and APX, which neutralize ROS and protect chloroplasts and mitochondria from oxidative damage (Colla et al., 2015). SWEs have antioxidant activity and scavenge free radicals, and foliar application of SWEs in wheat improves SOD and CAT activity while dramatically lowering lipid peroxidation (Shukla et al., 2018). Microbial biostimulants influence antioxidant activity. *Pseudomonas* and *Bacillus* spp. stimulate antioxidant responses by systemic acquired resistance (SAR) and ISR pathways (Pieterse et al., 2014). AMF enhances ROS-scavenging capacity under drought and salinity by improving the antioxidant enzyme pool (Ruiz-Lozano et al., 2012; Aroca et al., 2013).

1.3.5 Osmoprotection and maintenance of water balance

Cellular water balance is crucial for plant survival and productivity under drought or salinity stress, and biostimulants can increase osmoprotectant activity by stimulating the accumulation of solutes (Bulgari et al., 2019). These low-molecular-weight organic compounds act as osmotic buffers, protecting the photosynthetic apparatus and ensuring cellular homeostasis during dehydration or salt-induced water deficit. SWEs are effective in promoting osmotic adjustment due to their richness in osmoprotective molecules (Khan et al., 2009; Battacharyya et al., 2015). Microbial biostimulants contribute to drought and salinity tolerance through both structural and physiological mechanisms. AMF increases root hydraulic conductivity, root surface area, and upregulates the expression of aquaporin genes, which facilitate water transport across plasma membranes (Porcel et al., 2012; Ruiz-Lozano et al., 2016). AMF also enhances ABA signaling, improving stomatal regulation and water conservation under water-deficit conditions (Aroca et al., 2013). Similarly, *Pseudomonas*, *Bacillus*, and *Azospirillum* induce osmolyte biosynthesis leading to improved WUE (Vurukonda et al., 2016). Through these coordinated mechanisms, biostimulants enable plants to maintain favorable cellular water potential under adverse environmental conditions.

1.3.6 Activation of stress-responsive genes and signaling pathways

Biostimulants can activate molecular and signaling pathways that enhance plant tolerance to stress. Its treatments upregulate stress-responsive genes associated with osmotic adaptation, antioxidant activity, secondary metabolism, and hormonal signaling (Bulgari et al., 2015; Rouphael and Colla, 2020). Humic substances regulate the transcription of nitrate transporter (NRT) genes and H^+-ATPases to increase nutrient uptake efficiency under nutrient or water stress (Nardi et al., 2021; Nazir et al., 2023). Protein hydrolysates increased water and nutrient uptake by stimulating genes related to root morphogenesis and auxin and ethylene biosynthesis under stress conditions (Colla et al., 2015). Small signaling molecules like nitric oxide (NO), hydrogen sulfide (H_2S), and MT act as key intermediates in biostimulant-induced stress tolerance. NO and H_2S interact with ROS and phytohormones (ABA, JA, and SA) to help in stress (Kumar et al., 2022b; Khatoon et al., 2023). This molecular reprogramming results in a physiological condition known as priming, where plants develop enhanced responses to stress. Primed plants show faster and stronger osmolyte synthesis and higher ROS-scavenging capacity compared to nonprimed counterparts (Conrath et al., 2015; Rouphael and Colla, 2020).

1.3.7 Enhancement of secondary metabolite production

Biostimulants also play a role in secondary metabolism regulation, which leads to increased synthesis of flavonoids, phenolics, alkaloids, and terpenoids. These secondary metabolites act as antioxidants, antimicrobial agents, and structural components of plant tissues against both abiotic and biotic stresses (Garg et al., 2024; Bulgari et al., 2019). SWEs contain elicitor molecules that stimulate the phenylpropanoid pathway as a result of increasing the flavonoids, phenolic acids, and lignin concentration in plants (Sharma et al., 2019). The increased synthesis of these secondary metabolites reduced oxidative stress through ROS scavenging activity as well as strengthened cell wall integrity, thereby improving resistance to pathogen attack and abiotic stress (Khan et al., 2025). Microbial biostimulants are inducers of SAR and ISR, and these processes involve the accumulation of phytoalexins, PR proteins, phenylalanine ammonia-lyase (PAL), PODs, and polyphenol oxidases (Pieterse et al., 2014; Vurukonda et al., 2016). Such activation of defense pathways enhances the plant's ability to recognize and respond to pathogens and maintain the growth and productivity of plants.

1.3.7.1 *Synergistic effects and crosstalk between mechanisms*

Biostimulant activity is due to the synergistic interplay of numerous physiological and molecular pathways. Biostimulants affect plant metabolism at multiple levels, resulting in a complex network of signaling and metabolic changes (du Jardin, 2015; Rouphael and Colla, 2020). SWEs include auxin-like chemicals produced by PGPR, which increase root initiation and elongation (Craigie, 2011; Ertani et al., 2014). The simultaneous activation of various processes results in a synergistic outcome, in which the combined effect of many pathways outweighs the total of their separate contributions. This integration of nutrient absorption, hormonal control, and redox balance in plants works better under stress.

1.4 Stress-specific applications of biostimulants

Environmental stresses have a significant impact on plant function, reducing production and quality. Biostimulants, which change the physiological, biochemical, and molecular processes of plants, have become key tools for mitigating the harmful impacts of diverse conditions. Their efficacy has been tested in a variety of biotic (diseases and pests) and abiotic (drought, salt, extreme temperatures, nutrient deficits, and heavy metal toxicity) environments. Using examples from several crop production systems, this section focuses on the stress-specific functions of biostimulants in increasing plant resilience.

1.4.1 Biostimulants in mitigating drought stress

The most common stressor influencing agricultural productivity worldwide is still drought. By boosting osmoprotectant accumulation, increasing WUE, and improving root system architecture, biostimulants lessen the harm caused by drought. According to Canellas et al., 2015, humic compounds improve access to water reserves and allow for deeper soil penetration by increasing root length and branching. Plants develop and accumulate chemicals called osmolytes (glycine betaine and proline) to create tolerance to stress and to prevent or counteract the harmful effects of abiotic stresses. They offer ROS scavenging, antioxidant compound defense, protein stability, cell integrity preservation, and protection of biological components (Roychoudhury and Tripathi, 2019). Amino acid formulations and protein hydrolysates increase the production of proline and glycine betaine, which are molecules that stabilize proteins and membranes in the absence of water (Colla et al., 2015).

SWEs from *Ascophyllum nodosum* contain bioactive compounds such as betaines, CKs, and polyphenols that contribute to modulating stomatal closure and increasing leaf RWC (Shukla et al., 2018). Similarly, MT serves as a multifunctional signaling molecule in plants and is critical for enhancing drought resilience. MT enhances lateral root growth and lowers drought-induced oxidative stress by boosting antioxidant enzymes including CAT and SOD (Arnao and Hernández-Ruiz, 2019). Microbial inoculants such as AMF and PGPR boost water absorption and soil moisture retention by establishing hyphal networks (Rouphael et al., 2017). PGPR application can minimize oxidative stress in plants grown in salt- and water-deficient conditions by balancing phytohormone levels, optimizing antioxidant activity and osmoprotectant production, and correcting ion imbalance (Batool et al., 2020).

These biostimulants work together to increase drought tolerance through a number of signaling, metabolic, and structural processes.

1.4.2 Biostimulants in alleviating salinity stress

Around 20% of irrigated land globally is affected by salinity, which poses a major risk to food security. The main ways that biostimulants reduce salinity stress are via maintaining osmotic equilibrium, improving antioxidant defense, and controlling ion homeostasis. Amino acids and protein hydrolysates improve osmotic adjustment under salt stress by increasing proline and other suitable solutes (Colla et al., 2015). Since these algae contain hormones (such as auxins and CKs), polyphenols (such as phloroglucinol and eckol), polysaccharides (such as fucoidan, laminaran, carrageenan, and alginates), and kahydrin (a derivative of vitamin K), which are growth promoters and activators of the plant defense system (Ertani et al., 2018; Baltazar et al., 2021), humic substances and SWEs decrease Na^+ uptake while maintaining K^+/Na^+ balance, which is essential for enzymatic activity and membrane stability (Nardi et al., 2021; Nazir et al., 2023). Furthermore, seaweed-derived betaines and polyamines preserve photosynthetic apparatus by stabilizing thylakoid membranes during salt stress. PGPR strains such as *Pseudomonas fluorescens* and AMF increase salt tolerance by altering phytohormone balance and improving nutrient uptake, particularly phosphorus (Egamberdieva et al., 2019). By functioning as osmoprotectants and increasing the activity of antioxidant enzymes, chitosan-based formulations lessen lipid peroxidation in high-salinity conditions (Farouk et al., 2011). By improving cell wall rigidity and selective ion transport, emerging biostimulants like silicon nanoparticles (Si NPs) also reduce salt-induced ion toxicity (Alzahrani et al., 2018). Thus the synergistic application of biostimulants provides a sustainable means of maintaining agricultural productivity in saline-prone areas.

1.4.3 Biostimulants in heat and cold stress tolerance

Climate change is increasing the frequency of extreme temperature changes. Heat stress reduces pollen viability, inhibits photosynthesis, and accelerates the accumulation of ROS. SWEs and protein hydrolysates are vital for heat stress reduction because they promote antioxidant activity and maintain chlorophyll integrity (Colla et al., 2017; Anjos Neto et al., 2020)). MT and BRs, when administered exogenously, improve heat tolerance by boosting pollen fertility and upregulating heat shock proteins (HSPs) (Arnao and Hernández-Ruiz, 2019).

Cold stress, on the other hand, inhibits nutrient transport, reduces enzyme activity, and impairs membrane fluidity. Chitosan treatments produce the accumulation of soluble sugars and polyamines, which protects proteins and membranes against chilling damage. SWEs strong in phytohormones, such as CKs and ABA, boost antifreeze protein synthesis and photosynthetic efficiency. Apart from controlling development, auxins are essential for stress reactions, especially when it comes to modifying root architecture in harsh environments like cold temperatures (Tiwari et al., 2023).

Ornithine and arginine are the two amino acids that make up putrescine, a simple polyamine. Apart from its direct connection to plant growth and development, it also helps plants become more resilient to a range of abiotic stressors, including drought, salinity, and extreme temperatures (González-Hernández et al., 2022).

Furthermore, it has been demonstrated that AMF colonization increases cold tolerance by enhancing root metabolic activity and glucose allocation in cold climates (Zhu et al., 2016). According to these results, biostimulants offer multiple layers of defense against temperature fluctuations.

1.4.4 Biostimulants in nutrient deficiency stress

In intensive cropping systems, nutrient deficits are common, and biostimulants can help improve nutrient usage efficiency NUE. According to Canellas et al. (2015), humic compounds increase the cation exchange capacity (CEC) of soil, which facilitates the availability of micronutrients such as Fe, Zn, and Mn. Protein hydrolysates improve the solubility and absorption of micronutrients by acting as chelating agents (Ertani et al., 2009). As a biostimulant, yeast can alter plant physiological processes by enhancing seedling development and germination under stressful environments. *Oryza sativa* (rice) was used to test the effects of yeast extract, a sustainable biostimulant made from *Saccharomyces cerevisiae* (yeast) that contains all the nutrients, vitamins, amino acids, and growth factors. Therefore, in both tolerant and sensitive rice seedling kinds, yeast priming under nutrient-deficient circumstances preserved cell homeostasis and improved stress adaptation (Johnson and Puthur, 2022).

Indirectly boosting nutrient mobilization, SWEs increase rhizosphere microbial activity and root exudation. To enhance iron intake, microbial biostimulants, particularly PGPR, release siderophores, fix ambient N, and solubilize phosphates (Vessey, 2003). By controlling ion channels in root plasma membranes, chitosan-based coatings increase the effectiveness of nutrient absorption. Recent research has also linked MT to improved photosynthetic N usage efficiency and nutrient translocation (Liang et al., 2018). Therefore, by encouraging effective nutrient uptake and lowering reliance on chemical inputs, biostimulants provide sustainable substitutes for synthetic fertilizers.

1.4.5 Biostimulants in heavy metal stress

A rising problem in agricultural soils is heavy metal contamination that causes oxidative stress, interferes with cellular metabolism, and leads to low crop yields. Biostimulants use improved detoxification pathways, chelation, and sequestration to reduce metal toxicity. Toxic metals such as cadmium (Cd), lead (Pb), and nickel (Ni) are bound by humic acids, which decreases their toxicity and bioavailability (Aydin et al., 2012). As metal chelators, protein hydrolysates and amino acid complexes help the metals be compartmentalized into vacuoles. Wheat was treated with a commercial biostimulant (Fyto-fitness) in an environment contaminated with selenium (Se). It was shown that by changing the characteristics of the soil, the biostimulant increased biomass and affected plant metabolism. The fact that the Se (IV) accumulated more in the roots than in the shoots suggests that this could shield the plant from Se stress (Xiao et al., 2020). Magnesium (Mg), Mn, Cu, and Ca concentrations in wheat decreased when urea ammonium nitrate fertilizer was combined with another commercial biostimulant (Glutacetine) (Maignan et al., 2020). Moreover, a naturally derived biostimulant, IAA (Drobek et al., 2019), produced by AMF species (*Penicillium roqueforti*), depicted significant results on wheat to suppress Ni, Cd, Cu, Zn, and Pb that is grown in a heavy metal–rich environment (Ikram et al., 2018). By immobilizing metals in the rhizosphere and strengthening plants' antioxidant defense systems, chitosan nanoparticles have proven very successful in reducing the toxicity of heavy metals (Sobral et al., 2022). Additionally, SWEs increase the formation of phytochelatin and glutathione, both of which are essential for detoxification pathways. AMF reduces food chain pollution by immobilizing metals in fungal hyphae and stopping their transport to aerial plant parts (Herath et al., 2021). Likewise, reduced Mn, Co, and Ni levels were found in the AMF-inoculated grain in both wheat and barley, according to another AMF species (*Rhizophagus irregularis*) (Watts-Williams and Gilbert, 2021). These results demonstrate the potential of biostimulants to clean up contaminated soils and promote safer farming practices.

1.4.6 Biostimulants in biotic stress management

By inducing plant defensive responses, biostimulants, although mostly focused on abiotic stressors, also provide protection against diseases and pests. By triggering PR proteins, chitosan is known to generate SAR (Hadwiger, 2013). The activation of PR genes is in turn mediated by endogenous SA as a molecule involved in signal transduction (Legendré et al., 1992). According to Goñi et al. (2016), SWEs improve plant protection against bacterial and fungal diseases by activating the signaling pathways for SA and JA. Protein hydrolysates promote the accumulation of secondary metabolites, such as flavonoids and phenolics, which inhibit pathogen invasion and herbivory.

Trichoderma spp. and other microbial biostimulants improve plant immunity and stimulate growth by inducing ISR (Woo et al., 2014). The saprophytic fungi belonging to *Trichoderma* spp. can grow along the entire length of the root system, along which it establishes a barrier against pathogen attack (Harman et al., 2004). By altering redox signaling and triggering stress-responsive transcription factors, emerging compounds like MT and BRs enhance immune responses even further (Arnao and Hernández-Ruiz, 2019). As a result, biostimulants lessen dependency on artificial agrochemicals while supporting integrated approaches to pest and disease management.

1.5 Rhizosphere and soil health

The rhizosphere, the confined soil area immediately affected by root exudates and related microbes, is pivotal in influencing plant health and productivity. Biostimulants, due to their varied chemical and biological makeup, significantly affect the rhizosphere environment. They foster advantageous microbial communities, boost soil structure, and optimize nutrient cycling, ultimately establishing a more conducive ecology for plant growth. The incorporation of biostimulants into agricultural methods has garnered interest as a method to enhance plant stress resilience while simultaneously preserving long-term soil health and fertility.

1.5.1 Enhancement of soil microbial communities

One important family of biostimulants that directly alters rhizosphere dynamics are microbial inoculants, including *Trichoderma* spp., mycorrhizal fungi, and PGPR. In agricultural systems, microorganisms are crucial, especially those that plant growth-promoting microorganisms (PGPMs). The following three pathways are primarily responsible for the benefits of plant growth: (1) By supplying fixed N or other nutrients, PGPMs functioning as biofertilizers (such as phosphate-solubilizing bacteria and N-fixing bacteria) aid in the uptake of nutrients by plants (Kennedy and Islam, 2001); (2) microbes that express phytohormones, like *Azospirillum*, are known as phytostimulators. They can directly stimulate plant development by generating plant hormones (Spaepen et al., 2007; Glick et al., 2007). *Trichoderma*, *Pseudomonas*, and *Bacillus* are examples of biological control agents that shield plants against phytopathogenic organisms (Peighami-Ashnaei et al., 2009; Mohiddin et al., 2010; Dawar et al., 2010). By colonizing the rhizosphere and forming mutualistic relationships with plant roots, these microbes frequently enhance nutrient intake, water absorption, and disease defense. In addition to producing phytohormones like auxins and CKs that promote root growth, PGPR improves N fixation and solubilizes phosphorus (Rouphael and Colla, 2020). In a similar vein, AMF enhance the surface area of roots and promote effective uptake of immobile nutrients such as Zn and phosphorus (P). By modifying the activity of antioxidant enzymes, microbial inoculants have been demonstrated to lessen root oxidative stress in drought-stressed soils (Vishwakarma et al., 2019).

1.5.2 Impact on soil physicochemical properties

Humic and fulvic acids are examples of nonmicrobial biostimulants that have a major impact on soil health. Better nutrient availability is ensured by their capacity to boost CEC, improve soil aggregation, and increase water retention. Humic compounds help plants absorb nutrients by preventing nutrient leaching and facilitating their slow release through complex formation with micronutrients. When used as soil amendments, protein hydrolysates and SWEs increase microbial activity and the turnover of organic matter, improving soil fertility (du Jardin, 2015). This double impact on biological activity and soil physicochemical characteristics emphasizes how biostimulants work in concert to maintain the health of the rhizosphere.

1.5.3 Biostimulant-induced rhizosphere signaling

Biostimulants may function as signaling molecules in the rhizosphere, according to new research. For instance, chitosan activates rhizosphere microbes involved in N cycling and disease suppression in addition to improving plant defense responses (Pichyangkura and Chadchawan, 2015). Likewise, biostimulants originating from plants emit allelochemicals and secondary metabolites that affect the dynamics of microbial populations. The rhizosphere microbiome is modulated by signaling, which makes the soil ecosystem more robust and able to protect plants from biotic and abiotic challenges. During drought or salinity stress, certain microorganisms generate antioxidants and osmoprotectants that maintain the cellular integrity of plants (Nephali et al., 2020). Additionally, they can trigger the synthesis of hormones linked to stress, like ABA, which controls water loss and encourages root development. These microbes greatly improve soil structure, increase plant resilience, and improve water and nutrient uptake by interacting with plant roots (Rai et al., 2021).

1.5.4 Contribution to long-term soil sustainability

Biostimulants have the ability to enhance soil health over the long run in addition to providing short-term gains in crop output. Utilizing microbial inoculants and bioactive natural compounds can be a useful strategy to improve plant nutrient uptake and assimilation as well as soil nutrient availability (De Pascale et al., 2017). Their use reduces dependency on chemical pesticides and fertilizers, which can gradually damage soil quality. For economic and environmental reasons, it is critical to improve nutrient consumption efficiency, particularly for N and P (Barłóg 2023. Biostimulants improve beneficial microbial communities and promote the buildup of organic matter, hence preserving soil fertility, structure, and biodiversity. Integrating biostimulants into climate-smart agriculture is a sustainable technique for restoring degraded soils and ensuring crop stress tolerance.

1.6 Applications in sustainable agriculture

Biostimulants offer a possible path toward sustainable agricultural systems by reducing reliance on synthetic inputs while increasing crop yield and resilience. Their use increases water and nutrient use efficiency, as well as stress tolerance, both of which are critical for maintaining yields in resource-constrained and climate-stressed areas (Meena et al., 2025). Humic acid and protein hydrolysates increase nutrient availability, resulting in fewer fertilizer applications, whereas some microbial inoculants, such as AMF and PGPR, promote nutrient mobilization and soil fertility (Rouphael and Colla, 2020; Kumari et al., 2022). SWEs and plant-derived biostimulants are commonly used in organic and climate-smart farming systems to improve crop vigor, productivity, and yield stability, serving as sustainable alternatives to conventional chemical growth regulators and pesticides (Khoulati et al., 2025). Biostimulants also help with integrated pest and disease management by priming plant defense mechanisms and increasing the growth of beneficial rhizosphere bacteria that reduce pathogenic activity (Pichyangkura and Chadchawan, 2015; Carillo, 2025). Importantly, their involvement in altering root system design improves nutrient and water absorption efficiency, lowering input losses and reducing environmental pollution (Meena et al., 2025). Biostimulants promote sustainable soil ecosystems by hastening organic matter decomposition and cultivating a resilient, functionally varied microbial community. Incorporating biostimulants into precision agriculture frameworks promotes sustainable production by allowing for site-specific management, which lowers resource waste and maximizes physiological crop responses (Quille et al., 2025). As agriculture around the world strives to ensure food production while limiting negative environmental impacts, the use of biostimulants represents a possible path toward resilient and sustainable agroecosystems (Khoulati et al., 2025). Biostimulants—with wider adoption, confirmed by research, and supervised by regulatory frameworks—-could play a critical role in improving low-input, resilient agricultural systems capable of maintaining output under stress situations (Meena et al., 2025).

1.7 Challenges and future perspectives

Despite increased interest and documented benefits, biostimulants face various barriers to widespread application in agriculture. A key restriction is the absence of uniform definitions, formulations, and regulatory frameworks across countries. The term "biostimulant" encompasses a wide range of substances and organisms, which complicates classification and creates inconsistency in labeling, quality control, and marketing (du Jardin, 2015). Variability in source material, extraction processes, and environmental conditions often leads to inconsistent performance in field trials, making farmers hesitant to adopt them over conventional inputs. Another challenge is the limited mechanistic understanding of biostimulant action at the molecular and cellular levels. While many studies highlight improvements in nutrient uptake, stress tolerance, or root development, the precise signaling pathways, gene expression changes, and metabolomic shifts induced by biostimulants remain insufficiently characterized. This knowledge gap restricts the ability to design next-generation, targeted biostimulant formulations with predictable outcomes (De Diego and Spíchal, 2020). Scalability and cost-effectiveness are also important considerations, especially in resource-poor areas where affordability is a key predictor of adoption. Future perspectives include the use of multiomics methods, such as genomes, transcriptomics, proteomics, and metabolomics, to understand the mechanisms underpinning biostimulant effects and find biomarkers for efficacy prediction. Advances in biotechnology and synthetic biology may allow for the creation of engineered microbial consortia or bioactive compounds customized to specific crops and stress levels. Precision agriculture technologies, including remote sensing and AI-powered decision-making tools, can improve the timing and dose of biostimulant applications, increasing efficiency and sustainability. Furthermore, developing unified worldwide rules and certification systems will be critical for increasing farmer confidence and ensuring product consistency. Collaboration among academia, industry, and policymakers can boost innovation while resolving concerns about efficacy, safety, and environmental impact. Ultimately, biostimulants' future lies in their integration into holistic, climate-smart farming systems, where they serve as synergistic complements to traditional agronomic approaches rather than substitutes. Biostimulants have the potential to improve sustainable agriculture in the face of climate change and global food security challenges by connecting scientific knowledge to practical implementation.

1.8 Conclusion

Biostimulants have emerged as a transformative tool in modern agriculture, offering a sustainable pathway to improve crop productivity, resilience, and environmental stewardship. As global agriculture faces the dual challenges of climate change and increasing food demand, biostimulants provide a complementary approach to conventional fertilizers and pesticides by enhancing plant physiological functions, strengthening stress tolerance, and supporting soil and rhizosphere health. Unlike traditional agrochemicals, biostimulants act through multiple mechanisms, including improved nutrient uptake, modulation of hormonal signaling, enhancement of antioxidant defense systems, and stimulation of beneficial microbial interactions. These multifaceted actions position them as vital components of climate-smart and resource-efficient farming systems. The diversity of biostimulants ranging from humic substances, protein hydrolysates, and SWEs to microbial inoculants and plant-derived metabolites ensures a broad spectrum of applications across crops and environments. Their ability to improve WUE, reduce oxidative damage under stress, and promote root system development makes them particularly relevant in mitigating the adverse impacts of drought, salinity, and temperature extremes. Moreover, their contribution to soil microbial diversity and nutrient cycling aligns with the principles of sustainable intensification and regenerative agriculture.

However, realizing the full potential of biostimulants requires overcoming key challenges. The lack of standardized definitions, variable efficacy across environments, limited mechanistic insights, and fragmented regulatory frameworks remain significant hurdles. Addressing these issues will demand interdisciplinary research efforts, stronger collaboration between academia and industry, and the establishment of harmonized international guidelines for quality assurance and product validation. Integrating advanced technologies such as multiomics, synthetic biology, and precision farming tools can further accelerate the development of next-generation biostimulants with enhanced specificity and predictability. Looking ahead, the success of biostimulants will depend not only on scientific innovation but also on building trust and awareness among farmers, policymakers, and consumers. Demonstrating consistent field-level benefits, ensuring affordability, and aligning with sustainable development goals will be critical for widespread adoption. In this context, biostimulants should be viewed not as isolated inputs but as integral components of holistic, climate-resilient agricultural systems that balance productivity with ecological integrity. By bridging fundamental science with applied practice, biostimulants hold the promise of reshaping global agriculture, helping to secure food and nutritional security while safeguarding natural resources for future generations.

References

Almadi, L., Paoletti, A., Cinosi, N., Daher, E., Rosati, A., Di Vaio, C., et al., 2020. A biostimulant based on protein hydrolysates promotes the growth of young olive trees. Agriculture 10 (12), 618. Available at https://doi.org/10.3390/agriculture10120618.

Alzahrani, Y., Kuşvuran, A., Alharby, H.F., Kuşvuran, S., Rady, M.M., 2018. The defensive role of silicon in wheat against stress conditions induced by drought, salinity or cadmium. Ecotoxicology and Environmental Safety 154, 187–196. Available at https://doi.org/10.1016/j.ecoenv.2018.02.057.

Anjos Neto, A.P.dos, Oliveira, G.R.F., Mello, S.da C., Silva, M.S. da, Gomes-Junior, F.G., Novembre, A.D.daL.C., et al., 2020. Seed priming with seaweed extract mitigate heat stress in spinach: effect on germination, seedling growth and antioxidant capacity. Bragantia 79, 502–511. Available at https://doi.org/10.1590/1678-4499.20200127.

Arnao, M.B., Hernández-Ruiz, J., 2019. Melatonin: a new plant hormone and/or a plant master regulator? Trends in Plant Science 24 (1), 38–48. Available at https://doi.org/10.1016/j.tplants.2018.10.010.

Aroca, R., Ruiz-Lozano, J.M., Zamarreño, A.M., Paz, J.A., García-Mina, J.M., Pozo, M.J., et al., 2013. Arbuscular mycorrhizal symbiosis influences strigolactone production under salinity and alleviates salt stress in lettuce plants. Journal of Plant Physiology 170 (1), 47–55. Available at https://doi.org/10.1016/j.jplph.2012.08.020.

Ashraf, M., Foolad, M.R., 2007. Roles of glycine betaine and proline in improving plant abiotic stress resistance. Environmental and Experimental Botany 59 (2), 206–216. Available at https://doi.org/10.1016/j.envexpbot.2005.12.006.

Ayed, S., Bouhaouel, I., Jebari, H., Hamada, W., 2022. Use of biostimulants: towards sustainable approach to enhance durum wheat performances. Plants 11 (1), 133. Available at https://doi.org/10.3390/plants11010133.

Backer, R., Rokem, J.S., Ilangumaran, G., Lamont, J., Praslickova, D., Ricci, E., et al., 2018. Plant growth-promoting rhizobacteria: context, mechanisms of action, and roadmap to commercialization of biostimulants for sustainable agriculture. Frontiers in Plant Science 871. Available at https://doi.org/10.3389/fpls.2018.01473.

Baltazar, M., Correia, S., Guinan, K.J., Sujeeth, N., Bragança, R., Gonçalves, B., 2021. Recent advances in the molecular effects of biostimulants in plants: an overview. Biomolecules 11 (8), 1096. Available at https://doi.org/10.3390/biom11081096.

Barłóg, P., 2023. Improving Fertilizer Use Efficiency—Methods and Strategies for the Future. Plants 12, 3658. Available at https://doi.org/10.3390/plants12203658.

Batool, T., Ali, S., Seleiman, M.F., Naveed, N.H., Ali, A., Ahmed, K., et al., 2020. Plant growth promoting rhizobacteria alleviates drought stress in potato in response to suppressive oxidative stress and antioxidant enzymes activities. Scientific Reports 10 (1). Available at https://doi.org/10.1038/s41598-020-73489-z.

Battacharyya, D., Babgohari, M.Z., Rathor, P., Prithiviraj, B., 2015. Seaweed extracts as biostimulants in horticulture. Scientia Horticulturae 196, 39–48. Available at https://doi.org/10.1016/j.scienta.2015.09.012.

Bayat, H., Shafie, F., Aminifard, M.H., Daghighi, S., 2021. Comparative effects of humic and fulvic acids as biostimulants on growth, antioxidant activity and nutrient content of yarrow (*Achillea millefolium* L.). Scientia Horticulturae 279. Available at https://doi.org/10.1016/j.scienta.2021.109912.

Begum, N., Qin, C., Ahanger, M.A., Raza, S., Khan, M.I., Ashraf, M., et al., 2019. Role of arbuscular mycorrhizal fungi in plant growth regulation: implications in abiotic stress tolerance. Frontiers in Plant Science 10. Available at https://doi.org/10.3389/fpls.2019.01068.

Berg, G., Grube, M., Schloter, M., Smalla, K., 2014. The plant microbiome and its importance for plant and human health. Frontiers in Microbiology 5. Available at https://doi.org/10.3389/fmicb.2014.00491.

Bhattacharyya, P.N., Jha, D.K., 2012. Plant growth-promoting rhizobacteria (PGPR): emergence in agriculture. World Journal of Microbiology and Biotechnology 28 (4), 1327–1350. Available at https://doi.org/10.1007/s11274-011-0979-9.

Braziene, Z., Paltanavicius, V., Avizienytė, D., 2021. The influence of fulvic acid on spring cereals and sugar beets seed germination and plant productivity. Environmental Research 195, 110824. Available at https://doi.org/10.1016/j.envres.2021.110824.

Bulgari, R., Cocetta, G., Trivellini, A., Vernieri, P., Ferrante, A., 2015. Biostimulants and crop responses: a review. Biological Agriculture and Horticulture 31 (1), 1–17. Available at https://doi.org/10.1080/01448765.2014.964649.

Bulgari, R., Franzoni, G., Ferrante, A., 2019. Biostimulants application in horticultural crops under abiotic stress conditions. Agronomy 9 (6), 306. Available at https://doi.org/10.3390/agronomy9060306.

Calvo, P., Nelson, L., Kloepper, J.W., 2014. Agricultural uses of plant biostimulants. Plant and Soil 383 (1-2), 3–41. Available at https://doi.org/10.1007/s11104-014-2131-8.

Canellas, L.P., Olivares, F.L., Aguiar, N.O., Jones, D.L., Nebbioso, A., Mazzei, P., et al., 2015. Humic and fulvic acids as biostimulants in horticulture. Scientia Horticulturae 196, 15–27. Available at https://doi.org/10.1016/j.scienta.2015.09.013.

Capstaff, N.M., Morrison, F., Cheema, J., Brett, P., Hill, L., Muñoz-García, J.C., et al., 2020. Fulvic acid increases forage legume growth inducing preferential up-regulation of nodulation and signalling-related genes. Journal of Experimental Botany 71 (18), 5689–5704. Available at https://doi.org/10.1093/jxb/eraa283.

Carillo, P., 2025. Can biostimulants enhance plant resilience to heat and water stress in the Mediterranean hotspot? Plant Stress 16, 100802. Available at https://doi.org/10.1016/j.stress.2025.100802.

Casadesús, A., Polo, J., Munné-Bosch, S., 2019. Hormonal effects of an enzymatically hydrolyzed animal protein-based biostimulant (pepton) in water-stressed tomato plants. Frontiers in Plant Science 10. Available at https://doi.org/10.3389/fpls.2019.00758.

Chen, H., Zhou, Y., Li, J., Li, T., 2024. Plant biostimulant as an environmentally friendly alternative to modern agriculture. Journal of Agricultural and Food Chemistry 72 (1), 15–30.

Colla, G., Hoagland, L., Ruzzi, M., Cardarelli, M., Bonini, P., Canaguier, R., et al., 2017. Biostimulant Action of Protein Hydrolysates: Unraveling Their Effects on Plant Physiology and Microbiome. Frontiers in Plant Science 8. Available at https://doi.org/10.3389/fpls.2017.02202.

Colla, G., Nardi, S., Cardarelli, M., Ertani, A., Lucini, L., Canaguier, R., et al., 2015. Protein hydrolysates as biostimulants in horticulture. Scientia Horticulturae 196, 28–38. Available at https://doi.org/10.1016/j.scienta.2015.08.037.

Conrath, U., Beckers, G.J.M., Langenbach, C.J.G., Jaskiewicz, M.R., 2015. Priming for enhanced defense. Annual Review of Phytopathology 53, 97–119. Available at https://doi.org/10.1146/annurev-phyto-080614-120132.

Craigie, J.S., 2011. Seaweed extract stimuli in plant science and agriculture. Journal of Applied Phycology 23 (3), 371–393. Available at https://doi.org/10.1007/s10811-010-9560-4.

Cristofano, F., El-Nakhel, C., Rouphael, Y., 2021. Biostimulant substances for sustainable agriculture: origin, operating mechanisms and effects on cucurbits, leafy greens, and nightshade vegetables species. Biomolecules 11 (8), 1103. Available at https://doi.org/10.3390/biom11081103.

Dawar, S., Wahab, S., Tariq, M., Zaki, M.J., 2010. Application of *Bacillus* species in the control of root rot diseases of crop plants. Archives of Phytopathology and Plant Protection 43 (4), 412–418. Available at https://doi.org/10.1080/03235400701850870.

De Diego, N., Spíchal, L., 2022. Presence and future of plant phenotyping approaches in biostimulant research and development. Journal of Experimental Botany 73, 5199–5212. Available at https://doi.org/10.1093/jxb/erac275.

Drobek, M., Frąc, M., Cybulska, J., 2019. Plant biostimulants: importance of the quality and yield of horticultural crops and the improvement of plant tolerance to abiotic stress—a review. Agronomy 9 (6), 335. Available at https://doi.org/10.3390/agronomy9060335.

Egamberdieva, D., Wirth, S., Bellingrath-Kimura, S.D., Mishra, J., Arora, N.K., 2019. Salt-Tolerant Plant Growth Promoting Rhizobacteria for Enhancing Crop Productivity of Saline Soils. Frontiers in Microbiology 10. Available at https://doi.org/10.3389/fmicb.2019.02791.

Ertani, A., Cavani, L., Pizzeghello, D., Brandellero, E., Altissimo, A., Ciavatta, C., et al., 2009. Biostimulant activity of two protein hydrolyzates in the growth and nitrogen metabolism of maize seedlings. Journal of Plant Nutrition and Soil Science 172 (2), 237–244. Available at https://doi.org/10.1002/jpln.200800174.

Ertani, A., Pizzeghello, D., Francioso, O., Sambo, P., Sanchez-Cortes, S., Nardi, S., 2014. *Capsicum chinensis* L. growth and nutraceutical properties are enhanced by biostimulants in a long-term period: chemical and metabolomic approaches. Frontiers in Plant Science 5. Available at https://doi.org/10.3389/fpls.2014.00375.

Ertani, A., Francioso, O., Tinti, A., Schiavon, M., Pizzeghello, D., Nardi, S., 2018. Evaluation of seaweed extracts from laminaria and *Ascophyllum nodosum* spp. as *Biostimulants* in *Zea mays* l. using a combination of chemical, biochemical and morphological approaches. Frontiers in Plant Science 9. Available at https://doi.org/10.3389/fpls.2018.00428.

European Biostimulant Industry Council (EBIC), 2012. Representing plant biostimulants in policy and practice. European Biostimulant Industry Council, Brussels. Available at http://www.biostimulants.eu/.

FAO, 2017. The State of Food and Agriculture 2017. Leveraging food systems for inclusive rural transformation. FAO, Rome. Available at https://openknowledge.fao.org/handle/20.500.14283/i7658en FAO. 2017. The State of Food and Agriculture 2017. Leveraging food systems for inclusive rural transformation. Rome.

García, A.C., Van Tol De Castro, T.A., Santos, L.A., Tavares, O.C.H., Castro, R.N., Berbara, R.L.L., et al., 2019. Structure-property-function relationship of humic substances in modulating the root growth of plants: a review. Journal of Environmental Quality 48 (6), 1622–1632. Available at https://doi.org/10.2134/jeq2019.01.0027.

Garg, S., Nain, P., Kumar, A., Joshi, S., Punetha, H., Sharma, P.K., et al., 2024. Next generation plant biostimulants & genome sequencing strategies for sustainable agriculture development. Frontiers in Microbiology 15. Available at https://doi.org/10.3389/fmicb.2024.1439561.

Glick, B.R., 2012. Plant growth-promoting bacteria: mechanisms and applications. Scientifica 2012 (1), 963401.

Glick, B.R., Todorovic, B., Czarny, J., Cheng, Z., Duan, J., McConkey, B., 2007. Promotion of plant growth by bacterial ACC deaminase. Critical Reviews in Plant Sciences 26 (5-6), 227–242. Available at https://doi.org/10.1080/07352680701572966.

Godara, A., Bakshi, M., 2021. Effect of application of bio stimulants like humic acid and vermiwash on the growth, quality and yield of plant -a review. Plant Cell Biotechnology and Molecular Biology 22 (19-20), 22–30.

González-Hernández, A.I., Scalschi, L., Vicedo, B., Marcos-Barbero, E.L., Morcuende, R., Camañes, G., 2022. Putrescine: a key metabolite involved in plant development, tolerance and resistance responses to stress. International Journal of Molecular Sciences 23 (6), 2971. Available at https://doi.org/10.3390/ijms23062971.

Goñi, O., Fort, A., Quille, P., McKeown, P.C., Spillane, C., O'Connell, S., 2016. Comparative Transcriptome Analysis of Two *Ascophyllum nodosum* Extract Biostimulants: Same Seaweed but Different. Journal of Agricultural and Food Chemistry 64, 2980–2989. Available at https://doi.org/10.1021/acs.jafc.6b00621.

Harman, G.E., Lorito, M., Lynch, J.M., 2004. Uses of *Trichoderma* spp. to alleviate or remediate soil and water pollution. Advances in Applied Microbiology 56, 313–330. Available at https://doi.org/10.1016/S0065-2164(04)56010-0.

Herath, B.M.M.D., Madushan, K.W.A., Lakmali, J.P.D., Yapa, P.N., 2021. Arbuscular mycorrhizal fungi as a potential tool for bioremediation of heavy metals in contaminated soil. World Journal of Advanced Research and Reviews 10, 217–228. Available at https://doi.org/10.30574/wjarr.2021.10.3.0255.

Hasanuzzaman, M., Nahar, K., Alam, M.M., Roychowdhury, R., Fujita, M., 2013. Physiological, biochemical, and molecular mechanisms of heat stress tolerance in plants. International Journal of Molecular Sciences 14 (5), 9643–9684. Available at https://doi.org/10.3390/ijms14059643.

Hadwiger, L.A., 2013. Multiple effects of chitosan on plant systems: Solid science or hype. Plant Science 208, 42–49. Available at https://doi.org/10.1016/j.plantsci.2013.03.007.

Ikram, M., Ali, N., Jan, G., Jan, F.G., Rahman, I.U., Iqbal, A., et al., 2018. IAA producing fungal endophyte *Penicillium roqueforti* Thom., enhances stress tolerance and nutrients uptake in wheat plants grown on heavy metal contaminated soils. PLoS One 13 (11), e0208150. Available at https://doi.org/10.1371/journal.pone.0208150.

Iriti, M., Faoro, F., 2014. Chitosan as a MAMP, searching for a PRR. Plant Signaling & Behavior 4 (1), 66–68. Available at https://doi.org/10.4161/psb.4.1.7408.

Jacob, P., Hirt, H., Bendahmane, A., 2017. The heat-shock protein/chaperone network and multiple stress resistance. Plant Biotechnology Journal 15 (4), 405–414. Available at https://doi.org/10.1111/pbi.12659.

du Jardin, P., 2015. Plant biostimulants: definition, concept, main categories and regulation. Scientia Horticulturae 196, 3–14. Available at https://doi.org/10.1016/j.scienta.2015.09.021.

Jennings, R.C., 1968. Gibberellins as endogenous growth regulators in green and brown algae. Planta 80 (1), 34–42. Available at https://doi.org/10.1007/bf00387187.

Johnson, R., Puthur, J.T., 2022. Biostimulant priming in *Oryza sativa*: a novel approach to reprogram the functional biology under nutrient-deficient soil. Cereal Research Communications 50 (1), 45–52. Available at https://doi.org/10.1007/s42976-021-00150-4.

Kennedy, I.R., Islam, N., 2001. The current and potential contribution of asymbiotic nitrogen fixation to nitrogen requirements on farms: a review. Australian Journal of Experimental Agriculture 41 (3), 447–457. Available at https://doi.org/10.1071/ea00081.

Khan, W., Rayirath, U.P., Subramanian, S., Jithesh, M.N., Rayorath, P., Hodges, D.M., et al., 2009. Seaweed extracts as biostimulants of plant growth and development. Journal of Plant Growth Regulation 28 (4), 386–399. Available at https://doi.org/10.1007/s00344-009-9103-x.

Khan, A., Kanwal, F., Ullah, S., Fahad, M., Tariq, L., Altaf, M., et al., 2025. Plant Secondary Metabolites—Central Regulators Against Abiotic and Biotic Stresses. Metabolites 15, 276. Available at https://doi.org/10.3390/metabo15040276.

Khan, A., Kanwal, F., Ullah, S., Fahad, M., Tariq, L., Altaf, M., et al., 2025. Plant Secondary Metabolites—Central Regulators Against Abiotic and Biotic Stresses. Metabolites 15, 276. Available at https://doi.org/10.3390/metabo15040276.

Khatoon, S., Jain, K., Mahajan, M., Khan, M.I.R., 2023. Chapter 2 - Nitric oxide biosynthesis under stressful environments. In: Khan, M.I.R., Iqbal, N., Poór, P., Ferrante, A., Singh, V.P., Tripathi, D.K. (Eds.), Nitric Oxide in Developing Plant Stress Resilience. Academic Press, pp. 17–30. Available at https://doi.org/10.1016/B978-0-323-91209-9.00013-0.

Khoulati, A., Ouahhoud, S., Taibi, M., Ezrari, S., Mamri, S., Merah, O., et al., 2025. Harnessing biostimulants for sustainable agriculture: innovations, challenges, and future prospects. Discover Agriculture 3. Available at https://doi.org/10.1007/s44279-025-00177-9.

Khatoon, S., Mahajan, M., Khan, M.I.R., 2025. Ethylene–potassium interactions safeguard ionic balance, chlorophyll biosynthesis, and source–sink metabolism to sustain wheat growth and yield responses under salt stress. Plant Stress 18. Available at https://doi.org/10.1016/j.stress.2025.101049.

Khatoon, S., Mahajan, M., Kumari, S., Iqbal, N., Wahid, I., Iqbal, M., et al., 2024. Green-synthesized gold nanoparticles induce adaptation in photosynthetic responses, sugar and nitrogen metabolism, and seed yield of salt-stressed mustard plants. Clean Technologies and Environmental Policy. Available at https://doi.org/10.1007/s10098-024-02761-x.

Khoulati, A., Ouahhoud, S., Taibi, M., Ezrari, S., Mamri, S., Merah, O., et al., 2025. Harnessing biostimulants for sustainable agriculture: innovations, challenges, and future prospects. Discover Agriculture 3, 56. Available at https://doi.org/10.1007/s44279-025-00177-9.

Kumar, A., Dubey, A., 2020. Rhizosphere microbiome: engineering bacterial competitiveness for enhancing crop production. Journal of Advanced Research 24, 337–352. Available at https://doi.org/10.1016/j.jare.2020.04.014.

Kumar, A., Kumar, S., Venkatesh, K., Singh, N.K., Mandal, P.K., Sinha, S.K., 2022. Physio-molecular traits of contrasting bread wheat genotypes associated with 15N influx exhibiting homeolog expression bias in nitrate transporter genes under different external nitrate concentrations. Planta 255 (5), 104. Available at https://doi.org/10.1007/s00425-022-03890-7.

Kumar, V., Kumar, A., Tewari, K., Garg, N.K., Changan, S.S., Tyagi, A., 2022b. Isolation and characterization of drought and ABA responsive promoter of a transcription factor encoding gene from rice. Physiology and Molecular Biology of Plants 28 (10), 1813–1831. Available at https://doi.org/10.1007/s12298-022-01246-9.

Kumari, M., Swarupa, P., Kesari, K.K., Kumar, A., 2022. Microbial inoculants as plant biostimulants: a review on risk status. Life (Chicago, Ill.: 1978) 13 (1), 12. Available at https://doi.org/10.3390/life13010012.

Lesk, C., Rowhani, P., Ramankutty, N., 2016. Influence of extreme weather disasters on global crop production. Nature 529, 84–87. Available at https://doi.org/10.1038/nature16467.

Lotfi, R., Kalaji, H.M., Valizadeh, G.R., Khalilvand Behrozyar, E., Hemati, A., Gharavi-Kochebagh, P., et al., 2018. Effects of humic acid on photosynthetic efficiency of rapeseed plants growing under different watering conditions. Photosynthetica 56 (3), 962–970. Available at https://doi.org/10.1007/s11099-017-0745-9.

Liang, B., Ma, C., Zhang, Z., Wei, Z., Gao, T.g, Zhao, Q., et al., 2018. Long-term exogenous application of melatonin improves nutrient uptake fluxes in apple plants under moderate drought stress. Environmental and Experimental Botany 155, 650–661. Available at https://doi.org/10.1016/j.envexpbot.2018.08.016.

Ma, Y., Freitas, H., Dias, M.C., 2022. Strategies and prospects for biostimulants to alleviate abiotic stress in plants. Frontiers in Plant Science 13. Available at https://doi.org/10.3389/fpls.2022.1024243.

Mahajan, M., Nazir, F., Jahan, B., Siddiqui, M.H., Iqbal, N., Khan, M.I.R., 2023. Salicylic acid mitigates arsenic stress in rice (*Oryza sativa*) via modulation of nitrogen–sulfur assimilation, ethylene biosynthesis, and defense systems. Agriculture 13 (7), 1293.

Mahajan, M., Chaudhary, S., Khan, M.I.R., 2025a. Lanthanum (a rare earth element)-induced abscisic acid responses to modulate stomatal dynamics, nitrogen-carbon assimilation, metabolomic profile and yield traits in wheat. Journal of Hazardous Materials 497. Available at https://doi.org/10.1016/j.jhazmat.2025.139763.

Mahajan, M., Poor, P., Kaur, H., Aher, R.R., Palakolanu, S.R., Khan, M.I.R., 2025b. Salt stress tolerance and abscisic acid in plants: associating role of plant growth regulators and transcription factors. Plant Physiology and Biochemistry 228. Available at https://doi.org/10.1016/j.plaphy.2025.110303.

Mahajan, M., Khatoon, S., Prakash, S., Khan, M.I.R., 2025. Plant stress hormones: fostering drought stress resilience. In: Khan, M.I.R., Nazir, F., Siddique, K.H.M. (Eds.), Optimizing Plant Health Under Abiotic-Stress Environments: Ensuring Food and Nutritional Security. CRC Press, pp. 123–142.

Mahajan, M., Sahoo, R.N., Khan, M.I.R., 2025d. Lanthanum supplementation with abscisic acid modulates rhizosphere dynamics through changes in nitric oxide synthesis in wheat. Physiologia Plantarum 177 (1). Available at https://doi.org/10.1111/ppl.70126.

Maignan, V., Bernay, B., Géliot, P., Avice, J.C., 2020. Biostimulant effects of glutacetine® and its derived formulations mixed with n fertilizer on post-heading n uptake and remobilization, seed yield, and grain quality in winter wheat. Frontiers in Plant Science 11. Available at https://doi.org/10.3389/fpls.2020.607615.

Maignan, V., Géliot, P., Avice, J.C., 2021. Glutacetine® biostimulant applied on wheat under contrasting field conditions improves grain number leading to better yield, upgrades n-related traits and changes grain ionome. Plants 10 (3), 1–21. Available at https://doi.org/10.3390/plants10030456.

Markets and Markets. 2022. Biostimulants market by active ingredient (Humic substances, amino acids, seaweed extracts and microbial amendments), mode of application (foliar, soil treatment, seed treatment), form (liquid, and dry), crop type, and region - Global Forecast to 2027. AGI2920. https://www.marketsandmarkets.com/Market-Reports/biostimulant-market-1081.html

Meena, D.C., Birthal, P.S., Kumara, T.M.K., 2025. Biostimulants for sustainable development of agriculture: a bibliometric content analysis. Discover Agriculture 3, 2. Available at https://doi.org/10.1007/s44279-024-00149-5.

Mittler, R., 2017. ROS are good. Trends in Plant Science 22 (1), 11–19. Available at https://doi.org/10.1016/j.tplants.2016.08.002.

El Modafar, C., Elgadda, M., El Boutachfaiti, R., Abouraicha, E., Zehhar, N., Petit, E., et al., 2012. Induction of natural defence accompanied by salicylic acid-dependant systemic acquired resistance in tomato seedlings in response to bioelicitors isolated from green algae. Scientia Horticulturae 138, 55–63. Available at https://doi.org/10.1016/j.scienta.2012.02.011.

Mohiddin, F.A., Khan, M.R., Khan, S.M., Bhat, B.H., 2010. Why trichoderma is considered super hero (Super fungus) against the evil parasites? Plant Pathology Journal 9 (3), 92–102. Available at https://doi.org/10.3923/ppj.2010.92.102.

Munns, R., Tester, M., 2008. Mechanisms of salinity tolerance. Annual Review of Plant Biology 59 (1), 651–681. Available at https://doi.org/10.1146/annurev.arplant.59.032607.092911.

Nadeem, S.M., Ahmad, M., Zahir, Z.A., Javaid, A., Ashraf, M., 2014. The role of mycorrhizae and plant growth promoting rhizobacteria (PGPR) in improving crop productivity under stressful environments. Biotechnology Advances 32 (2), 429–448. Available at https://doi.org/10.1016/j.biotechadv.2013.12.005.

Nardi, S., Schiavon, M., Francioso, O., 2021. Chemical structure and biological activity of humic substances define their role as plant growth promoters. Molecules (Basel, Switzerland) 26 (8), 2256.

Nazir, F., Mahajan, M., Khatoon, S., Albaqami, M., Ashfaque, F., Chhillar, H., et al., 2023. Sustaining nitrogen dynamics: a critical aspect for improving salt tolerance in plants. Frontiers in Plant Science 14, 1087946. Available at https://doi.org/10.3389/fpls.2023.1087946.

Nephali, L., Piater, L.A., Dubery, I.A., Patterson, V., Huyser, J., Burgess, K., et al., 2020. Biostimulants for plant growth and mitigation of abiotic stresses: a metabolomics perspective. Metabolites 10 (12), 505. Available at https://doi.org/10.3390/metabo10120505.

Nolan, T., Chen, J., Yin, Y., 2017. Cross-talk of brassinosteroid signaling in controlling growth and stress responses. Biochemical Journal 474 (16), 2641–2661. Available at https://doi.org/10.1042/BCJ20160633.

Omidbakhshfard, M.A., Sujeeth, N., Gupta, S., Omranian, N., Guinan, K.J., Brotman, Y., et al., 2020. A biostimulant obtained from the seaweed ascophyllum nodosum protects arabidopsis thaliana from severe oxidative stress. International Journal of Molecular Sciences 21 (2), 474. Available at https://doi.org/10.3390/ijms21020474.

Van Oosten, M.J., Pepe, O., De Pascale, S., Silletti, S., Maggio, A., 2017. The role of biostimulants and bioeffectors as alleviators of abiotic stress in crop plants. Chemical and Biological Technologies in Agriculture 4, 5. Available at https://doi.org/10.1186/s40538-017-0089-5.

De Pascale, S., Rouphael, Y., Colla, G., 2017. Plant biostimulants: innovative tool for enhancing plant nutrition in organic farming. European Journal of Horticultural Science 82 (6), 277–285. Available at https://doi.org/10.17660/eJHS.2017/82.6.2.

Paul, K., Sorrentino, M., Lucini, L., Rouphael, Y., Cardarelli, M., Bonini, P., et al., 2019. Understanding the biostimulant action of vegetal-derived protein hydrolysates by high-throughput plant phenotyping and metabolomics: a case study on tomato. Frontiers in Plant Science 10. Available at https://doi.org/10.3389/fpls.2019.00047.

Peighami-Ashnaei, S., Sharifi-Tehrani, A., Ahmadzadeh, M., Behboudi, K., 2009. Interaction of different media on production and biocontrol efficacy of Pseudomonas fluorescens P-35 and bacillus subtilis B-3 against grey mould of apple. Journal of Plant Pathology 91 (1), 65–70. Available at https://doi.org/10.4454/jpp.v91i1.625.

Pichyangkura, R., Chadchawan, S., 2015. Biostimulant activity of chitosan in horticulture. Scientia Horticulturae 196, 49–65. Available at https://doi.org/10.1016/j.scienta.2015.09.031.

Pieterse, C.M.J., Zamioudis, C., Berendsen, R.L., Weller, D.M., Van Wees, S.C.M., Bakker, P.A.H.M., 2014. Induced systemic resistance by beneficial microbes. Annual Review of Phytopathology 52, 347–375. Available at https://doi.org/10.1146/annurev-phyto-082712-102340.

Porcel, R., Aroca, R., Ruiz-Lozano, J.M., 2012. Salinity stress alleviation using arbuscular mycorrhizal fungi. a review. Agronomy for Sustainable Development 32 (1), 181–200. Available at https://doi.org/10.1007/s13593-011-0029-x.

Qin, K., Leskovar, D.I., 2018. Lignite-derived humic substances modulate pepper and soil-biota growth under water deficit stress. Journal of Plant Nutrition and Soil Science 181 (5), 655–663. Available at https://doi.org/10.1002/jpln.201800078.

Quille, P., Kacprzyk, J., O'Connell, S., Ng, C.K.Y., 2025. Reducing fertiliser inputs: plant biostimulants as an emerging strategy to improve nutrient use efficiency. Discover Sustainability 6 (1). Available at https://doi.org/10.1007/s43621-025-00910-w.

Rai, N., Rai, S.P., Sarma, B.K., 2021. Prospects for abiotic stress tolerance in crops utilizing phyto- and biostimulants. Frontiers in Sustainable Food Systems 5. Available at https://doi.org/10.3389/fsufs.2021.754853.

Rentel, M.C., Lecourieux, D., Ouaked, F., Usher, S.L., Petersen, L., Okamoto, H., et al., 2004. OXI1 kinase is necessary for oxidative burst-mediated signalling in *Arabidopsis*. Nature 427 (6977), 858–861. Available at https://doi.org/10.1038/nature02353.

Rouphael, Y., Colla, G., 2020. Toward a sustainable agriculture through plant biostimulants: from experimental data to practical applications. Agronomy 10 (10), 1461. Available at https://doi.org/10.3390/agronomy10101461.

Rouphael, Y., Colla, G., Giordano, M., El-Nakhel, C., Kyriacou, M.C., De Pascale, S., 2017. Foliar applications of a legume-derived protein hydrolysate elicit dose-dependent increases of growth, leaf mineral composition, yield and fruit quality in two greenhouse tomato cultivars. Scientia Horticulturae 226, 353–360. Available at https://doi.org/10.1016/j.scienta.2017.09.007.

Rouphael, Y., Giordano, M., Cardarelli, M., Cozzolino, E., Mori, M., Kyriacou, M.C., et al., 2018. Plant-and seaweed-based extracts increase yield but differentially modulate nutritional quality of greenhouse spinach through biostimulant action. Agronomy 8 (7), 126. Available at https://doi.org/10.3390/agronomy8070126.

Roychoudhury, A., Tripathi, D., 2019. Osmolytes and their role in abiotic stress tolerance in plants. In: Roychoudhury, A., Tripathi, D. (Eds.), Molecular Plant Abiotic Stress: Biology and Biotechnology. Wiley, pp. 91–104. Available at https://doi.org/10.1002/9781119463665.ch5.

Ruiz-Lozano, J.M., Porcel, R., Bárzana, G., Azcón, R., Aroca, R., 2012. Contribution of arbuscular mycorrhizal symbiosis to plant drought tolerance: state of the art. In: Aroca, R. (Ed.), Plant Responses to Drought Stress: From Morphological to Molecular Features. Springer, Berlin, Heidelberg, pp. 335–362. Available at https://doi.org/10.1007/978-3-642-32653-0_13.

Ruiz-Lozano, J.M., Aroca, R., Zamarreño, Á.M., Molina, S., Andreo-Jiménez, B., Porcel, R., et al., 2016. Arbuscular mycorrhizal symbiosis induces strigolactone biosynthesis under drought and improves drought tolerance in lettuce and tomato. Plant Cell and Environment 39 (2), 441–452. Available at https://doi.org/10.1111/pce.12631.

Santi, C., Zamboni, A., Varanini, Z., Pandolfini, T., 2017. Growth stimulatory effects and genome-wide transcriptional changes produced by protein hydrolysates in maize seedlings. Frontiers in Plant Science 8. Available at https://doi.org/10.3389/fpls.2017.00433.

Santoyo, G., Moreno-Hagelsieb, G., del Carmen Orozco-Mosqueda, M., Glick, B.R., 2016. Plant growth-promoting bacterial endophytes. Microbiological Research 183, 92–99. Available at https://doi.org/10.1016/j.micres.2015.11.008.

Savci, S., 2012. Investigation of effect of chemical fertilizers on environment. APCBEE Procedia 1, 287–292. Available at https://doi.org/10.1016/j.apcbee.2012.03.047.

Selmar, D., Kleinwächter, M., 2013. Influencing the product quality by deliberately applying drought stress during the cultivation of medicinal plants. Industrial Crops and Products 42 (1), 558–566. Available at https://doi.org/10.1016/j.indcrop.2012.06.020.

Seo, P.J., Lee, A.-K., Xiang, F., Park, C.-M., 2008. Molecular and functional profiling of *Arabidopsis* pathogenesis-related genes: insights into their roles in salt response of seed germination. Plant and Cell Physiology 49 (3), 334–344. Available at https://doi.org/10.1093/pcp/pcn011.

Sestili, F., Rouphael, Y., Cardarelli, M., Pucci, A., Bonini, P., Canaguier, R., et al., 2018. Protein hydrolysate stimulates growth in tomato coupled with N-dependent gene expression involved in N assimilation. Frontiers in Plant Science 9. Available at https://doi.org/10.3389/fpls.2018.01233.

Sharma, A., Shahzad, B., Rehman, A., Bhardwaj, R., Landi, M., Zheng, B., 2019. Response of phenylpropanoid pathway and the role of polyphenols in plants under abiotic stress. Molecules (Basel, Switzerland) 24 (13), 2452. Available at https://doi.org/10.3390/molecules24132452.

Shukla, P.S., Borza, T., Critchley, A.T., Prithiviraj, B., 2016. Carrageenans from red seaweeds as promoters of growth and elicitors of defense response in plants. Frontiers in Marine Science 3 (MAY). Available at https://doi.org/10.3389/fmars.2016.00081.

Shukla, P.S., Shotton, K., Norman, E., Neily, W., Critchley, A.T., Prithiviraj, B., 2018. Seaweed extract improve drought tolerance of soybean by regulating stress-response genes. AoB plants 10 (1), plx051.

Sinha, S.K., Kumar, A., Tyagi, A., Venkatesh, K., Paul, D., Singh, N.K., et al., 2020. Root architecture traits variation and nitrate-influx responses in diverse wheat genotypes under different external nitrogen concentrations. Plant Physiology and Biochemistry 148, 246–259. Available at https://doi.org/10.1016/j.plaphy.2020.01.018.

Smith, S.E., Smith, F.A., 2011. Roles of arbuscular mycorrhizas in plant nutrition and growth: new paradigms from cellular to ecosystem scales. Annual Review of Plant Biology 62, 227–250. Available at https://doi.org/10.1146/annurev-arplant-042110-103846.

Spaepen, S., Vanderleyden, J., 2011. Auxin and Plant-microbe interactions. Cold Spring Harbor Perspectives in Biology 3 (4), a001438. Available at https://doi.org/10.1101/cshperspect.a001438.

Spaepen, S., Vanderleyden, J., Remans, R., 2007. Indole-3-acetic acid in microbial and microorganism-plant signaling. FEMS Microbiology Reviews 31 (4), 425–448. Available at https://doi.org/10.1111/j.1574-6976.2007.00072.x.

Stirk, W.A., Novák, O., Hradecká, V., Pěnčík, A., Rolčík, J., Strnad, M., et al., 2009. Endogenous cytokinins, auxins and abscisic acid in *Ulva fasciata* (Chlorophyta) and *Dictyota humifusa* (Phaeophyta): towards understanding their biosynthesis and homoeostasis. European Journal of Phycology 44 (2), 231–240. Available at https://doi.org/10.1080/09670260802573717.

Sun, J., Qiu, C., Ding, Y., Wang, Y., Sun, L., Fan, K., et al., 2020. Fulvic acid ameliorates drought stress-induced damage in tea plants by regulating the ascorbate metabolism and flavonoids biosynthesis. BMC Genomics 21, 411. Available at https://doi.org/10.1186/s12864-020-06815-4.

Taskos, D., Stamatiadis, S., Yvin, J.C., Jamois, F., 2019. Effects of an *Ascophyllum nodosum* (L.) Le Jol. extract on grapevine yield and berry composition of a *Merlot vineyard*. Scientia Horticulturae 250, 27–32. Available at https://doi.org/10.1016/j.scienta.2019.02.030.

Tester, M., Langridge, P., 2010. Breeding technologies to increase crop production in a changing world. Science (New York, N.Y.) 327 (5967), 818–822. Available at https://doi.org/10.1126/science.1183700.

Tiwari, M., Kumar, R., Subramanian, S., Doherty, C.J., Jagadish, S.V.K., 2023. Auxin–cytokinin interplay shapes root functionality under low-temperature stress. Trends in Plant Science 28 (4), 447–459. Available at https://doi.org/10.1016/j.tplants.2022.12.004.

Tyagi, K., V, P., Tyagi, P., Kumari, A., Pandey, R., Meena, N.L., et al., 2023. Seed priming with melatonin induces rhizogenesis and modulates physio-biochemical traits in high-yielding rice (*Oryza sativa* L.) genotypes. South African Journal of Botany 163, 191–200. Available at https://doi.org/10.1016/j.sajb.2023.10.043.

Vishwakarma, K., Mishra, M., Patil, G., Mulkey, S., Ramawat, N., Pratap Singh, V., et al., 2019. Avenues of the membrane transport system in adaptation of plants to abiotic stresses. Critical Reviews in Biotechnology 39, 861–883. Available at https://doi.org/10.1080/07388551.2019.1616669.

Vriet, C., Russinova, E., Reuzeau, C., 2013. From squalene to brassinolide: the steroid metabolic and signaling pathways across the plant kingdom. Molecular Plant 6 (6), 1738–1757. Available at https://doi.org/10.1093/mp/sst096.

Vurukonda, S.S.K.P., Vardharajula, S., Shrivastava, M., SkZ, A., 2016. Enhancement of drought stress tolerance in crops by plant growth promoting rhizobacteria. Microbiological Research 184, 13–24. Available at https://doi.org/10.1016/j.micres.2015.12.003.

Woo, S.L., Ruocco, M., Vinale, F., Nigro, M., Marra, R., Lombardi, N., et al., 2014. Trichoderma-based Products and their Widespread Use in Agriculture. The Open Mycology Journal 8, 71–126. Available at https://doi.org/10.2174/1874437001408010071.

Watts-Williams, S.J., Gilbert, S.E., 2021. Arbuscular mycorrhizal fungi affect the concentration and distribution of nutrients in the grain differently in barley compared with wheat. Plants People Planet 3 (5), 567–577. Available at https://doi.org/10.1002/ppp3.10090.

Xiao, T., Boada, R., Marini, C., Llugany, M., Valiente, M., 2020. Influence of a plant biostimulant on the uptake, distribution and speciation of Se in Se-enriched wheat (*Triticum aestivum* L. cv. Pinzón). Plant and Soil 455 (1-2), 409–423. Available at https://doi.org/10.1007/s11104-020-04686-y.

Xing, K., Zhu, X., Peng, X., Qin, S., 2015. Chitosan antimicrobial and eliciting properties for pest control in agriculture: a review. Agronomy for Sustainable Development 35 (2), 569–588. Available at https://doi.org/10.1007/s13593-014-0252-3.

Xu, L., Geelen, D., 2018. Developing biostimulants from agro-food and industrial by-products. Frontiers in Plant Science 9. Available at https://doi.org/10.3389/fpls.2018.01567.

Yakhin, O.I., Lubyanov, A.A., Yakhin, I.A., Brown, P.H., 2017. Biostimulants in plant science: a global perspective. Frontiers in Plant Science 7. Available at https://doi.org/10.3389/fpls.2016.02049.

Zandonadi, D.B., Santos, M.P., Busato, J.G., Peres, L.E.P., Façanha, A.R., 2013. Plant physiology as affected by humified organic matter. Theoretical and Experimental Plant Physiology 25 (1), 13–25. Available at https://doi.org/10.1590/S2197-00252013000100003.

Zhang, N., Zhao, B., Zhang, H.J., Weeda, S., Yang, C., Yang, Z.C., et al., 2013. Melatonin promotes water-stress tolerance, lateral root formation, and seed germination in cucumber (*Cucumis sativus* L.). Journal of Pineal Research 54 (1), 15–23. Available at https://doi.org/10.1111/j.1600-079X.2012.01015.x.

Zhu, C', Ling, N', Guo, J', Wang, M', Guo, S', Shen, Q', 2016. Impacts of Fertilization Regimes on Arbuscular Mycorrhizal Fungal (AMF) Community Composition Were Correlated with Organic Matter Composition in Maize Rhizosphere Soil. Frontiers in Microbiology 7. Available at https://doi.org/10.3389/fmicb.2016.01840.

Zulfiqar, F., Moosa, A., Ali, H.M., Bermejo, N.F., Munné-Bosch, S., 2024. Biostimulants: a sufficiently effective tool for sustainable agriculture in the era of climate change? Plant Physiology and Biochemistry 211, 108699. Available at https://doi.org/10.1016/j.plaphy.2024.108699.

CHAPTER

2

Utilizing plant growth regulators for sustainable agricultural intensification under abiotic stress

Anshu Jha, Ragini Singh, Preeti Singh, Shahid Umar, and Noushina Iqbal

Department of Botany, School of Chemical and Life Sciences, Jamia Hamdard, New Delhi, Delhi, India

2.1 Introduction

Industrialization and increasing population lead to climate change and an increase in food demand, altering the global environment (Malhi et al., 2020). Abiotic stress caused due to climate change directly and indirectly affects the crop growth and productivity of crops. Drought, heat, salinity, and heavy metals have an alarming impact on food security worldwide (Cakmak, 2005; Mahajan et al., 2025a,b,c,d). The shift in environmental conditions intensifies the effects of abiotic stressors, which has a negative role on the agricultural ecosystems and human well-being (Arunanondchai et al., 2018). Additionally, crop performance is disturbed by diminishing soil fertility, air pollution, and limited water availability (Noya et al., 2018). Over the past few decades, the indiscriminate use of chemical fertilizers and pesticides has indeed helped boost crop yields but has also caused severe ecological imbalances by degrading soil fertility, polluting water bodies, and disrupting beneficial microbial communities.

Phytohormones serve as the plant's internal communication system and biochemical messengers that coordinate countless physiological and metabolic activities. They help plants adjust and thrive even when faced with challenging environmental conditions. Phytohormones such as auxins (AX), gibberellins (GAs), cytokinins (CKs), abscisic acid (ABA), ethylene (ET), salicylic acid (SA), jasmonates (JAs), brassinosteroids (BRs), and strigolactones (SLs) work together through intricate signaling networks, guiding essential processes like cell division, growth, aging, and responses to abiotic stress (Rademacher, 2015; Mahajan et al., 2023; Patyal et al., 2024). Application of plant growth regulators (PGRs) to improve crop resilience is a significant way to overcome the challenges faced by abiotic stress and ensure sustainability (Patyal et al., 2024). PGRs are naturally occurring or synthetic compounds that, at low dosage,

Harnessing Plant-Stimulating Molecules
DOI: https://doi.org/10.1016/B978-0-443-44994-9.00014-6

control the developmental and metabolic processes such as flowering, fruit formation, and ripening (Rademacher, 2015). By modulating hormonal balance and signaling pathways, plants could better utilize the available nutrients, maintain osmotic homeostasis, enhance antioxidant defenses, and sustain productivity under abiotic stress conditions such as drought, salinity, and heat stress. The present chapter focuses on the role of PGRs in abiotic stress tolerance through various approaches.

2.2 Plant growth regulators: an introduction to major classes and their functional roles in plant growth and stress tolerance

PGRs can be used for modern agriculture as a provision for sustainable agriculture. They could enhance crop yield by 10–15 million tons per year, thereby reducing the need for chemical fertilizers and pesticides, leading to environmental destruction (Rademacher, 2015; Zhang et al., 2024). They could be used as synonyms—growth hormones, phytohormones, plant hormones, and bioregulators (Sabagh et al., 2021). They could act as thermoprotectants, reduce oxidative stress, improve photosynthesis, accumulate of stress-responsive proteins, and perform many other regulatory functions necessary in mitigating abiotic stress conditions (Sharma et al., 2020). PGRs are synthesized endogenously by plants and modulate a wide range of physiological functions. Their synthetic counterparts could initiate diverse biochemical and physiological pathways that take part in plant growth and development (Sabagh et al., 2021; Mahajan et al., 2025a). They regulate cell division, enlargement, and differentiation, which provide control of dormancy, senescence, and organogenesis from the germination stage until fructification. They could also control physiological processes such as sugar metabolism, plant stress responses, and stomata aperture control (Lopez-Lauri, 2016). Phytohormones antagonistically or synergistically work together in regulation of plant defense and growth (Zhang et al., 2022). The most important natural PGRs are ABA, ET, CKs, AX, and GAs. Other PGRs identified are JA, SA, and BRs and SLs.

2.2.1 Auxin

In plants, AX was first discovered as a mobile signal mediating the phototropic growth of coleoptiles (Darwin, 2017). Indole-3-acetic acid (IAA), the most abundant plant hormone from the AX family of indole derivatives, is involved in the regulation of plant growth and development, such as cell division, elongation, fruit development, and senescence (Grossmann, 2010; McSteen, 2010). It has significance in plant defense against abiotic stress (Bianco and Defez, 2009; Nazir et al., 2023b). AX plays a role in numerous developmental processes such as vascular tissue differentiation, axillary bud formation, apical dominance, root growth, and floral organ development (Zhao, 2010).

2.2.2 Cytokinin

CK is one of the key phytohormones that affect cell division, which in turn affects plant growth (Chen et al., 1985; Binns, n.d.; Wybouw and De Rybel, 2019). It also stimulates lateral buds and inhibits the aging of organs and plant tissues (Wybouw and De Rybel, 2019). It plays a role in a variety of processes, including fruit and leaf abscission, morphogenesis, chloroplast biogenesis, seed germination, assimilate mobilization and maintenance, stomatal functioning regulation, and root-to-shoot signaling in stressed plants (Chen et al., 1985; Binns, n.d.). N6-adenine derivatives known as CK were initially identified by van Overbeek et al. (1941) in the milk of coconuts (*Cocos nucifera*) (Sakakibara, 2006). The most prominent CKs present in higher plants are isopentenyladenine (iP), zeatin (Z), and dihydrozeatin (DZ). The biologically active substances are believed to be free bases and their ribosides (iPR, ZR, and DZR). Glycosidic conjugates contribute to the transit of CK, defense against degradation, and both irreversible and reversible inactivation (Mok, 2019). The synthesis of CK mainly takes place in root apical meristems, but it can also be produced in fruits and young leaves (Šmehilová et al., 2009).

2.2.3 Gibberellin

GAs form a large group of carboxylic acids and are tetracyclic and diterpenoid phytohormones (Gupta and Chakrabarty, 2013; Camara et al., 2018). GA has an important role in the promotion of organ growth, cell elongation and/or division, and activation of most developmental processes (i.e., seed germination, induction of flowering, and maturation) in many plant species (Gupta and Chakrabarty, 2013; Nazir et al., 2024). GA was first discovered in the fungus *Gibberella fujikuroi,* which stimulated growth in infected vascular plants. GAs are numbered in order of their discovery, starting with GA1 (Takahashi, 1998). The biologically active GAs in plants are GA1, GA3, GA4, and GA7. These differ from each other according to the presence of a hydroxyl group in C-3β, a carboxyl group in C-6, and a lactone between C-4 and C-10. The 3β-hydroxyl group can be replaced by other functional groups at positions C-2 and/or C-3. However, GA5 and GA6 are bioactive forms of GAs that lack a hydroxyl group at C-3β (Camara et al., 2018).

2.2.4 Abscisic acid

ABA is involved in many cellular processes, including seed development, dormancy, germination, vegetative growth, and responses to environmental stresses (Xiong and Zhu, 2003; Mahajan et al., 2025d). The 15-carbon sesquiterpene ABA is produced in chloroplasts and other plastids from carotenoids. The essential enzyme for the production of ABA is 9-cis-epoxycarotenoid dioxygenase (NCED) (Seo and Koshiba, 2002). ABA is present in high amounts in young tissues (Cardoso et al., 2020). In general, nutrient deficiencies result in higher ABA concentrations, while temperatures above 40°C cause them to drop (Cardoso et al., 2020). ABA concentrations in root tissue are typically lower than those in leaves; dehydration of detached roots of different species, ages, and branching orders also promotes the synthesis of ABA (Davies, 1995). In plants, ABA is primarily synthesized in mature leaves and roots via the carotenoid pathway (Chen et al., 2020).

2.2.5 Ethylene

ET is a volatile phytohormone, which controls many developmental and physiological activities, including seed dormancy and germination, growth of vegetation, flowering, maturation of climacteric fruit, and aging (Watkins, 2002; Khatoon et al., 2025). Additionally, ET provides protection to plants from biological and nonbiological stressors. It causes leaf abscission and senescence in response to stressor developmental signals (Rademacher, 2015). ET treatments are used for the ripening of climacteric fruits, such as peppers and tomatoes, for maintaining consistency in harvest (Watkins, 2002). Since ET is a gaseous phytohormone, it could easily pass through from its production site due to biochemical and/or metabolic processes, which makes sensing very prompt (Pattyn et al., 2021; Sharma et al., 2023).

2.2.6 Brassinosteroid

BRs are polyhydroxylated steroid phytohormones, which have a significant role in morphological and physiological responses toward resistance to abiotic and biotic stress (Yao et al., 2023). BRs have a crucial role in various developmental processes such as fruit formation, floral initiation, and shoot and root growth. BRs provide antioxidant defense mechanisms, which in turn reduces the accumulation of reactive oxygen species (ROS) and enhances photosystem protection in plants against abiotic stress (Vardhini and Anjum, 2015; Sharma et al., 2017). Additionally, to maintain membrane function in abiotic stress, BRs can alter the structure of cellular walls and cellular membranes. Furthermore, BR signaling has an essential role in promoting the expression of various enzymes linked to the mechanisms involved in abiotic stress tolerance, including loosening and extension of cellular walls (Guo et al., 2009), and increasing the expression of pectin methyl esterase (Yang et al., 2014).

2.2.7 Strigolactone

SLs are carotenoid-derived terpenoids, which were discovered as a germination catalyst for root parasitic plant, witchweed (*Striga* spp.). SL has a role in seed germination, shoot branching, leaf senescence, root development, and protection against abiotic stress (e.g., light stress and high-temperature stress) (Wu et al., 2022). They can regulate interactions between arbuscular mycorrhizal fungi and root parasitic plants, thereby known as rhizosphere chemical signals (Soliman et al., 2022). There is growing evidence of a crosstalk between SL signaling and function; SLs' signaling and their impact on plant architecture suggest that SLs interact with other PGRs, specifically AXs and ABA (Cheng et al., 2013).

2.2.8 Jasmonic acid

JAs originate from oxylipins and are involved in plant development and stress responses (Katsir et al., 2008). It mediates a variety of biotic and abiotic stress processes in plants, including ultraviolet (UV) radiation, wounding, fruit ripening, and seed germination (Gomi, 2020). The first indication that JA promotes senescence came from a component that was isolated from wormwood and caused oats to rapidly lose their chlorophyll (Kelley and Estelle, 2012). The metabolites of JA have been derived from various reactions involving esterification, methylation, sulfation, glycosylation, conjugation, decarboxylation, hydroxylation, and carboxylation. Methyl jasmonate (MeJA) is a volatile phytohormone, which is metabolized from JA (Wasternack and Strnad, 2018). JA and MeJA are two of the major phytohormones that regulate both defense responses and the development processes (Wasternack et al., 2013). Additionally, volatile oil from plants can be promoted with the help of JA, thereby enhancing the antibacterial qualities. It has been observed that JA has a significant role in controlling the life cycle of plants (Gomi, 2020).

2.2.9 Salicylic acid

SA is a noteworthy phytohormone participating in plant growth and development (Larkindale and Knight, 2002). Heat production in plants is initiated by SA by initiating an alternative respiration pathway, which releases foul smell produced by compounds that attract pollinators (Arif et al., 2020). SA acts as a messenger and manages the expression of plant-pathogen-related genes providing disease resistance (Mohamed et al., 2020). It provides a defense system toward abiotic stress factors, such as heat, saline conditions, and oxidative stress. SA plays a major role in regulating photosynthesis, respiration, vegetative growth, seed germination, flowering, and aging (Arif et al., 2020). In plants, SA fuels the SAR mechanism, which in turn triggers the pathogenesis-related proteins that fight against phytopathogens, including bacteria and fungi (Mohamed et al., 2020). Additionally, it regulates the interaction between plants and water, hence controls many metabolic activities in plants (Mohamed et al., 2020).

2.3 Role of plant growth regulators in abiotic stress tolerance in plants

2.3.1 Impact of plant growth regulators on drought stress tolerance

Among the major environmental factors limiting plant productivity is water scarcity. Due to the severity and duration of stress, crop output faces more loss than any other source of stress (Gull et al., 2019). The severity of the drought depends on the availability and distribution of rainfall, evaporation requirements, and the moisture-storing capacity of soils (Hazel, 1995; Mahajan et al., 2025c). As temperatures and atmospheric carbon dioxide (CO_2) levels continue to rise, the climate has changed globally in recent years. Climate change has led to an uneven distribution of rainfall due to changes in temperature and atmospheric CO_2, which acts as a significant stressor in the form of drought. Crop plants stop growing after being subjected to drought stress. Under drought stress conditions, plants reduce their metabolic needs and shoot growth (Grime, 1977). To enhance the metabolic rates and increase the plant resistance to drought stress conditions, following phytohormones work individually and in combination.

AX is involved in cell dedifferentiation and differentiation, root morphology or architecture, geotropism, root growth, floral organ development, and seed dormancy (Khosravi-nejad et al., 2022). AX homeostasis regulates ABA production in response to drought stress (Farhangi-Abriz and Torabian, 2018). In drought-stressed rice (*Oryza sativa*), transcription of IAA-conjugating genes is increased (Ronzan et al., 2019). AX controls hydrotropism responses, which allow plant roots to detect and react to the moisture content of the soil (Kaneyasu et al., 2007). It has been observed that AX is required for the *Arabidopsis thaliana* (*Arabidopsis*) roots to undergo the transition from cell division to cell differentiation (Di Mambro et al., 2017). Hydrotropism and gravitropism take place simultaneously in the soil when AX and ABA work together to encourage the formation of root hair (Cassab et al., 2013). AX has a smaller role in hydrotropism than in gravitropism (Cassab et al., 2013).

The use of ABA reduces abiotic stressors, such as moisture deficiency (Keskin et al., 2010). To ensure resistance to moisture stress, high cellular ABA stimulates changes in stomatal conductance, root hydraulic conductivity, photosynthesis, biomass distribution between roots and shoots, osmolyte formation, and the synthesis of stress-responsive proteins and genes (Finkelstein et al., 2008; Lim et al., 2015). Many plants' leaves accumulate ABA when they are under water stress (Upreti and Murti, 1999). After the stomata are closed, the increase in ABA enables plants to limit water loss through transpiration, and when the underlying hydraulic conductivity rises, the plant's water uptake improves (Thompson et al., 2007). According to Zhang et al. (2006), ABA also regulates stomatal conductance, allele expression, and catabolic changes in conditions of moisture deficit, all of which are crucial for the movement of nutrients from roots to stems. The 9-cis-epoxycarotenoid dioxygenase (NCED) is an enzyme used in ABA production and is found to be a dominant contributor in ABA production in Arabidopsis under drought stress. Under drought stress, NCED gets upregulated in Arabidopsis (Burbidge et al., 1999), tomato (Tan et al., 1997), and maize (Chernys and Zeevaart, 2000). NCED3 increases the ABA level, which promotes stomatal closure (Takahashi et al., 2018).

On the other hand, elevated CK levels in plants have been shown to improve proline levels, delaying leaf senescence and improving plant viability in water-stressed environments (Alvarez et al., 2008). Increased CK levels increased plant survival in conditions of water deprivation (Rivero et al., 2007). SARK, a promoter triggered by stress and maturation, is used upstream of isopentenyltransferase (IPT) to produce drought tolerance in monocots and dicots by overexpressing it (Rivero et al., 2007; Qin et al., 2011). Overexpression of CK oxidase/dehydrogenase (CKX) may lead to downregulation of CK content (Werner et al., 2010). Elevated content of protective compounds can be achieved by constitutive CKX expression, which leads to increased drought tolerance in Arabidopsis (Werner et al., 2001; Nishiyama et al., 2011), tobacco (Macková et al., 2013), and barley (Pospíšilová et al., 2016). It has been shown in Akter et al.'s (2014) study that at the reproductive stage, the drought-stressed maize plants when treated with CK showed an 80.9% increased stem diameter relative to untreated plants, plant height was also observed to be increased. The dry weight of the plant increased with the increase of CK concentration at the vegetative stage. Yield was found to improve with increased CK concentration.

To adapt to water scarcity, changes in root architecture are essential. A crosstalk between GA and IAA in roots in response to drought is associated with lateral root extension and decreased GA2 oxidase expression in the roots (Gou et al., 2010; Krugman et al., 2011). Signals from the roots sustain root growth and redistribute photo-assimilates to regulate shoot and root growth. Water status can be elevated by GA3 treatments of water-stressed maize plants. Increase in dry weight, chlorophyll content, and seedling growth was observed in plants treated with GA3. It can reduce the proline accumulation caused by the drought stress (Kaya et al., 2006). Crosstalk between ET and GA appears to be the cause of ABA-mediated root signals that result in stomatal closure (Coelho Filho et al., 2013). The *DELLA* and *XERICO* genes, which impede growth but promote plant survival, were expressed more in Arabidopsis under drought stress (Zentella et al., 2007). Water deficit has been linked to extended distance signaling of 1-aminocyclopropane-1-carboxylic acid (ACC), where the increased ACC produced high levels of endogenous ET in the plants (Mayak et al., 2004). This led to abscission, early senescence, growth hindrance, and ultimately a yield penalty in important crops (Dodd, 2005). In plants, ET causes leaf abscission, which reduces water loss, and improves water usage efficiency during dry conditions (Apelbaum and Yang, 1981; Zhang et al., 2009b) showed that tolerance to drought stress can be executed in transgenic tobacco by expressing the drought- and ET-inducible soybean *GmERF3* gene due to a higher proline and soluble sugar accumulation. According to Scarpeci et al. (2017), overexpressing the *AtERF019* gene in Arabidopsis provided drought tolerance through reducing stomatal conductance and decreasing the transpiration rate. Kentucky bluegrass (*Poa pratensis*) when pretreated with ethephon promotes antioxidant enzyme activity, and the accumulation of proline that leads to an osmotic enhancement when exposed to drought stress (Zhang et al., 2021).

BRs have been observed to increase tolerance against high temperatures in canola (*Brassica napus*) (Singh and Shono, 2005) and brome grass. With the help of BRs, the plants gain resistance due to the increased activity of the antioxidant pathway. BRs alter the antioxidant capacity of plants under stress conditions as they act as secondary messengers (Wilen et al., 1995). According to Sairam (1994), the enhancement of proteins in wheat plants is by homobrassinolide under moisture stress conditions. SLs applied exogenously have been observed to improve the shoot and root length under drought stress. It also enhances the shoot and root dry weight under moisture conditions. The chlorophyll a (21%), chlorophyll b (68%), and chlorophyll a+b (31%) increased in drought-stressed plants under application of SLs. It significantly enhances the catalase (CAT) activity by 43% and ascorbate peroxidase (APX) activity by 68% in drought-stressed plants as compared to control (Sattar et al., 2022). Plant hormones interact with one another at different stages to control environmental and developmental pathways (Zaid and Mohammad, 2018). Under drought stress, JAZ proteins are destroyed, resulting in active transcription factors such as MYC2, which upregulate genes associated with abiotic stress tolerance (Ghorbel et al., 2021). There is much evidence suggesting SA may have a role in drought stress by modifying regulating drought-related genes through transcriptional regulation and stomatal aperture (Yadav et al., 2021). In green olive (*Phillyrea angustifolia*), endogenous SA increased during water shortage (Llanes et al., n.d.; Álvarez et al., 2019). Proline production is controlled by SA, which also maintains cellular redox state in turnip (*Brassica rapa*) plants (La et al., 2019).

2.3.2 Role of plant growth regulators in salt stress tolerance

Salt stress is one of the most significant limiting factors for crop productivity and growth (Khatoon et al., 2024). Salt stress lowers crop yields and production, thus making it a global threat to global agriculture. It may be caused by natural processes or human activity, such as the weathering of rocks, irrigation schemes that use salt-rich irrigation water, or inadequate drainage (He et al., 2018). Crop growth and yield are decreased by salt stress in a number of ways. Osmotic stress and ion toxicity are the two primary impacts of salt stress on crop plants. Some downstream effects of salt stress include decreased cytosolic metabolism, assimilate synthesis, and impaired cell growth and membrane function (Fahad et al., 2017).

It has been widely established how PGRs control a variety of physiological and biochemical processes as well as the physiological traits of several crops grown under salt stress (Fatma et al., 2014). An efficient and useful method for granting crop salt resistance may be the exogenous application of PGRs (Amjad et al., 2014). Increase in salt stress leads to inhibition of the formation of roots, but mild salt stress causes less elongation and larger growth of lateral roots (Zolla et al., 2010; Nazir et al., 2023a).

Phytohormones and many signal molecules can together interact to control salt stress, with AX serving as a crucial mediator (Ribba et al., 2020). Under salt stress, the lateral root development is improved by the salt-responsive AX signaling genes *AXR1, AXR4, IAR4,* and *TIR1*. Basipetal AX transport modifications can help in the circulation of AX efflux transporters (PIN2) and modulate the gravitropism responses, which may contribute to minimizing the effects of salt stress in plants (Vanneste and Friml, 2009; Galvan-Ampudia and Testerink, 2011). Increase in osmolarity of the shoot sap, the regularity of shoot turgor, and growth are maintained in salt-resistant hybrids under the control of different AX concentrations (De Costa et al., 2007). Indole-3-butyric acid (IBA), an endogenous AX, is found in maize. Also, IAA makes resistant maize genotypes under salt stress by maintaining the osmolarity (Epstein and Ludwig, 1993). According to Skirycz et al. (2011) and Claeys et al. (2012), GA mediates osmotic changes in response to salt stress. Exogenous application of GA_3, a bioactive GA, enhanced plant water availability at low salt stress conditions in tomato plants (Maggio et al., 2010). GA_3 can mitigate the disadvantages of salt stress in mustard plants. It can restore the normal chlorophyll level by enhancing ultrastructural morphogenesis of plastids, which is coupled with retention of chlorophyll and delay of senescence. The application of exogenous GA_3 was reported to improve the seed germination rate under salt stress (Arteca, 1996; Shah, 2007). Growth and grain yield can be increased by seed treatment with GA3 (Javid et al., 2011). Furthermore, when Arabidopsis is exposed to salt stress, ET and GA work together to control the cell cycle and proliferation (Skirycz et al., 2011; Claeys et al., 2012).

ET, also known as stress phytohormone, is involved in salt stress response, though it has roles in the regulation of plant growth and development (Abeles et al., 2012). ACC and ET can be induced by salt and other abiotic stresses in many plant species (Morgan and Drew, 1997). Under salt stress, some plants like *Arabidopsis, Cakile maritima,* and *Thellungiella salsuginea* accumulate more ACC in both leaves and roots (Ellouzi et al., 2014). Tolerance of plants to high salt can be increased by the application of ET or ACC (Cao et al., 2007) or by enhancing the expression of ROS scavengers (Peng et al., 2014). Endogenous ET can increase resistance to salt stress tolerance in various plants such as Arabidopsis and maize (Riyazuddin et al., 2020). Ion toxicity and abnormal plant growth may take place in salt stress due to the accumulation of sodium (Na^+) and chloride (Cl^-) ions and the production of ROS. Antiporters can facilitate the efflux of sodium from the cytosol; ROS scavenging enzymes such as superoxide dismutase (SOD), peroxidase (POX), and CAT with the help of ET, which may also help in the accumulation of osmolytes such as proline and glycine betaine to prevent dehydration (Kukreja et al., 2005; Savvides et al., 2016; Li et al., 2019; Lang et al., 2020; Riyazuddin et al., 2020).

CKs are involved in making the plant salt tolerant (Iqbal et al., 2006; Javid et al., 2011). CK biosynthesis reduces in the root system and the supply to shoots also reduces under salt stress, which leads to alteration in the global gene expression network that elicits the responses required to resist the salt stress (Hare et al., 1997; Nishiyama et al., 2012). Kinetin can break the stress-induced seed dormancy of crop plants such as tomato, barley, and cotton when applied exogenously (Javid et al., 2011). Kinetin acts as a direct free radical scavenger or may be involved in an antioxidative mechanism, which avoids purine breakdown (Javid et al., 2011). Additionally, CKs enhance the capacity of antioxidant defense systems to protect their cells from the accumulation of ROS triggered by stress, preserve chloroplast integrity (Rivero et al., 2007), reduce electrolyte leakage, and/or enhance malondialdehyde (MDA) levels (Liao et al., 2017; Xu et al., 2017). In wheat, kinetin used in seed treatment was shown to be effective in increasing germination rate in the salt-intolerant cultivar under salt stress. It also increased shoot fresh, dry weights, and root fresh weight in the salt-tolerant cultivar under salt stress (Iqbal et al., 2006).

According to Duan et al. (2013), ABA is also thought to be a crucial signaling molecule that promotes the inhibition of lateral root growth in plants under salt stress. When lateral root cells are under salt stress, the ABA level elevates completely, causing postemergence lateral roots to become dormant. To reduce the passage of Na^+ ions through the endodermis, lateral roots produce a thick and well-developed Casparian strip during the dormant season. In the presence of Na^+ ions, ABA signaling is triggered within endodermal cells, preventing lateral root expansion in conditions of high salt. There is some evidence of crosstalk between ABA and JA for regulating salt stress (Moons et al., 1997; Shinozaki and Yamaguchi-Shinozaki, 2007). Elongation and increased biomass production in shoots of Kharchia were observed with ABA treatment under salt stress. ABA treatment can induce proline accumulation, thereby increasing the activities of synthesizing enzymes, i.e., glutathione dehydrogenase (GDH) and pyrroline-5-carboxylate synthase (P5CS) (Kaur and Asthir, 2020).

High endogenous JA levels are reported in salt-tolerant cultivar crops compared to the salt-sensitive cultivar (Kang et al., 2005). Additionally, biologically active JA is found to be increased by high salt stress in plants (Moons et al., 1997), which indicates that the higher JA could act as an effective protector against salt stress. In salt-tolerant rice, exogenous application of JA can effectively reduce Na^+ (Kang et al., 2005). Moreover, JA treatment can recover seedling development and photosynthetic activity in several cultivar crops that were disrupted due to salt stress (Yoon et al., 2009; Javid et al., 2011).

BRs application helps in making the salt-tolerant plants (Ali et al., 2007). Seed dormancy caused by salt stress can be eliminated by BRs (Kagale et al., 2007). The effect of salt-induced inhibition of photosynthesis can also be improved by BRs (Ali et al., 2007; Hayat et al., 2007). BRs can activate ribulose-1,5-bisphosphate carboxylase/oxygenase (RuBisCo), which is important for efficient photosynthesis, and it is protected by BRs along with other photosynthetic apparatus (Takahama and Oniki, 1992; Yamori et al., 2006). The quantum yield of photosystem II (PSII) in wheat is enhanced with the application of BRs (Shahbaz et al., 2008). BRs mitigate the inhibitory effect of salt stress on photosynthetic pigments (Anuradha and Ram Rao, 2003).

Under salt stress, foliar applications of SA results in increased fresh and dry weight of roots and shoots, shoot diameter, leaf number, and leaf relative water content (RWC) in cucumber plants. SA foliar spray lowered the electrolyte leakage, which was increased under salt stress and led to an increase in nutrient content under salt stress in cucumber plants (Yildirim et al., 2008). The SL treatment on rice seedlings increases plant height and root length with high-salt-stress treatment. The increased root system is due to the enhancement in absorbing more nutrients and water. Due to salt stress the aboveground and underground plant parts are suppressed, which can be mitigated by synthetic SL GR24. SL improves the photosynthesis in the rice, which is an advantage under adverse stress (Ling et al., 2020). Similarly, in chili plants, SL can increase plant height, dry weight, fruit length, fruit girth, and fruit yield under salt stress. Chlorophyll a, chlorophyll b, and total chlorophyll are also observed to be improved under salt stress on application of SLs (Danish et al., 2024).

2.3.3 Role of plant growth regulators in heat stress tolerance

Rising global temperature is becoming a serious issue that impacts plant productivity and development, especially in crops. The most significant factor affecting crop growth and food security is heat stress (Liliane and Charles, 2020). Heat stress harms plants' yield, photosynthetic efficiency, and seed germination rate (Liu et al., 2018). AX helps seedlings elevate the photosynthetic and meristematic tissues away from heat-absorbing soil by elongating hypocotyls, thereby providing the cooling effect of moving air (Gray et al., 1998). Under heat stress, exogenous application of auxin induces anther cell proliferation and can reverse male sterility caused by high temperatures (Sakata et al., 2010; Oshino et al., 2011). AX is also involved in reproductive development; its effect was seen in the grain yield of five wheat lines that were exposed to different levels of heat stress for 6 hours per day during early flowering. A semidwarf cultivar of wheat, 'CDC Go', responded to AX application (1 µM) and produced increased yield and had higher grain number (Abeysingha et al., 2021).

CK plays an important role in minimizing heat injury by leaf cooling due to an increase in transpiration rate. CK stimulates the opening of the stomata that facilitates transpiration. Heat shock protein, glycine-rich proteins, and many other heat-responsive proteins are induced by CK under heat stress. Endogenous level of CK can be increased by over-expression of the CK biosynthetic gene *isopentenyl ltransferase* (*IPT*) that enhances heat stress tolerance in bentgrass (Xing et al., 2009). Injuries caused by heat stress in creeping bentgrass can be alleviated by injecting CK to the root zone. Zeatin riboside, a bioactive cytokine, when applied exogenously was effective in slowing leaf senescence and alleviating heat-induced lipid peroxidation of cell membranes (Liu and Huang, 2002). CK and GA act antagonistically in numerous developmental processes. In tomato, GA-dependent hypocotyl elongation as well as leaf serration is inhibited by CK (Fleishon et al., 2011).

Exogenous application of GA_3 in Arabidopsis reversed the inhibition in the seed germination and seedling growth (Alonso-Ramírez et al., 2009). During heat stress, there is rapid elongation of the stem. However, when biosynthesis of GA is suppressed, induced hypocotyl elongation in Arabidopsis is reduced (Stavang et al., 2009). Flowering time is not only important for shaping the life cycles of plants but also that of the pollinator species. GAs can influence flowering as they stimulate degradation of DELLA protein that influences PIF4, which can activate flowering locus T (FT). Heat stress may accelerate floral transition, which is dependent on FT. FT, on the other hand, is activated by PIF4, thereby GAs control flowering (Kumar et al., 2012). Foliar spray of GA3 was observed to have a favorable impact on the morphological, physiological, and biochemical characteristics of tomato. GA3 helps to alleviate the heat stress and increase the shoot length in tomato. 75 mg L^{-1} of GA3 spray produces maximum shoot length in high temperatures. Reduced electrolyte leakage, MDA, and hydrogen peroxide (H_2O_2) contents and increased antioxidant defense system can be obtained by exogenous application of GA3 in a concentration-dependent manner (Guo et al., 2022).

ET plays an important role in thermotolerance by activating several stress-related proteins involved in maintaining the functional integrity and stability of plant cells, when applied exogenously (Jegadeesan et al., 2018). ET signaling helps in reducing oxidative stress and maintaining chlorophyll to confer heat stress tolerance. ET signaling also helps in activating heat shock protein (HSPs) (Wu and Yang, 2019). Ethephon, an ET releasing compound, when applied exogenously shows improved kernel set and grain yield during heat stress conditions in maize (Cicchino et al., 2013). Ethephon application also improves the pollen viability and pollen germination in heat stress (Jegadeesan et al., 2018).

According to Hayat et al. (2010), SA is a universal phytohormone that regulates different physiological responses when exposed to abiotic stress. According to Mohammed and Tarpley (2009), applying SA to rice could improve the dry count distribution by up to 16%, reducing the negative impacts of nighttime high temperatures of 32°C. Rice's ability to withstand heat stress was enhanced by foliar spraying it with SA (0.1 mM) (Chandrakala et al., 2013). Applying SA through foliar spray improved the proline, sugar, and antioxidant enzymes (CAT, POD, and APX) as well as phytohormone concentrations (ABA, IAA, and GA3) of rice spikelets, which in turn enhanced spikelet numbers, seed setting rate, and crop yield (Zhang et al., 2017).

According to Kobayasi et al. (2009), certain growth regulators, such as MeJA, can extend the flowering period till the early morning and may be utilized to alleviate heat stress. According to Kobayasi and Atsuta (2009), the application of MeJA led to both an improvement in the quantity of flowers and an early opening of the flowers. Under heat stress, JA can lead to heat shock–induced enhancement in secondary metabolism in *Angelica sinensis* (Dang Gui) (Ahammed et al., 2016).

ABA levels transiently increase when plants are exposed to heat stress in *Pisum sativum* (pea), which might indicate its involvement in heat sensing and acclimation (Liu et al., 2006). Under heat stress, exogenous application of ABA can promote starch accumulation by increasing the activities of glutamate-oxaloacetate transaminase and glutamate-pyruvate transaminase (Asthir and Bhatia, 2014). In the tobacco plant, stomatal conductance increases under heat stress, which is mediated by ABA. During high transpiration to decrease the leaf temperature, there is excess water loss, which is reduced by stomatal closure induced by ABA (Macková et al., 2013). By improving photosynthetic activity, water use efficiency (WUE), and crop output, foliar administration of ascorbic acid reduced heat stress (Fahad et al., 2016).

BR helps in maintaining the thermotolerance in *B. napus* (rapeseed), which is associated with increased synthesis and accumulation of HSPs (Dhaubhadel et al., 2002). Exogenous application of BR helps to increase net photosynthetic rate, stomatal conductance, and maximum carboxylation rate of RuBisCO in tomato leaves under heat stress (Ogweno et al., 2008). Under heat stress, BR upregulated genes are linked to plant growth and development processes, which target genes encoding cell elongation and cell wall modification enzymes, AX-responsive factors, and TFs (Goda et al., 2002; Müssig and Altmann, 2003; Vert et al., 2005).

SL biosynthesis or signaling is crucial for the temperature tolerance of plants. Application of SLs is recommended for heat suppression at leaf elongation, which is associated with the increase in transcript levels of cell cycle–related genes and decrease in transcript levels of genes involved in AX transport in elongating leaves of tall fescue (Hu et al., 2019). SL biosynthesis is induced at an early stage of heat stress. It accumulates in the roots and plays a positive role in the heat stress response and HSP70 protein accumulation. SLs help in the induction of ABA biosynthesis and ABA-dependent transcriptional responses under heat stress. They are generally involved in the regulation of the enzymatic antioxidant response under both optimal and stress conditions in tomato plants (Chi et al., 2021). Table 2.1 shows the roles of various PGRs in abiotic stress tolerance.

2.3.4 Role of plant growth regulators in alleviation of heavy metal stress

Due to increasing industrialization, sewage wastewater irrigation, and a greater dependency on chemical fertilizers, toxic metals are added to agricultural soils, which negatively impacts the soil–plant environment system (Nievola et al., 2017). These toxic heavy metals inhibit plants' uptake of nutrients and induce cellular damage in plants, subsequently affecting plant development and plant productivity. Over time, these metals bioaccumulate and gradually infiltrate plants through the air, water, and food chain (Rauwane and Ntushelo, 2020). PGRs are employed to increase crop plants' biomass production by reducing the detrimental effects of metal stress (Ouzounidou and Ilias, 2005). According to Peleg and Blumwald (2011) and Vriet et al. (2012), PGRs—specifically, ABA, GAs, AXs, CKs, ET, JAs, BRs, and SA—are also crucial for helping plants during heavy metal toxicity.

AX, under metal stress, responds by change of auxin homeostasis including auxin redistribution, AX transport, and AX stability (Potters et al., 2009). ROS and auxin avoid deleterious effects of stress on plant growth under metal stress. When Arabidopsis seedlings were exposed to oxidative stress-inducing agents, AX homeostasis and physiological responses were altered, suggesting possible crosstalk (Pasternak et al., 2005). AX application under cadmium stress reduced the damage of plants (MingMing et al., 2010). Cadmium toxicity on maize plants was minimized by treatment of maize coleoptile segments with IAA (Karcz and Kurtyka, 2007).

ABA provides resistance to plants against toxic metals such as arsenic (As), cadmium (Cd), and lead (Pb) (Maestri and Marmiroli, 2012; Vishwakarma et al., 2017). ABA content increased in rice, potato (*Solanum tuberosum*), oilseed rape (*B. napus*), apple (*Malus hupehensis*), Alfred's Stonecrop (*Sedum alfredii*), and other plants when exposed to Cd (Hsu and Kao, 2007; Stroiński et al., 2013; Yan et al., 2016; Shi et al., 2019; Zhang et al., 2019; Lu et al., 2020). ABA production peaked in the roots and leaves of Cd-tolerant rice cultivar than those in the Cd-sensitive genotype (Hsu and Kao, 2007), which indicates the positive correlation between endogenous ABA content and Cd tolerance. Pretreatment with ABA in the leaves of Indian mustard may reduce Cd accumulation (Salt et al., 1995). Moreover, exposing germinating chickpea (*Cicer arietinum*) and leaves of pea with Pb leads to increased endogenous ABA levels (Parys et al., 1998; Atici et al., 2005). It was observed that ABA biosynthesis was upregulated in the roots and shoots of Arabidopsis (Pb sensitive) and hoary mustard (*Hirschfeldia incana*) (Pb tolerant) (Auguy et al., 2016).

CKs provide plant resistance to heavy metals through different signaling pathways. Exogenous application of CKs elevates plant resistance to metal stress (Fahad et al., 2015;

TABLE 2.1 Role of plant growth regulators in various agricultural crops in abiotic stress tolerance.

PGRs	Plants	Abiotic stress	Impacts on plant	References
Auxin	Rice	Drought	Transcription of IAA-conjugating genes is increased.	Ronzan et al. (2019)
	Maize	Salt	IAA maintains the osmolarity of the shoot sap.	Epstein and Ludwig (1993)
	Wheat	Heat	Increases the grain yield and number.	Abeysingha et al. (2021)
	Maize	Heavy metal	IAA makes plant resistant for cadmium toxicity.	Karcz and Kurtyka (2007)
Abscisic acid	Arabidopsis	Drought	*NCED* gets upregulated in *Arabidopsis* and increases the ABA level, which promotes stomatal closure.	Burbidge et al. (1999), Takahashi et al. (2018)
	Kharchia	Salt	Elongation and increased biomass production in shoots.	Kaur and Asthir (2020)
	Tobacco	Heat	Stomatal conductance increases and water loss is reduced by stomatal closure.	Macková et al. (2013)
	Mustard	Heavy metal	Reduces cadmium accumulation in pretreated leaves of mustard with ABA.	Salt et al. (1995)
Ethylene	Kentucky bluegrass	Drought	Pretreatment with ethephon promotes antioxidant enzyme activity and the accumulation of proline that leads to an osmotic enhancement.	Zhang et al. (2021)
	Arabidopsis	Salt	Ethylene helps in activation of antiporters that can facilitate the efflux of sodium from the cytosol.	Kukreja et al. (2005)
	Maize	Heat	Ethephon, when applied exogenously shows improved kernel set and grain yield.	Cicchino et al. (2013)
	Pea	Heavy metal	Ethylene levels are elevated at high cadmium concentrations.	Vassilev et al. (2004)
Cytokinin	Maize	Drought	Increased stem diameter, plant height, dry weight, and yield attributes.	Akter et al. (2014)
	Wheat	Salt	Kinetin increases germination rate in salt-intolerant cultivar and increased shoot fresh, dry weights, and root fresh weight in the salt-tolerant cultivar.	Iqbal et al. (2006)
	Bentgrass	Heat	Zeatin riboside slows leaf senescence and alleviates heat-induced lipid peroxidation of cell membranes.	Liu and Huang (2002)
	Tomato	Heavy metal	Kinetin application ameliorates the cadmium-induced toxicity.	Singh et al. (2018)

(*Continued*)

TABLE 2.1 (Continued)

PGRs	Plants	Abiotic stress	Impacts on plant	References
Gibberellins	Tomato	Drought	Foliar spray of GA3 has a favorable impact on the morphological, physiological, and biochemical characteristics of tomato.	Guo et al. (2022)
	Mustard	Salt	GA3 can restore the normal chlorophyll level delay of senescence. Seed germination rate can be improved.	Arteca (1996), Shah (2007)
	Maize	Heat	GA3 can elevate the proline accumulation and improve physiological parameters.	Kaya et al. (2006)
	Pea	Heavy metal	Exogenous GA application lessens the toxicity of chromium on pea seed growth by controlling ROS formation.	Gangwar et al. (2011).

PGRs, Plant growth regulators; *IAA*, indole-3-acetic acid; *NCED*, 9-cis-epoxycarotenoid dioxygenase; *ABA*, abscisic acid; *GA*, gibberellins; *ROS*, reactive oxygen species.

Mohan et al., 2016). The kinetin group present in the molecular structure of CK has the potential to induce a plant antioxidant defense system, which provides the protective roles (Hönig et al., 2018). CKs interact with other phytohormones to reduce heavy metal uptake. It alters the expression of metal transporter genes or modifies the root architecture to reduce the absorption of metals (Bhardwaj and Garg, 2022). Moreover, CKs leads to improved antioxidant defense system in plants, which reduces oxidative stress caused by heavy metals (HMs) (Zhang et al., 2024) in green algae, including *Chlorella vulgaris* (Piotrowska-Niczyporuk et al., 2012) and *Acutodesmus obliquus* (Piotrowska-Niczyporuk et al., 2018). Under heavy metal stress, CKs promote cell division, growth, and tissue differentiation. Kinetin application ameliorates the Cd-induced toxicity in tomato (Singh et al., 2018) and pea seedlings (Al-Hakimi, 2007). The effects of AX (IAA, IBA, and phenylacetic acid [PAA]) and CKs (trans zeatin (tZ), kinetin (Kin), and N,N′-diphenylurea (DPU)) on the growth, oxidative damage, level of antioxidants, and the activity of antioxidant enzymes as well as the contents of proteins and photosynthetic pigments in *A. obliquus* were investigated under 100 μM Pb stress. Exogenous application of AX and CKs helps reduce the absorption of Pb by algal cells, which in turn alleviates the accumulation of proteins, chlorophylls, and carotenoids. Moreover, the combined effect of these phytohormones also activates the xanthophyll cycle involved in the protection of the photosynthetic apparatus under stress. The oxidative stress is reduced in algal cultures treated with phytohormones in the presence of toxic Pb because of the antioxidant defense system (Piotrowska-Niczyporuk et al., 2018).

One of the responses toward heavy metal stress in plants is the enhanced expression of ET-related biosynthetic genes, which leads to increased ET production (Asgher et al., 2014; Khan and Khan, 2014; Khan et al., 2015). ABA and ET under Cd stress have different effects on physiological and biochemical responses of plants, which include Cd transport, redox homeostasis, and Cd chelation. They can regulate the formation of apoplastic barrier, thereby reducing the accumulation of Cd in plants (Tao et al., 2019; Liu et al., 2021). ET levels are elevated at high cadmium concentrations in pea and barley (Vassilev et al., 2004; Rodríguez-Serrano et al., 2006). It has been observed that there is increased regulation of two of the six ACS gene family members (*ACS1* and *ACS2*) and two of the seven ACO gene family members in rice (*ACO4* and *ACO5*) (Rzewuski and Sauter, 2008). This indicates chromium (Cr) signaling involves ET biosynthesis (Trinh et al., 2014).

The reduced endogenous JA led to increased sensitivity to Cd in tomato seedlings results. In soybean (*Glycine max*) seedlings, nickel (Ni^{2+}) stress tolerance was observed if the seedling was treated with JA before the presence of nickel chloride ($NiCl_2$) contamination (Sirhindi et al., 2015). Furthermore, by controlling the antioxidant machinery and protecting DNA synthesis of total proteins, JA can protect the seedlings (Sirhindi et al., 2015). The extreme effects of oxidative stress on growth, biomass production, and protein concentrations in Ni-stressed plants can be alleviated with the help of JA (Azeem, 2018). Cd accumulation rates can be decreased by the application of JA in *Vicia faba* (faba bean) roots, shoots, and leaves by enhancing osmotic and antioxidant activity and by inhibiting H_2O_2 and MDA accumulation (Noriega et al., 2012). With the increasing Pb concentration, a decrease in plant growth was observed, which was improved with the application of JA. Additionally, under Pb stress, the foliar application of JA, SA, proline, and their combination led to increased growth traits of maize compared to untreated plants. SOD, POX, and CAT activities were reduced in Pb-stressed plants treated with JA, SA, and proline compared with untreated plants. The nitrogen (N), phosphorus (P), and potassium (K) concentrations in maize grains also increased on exogenous application of JA, SA, proline, and their combination (Hashem and El-Sherif, 2019).

SA essentially modifies metabolic pathways like antioxidant activities; its application to crops produces an adaptive impact and subsequently boosts resistance to heavy metal stress. Previous research showed that copper toxicity in cucumber and tobacco caused defensive mechanisms to be developed by SA treatment (Strobel and Kuc, 1995). However, foliar-applied SA was later proposed to help protect against heavy metal stress. The application of SA reduced the toxicity of mercury in alfalfa and increased the potency of APX, POD, and nicotinamide adenine dinucleotide phosphate (NADPH). It highlights how ROS signaling plays an integral role in this process for enhancement (Zhou et al., 2009). One of the crucial PGRs that can control ionic absorption and effectively lessen the toxicity of heavy metals in plants is BR. Cd is a very toxic heavy metal that inhibits the synthesis of chlorophyll, disrupts the function of several enzymes, and reduces photosynthetic efficiency by using less energy (Vassilev and Yordanov, 1997). However, the use of BRs can reduce the toxicity of Cd. The use of epibrassinolide (EBL) reduced the severe effects of Cd toxicity (Janeczko et al., 2005).

According to numerous researches, using GA reduces abiotic stressors, particularly the resistance to heavy metal stress. Zhu et al. (2012) investigated how applying GA at 5 μM decreased lipid peroxidation and Cd uptake. GA has no effect on the lowering of Cd uptake in an IRT1 knockout mutant irt1, which supports the idea that GAs lower nitrogen oxide levels, which in turn lower the expression of the IRT1 gene, an iron (Fe) transporter implicated in Cd uptake (Hamayun et al., 2010). According to Siddiqui et al. (2011), Ni has been demonstrated to decrease growth, chlorophyll concentrations, and carbonic anhydrase activity by increasing the formation of ROS; however, the detrimental effects of Ni are lessened when GAs are applied. By controlling ROS formation and an antioxidant defense mechanism, exogenous GA application lessens the toxicity of chromium on pea seed growth and ammonium absorption (Gangwar et al., 2011). According to Meng et al. (2009), exogenous GA treatment reduces the negative effects of Cd on oxidative damage and seedling establishment. Additionally, by controlling the activities of proteases, CAT, and POD, GAs reduce the detrimental effects of Cd and molybdenum on broad bean and lupin plants (Sharaf et al., 2009). The foliar application of SLs can decrease the growth inhibition caused by Cd, lessen the plant Cd contents, shield chlorophyll pigments, and decrease Cd-induced oxidative stress in wheat. Moreover, in wheat, enhanced shoot and root nitric oxide (NO) content was observed when SL applied on wheat foliage. Additionally, it can overcome the suppression of SOD and POD caused by Cd stress by elevating antioxidant defense systems (Shah et al., 2023).

2.4 Crosstalk between plant growth regulators for abiotic stress tolerance

Hormonal crosstalk allows for the wider remediation strategies and regulation against abiotic stress (Pieterse et al., 2012). Phytohormones not only act in isolation but also by synergistic or antagonistic crosstalk (Peleg and Blumwald, 2011). In drought stress conditions, SA and ABA regulate the osmotic stability, stomatal conductance, and provision of photo-assimilates. ABA also interacts with ET under drought stress by regulating the stomatal conductance along with shoot and root growth (Gazzarrini and Mccourt, 2003). ABA is an important phytohormone for regulating stomatal genes, along with which JA and BR also trigger the stomatal closure under drought stress (Munemasa et al., 2011). Under drought stress, MeJA and ABA function together in symphony and the application of either of the phytohormones to a detached leaf can lead to closed stomata (Salvi et al., 2021). However, according to Hossain et al. (2011), a sufficient amount of ABA is required for MeJA to initiate the stomatal closure. Enhanced MeJA level in rice maintains the higher level of ABA biosynthesis, which is essential for plant resistance against drought (Kim et al., 2009). CK degradation in rice during drought stress is initiated by the increase in cytokinin oxidase/dehydrogenase (OsCKX9) promoted by SL (Duan et al., 2019). In lotus (*Nelumbo nucifera*), tomato, and Arabidopsis, the increase in ABA level led to a sudden reduction in the SL level (Ha et al., 2014; Liu et al., 2015; Visentin et al., 2016). Their interaction is important for regulation of stomatal development (Ha et al., 2014). Elevated ABA level may lead to less resistance toward drought stress in SL biosynthetic mutants of tomato and lotus (Liu et al., 2015). AX works in concert with BR to improve the drought condition in plants, their interaction helps in hypocotyl elongation in numerous plants. BR modulates the endogenous AX level (Mandava, 1988). An improvement in resistance to drought can be possible in rice plants by the expression of an IPT gene (a CK biosynthetic gene) driven by SARK promoter (stress and senescence-inducible promoter), which can increase BR-associated genes leading to decrease in the JA-related genes. This crosstalk can enhance grain yield in rice (Peleg et al., 2011). ET and GA interact in antagonism and synergism with each other by providing a defense system and increasing plants' growth against salt stress. GA and ET affect each other's synthesis (Iqbal et al., 2014). GA and ET interact through aspartate-glutamate-leucine-leucine-alanine (DELLA) proteins (Achard et al., 2003). DELLA proteins can be increased by ET, reducing the GA action, thereby showing better resistance to salinity (Achard et al., 2006). ET acts synergistically with melatonin to alleviate salt stress in wheat through upregulation of antioxidants' activity, which detoxifies ROS (Khan et al., 2022). The interaction between EIN3/EIL1 and JAZ-MYC2 acts as a key node connecting various signaling and biosynthetic pathways responsible for defense via enhancement in secondary metabolites' production (Pérez-Llorca et al., 2023). ET and JA may either work synergistically or antagonistically in stress conditions. Synergistic act has been seen via EIN3/EIL1 along with JAZs-MYC2, which can reduce leaf growth and expansion by regulating AUXs, controlling JA synthesis (Ludwig et al., 2005; Yang et al., 2019; Singh et al., 2023) showed that SlWRKY23 was induced in tomato by osmotic stress, and both ET and AX were involved in imparting tolerance to mannitol and sodium chloride (NaCl) stress through activation of WRKY. In salt stress, the exogenous application of CK downregulates ABA genes like mitogen-activated protein kinase kinase 9 (MKK9), ACC synthase 6 (ACS6), and ERF1 and ERF5, involved in stress response, suggesting the interrelation between CK with ABA and ET (Golan et al., 2016). Under salt stress, EIN2 alteration leads to increased ABA production (Dong et al., 2011). Upregulation of stress-responsive genes involved in ABA signaling is achieved by ACS7 mutation, enhancing salt tolerance in plants (Wang et al., 2007). ET level can be increased under salt stress by exogenous application of BRs, which can also increase alternative oxidase pathway (AOX) components in cucumber seedlings. This synergistic effect can eliminate oxidative stress as ET activates AOX (Jiroutova et al., 2018). During heat stress, the increase in ABA level in seed leads to the upregulation of synthetic ABA enzyme genes *ZEP*, *NCED2*, *NCED5*, and *NCED9*. On the other hand, decreased GA level is the contribution of the decrease in synthetic GA enzyme genes *GA20ox1*, *GA20ox2*, *GA20ox3*, *GA3ox1*, and *GA3ox2* (Toh et al., 2008). To delay the germination during heat stress, FUS3 leads to the activation of ABA metabolic genes and restrains the GA metabolic genes (Chiu et al., 2012). Under heat stress, in Arabidopsis, GA and ABA metabolic genes are regulated by SOM, which is activated by DELLA proteins (Lim et al., 2013). It was observed in the Yunnan pine (*Pinus yunnanensis*) under heat and drought stress, the combined treatment of MeJA and GA decreased the wilting and increased the chlorophyll content. Both the phytohormones are antagonists to each other, which maintains a balance between enhancing the growth and adapting to the stress, suggesting the growth and defense trade-offs at an early stage (Pan et al., 2025). Under heat stress, GA_3 and silicon (Si) combined treatment in date palm (*Phoenix dactylifera*) increased the accumulation of SA, which in turn regulated numerous biochemical and physiological processes in plants for thermotolerance (Snyman and Cronjé, 2008; Sharma et al., 2020). Endogenous SA accumulation activates proline biosynthesis to maintain osmotic potential and increase antioxidant enzymes (Sharma et al., 2020). Si and GA3 treatment led to the antagonistic relation between SA and ABA (Lee et al., 2019). Increase in endogenous SA by the treatment with GA3 and Si helps in the accumulation of heat shock transcription factor genes such as *Hsf3* that triggers plant defense systems under heat stress (Snyman and Cronjé, 2008; Wei et al., 2018). Cd stress can be avoided by a crosstalk between IAA and ET, leading to activation of the antioxidant system. In diageotropica (dgt), a mutant tomato plant, which has decreased sensitivity to IAA, and a disrupted auxin induced ET biosynthesis, a higher Cd tolerance was observed (Gratão et al., 2012). Total phenolic content can be increased in the presence of Pb and zinc (Zn) by the interaction between ABA and GA3; however, in the presence of Cd, phenolic compounds decreased (Ahmad et al., 2015). ABA suppresses the synthesis of BR under metal toxicity and has a role in interaction between CKs and BR (Zhang et al., 2009a). Cd stress in wheat can be regulated by the exogenous application of SA by an increase in the GSH amount, which helps in the removal of metals and reduces the oxidative stress. The supplementation of SA can also increase the ABA level in wheat when introduced with Cd stress. On the other hand, endogenous ABA controls the SA-mediated regulation of dehydrin proteins under heavy metal stress in wheat (Kovács et al., 2014; Shakirova et al., 2016).

2.5 Potential role of plant growth regulators: future perspectives

Scientific and technological developments offer promising opportunities to improve PGRs' shortcomings and emphasize their significance in sustainable agriculture. For instance, biotechnology breakthroughs have enabled the engineering of plants to produce the optimal amount of endogenous PGRs (Gupta et al., 2024). This implies that there will be no need for exogenous PGR applications. To enhance the nutrient uptake and resistance of plants to abiotic stress, crop plants can be designed to overexpress the genes involved.

Research development on biostimulants offers economical and eco-friendly alternatives (Malik et al., 2020). Formulation science can be developed to create the possibility to alter PGR products for desirable crops, environments, or phases of growth. For example, encapsulated PGRs or controlled-release formulations can lead to distribution of active components over an extended period of time, assuring the intended impact without producing waste. Research can be done on the enduring impacts of PGRs on biodiversity and soil health. Soil microbial communities may interact with PGRs that may serve as the basis for formulating products to improve soil fertility. PGRs have the potential to be incorporated into crop management plans (Small and Degenhardt, 2018). Environmental stability can be preserved by combining all the sustainable practices like crop rotation, organic farming, and integrated pest management (Ali et al., 2024). PGR biosynthesis pathways can be optimized by genetic engineering techniques, such as CRISPR/Cas9 and other genome-editing technologies (Sims, n.d.), or the plant's response to stress can be improved by increasing receptor activity. Omic(s) techniques, including transcriptomics, proteomics, and metabolomics, can shed light on the ways in which PGRs engage with plant signaling networks. The development of synthetic analogy or novel delivery methods for PGRs to improve their stability, bioavailability, and focused activity is made possible by synthetic biology. Additionally, predictive models for optimizing PGR activity under a variety of environmental conditions might be created using systems biology. PGRs can play a more significant role in attaining sustainable production of highly productive vegetable crops in an ecological-friendly setting if given the correct challenges and opportunities for the future.

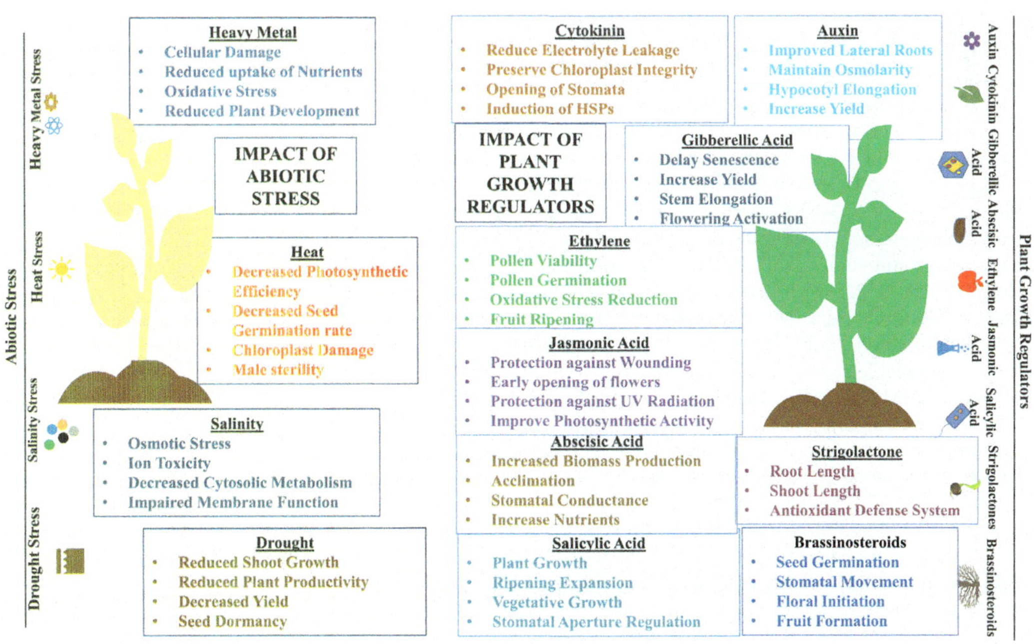

FIGURE 2.1 The role of phytohormones in improving plant tolerance against multiple abiotic stresses.

2.6 Conclusion

Alongside population expansion, the demand for food security has increased globally in recent decades. However, biotic and abiotic stressors are becoming more and more detrimental to seed germination, growth, and yields. PGRs, which function as chemical messengers, are essential for controlling how plants develop and react to environmental stressors. Important PGRs such as AX, CK, GA, ET, SA, JA, and ABA are crucial for crops' reactions to stress. These PGRs provide plants the ability to detect and adjust to harsh environmental factors like salt, drought, and extreme temperatures. Plants increase their resistance to stress by producing regulators and activating enzymes. Either independently or by developing a crosstalk within them, the PGRs activate a cascade event that ultimately attracts different transcription factors that either stimulate or inhibit a range of genes. In conclusion, PGRs have strong potential to reduce the negative effects of abiotic stress on crops and its production while maintaining their potential for future sustainable agricultural practices. Fig. 2.1 shows the role of PGRs in different abiotic stresses and the mechanisms involved.

References

Abeles, F.B., Morgan, P.W., Saltveit Jr, M.E., 2012. Ethylene in Plant Biology. Academic Press.

Abeysingha, D.N., Ozga, J.A., Strydhorst, S., Doyle, P., Iqbal, M., Yang, R.C., et al., 2021. The effect of auxins on amelioration of heat stress-induced wheat (*Triticum aestivum* L.) grain loss. Journal of Agronomy and Crop Science 207 (6), 970–983. Available at https://doi.org/10.1111/jac.12555, http://onlinelibrary.wiley.com/journal/10.1111/(ISSN)1439-037X.

Achard, P., Vriezen, W.H., Van Der Straeten, D., Harberd, N.P., 2003. Ethylene regulates *Arabidopsis* development via the modulation of DELLA Protein Growth Repressor Function. The Plant Cell 15 (12), 2816–2825. Available at https://doi.org/10.1105/tpc.015685, http://www.plantcell.org/.

Achard, P., Cheng, H., De Grauwe, L., Decat, J., Schoutteten, H., Moritz, T., et al., 2006. Integration of plant responses to environmentally activated phytohormonal signals. Science (New York, N.Y.) 311 (5757), 91–94. Available at https://doi.org/10.1126/science.1118642.

Ahammed, G.J., Li, X., Zhou, J., Zhou, Y.H., Yu, J.Q., 2016. Role of hormones in plant adaptation to heat stress. Plant Hormones under Challenging Environmental Factors. Springer Netherlands, China, pp. 1–21. Available at http://doi.org/10.1007/978-94-017-7758-2, https://doi.org/10.1007/978-94-017-7758-2_1.

Ahmad, A., Hadi, F., Ali, N., 2015. Effective Phytoextraction Of Cadmium (Cd) with increasing concentration of total phenolics and free proline in *Cannabis sativa* (L) plant under various treatments of fertilizers, plant growth regulators and sodium salt. International Journal of Phytoremediation 17 (1), 56–65. Available at https://doi.org/10.1080/15226514.2013.828018, http://www.tandf.co.uk/journals/titles/15226514.asp.

Akter, N., Rafiqul Islam, M., Abdul Karim, M., Hossain, T., 2014. Alleviation of drought stress in maize by exogenous application of gibberellic acid and cytokinin. Journal of Crop Science and Biotechnology 17 (1), 41–48. Available at https://doi.org/10.1007/s12892-013-0117-3.

Al-Hakimi, A.M.A., 2007. Modification of cadmium toxicity in pea seedlings by kinetin. Plant, Soil and Environment 53 (3), 129–135. Available at https://doi.org/10.17221/2228-pse.

Ali, B., Hayat, S., Ahmad, A., 2007. 28-Homobrassinolide ameliorates the saline stress in chickpea (*Cicer arietinum* L). Environmental and Experimental Botany 59 (2), 217–223. Available at https://doi.org/10.1016/j.envexpbot.2005.12.002.

Ali, A., Niu, G., Masabni, J., Ferrante, A., Cocetta, G., 2024. Integrated nutrient management of fruits, vegetables, and crops through the use of biostimulants, soilless cultivation, and traditional and modern approaches—a mini review. Agriculture 14 (8), 1330. Available at https://doi.org/10.3390/agriculture14081330.

Alonso-Ramírez, A., Rodríguez, D., Reyes, D., Jiménez, J.A., Nicolás, G., López-Climent, M., et al., 2009. Evidence for a role of gibberellins in salicylic acid-modulated early plant responses to abiotic stress in *Arabidopsis* seeds. Plant Physiology 150 (3), 1335–1344. Available at https://doi.org/10.1104/pp.109.139352.

Alvarez, S., Marsh, E.L., Schroeder, S.G., Schachtman, D.P., 2008. Metabolomic and proteomic changes in the xylem sap of maize under drought. Plant, Cell and Environment 31 (3), 325–340. Available at https://doi.org/10.1111/j.1365-3040.2007.01770.x.

Álvarez, S., Gómez-Bellot, M.J., Acosta-Motos, J.R., Sánchez-Blanco, M.J., 2019. Application of deficit irrigation in *Phillyrea angustifolia* for landscaping purposes. Agricultural Water Management 218, 193–202. Available at https://doi.org/10.1016/j.agwat.2019.03.049.

Amjad, M., Akhtar, J., Anwar-ul-Haq, M., Yang, A., Akhtar, S.S., Jacobsen, S.E., 2014. Integrating role of ethylene and ABA in tomato plants adaptation to salt stress. Scientia Horticulturae 172, 109–116. Available at https://doi.org/10.1016/j.scienta.2014.03.024, www.elsevier.com/inca/publications/store/5/0/3/3/1/6.

Anuradha, S., Ram Rao, S.S., 2003. Application of brassinosteroids to rice seeds (*Oryza sativa* L.) reduced the impact of salt stress on growth, prevented photosynthetic pigment loss and increased nitrate reductase activity. Plant Growth Regulation 40 (1), 29–32. Available at https://doi.org/10.1023/a:1023080720374.

Apelbaum, A., Yang, S.F., 1981. Biosynthesis of stress ethylene induced by water deficit. Plant Physiology 68 (3), 594–596. Available at https://doi.org/10.1104/pp.68.3.594.

Arif, Y., Sami, F., Siddiqui, H., Bajguz, A., Hayat, S., 2020. Salicylic acid in relation to other phytohormones in plant: a study towards physiology and signal transduction under challenging environment. Environmental and Experimental Botany 175, 104040. Available at https://doi.org/10.1016/j.envexpbot.2020.104040.

Arteca, R.N., 1996. Plant Growth Substances: Principles and Applications. Springer US. Available at https://doi.org/10.1007/978-1-4757-2451-6.

Arunanondchai, P., Fei, C., Fisher, A., McCarl, B.A., Wang, W., Yang, Y., 2018. How Does Climate Change Affect Agriculture? The Routledge Handbook of Agricultural Economics. Taylor and Francis, United States, pp. 191–210. Available at https://www.routledge.com/The-Routledge-Handbook-of-Agricultural-Economics/Cramer-Paudel-Schmitz/p/book/9781315623351.

Asgher, M., Khan, N.A., Khan, M.I.R., Fatma, M., Masood, A., 2014. Ethylene production is associated with alleviation of cadmium-induced oxidative stress by sulfur in mustard types differing in ethylene sensitivity. Ecotoxicology and Environmental Safety 106, 54–61. Available at https://doi.org/10.1016/j.ecoenv.2014.04.017, http://www.elsevier.com/inca/publications/store/6/2/2/8/1/9/index.htt.

Asthir, B., Bhatia, S., 2014. In vivo studies on artificial induction of thermotolerance to detached panicles of wheat (*Triticum aestivum* L) cultivars under heat stress. Journal of Food Science and Technology 51 (1), 118–123. Available at https://doi.org/10.1007/s13197-011-0458-1.

Atici, Ö., Ağar, G., Battal, P., 2005. Changes in phytohormone contents in chickpea seeds germinating under lead or zinc stress. Biologia plantarum 49 (2), 215–222. Available at https://doi.org/10.1007/s10535-005-5222-9.

Auguy, F., Fahr, M., Moulin, P., El Mzibri, M., Smouni, A., Filali-Maltouf, A., et al., 2016. Transcriptome changes in *Hirschfeldia incana* in response to lead exposure. Frontiers in Plant Science 6. Available at https://doi.org/10.3389/fpls.2015.01231 (2016).

Azeem, U., 2018. Ameliorating nickel stress by jasmonic acid treatment in *Zea mays* L. Russian Agricultural Sciences 44 (3), 209–215. Available at https://doi.org/10.3103/s1068367418030035.

Bhardwaj, I., Garg, N., 2022. Cereals and phytohormones under heavy metal stress. Sustainable Remedies for Abiotic Stress in Cereals. Springer Nature, India, pp. 369–393. Available at https://doi.org/10.1007/978-981-19-5121-3_15, https://link.springer.com/book/10.1007/978-981-19-5121-3.

Bianco, C., Defez, R., 2009. Medicago truncatula improves salt tolerance when nodulated by an indole-3-acetic acid-overproducing *Sinorhizobium meliloti* strain. Journal of Experimental Botany 60 (11), 3097–3107. Available at https://doi.org/10.1093/jxb/erp140.

Binns A.N., n.d. Cytokinin accumulation and action: biochemical, genetic, and molecular approaches.

Burbidge, A., Grieve, T.M., Jackson, A., Thompson, A., McCarty, D.R., Taylor, I.B., 1999. Characterization of the ABA-deficient tomato mutant notabilis and its relationship with maize Vp14. Plant Journal 17 (4), 427–431. Available at https://doi.org/10.1046/j.1365-313X.1999.00386.x.

Cakmak, I., 2005. Role of mineral nutrients in tolerance of crop plants to environmental stress factors. In Proceedings from the international symposium on fertigation–optimizing the utilization of water and nutrients (pp. 35–48).

Camara, M.C., Vandenberghe, L.P.S., Rodrigues, C., de Oliveira, J., Faulds, C., Bertrand, E., et al., 2018. Current advances in gibberellic acid (GA3) production, patented technologies and potential applications. Planta 248 (5), 1049–1062. Available at https://doi.org/10.1007/s00425-018-2959-x, link.springer.de/link/service/journals/00425/index.htm.

Cao, W.H., Liu, J., He, X.J., Mu, R.L., Zhou, H.L., Chen, S.Y., et al., 2007. Modulation of ethylene responses affects plant salt-stress responses. Plant Physiology 143 (2), 707–719. Available at https://doi.org/10.1104/pp.106.094292, http://www.plantphysiol.org/cgi/reprint/143/2/707.pdf.

Cardoso, A.A., Gori, A., Da-Silva, C.J., Brunetti, C., 2020. Abscisic acid biosynthesis and signaling in plants: key targets to improve water use efficiency and drought tolerance. Applied Sciences 10 (18), 6322. Available at https://doi.org/10.3390/app10186322.

Cassab, G.I., Eapen, D., Campos, M.E., 2013. Root hydrotropism: an update. American Journal of Botany 100 (1), 14–24. Available at https://doi.org/10.3732/ajb.1200306Mexico, http://www.amjbot.org/content/100/1/14.full.pdf+html.

Chandrakala, J.U., Chaturvedi, A.K., Ramesh, K.V., Rai, P., Khetarpal, S., Pal, M., 2013. Acclimation response of signalling molecules for high temperature stress on photosynthetic characteristics in rice genotypes. Indian Journal of Plant Physiology 18 (2), 142–150. Available at https://doi.org/10.1007/s40502-013-0021-3.

Chen, C.M., Ertl, J.R., Leisner, S.M., Chang, C.C., 1985. Localization of cytokinin biosynthetic sites in pea plants and carrot roots. Plant Physiology 78 (3), 510–513. Available at https://doi.org/10.1104/pp.78.3.510, http://www.plantphysiol.org/content/78/3/510.full.pdf.

Chen, K., Li, G.-J., Bressan, R.A., Song, C.-P., Zhu, J.-K., Zhao, Y., 2020. Abscisic acid dynamics, signaling, and functions in plants. Journal of Integrative Plant Biology 62 (1), 25–54. Available at https://doi.org/10.1111/jipb.12899.

Cheng, X., Ruyter-Spira, C., Bouwmeester, H., 2013. The interaction between strigolactones and other plant hormones in the regulation of plant development. Frontiers in Plant Science 4. Available at https://doi.org/10.3389/fpls.2013.00199.

Chernys, J.T., Zeevaart, J.A.D., 2000. Characterization of the 9-cis-epoxycarotenoid dioxygenase gene family and the regulation of abscisic acid biosynthesis in avocado. Plant Physiology 124 (1), 343–354. Available at https://doi.org/10.1104/pp.124.1.343.

Chi, C., Xu, X., Wang, M., Zhang, H., Fang, P., Zhou, J., et al., 2021. Strigolactones positively regulate abscisic acid-dependent heat and cold tolerance in tomato. Horticulture Research 8 (1). Available at https://doi.org/10.1038/s41438-021-00668-y.

Chiu, R.S., Nahal, H., Provart, N.J., Gazzarrini, S., 2012. The role of the *Arabidopsis* FUSCA3 transcription factor during inhibition of seed germination at high temperature. BMC Plant Biology 12. Available at https://doi.org/10.1186/1471-2229-12-15Canada, http://www.biomedcentral.com/1471-2229/12/15.

Cicchino, M.A., Rattalino Edreira, J.I., Otegui, M.E., 2013. Maize physiological responses to heat stress and hormonal plant growth regulators related to ethylene metabolism. Crop Science 53 (5), 2135–2146. Available at https://doi.org/10.2135/cropsci2013.03.0136Argentina, https://www.crops.org/publications/cs/pdfs/53/5/2135.

Claeys, H., Skirycz, A., Maleux, K., Inzé, D., 2012. DELLA signaling mediates stress-induced cell differentiation in *Arabidopsis* leaves through modulation of anaphase-promoting complex/cyclosome activity. Plant Physiology 159 (2), 739–747. Available at https://doi.org/10.1104/pp.112.195032, http://www.plantphysiol.org/content/159/2/739.full.pdf+html.

Coelho Filho, M.A., Colebrook, E.H., Lloyd, D.P.A., Webster, C.P., Mooney, S.J., Phillips, A.L., et al., 2013. The involvement of gibberellin signalling in the effect of soil resistance to root penetration on leaf elongation and tiller number in wheat. Plant and Soil 371 (1-2), 81–94. Available at https://doi.org/10.1007/s11104-013-1662-8.

Danish, S., Hareem, M., Dawar, K., Naz, T., Iqbal, M.M., Ansari, M.J., et al., 2024. The role of strigolactone in alleviating salinity stress in chili pepper. BMC Plant Biology 24 (1). Available at https://doi.org/10.1186/s12870-024-04900-4, https://bmcplantbiol.biomedcentral.com/.

Darwin, C., 2017. The power of movement in plants by. Charles Darwin-Delphi classics (Illustrated). Delphi Classics.

Davies, P.J., 1995. The plant hormones: their nature, occurrence, and functions. Plant Hormones: Physiology, Biochemistry and Molecular Biology. Springer Netherlands, pp. 1–12.

De Costa, W., Zörb, C., Hartung, W., Schubert, S., 2007. Salt resistance is determined by osmotic adjustment and abscisic acid in newly developed maize hybrids in the first phase of salt stress. Physiologia Plantarum 131 (2), 311–321. Available at https://doi.org/10.1111/j.1399-3054.2007.00962.x.

Dhaubhadel, S., Browning, K.S., Gallie, D.R., Krishna, P., 2002. Brassinosteroid functions to protect the translational machinery and heat-shock protein synthesis following thermal stress. Plant Journal 29 (6), 681–691. Available at https://doi.org/10.1046/j.1365-313X.2002.01257.x.

Di Mambro, R., De Ruvo, M., Pacifici, E., Salvi, E., Sozzani, R., Benfey, P.N., et al., 2017. Auxin minimum triggers the developmental switch from cell division to cell differentiation in the *Arabidopsis* root. Proceedings of the National Academy of Sciences 114 (36). Available at https://doi.org/10.1073/pnas.1705833114.

Dodd, I.C., 2005. Root-to-shoot signalling: assessing the roles of 'up' in the up and down world of long-distance signalling in planta. Plant and Soil 274 (1-2), 251–270. Available at https://doi.org/10.1007/s11104-004-0966-0.

Dong, H., Zhen, Z., Peng, J., Chang, L., Gong, Q., Wang, N.N., 2011. Loss of ACS7 confers abiotic stress tolerance by modulating ABA sensitivity and accumulation in *Arabidopsis*. Journal of Experimental Botany 62 (14), 4875–4887. Available at https://doi.org/10.1093/jxb/err143.

Duan, L., Dietrich, D., Ng, C.H., Chan, P.M.Y., Bhalerao, R., Bennett, M.J., et al., 2013. Endodermal ABA signaling promotes lateral root quiescence during salt stress in *Arabidopsis* seedlings. The Plant Cell 25 (1), 324–341. Available at https://doi.org/10.1105/tpc.112.107227, http://www.plantcell.org/content/25/1/324.full.pdf +html.

Duan, J., Yu, H., Yuan, K., Liao, Z., Meng, X., Jing, Y., et al., 2019. Strigolactone promotes cytokinin degradation through transcriptional activation of CYTOKININ OXIDASE/DEHYDROGENASE 9 in rice. Proceedings of the National Academy of Sciences 116 (28), 14319–14324. Available at https://doi.org/10.1073/pnas.1810980116.

Ellouzi, H., Hamed, K.B., Hernández, I., Cela, J., Müller, M., Magné, C., et al., 2014. A comparative study of the early osmotic, ionic, redox and hormonal signaling response in leaves and roots of two halophytes and a glycophyte to salinity. Planta 240 (6), 1299–1317. Available at https://doi.org/10.1007/s00425-014-2154-7, link. springer.de/link/service/journals/00425/index.htm.

Epstein, E., Ludwig, J., 1993. Indole-3-butyric acid in plants: occurrence, synthesis, metabolism and transport. Physiologia Plantarum 88 (2), 382–389. Available at https://doi.org/10.1034/j.1399-3054.1993.880224.x.

Fahad, S., Hussain, S., Matloob, A., Khan, F.A., Khaliq, A., Saud, S., et al., 2015. Phytohormones and plant responses to salinity stress: a review. Plant Growth Regulation 75 (2), 391–404. Available at https://doi.org/10.1007/s10725-014-0013-y, http://www.springeronline.com/sgw/cda/frontpage/0,11855,1-40109-70-35547532-0,00. html.

Fahad, et al., 2016. Exogenously applied plant growth regulators enhance the morpho-physiological growth and yield of rice under high temperature. Frontiers in Plant Science 7.

Fahad, S., Bajwa, A.A., Nazir, U., Anjum, S.A., Farooq, A., Zohaib, A., et al., 2017. Crop production under drought and heat stress: plant responses and management options. Frontiers in Plant Science 8. Available at https://doi.org/10.3389/fpls.2017.01147, http://journal.frontiersin.org/article/10.3389/fpls.2017.01147/full.

Farhangi-Abriz, S., Torabian, S., 2018. Biochar increased plant growth-promoting hormones and helped to alleviates salt stress in common bean seedlings. Journal of Plant Growth Regulation 37 (2), 591–601. Available at https://doi.org/10.1007/s00344-017-9756-9.

Fatma, M., Asgher, M., Masood, A., Khan, N.A., 2014. Excess sulfur supplementation improves photosynthesis and growth in mustard under salt stress through increased production of glutathione. Environmental and Experimental Botany 107, 55–63. Available at https://doi.org/10.1016/j.envexpbot.2014.05.008, www.elsevier. com/inca/publications/store/2/6/7.

Finkelstein, R., Reeves, W., Ariizumi, T., Steber, C., 2008. Molecular aspects of seed dormancy. Annual Review of Plant Biology 59 (1), 387–415. Available at https://doi.org/10.1146/annurev.arplant.59.032607.092740.

Fleishon, S., Shani, E., Ori, N., Weiss, D., 2011. Negative reciprocal interactions between gibberellin and cytokinin in tomato. New Phytologist 190 (3), 609–617. Available at https://doi.org/10.1111/j.1469-8137.2010.03616.x.

Galvan-Ampudia, C.S., Testerink, C., 2011. Salt stress signals shape the plant root. Current Opinion in Plant Biology 14 (3), 296–302. Available at https://doi.org/10.1016/j.pbi.2011.03.019.

Gangwar, S., Singh, V.P., Srivastava, P.K., Maurya, J.N., 2011. Modification of chromium (VI) phytotoxicity by exogenous gibberellic acid application in *Pisum sativum* (L.) seedlings. Acta Physiologiae Plantarum 33 (4), 1385–1397. Available at https://doi.org/10.1007/s11738-010-0672-x.

Gazzarrini, S., Mccourt, P., 2003. Cross-talk in plant hormone signalling: what *Arabidopsis* mutants are telling us. Annals of Botany 91 (6), 605–612. Available at 10.1093/aob/mcg064.

Ghorbel, M., Brini, F., Sharma, A., Landi, M., 2021. Role of jasmonic acid in plants: the molecular point of view. Plant Cell Reports 40 (8), 1471–1494. Available at https://doi.org/10.1007/s00299-021-02687-4.

Goda, H., Shimada, Y., Asami, T., Fujioka, S., Yoshida, S., 2002. Microarray analysis of brassinosteroid-regulated genes in *Arabidopsis*. Plant Physiology 130 (3), 1319–1334. Available at https://doi.org/10.1104/pp.011254.

Golan, Y., Shirron, N., Avni, A., Shmoish, M., Gepstein, S., 2016. Cytokinins induce transcriptional reprograming and improve *Arabidopsis* plant performance under drought and salt stress conditions. Frontiers in Environmental Science 4. Available at https://doi.org/10.3389/fenvs.2016.00063.

Gomi, K., 2020. Jasmonic acid: an essential plant hormone. International Journal of Molecular Sciences 21 (4), 1261. Available at https://doi.org/10.3390/ijms21041261.

Gou, J., Strauss, S.H., Tsai, C.J., Fang, K., Chen, Y., Jiang, X., et al., 2010. Gibberellins regulate lateral root formation in Populus through interactions with auxin and other hormones. The Plant Cell 22 (3), 623–639. Available at https://doi.org/10.1105/tpc.109.073239, http://www.plantcell.org/cgi/reprint/22/3/623.

Gratão, P.L., Monteiro, C.C., Carvalho, R.F., Tezotto, T., Piotto, F.A., Peres, L.E.P., et al., 2012. Biochemical dissection of diageotropica and Never ripe tomato mutants to Cd-stressful conditions. Plant Physiology and Biochemistry 56, 79–96. Available at https://doi.org/10.1016/j.plaphy.2012.04.009.

Gray, W.M., Östin, A., Sandberg, G., Romano, C.P., Estelle, M., 1998. High temperature promotes auxin-mediated hypocotyl elongation in *Arabidopsis*. Proceedings of the National Academy of Sciences 95 (12), 7197–7202. Available at https://doi.org/10.1073/pnas.95.12.7197.

Grime, J.P., 1977. Evidence for the existence of three primary strategies in plants and its relevance to ecological and evolutionary theory. The American Naturalist 111 (982), 1169–1194. Available at https://doi.org/10.1086/283244.

Grossmann, K., 2010. Auxin herbicides: current status of mechanism and mode of action. Pest Management Science 66 (2), 113–120. Available at https://doi.org/10.1002/ps.1860.

Gull, A., Lone, A.A., Wani, N.U., 2019. Biotic and abiotic stresses in plants. Abiotic and Biotic Stress in Plants. IntechOpen.

Guo, H., Li, L., Ye, H., Yu, X., Algreen, A., Yin, Y., 2009. Three related receptor-like kinases are required for optimal cell elongation in *Arabidopsis thaliana*. Proceedings of the National Academy of Sciences 106 (18), 7648–7653. Available at https://doi.org/10.1073/pnas.0812346106.

Guo, T., Gull, S., Ali, M.M., Yousef, A.F., Ercisli, S., Kalaji, H.M., et al., 2022. Heat stress mitigation in tomato (*Solanum lycopersicum* L.) through foliar application of gibberellic acid. Scientific Reports 12 (1). Available at https://doi.org/10.1038/s41598-022-15590-z, www.nature.com/srep/index.html.

Gupta, R., Chakrabarty, S.K., 2013. Gibberellic acid in plant: still a mystery unresolved. Plant Signaling and Behavior 8 (9). Available at https://doi.org/10.4161/psb.25504, https://www.landesbioscience.com/journals/psb/2013PSB0185.pdf.

Gupta, D., Saini, A., van der Vyver, C., Panda, S.K., 2024. Gene editing: paving the way for enhancing plant tolerance to abiotic stresses-mechanisms, breakthroughs, and future prospects. Journal of Plant Growth Regulation 43 (11), 3986–4002. Available at https://doi.org/10.1007/s00344-024-11395-8, https://www.springer.com/journal/344.

Ha, C.V., Leyva-González, M.A., Osakabe, Y., Tran, U.T., Nishiyama, R., Watanabe, Y., et al., 2014. Positive regulatory role of strigolactone in plant responses to drought and salt stress. Proceedings of the National Academy of Sciences 111 (2), 851–856. Available at https://doi.org/10.1073/pnas.1322135111.

Hamayun, M., Khan, S.A., Khan, A.L., Shin, J.H., Ahmad, B., Shin, D.H., et al., 2010. Exogenous gibberellic acid reprograms soybean to higher growth and salt stress tolerance. Journal of Agricultural and Food Chemistry 58 (12), 7226–7232. Available at https://doi.org/10.1021/jf101221t.

Hare, P.D., Cress, W.A., van Staden, J., 1997. The involvement of cytokinins in plant responses to environmental stress. Plant Growth Regulation 23 (1-2), 79–103. Available at https://doi.org/10.1023/a:1005954525087.

Hashem, H.A., El-Sherif, N.A., 2019. Exogenous jasmonic acid induces lead stress tolerance in kidney bean (*Phaseolus vulgaris* L.) by changing amino acid profile and stimulating antioxidant defense system. Jordan Journal of Biological Sciences 12 (3), 345–353. Available at http://jjbs.hu.edu.jo/files/vol12/n3/Paper%20number%2013.pdf.

Hayat, S., Ali, B., Aiman Hasan, S., Ahmad, A., 2007. Brassinosteroid enhanced the level of antioxidants under cadmium stress in *Brassica juncea*. Environmental and Experimental Botany 60 (1), 33–41. Available at https://doi.org/10.1016/j.envexpbot.2006.06.002.

Hayat, Q., Hayat, S., Irfan, M., Ahmad, A., 2010. Effect of exogenous salicylic acid under changing environment: a review. Environmental and Experimental Botany 68 (1), 14–25. Available at https://doi.org/10.1016/j.envexpbot.2009.08.005.

Hazel, J.R., 1995. Thermal Adaptation In Biological Membranes: Is Homeoviscous Adaptation The Explanation? Annual Review of Physiology 57 (1), 19–42. Available at https://doi.org/10.1146/annurev.ph.57.030195.000315.

He, M., He, C.Q., Ding, N.Z., 2018. Abiotic stresses: general defenses of land plants and chances for engineering multistress tolerance. Frontiers in Plant Science 871. Available at https://doi.org/10.3389/fpls.2018.01771, https://www.frontiersin.org/articles/10.3389/fpls.2018.01771/pdf.

Hönig, M., Plíhalová, L., Husičková, A., Nisler, J., Doležal, K., 2018. Role of cytokinins in senescence, antioxidant defence and photosynthesis. International Journal of Molecular Sciences 19 (12), 4045. Available at https://doi.org/10.3390/ijms19124045.

Hossain, M.A., Munemasa, S., Uraji, M., Nakamura, Y., Mori, I.C., Murata, Y., 2011. Involvement of endogenous abscisic acid in methyl jasmonate-induced stomatal closure in *Arabidopsis*. Plant Physiology 156 (1), 430–438. Available at https://doi.org/10.1104/pp.111.172254, http://www.plantphysiol.org/content/156/1/430.full.pdf+html.

Hsu, Y.T., Kao, C.H., 2007. Heat shock-mediated H_2O_2 accumulation and protection against Cd toxicity in rice seedlings. Plant and Soil 300 (1-2), 137–147. Available at https://doi.org/10.1007/s11104-007-9396-0.

Hu, Q., Zhang, S., Huang, B., 2019. Strigolactones promote leaf elongation in tall fescue through upregulation of cell cycle genes and downregulation of auxin transport genes in tall fescue under different temperature regimes. International Journal of Molecular Sciences 20 (8), 1836. Available at https://doi.org/10.3390/ijms20081836.

Iqbal, M., Ashraf, M., Jamil, A., 2006. Seed enhancement with cytokinins: changes in growth and grain yield in salt stressed wheat plants. Plant Growth Regulation 50 (1), 29–39. Available at https://doi.org/10.1007/s10725-006-9123-5.

Iqbal, N., Masood, A., Khan, N.A., 2014. Phytohormones in salinity tolerance: ethylene and gibberellins cross talk. Phytohormones and Abiotic Stress Tolerance in Plants. Springer-Verlag, Berlin Heidelberg, India, pp. 77–98. Available at http://doi.org/10.1007/978-3-642-25829-9, https://doi.org/10.1007/978-3-642-25829-9_3.

Janeczko, A., et al., 2005. Protection of winter rape photosystem 2 by 24-epi brassinolide under cadmium stress. Photosynthetica 43 (2), 293–298.

Javid, M.G., Sorooshzadeh, A., Moradi, F., Sanavy, S.A.M.M., Allahdadi, I., 2011. The role of phytohormones in alleviating salt stress in crop plants. Australian Journal of Crop Science 5 (6), 726–734. Available at http://www.cropj.com/javid_5_6_2011_726_734.pdf Iran.

Jegadeesan, S., Chaturvedi, P., Ghatak, A., Pressman, E., Meir, S., Faigenboim, A., et al., 2018. Proteomics of heat-stress and ethylene-mediated thermotolerance mechanisms in tomato pollen grains. Frontiers in Plant Science 9. Available at https://doi.org/10.3389/fpls.2018.01558.

Jiroutova, P., Oklestkova, J., Strnad, M., 2018. Crosstalk between brassinosteroids and ethylene during plant growth and under abiotic stress conditions. International Journal of Molecular Sciences 19 (10), 3283. Available at https://doi.org/10.3390/ijms19103283.

Kagale, S., Divi, U.K., Krochko, J.E., Keller, W.A., Krishna, P., 2007. Brassinosteroid confers tolerance in *Arabidopsis thaliana* and Brassica napus to a range of abiotic stresses. Planta 225 (2), 353–364. Available at https://doi.org/10.1007/s00425-006-0361-6.

Kaneyasu, T., Kobayashi, A., Nakayama, M., Fujii, N., Takahashi, H., Miyazawa, Y., 2007. Auxin response, but not its polar transport, plays a role in hydrotropism of *Arabidopsis* roots. Journal of Experimental Botany 58 (5), 1143–1150. Available at https://doi.org/10.1093/jxb/erl274.

Kang, D.J., Seo, Y.J., Lee, J.D., Ishii, R., Kim, K.U., Shin, D.H., et al., 2005. Jasmonic acid differentially affects growth, ion uptake and abscisic acid concentration in salt-tolerant and salt-sensitive rice cultivars. Journal of Agronomy and Crop Science 191 (4), 273–282. Available at https://doi.org/10.1111/j.1439-037X.2005.00153.x.

Karcz, W., Kurtyka, R., 2007. Effect of cadmium on growth, proton extrusion and membrane potential in maize coleoptile segments. Biologia plantarum 51 (4), 713–719. Available at https://doi.org/10.1007/s10535-007-0147-0.

Katsir, L., Chung, H.S., Koo, A.J., Howe, G.A., 2008. Jasmonate signaling: a conserved mechanism of hormone sensing. Current Opinion in Plant Biology 11 (4), 428–435. Available at https://doi.org/10.1016/j.pbi.2008.05.004.

Kaur, G., Asthir, B., 2020. Impact of exogenously applied ABA on proline metabolism conferring drought and salinity stress tolerance in wheat genotypes. Cereal Research Communications 48 (3), 309–315. Available at https://doi.org/10.1007/s42976-020-00041-0.

Kaya, C., Levent Tuna, A., Alfredo, A.C.A., 2006. Gibberellic acid improves water deficit tolerance in maize plants. Acta Physiologiae Plantarum 28 (4), 331–337. Available at https://doi.org/10.1007/s11738-006-0029-7.

Kelley, D.R., Estelle, M., 2012. Ubiquitin-mediated control of plant hormone signaling. Plant Physiology 160 (1), 47–55. Available at https://doi.org/10.1104/pp.112.200527, http://www.plantphysiol.org/content/160/1/47.full.pdf+html.

Keskin, B.C., Sarikaya, A.T., Yüksel, B., Memon, A.R., 2010. Abscisic acid regulated gene expression in bread wheat (*Triticum aestivum* L.). Australian Journal of Crop Science 4 (8), 617–625. Available at http://www.cropj.com/keskin_4_8_2010_617_625.pdf Turkey.

Khan, M.I., Khan, N.A., 2014. Ethylene reverses photosynthetic inhibition by nickel and zinc in mustard. Protoplasma 251 (5), 1007–1019.

Khan, M.I.R., Nazir, F., Asgher, M., Per, T.S., Khan, N.A., 2015. Selenium and sulfur influence ethylene formation and alleviate cadmium-induced oxidative stress by improving proline and glutathione production in wheat. Journal of Plant Physiology 173, 9–18. Available at https://doi.org/10.1016/j.jplph.2014.09.011, www.urbanfischer.de/journals/jpp/p_physio.htm.

Khan, S., et al., 2022. Involvement of ethylene in melatonin-modified photosynthetic N use efficiency. Physiologia Plantarum 174 (6), e13832.

Khatoon, S., et al., 2024. Green-synthesized gold nanoparticles induce adaptation. Clean Technologies and Environmental Policy 1–17.

Khatoon, S., Mahajan, M., Khan, M.I.R., 2025. Ethylene–potassium interactions safeguard ionic balance, chlorophyll biosynthesis, and source–sink metabolism to sustain wheat growth and yield responses under salt stress. Plant Stress 18. Available at https://doi.org/10.1016/j.stress.2025.101049, https://www.sciencedirect.com/science/journal/2667064X.

Khosravi-nejad, F., Khavari-nejad, R.A., Moradi, F., Najafi, F., 2022. Cytokinin and abscisic acid alleviate drought stress through changing organic acids profile, ion immolation, and fatty acid profile to improve yield of wheat (*Triticum aestivum* L.) cultivars. Physiology and Molecular Biology of Plants 28 (5), 1119–1129. Available at https://doi.org/10.1007/s12298-022-01173-9, https://www.springer.com/journal/12298.

Kim, E.H., Kim, Y.S., Park, S.H., Koo, Y.J., Choi, Y.D., Chung, Y.Y., et al., 2009. Methyl jasmonate reduces grain yield by mediating stress signals to alter spikelet development in rice1[W][OA]. Plant Physiology 149 (4), 1751–1760. Available at https://doi.org/10.1104/pp.108.134684, http://www.plantphysiol.org/cgi/reprint/149/4/1751.

Kobayasi, K., Atsuta, Y., 2009. Sterility and poor pollination due to early flower opening induced by methyl jasmonate. Plant Production Science 13 (1), 29–36. Available at https://doi.org/10.1626/pps.13.29, http://www.jstage.jst.go.jp/article/pps/13/1/29/_pdf.

Kobayasi, K., Masui, H., Atsuta, Y., Matsui, T., Yoshimoto, M., Hasegawa, T., 2009. Flower opening time in rice–Cultivar difference and effect of weather factors–. In: Proceedings of MARCO symposium.

Kovács, V., Gondor, O.K., Szalai, G., Darkó, E., Majláth, I., Janda, T., et al., 2014. Synthesis and role of salicylic acid in wheat varieties with different levels of cadmium tolerance. Journal of Hazardous Materials 280, 12–19. Available at https://doi.org/10.1016/j.jhazmat.2014.07.048, www.elsevier.com/locate/jhazmat.

Krugman, T., Peleg, Z., Quansah, L., Chagué, V., Korol, A.B., Nevo, E., et al., 2011. Alteration in expression of hormone-related genes in wild emmer wheat roots associated with drought adaptation mechanisms. Functional and Integrative Genomics 11 (4), 565–583. Available at https://doi.org/10.1007/s10142-011-0231-6.

Kukreja, S., Nandwal, A.S., Kumar, N., Sharma, S.K., Sharma, S.K., Unvi, V., et al., 2005. Plant water status, H_2O_2 scavenging enzymes, ethylene evolution and membrane integrity of *Cicer arietinum* roots as affected by salinity. Biologia Plantarum 49 (2), 305–308. Available at https://doi.org/10.1007/s10535-005-5308-4.

Kumar, S.V., Lucyshyn, D., Jaeger, K.E., Alós, E., Alvey, E., Harberd, N.P., et al., 2012. Transcription factor PIF4 controls the thermosensory activation of flowering. Nature 484 (7393), 242–245. Available at https://doi.org/10.1038/nature10928.

La, V.H., et al., 2019. Salicylic acid improves drought-stress tolerance. Horticulture, Environment and Biotechnology 60 (1), 31–40.

Lang, T., Deng, C., Yao, J., Zhang, H., Wang, Y., Deng, S., 2020. A Salt-signaling Network Involving Ethylene, Extracellular ATP, hydrogen peroxide, and calcium mediates K+/Na+ homeostasis in *Arabidopsis*. International Journal of Molecular Sciences 21 (22), 8683. Available at https://doi.org/10.3390/ijms21228683.

Larkindale, J., Knight, M.R., 2002. Protection against heat stress-induced oxidative damage. Plant Physiology 128 (2), 682.

Lee, B.R., et al., 2019. Antagonistic shifting from abscisic acid to salicylic acid-mediated sucrose accumulation. Environmental and Experimental Botany 162, 38–47.

Li, Y., Zhang, H., Zhang, Q., Liu, Q., Zhai, H., Zhao, N., et al., 2019. An AP2/ERF gene, IbRAP2-12, from sweetpotato is involved in salt and drought tolerance in transgenic *Arabidopsis*. Plant Science 281, 19–30. Available at https://doi.org/10.1016/j.plantsci.2019.01.009.

Liao, X., Guo, X., Wang, Q., Wang, Y., Zhao, D., Yao, L., et al., 2017. Overexpression of Ms DREB 6.2 results in cytokinin-deficient developmental phenotypes and enhances drought tolerance in transgenic apple plants. The Plant Journal 89 (3), 510–526. Available at https://doi.org/10.1111/tpj.13401.

Liliane, T.N., Charles, M.S., 2020. Factors affecting yield of crops. Agronomy-climate Change & Food Security 9, 9–24.

Lim, S., Park, J., Lee, N., Jeong, J., Toh, S., Watanabe, A., et al., 2013. ABA-insensitive3, ABA-insensitive5, and DELLAs interact to activate the expression of SOMNUS and other high-temperature-inducible genes in imbibed seeds in *Arabidopsis*. The Plant Cell 25 (12), 4863–4878. Available at https://doi.org/10.1105/tpc.113.118604, http://www.plantcell.org/content/25/12/4863.full.pdf.

Lim, C.W., Baek, W., Jung, J., Kim, J.H., Lee, S.C., 2015. Function of ABA in stomatal defense against biotic and drought stresses. International Journal of Molecular Sciences 16 (7), 15251–15270. Available at https://doi.org/10.3390/ijms160715251, http://www.mdpi.com/1422-0067/16/7/15251/pdf.

Ling, F., Su, Q., Jiang, H., Cui, J., He, X., Wu, Z., et al., 2020. Effects of strigolactone on photosynthetic and physiological characteristics in salt-stressed rice seedlings. Scientific Reports 10 (1). Available at https://doi.org/10.1038/s41598-020-63352-6.

Liu, X., Huang, B., 2002. Cytokinin effects on creeping bentgrass response to heat stress: II. Leaf senescence and antioxidant metabolism. Crop Science 42 (2), 466–472. Available at https://doi.org/10.2135/cropsci2002.4660.

Liu, H.Y., Sun, W.N., Su, W.A., Tang, Z.C., 2006. Co-regulation of water channels and potassium channels in rice. Physiologia Plantarum 128 (1), 58–69. Available at https://doi.org/10.1111/j.1399-3054.2006.00709.x.

Liu, J., He, H., Vitali, M., Visentin, I., Charnikhova, T., Haider, I., et al., 2015. Osmotic stress represses strigolactone biosynthesis in Lotus japonicus roots: exploring the interaction between strigolactones and ABA under abiotic stress. Planta 241 (6), 1435–1451. Available at https://doi.org/10.1007/s00425-015-2266-8, link.springer.de/link/service/journals/00425/index.htm.

Liu, X., Zhou, Y., Xiao, J., Bao, F., 2018. Effects of chilling on the structure, function and development of chloroplasts. Frontiers in Plant Science 9. Available at https://doi.org/10.3389/fpls.2018.01715.

Liu, Y., et al., 2021. Ethylene-mediated apoplastic barriers development. Journal of Hazardous Materials 403, 123729.

Llanes A.S. et al., n.d. Alterations of endogenous hormonal levels in plants under drought and salinity.

Lopez-Lauri, F., 2016. Plant growth regulators. Postharvest Management Approaches 125–139.

Lu, Q., Chen, S., Li, Y., Zheng, F., He, B., Gu, M., 2020. Exogenous abscisic acid (ABA) promotes cadmium (Cd) accumulation in Sedum alfredii Hance by regulating the expression of Cd stress response genes. Environmental Science and Pollution Research 27 (8), 8719–8731. Available at https://doi.org/10.1007/s11356-019-07512-w.

Ludwig, A.A., Saitoh, H., Felix, G., Freymark, G., Miersch, O., Wasternack, C., et al., 2005. Ethylene-mediated cross-talk between calcium-dependent protein kinase and MAPK signaling controls stress responses in plants. Proceedings of the National Academy of Sciences 102 (30), 10736–10741. Available at https://doi.org/10.1073/pnas.0502954102.

Macková, H., Hronková, M., Dobrá, J., Turečková, V., Novák, O., Lubovská, Z., et al., 2013. Enhanced drought and heat stress tolerance of tobacco plants with ectopically enhanced cytokinin oxidase/dehydrogenase gene expression. Journal of Experimental Botany 64 (10), 2805–2815. Available at https://doi.org/10.1093/jxb/ert131.

Maestri, E., Marmiroli, M., 2012. Genetic and molecular aspects of metal tolerance and hyperaccumulation. Metal Toxicity in Plants: Perception, Signaling and Remediation 41–63. Available at https://doi.org/10.1007/978-3-642-22081-4_3, http://doi.org/10.1007/978-3-642-22081-4.

Maggio, A., Barbieri, G., Raimondi, G., De Pascale, S., 2010. Contrasting effects of GA3 Treatments On Tomato Plants Exposed To Increasing Salinity. Journal of Plant Growth Regulation 29 (1), 63–72. Available at https://doi.org/10.1007/s00344-009-9114-7.

Mahajan, M., et al., 2023. Salicylic acid mitigates arsenic stress in rice. Agriculture 13 (7), 1293.

Mahajan, M., Sahoo, R.N., Khan, M.I.R., 2025a. Lanthanum supplementation with abscisic acid. Physiologia Plantarum 177 (1), e70126.

Mahajan, M., et al., 2025b. Salt stress tolerance and abscisic acid in plants. Plant Physiology and Biochemistry, 110303.

Mahajan, M., Khatoon, S., Prakash, S., Khan, M.I.R., 2025c. Plant stress hormones: fostering drought stress resilience. Optimizing Plant Health under Abiotic-Stress Environments: Ensuring Food and Nutritional Security. CRC Press, India, pp. 123–142. Available at https://doi.org/10.1201/9781032630380-7, https://www.taylorfrancis.com/books/edit/10.1201/9781032630380/.

Mahajan, M., Chaudhary, S., Khan, M.I.R., 2025d. Lanthanum-induced abscisic acid responses. Journal of Hazardous Materials 497, 139763.

Malhi, Y., Franklin, J., Seddon, N., Solan, M., Turner, M.G., Field, C.B., et al., 2020. Climate change and ecosystems: threats, opportunities and solutions. Philosophical Transactions of the Royal Society B: Biological Sciences 375 (1794), 20190104. Available at https://doi.org/10.1098/rstb.2019.0104.

Malik, A., et al., 2020. Biostimulant-treated seedlings under sustainable agriculture. Agronomy 11 (1), 14.

Mandava, N., 1988. Plant growth-promoting brassinosteroids. Annual Review of Plant Physiology and Plant Molecular Biology 39 (1), 23–52. Available at https://doi.org/10.1146/annurev.arplant.39.1.23.

MingMing, H., Fan, H., Kai, W., FengYun, Z., 2010. Effects of different kinds of exogenous auxin on the growth of rice roots under cadmium stress. Agricultural Science & Technology - Hunan 11, 45–48.

Mayak, S., Tirosh, T., Glick, B.R., 2004. Plant growth-promoting bacteria that confer resistance to water stress in tomatoes and peppers. Plant Science 166 (2), 525–530. Available at https://doi.org/10.1016/j.plantsci.2003.10.025, http://www.journals.elsevier.com/plant-science/.

McSteen, P., 2010. Auxin and monocot development. Cold Spring Harbor Perspectives in Biology 2 (3), a001479. Available at https://doi.org/10.1101/cshperspect.a001479.

Meng, H., Hua, S., Shamsi, I.H., Jilani, G., Li, Y., Jiang, L., 2009. Cadmium-induced stress on the seed germination and seedling growth of *Brassica napus* L., and its alleviation through exogenous plant growth regulators. Plant Growth Regulation 58 (1), 47–59. Available at https://doi.org/10.1007/s10725-008-9351-y.

Mohamed, H.I., El-Shazly, H.H., Badr, A., 2020. Role of salicylic acid in biotic and abiotic stress tolerance in plants. Plant Phenolics in Sustainable Agriculture: Volume 1 1, 533–554. Available at https://doi.org/10.1007/978-981-15-4890-1_23, https://link.springer.com/book/10.1007/978-981-15-4890-1.

Mohammed, A.R., Tarpley, L., 2009. Impact of high nighttime temperature on respiration, membrane stability, antioxidant capacity, and yield of rice plants. Crop Science 49 (1), 313–322. Available at https://doi.org/10.2135/cropsci2008.03.0161, http://crop.scijournals.org/cgi/reprint/49/1/313.

Mohan, T.C., Castrillo, G., Navarro, C., Zarco-Fernández, S., Ramireddy, E., Mateo, C., et al., 2016. Cytokinin determines thiol-mediated arsenic tolerance and accumulation. Plant Physiology 171 (2), 1418–1426. Available at https://doi.org/10.1104/pp.16.00372, http://www.plantphysiol.org/content/171/2/1418.full.pdf.

Mok, D.W., 2019. Cytokinins: Chemistry, Activity, and Function. CRC press.

Moons, A., Prinsen, E., Bauw, G., Van Montagu, M., 1997. Antagonistic effects of abscisic acid and jasmonates on salt stress-inducible transcripts in rice roots. The Plant Cell 9 (12), 2243–2259. Available at https://doi.org/10.1105/tpc.9.12.2243.

Morgan, P.W., Drew, M.C., 1997. Ethylene and plant responses to stress. Physiologia Plantarum 100 (3), 620–630. Available at https://doi.org/10.1034/j.1399-3054.1997.1000325.x.

Munemasa, S., Mori, I.C., Murata, Y., 2011. Methyl jasmonate signalling and signal crosstalk. Plant Signaling & Behavior 6 (7), 939–941.

Müssig, C., Altmann, T., 2003. Genomic brassinosteroid effects. Journal of Plant Growth Regulation 22 (4), 313–324. Available at https://doi.org/10.1007/s00344-003-0061-4.

Nazir, F., et al., 2023a. Sustaining nitrogen dynamics. Frontiers in Plant Science 14, 1087946.

Nazir, F., et al., 2023b. An explicit story of plant abiotic stress resilience. Plant and Soil 486 (1), 135–163.

Nazir, F., Khatoon, S., Mahajan, M., Kumari, S., AlAjmi, M.F., Rehman, M.T., et al., 2024. Enhanced arsenic stress tolerance in landrace and improved rice cultivars through modulation of gibberellic acid (GA3) synthesis and antioxidant metabolism via phosphorus and silicon supplementation. Plant Stress 13. Available at https://doi.org/10.1016/j.stress.2024.100511, https://www.sciencedirect.com/science/journal/2667064X.

Nievola, C.C., Carvalho, C.P., Carvalho, V., Rodrigues, E., 2017. Rapid responses of plants to temperature changes. Temperature 4 (4), 371–405. Available at https://doi.org/10.1080/23328940.2017.1377812, https://tandfonline.com/action/journalInformation?journalCode=ktmp20.

Nishiyama, R., Watanabe, Y., Fujita, Y., Le, D.T., Kojima, M., Werner, T., et al., 2011. Analysis of cytokinin mutants and regulation of cytokinin metabolic genes reveals important regulatory roles of cytokinins in drought, salt and abscisic acid responses, and abscisic acid biosynthesis. The Plant Cell 23 (6), 2169–2183. Available at https://doi.org/10.1105/tpc.111.087395, http://www.plantcell.org/content/23/6/2169.full.pdf+html.

Nishiyama, R., et al., 2012. Transcriptome analyses of a salt-tolerant cytokinin-deficient mutant. PLoS One 7 (2), e32124.

Noriega, G., Cruz, D.S., Batlle, A., Tomaro, M., Balestrasse, K., 2012. Heme oxygenase is involved in the protection exerted by jasmonic acid against cadmium stress in soybean roots. Journal of Plant Growth Regulation 31 (1), 79–89. Available at https://doi.org/10.1007/s00344-011-9221-0.

Noya, I., González-García, S., Bacenetti, J., Fiala, M., Moreira, M.T., 2018. Environmental impacts of the cultivation-phase associated with agricultural crops for feed production. Journal of Cleaner Production 172, 3721–3733. Available at https://doi.org/10.1016/j.jclepro.2017.07.132.

Ogweno, J.O., Song, X.S., Shi, K., Hu, W.H., Mao, W.H., Zhou, Y.H., et al., 2008. Brassinosteroids alleviate heat-induced inhibition of photosynthesis by increasing carboxylation efficiency and enhancing antioxidant systems in *Lycopersicon esculentum*. Journal of Plant Growth Regulation 27 (1), 49–57. Available at https://doi.org/10.1007/s00344-007-9030-7.

Oshino, T., Miura, S., Kikuchi, S., Hamada, K., Yano, K., Watanabe, M., et al., 2011. Auxin depletion in barley plants under high-temperature conditions represses DNA proliferation in organelles and nuclei via transcriptional alterations. Plant, Cell & Environment 34 (2), 284–290. Available at https://doi.org/10.1111/j.1365-3040.2010.02242.x.

Ouzounidou, G., Ilias, I., 2005. Hormone-induced protection of sunflower photosynthetic apparatus against copper toxicity. Biologia Plantarum 49 (2), 223–228. Available at https://doi.org/10.1007/s10535-005-3228-y.

Pan, Z., Yang, B., Li, S., Cai, N., Xu, Y., 2025. JA regulates the impact of combined heat stress and drought stress on the GA-mediated response of Pinus yunnanensis. Industrial Crops and Products 235, 121826. Available at https://doi.org/10.1016/j.indcrop.2025.121826.

Parys, E., Romanowska, E., Siedlecka, M., Poskuta, J.W., 1998. The effect of lead on photosynthesis and respiration in detached leaves and in mesophyll protoplasts of *Pisum sativum*. Acta Physiologiae Plantarum 20 (3), 313–322. Available at https://doi.org/10.1007/s11738-998-0064-7, https://rd.springer.com/journal/11738.

Pasternak, T., et al., 2005. Complementary interactions between oxidative stress and auxins. Journal of Experimental Botany 56 (418), 1991–2001.

Pattyn, J., Vaughan-Hirsch, J., Van de Poel, B., 2021. The regulation of ethylene biosynthesis: a complex multilevel control circuitry. New Phytologist 229 (2), 770–782. Available at https://doi.org/10.1111/nph.16873.

Patyal, U., Khushboo, K., Bala, R., Narwat, M., Kaur, M., 2024. The potential for plant growth regulators to impact crop productivity in future agricultural system. Plant Growth Regulators: Resilience for Sustainable Agriculture. Springer Nature, India, pp. 223–242. Available at https://doi.org/10.1007/978-981-97-2918-0_12, https://link.springer.com/book/10.1007/978-981-97-2918-0#:~:text=In%20plants%2C%20plant%20growth%20regulators,high%20yields%20in%20crop%20plants.

Peleg, Z., Blumwald, E., 2011. Hormone balance and abiotic stress tolerance in crop plants. Current Opinion in Plant Biology 14 (3), 290–295. Available at https://doi.org/10.1016/j.pbi.2011.02.001.

Peleg, Z., Reguera, M., Tumimbang, E., Walia, H., Blumwald, E., 2011. Cytokinin-mediated source/sink modifications improve drought tolerance and increase grain yield in rice under water-stress. Plant Biotechnology Journal 9 (7), 747–758. Available at https://doi.org/10.1111/j.1467-7652.2010.00584.x.

Peng, J., Li, Z., Wen, X., Li, W., Shi, H., Yang, L., et al., 2014. Salt-induced stabilization of EIN3/EIL1 confers salinity tolerance by deterring ROS accumulation in *Arabidopsis*. PLoS Genetics 10 (10), e1004664. Available at https://doi.org/10.1371/journal.pgen.1004664.

Pérez-Llorca, M., Pollmann, S., Müller, M., 2023. Ethylene and jasmonates signaling network mediating secondary metabolites under abiotic stress. International Journal of Molecular Sciences 24 (6), 5990. Available at https://doi.org/10.3390/ijms24065990.

Pieterse, C.M.J., Van Der Does, D., Zamioudis, C., Leon-Reyes, A., Van Wees, S.C.M., 2012. Hormonal modulation of plant immunity. Annual Review of Cell and Developmental Biology 28, 489–521. Available at https://doi.org/10.1146/annurev-cellbio-092910-154055.

Piotrowska-Niczyporuk, A., et al., 2012. Phytohormones as regulators of heavy metal biosorption. Plant Physiology and Biochemistry 52, 52–65.

Piotrowska-Niczyporuk, A., et al., 2018. Exogenously applied auxins and cytokinins ameliorate lead toxicity. Plant Physiology and Biochemistry 132, 535–546.

Pospíšilová, H., Jiskrová, E., Vojta, P., Mrízová, K., Kokáš, F., Čudejková, M.M., et al., 2016. Transgenic barley overexpressing a cytokinin dehydrogenase gene shows greater tolerance to drought stress. New Biotechnology 33 (5), 692–705. Available at https://doi.org/10.1016/j.nbt.2015.12.005, http://www.elsevier.com.

Potters, G., Pasternak, T.P., Guisez, Y., Jansen, M.A.K., 2009. Different stresses, similar morphogenic responses: integrating a plethora of pathways. Plant, Cell & Environment 32 (2), 158–169. Available at https://doi.org/10.1111/j.1365-3040.2008.01908.x.

Qin, H., Gu, Q., Zhang, J., Sun, L., Kuppu, S., Zhang, Y., et al., 2011. Regulated expression of an isopentenyltransferase gene (IPT) in peanut significantly improves drought tolerance and increases yield under field conditions. Plant and Cell Physiology 52 (11), 1904–1914. Available at https://doi.org/10.1093/pcp/pcr125.

Rademacher, W., 2015. Plant growth regulators: backgrounds and uses in plant production. Journal of Plant Growth Regulation 34 (4), 845–872. Available at https://doi.org/10.1007/s00344-015-9541-6.

Rauwane, M., Ntushelo, K., 2020. Understanding biotic stress and hormone signalling in cassava (*Manihot esculenta*): potential for using hyphenated analytical techniques. Applied Sciences 10 (22), 8152. Available at https://doi.org/10.3390/app10228152.

Ribba, T., Garrido-Vargas, F., O'Brien, J.A., 2020. Auxin-mediated responses under salt stress: from developmental regulation to biotechnological applications. Journal of Experimental Botany 71 (13), 3843–3853. Available at https://doi.org/10.1093/jxb/eraa241, http://jxb.oxfordjournals.org/.

Rivero, R.M., Kojima, M., Gepstein, A., Sakakibara, H., Mittler, R., Gepstein, S., et al., 2007. Delayed leaf senescence induces extreme drought tolerance in a flowering plant. Proceedings of the National Academy of Sciences 104 (49), 19631–19636. Available at https://doi.org/10.1073/pnas.0709453104.

Riyazuddin, R., Verma, R., Singh, K., Nisha, N., Keisham, M., Bhati, K.K., et al., 2020. Ethylene: a master regulator of salinity stress tolerance in plants. Biomolecules 10 (6), 1–22. Available at https://doi.org/10.3390/biom10060959, https://www.mdpi.com/2218-273X/10/6/959/pdf.

Rodríguez-Serrano, M., Romero-Puertas, M.C., Zabalza, A., Corpas, F.J., Gómez, M., Del Río, L.A., et al., 2006. Cadmium effect on oxidative metabolism of pea (*Pisum sativum* L.) roots. Imaging of reactive oxygen species and nitric oxide accumulation in vivo. Plant, Cell and Environment 29 (8), 1532–1544. Available at https://doi.org/10.1111/j.1365-3040.2006.01531.x.

Ronzan, M., Piacentini, D., Fattorini, L., Federica, D.R., Caboni, E., Eiche, E., et al., 2019. Auxin-jasmonate crosstalk in *Oryza sativa* L. root system formation after cadmium and/or arsenic exposure. Environmental and Experimental Botany 165, 59–69. Available at https://doi.org/10.1016/j.envexpbot.2019.05.013, www.elsevier.com/inca/publications/store/2/6/7.

Rzewuski, G., Sauter, M., 2008. Ethylene biosynthesis and signaling in rice. Plant Science 175 (1-2), 32–42. Available at https://doi.org/10.1016/j.plantsci.2008.01.012.

Sabagh, A.E.L., Mbarki, S., Hossain, A., Iqbal, M.A., Islam, M.S., Raza, A., et al., 2021. Potential role of plant growth regulators in administering crucial processes against abiotic stresses. Frontiers in Agronomy 3. Available at https://doi.org/10.3389/fagro.2021.648694, www.frontiersin.org/journals/agronomy#.

Sairam, R.K., 1994. Effects of homobrassinolide application on plant metabolism and grain yield under irrigated and moisture-stress conditions of two wheat varieties. Plant Growth Regulation 14 (2), 173–181. Available at https://doi.org/10.1007/bf00025220.

Sakakibara, H., 2006. CYTOKININS: activity, biosynthesis, and translocation. Annual Review of Plant Biology 57 (1), 431–449. Available at https://doi.org/10.1146/annurev.arplant.57.032905.105231.

Sakata, T., Oshino, T., Miura, S., Tomabechi, M., Tsunaga, Y., Higashitani, N., et al., 2010. Auxins reverse plant male sterility caused by high temperatures. Proceedings of the National Academy of Sciences 107 (19), 8569–8574. Available at https://doi.org/10.1073/pnas.1000869107.

Salt, D.E., Prince, R.C., Pickering, I.J., Raskin, I., 1995. Mechanisms of cadmium mobility and accumulation in Indian mustard. Plant Physiology 109 (4), 1427–1433. Available at https://doi.org/10.1104/pp.109.4.1427.

Salvi, P., Manna, M., Kaur, H., Thakur, T., Gandass, N., Bhatt, D., et al., 2021. Phytohormone signaling and crosstalk in regulating drought stress response in plants. Plant Cell Reports 40 (8), 1305–1329. Available at https://doi.org/10.1007/s00299-021-02683-8.

Sattar, A., Ul-Allah, S., Ijaz, M., Sher, A., Butt, M., Abbas, T., et al., 2022. Exogenous application of strigolactone alleviates drought stress in maize seedlings by regulating the physiological and antioxidants defense mechanisms. Cereal Research Communications 50 (2), 263–272. Available at https://doi.org/10.1007/s42976-021-00171-z, https://link.springer.com/journal/42976.

Savvides, A., Ali, S., Tester, M., Fotopoulos, V., 2016. Chemical priming of plants against multiple abiotic stresses: mission possible? Trends in Plant Science 21 (4), 329–340. Available at https://doi.org/10.1016/j.tplants.2015.11.003.

Scarpeci, T.E., Frea, V.S., Zanor, M.I., Valle, E.M., 2017. Overexpression of AtERF019 delays plant growth and senescence, and improves drought tolerance in *Arabidopsis*. Journal of Experimental Botany 68 (3), 673–685. Available at https://doi.org/10.1093/jxb/erw429, http://jxb.oxfordjournals.org/.

Seo, M., Koshiba, T., 2002. Complex regulation of ABA biosynthesis in plants. Trends in Plant Science 7 (1), 41–48. Available at https://doi.org/10.1016/s1360-1385(01)02187-2.

Shah, S.H., 2007. Effects of salt stress on mustard as affected by gibberellic acid application. Genetics and Applied Plant Physiology 33 (1–2), 97–106.

Shah, T., Asad, M., Khan, Z., Amjad, K., Alsahli, A.A., D'amato, R., 2023. Strigolactone decreases cadmium concentrations by regulating cadmium localization and glyoxalase defense system: effects on nodules organic acids and soybean yield. Chemosphere 335. Available at https://doi.org/10.1016/j.chemosphere.2023.139028, www.elsevier.com/locate/chemosphere.

Shahbaz, M., Ashraf, M., Athar, H.-ur-R., 2008. Does exogenous application of 24-epibrassinolide ameliorate salt induced growth inhibition in wheat (*Triticum aestivum* L.)? Plant Growth Regulation 55 (1), 51–64. Available at https://doi.org/10.1007/s10725-008-9262-y.

Shakirova, F.M., Allagulova, C.R., Maslennikova, D.R., Klyuchnikova, E.O., Avalbaev, A.M., Bezrukova, M.V., 2016. Salicylic acid-induced protection against cadmium toxicity in wheat plants. Environmental and Experimental Botany 122, 19–28. Available at https://doi.org/10.1016/j.envexpbot.2015.08.002, www.elsevier.com/inca/publications/store/2/6/7.

Sharaf, A.E., Farghal, I.I., Sofy, M.R., 2009. Role of gibberellic acid in abolishing the detrimental effects of Cd and Pb on broad bean and lupin plants. Research Journal of Agricultural and Biological Sciences 5 (5), 668–673.

Sharma, I., Kaur, N., Pati, P.K., 2017. Brassinosteroids: a promising option in deciphering remedial strategies for abiotic stress tolerance in rice. Frontiers in Plant Science 8. Available at https://doi.org/10.3389/fpls.2017.02151, http://journal.frontiersin.org/article/10.3389/fpls.2017.02151/full.

Sharma, L., Priya, M., Kaushal, N., Bhandhari, K., Chaudhary, S., Dhankher, O.P., et al., 2020. Plant growth-regulating molecules as thermoprotectants: functional relevance and prospects for improving heat tolerance in food crops. Journal of Experimental Botany 71 (2), 569–594. Available at https://doi.org/10.1093/jxb/erz333, http://jxb.oxfordjournals.org/.

Sharma, M., Negi, S., Kumar, P., Srivastava, D.K., Choudhary, M.K., Irfan, M., 2023. Fruit ripening under heat stress: the intriguing role of ethylene-mediated signaling. Plant Science 335. Available at https://doi.org/10.1016/j.plantsci.2023.111820, http://www.journals.elsevier.com/plant-science/.

Shi, W.G., Liu, W., Yu, W., Zhang, Y., Ding, S., Li, H., et al., 2019. Abscisic acid enhances lead translocation from the roots to the leaves and alleviates its toxicity in *Populus* × *canescens*. Journal of Hazardous Materials 362, 275–285. Available at https://doi.org/10.1016/j.jhazmat.2018.09.024, www.elsevier.com/locate/jhazmat.

Shinozaki, K., Yamaguchi-Shinozaki, K., 2007. Gene networks involved in drought stress response and tolerance. Journal of Experimental Botany 58 (2), 221–227. Available at https://doi.org/10.1093/jxb/erl164.

Siddiqui, M.H., Al-Whaibi, M.H., Basalah, M.O., 2011. Interactive effect of calcium and gibberellin on nickel tolerance in relation to antioxidant systems in *Triticum aestivum* L. Protoplasma 248 (3), 503–511. Available at https://doi.org/10.1007/s00709-010-0197-6.

Sims, K., n.d. Genetic transformation and regeneration in the Galena hop variety. Doctoral dissertation, Lincoln University.

Singh, I., Shono, M., 2005. Physiological and molecular effects of 24-epibrassinolide, a brassinosteroid on thermotolerance of tomato. Plant Growth Regulation 47 (2-3), 111–119. Available at https://doi.org/10.1007/s10725-005-3252-0.

Singh, S., Singh, A., Srivastava, P.K., Prasad, S.M., 2018. Cadmium toxicity and its amelioration by kinetin in tomato seedlings vis-à-vis ascorbate-glutathione cycle. Journal of Photochemistry and Photobiology B: Biology 178, 76–84. Available at https://doi.org/10.1016/j.jphotobiol.2017.10.025, www.elsevier.com/locate/jphotobiol.

Singh, D., Debnath, P., Sane, A.P., Sane, V.A., 2023. Tomato (*Solanum lycopersicum*) WRKY23 enhances salt and osmotic stress tolerance by modulating the ethylene and auxin pathways in transgenic *Arabidopsis*. Plant Physiology and Biochemistry 195, 330–340. Available at https://doi.org/10.1016/j.plaphy.2023.01.002, http://www.journals.elsevier.com/plant-physiology-and-biochemistry/.

Sirhindi, G., Mir, M.A., Sharma, P., Gill, S.S., Kaur, H., Mushtaq, R., 2015. Modulatory role of jasmonic acid on photosynthetic pigments, antioxidants and stress markers of *Glycine max* L. under nickel stress. Physiology and Molecular Biology of Plants 21 (4), 559–565. Available at https://doi.org/10.1007/s12298-015-0320-4, http://www.springerlink.com/content/120962/.

Skirycz, A., Claeys, H., de Bodt, S., Oikawa, A., Shinoda, S., Andriankaja, M., et al., 2011. Pause-and-stop: the effects of osmotic stress on cell proliferation during early leaf development in *Arabidopsis* and a role for ethylene signaling in cell cycle arrest. The Plant Cell 23 (5), 1876–1888. Available at https://doi.org/10.1105/tpc.111.084160, http://www.plantcell.org/content/23/5/1876.full.pdf+html.

Small, C.C., Degenhardt, D., 2018. Plant growth regulators for enhancing revegetation success in reclamation: a review. Ecological Engineering 118, 43–51. Available at https://doi.org/10.1016/j.ecoleng.2018.04.010, www.elsevier.com/inca/publications/store/5/2/2/7/5/1.

Šmehilová, M., Galuszka, P., Bilyeu, K.D., Jaworek, P., Kowalska, M., Šebela, M., et al., 2009. Subcellular localization and biochemical comparison of cytosolic and secreted cytokinin dehydrogenase enzymes from maize. Journal of Experimental Botany 60 (9), 2701–2712. Available at https://doi.org/10.1093/jxb/erp126.

Snyman, M., Cronjé, M.J., 2008. Modulation of heat shock factors accompanies salicylic acid-mediated potentiation. Journal of Experimental Botany 59 (8), 2125–2132.

Soliman, S., Wang, Y., Han, Z., Pervaiz, T., El-kereamy, A., 2022. Strigolactones in plants and their interaction with the ecological microbiome in response to abiotic stress. Plants 11 (24), 3499. Available at https://doi.org/10.3390/plants11243499.

Stavang, J.A., Gallego-Bartolomé, J., Gómez, M.D., Yoshida, S., Asami, T., Olsen, J.E., et al., 2009. Hormonal regulation of temperature-induced growth in *Arabidopsis*. Plant Journal 60 (4), 589–601. Available at https://doi.org/10.1111/j.1365-313X.2009.03983.x.

Strobel, N.E., Kuc, J.A., 1995. Chemical and biological inducers of systemic resistance to pathogens protect cucumber and tobacco plants from damage caused by paraquat and cupric chloride. Phytopathology 85 (10), 1306–1310. Available at https://doi.org/10.1094/Phyto-85-1306.

Stroiński, A., Giżewska, K., Zielezińska, M., 2013. Abscisic acid is required in transduction of cadmium signal. Biologia Plantarum 57 (1), 121–127.

Takahama, U., Oniki, T., 1992. Regulation of peroxidase-dependent oxidation of phenolics in the apoplast of spinach leaves by ascorbate. Plant and Cell Physiology 33 (4), 379–387.

Takahashi, N., 1998. Discovery of gibberellin. Discoveries in Plant Biology: Volume I. World Scientific, pp. 17–32.

Takahashi, F., et al., 2018. A small peptide modulates stomatal control via abscisic acid. Nature 556 (7700), 235–238.

Tan, B.C., Schwartz, S.H., Zeevaart, J.A.D., McCarty, D.R., 1997. Genetic control of abscisic acid biosynthesis in maize. Proceedings of the National Academy of Sciences 94 (22), 12235–12240. Available at https://doi.org/10.1073/pnas.94.22.12235.

Tao, Q., Jupa, R., Liu, Y., Luo, J., Li, J., Kováč, J., et al., 2019. Abscisic acid-mediated modifications of radial apoplastic transport pathway play a key role in cadmium uptake in hyperaccumulator *Sedum alfredii*. Plant, Cell & Environment 42 (5), 1425–1440. Available at https://doi.org/10.1111/pce.13506.

Thompson, A.J., Andrews, J., Mulholland, B.J., McKee, J.M.T., Hilton, H.W., Horridge, J.S., et al., 2007. Overproduction of abscisic acid in tomato increases transpiration efficiency and root hydraulic conductivity and influences leaf expansion. Plant Physiology 143 (4), 1905–1917. Available at https://doi.org/10.1104/pp.106.093559, http://www.plantphysiol.org/.

Toh, S., Imamura, A., Watanabe, A., Nakabayashi, K., Okamoto, M., Jikumaru, Y., et al., 2008. High temperature-induced abscisic acid biosynthesis and its role in the inhibition of gibberellin action in *Arabidopsis* seeds. Plant Physiology 146 (3), 1368–1385. Available at https://doi.org/10.1104/pp.107.113738, http://www.plantphysiol.org/cgi/reprint/146/3/1368.

Trinh, N.N., Huang, T.L., Chi, W.C., Fu, S.F., Chen, C.C., Huang, H.J., 2014. Chromium stress response effect on signal transduction and expression of signaling genes in rice. Physiologia Plantarum 150 (2), 205–224. Available at https://doi.org/10.1111/ppl.12088.

Upreti, K.K., Murti, G.S., 1999. Water stress induced changes in common polyamines and abscisic acid in French bean. Plant Growth Regulation 27 (3), 159–167.

van Overbeek, J., Conklin, M.E., Blakeslee, A.F., 1941. Factors in coconut milk essential for growth and development. Science (New York, N.Y.) 94 (2441), 350–351.

Vanneste, S., Friml, J., 2009. Auxin: a trigger for change in plant development. Cell 136 (6), 1005–1016. Available at https://doi.org/10.1016/j.cell.2009.03.001.

Vardhini, B.V., Anjum, N.A., 2015. Brassinosteroids make plant life easier under abiotic stresses mainly by modulating major components of antioxidant defense system. Frontiers in Environmental Science 2. Available at https://doi.org/10.3389/fenvs.2014.00067, http://journal.frontiersin.org/article/10.3389/fenvs.2014.00067/pdf.

Vassilev, A., Yordanov, I., 1997. Reductive analysis of factors limiting growth of cadmium-treated plants: a review. Bulgarian Journal of Plant Physiology 23 (3–4), 114–133.

Vassilev, A., Lidon, F., Scotti, P., Da Graca, M., Yordanov, I., 2004. Cadmium-induced changes in chloroplast lipids and photosystem activities in barley plants. Biologia plantarum 48 (1), 153–156. Available at https://doi.org/10.1023/b:biop.0000024295.27419.89.

Vert, G., Nemhauser, J.L., Geldner, N., Hong, F., Chory, J., 2005. Molecular mechanisms of steroid hormone signaling in plants. Annual Review of Cell and Developmental Biology 21, 177–201. Available at https://doi.org/10.1146/annurev.cellbio.21.090704.151241.

Visentin, I., Vitali, M., Ferrero, M., Zhang, Y., Ruyter-Spira, C., Novák, O., et al., 2016. Low levels of strigolactones in roots as a component of the systemic signal of drought stress in tomato. New Phytologist 212 (4), 954–963. Available at https://doi.org/10.1111/nph.14190.

Vishwakarma, K., Upadhyay, N., Kumar, N., Yadav, G., Singh, J., Mishra, R.K., et al., 2017. Abscisic acid signaling and abiotic stress tolerance in plants: a review on current knowledge and future prospects. Frontiers in Plant Science 8. Available at https://doi.org/10.3389/fpls.2017.00161, http://journal.frontiersin.org/article/10.3389/fpls.2017.00161/full.

Vriet, C., Russinova, E., Reuzeau, C., 2012. Boosting crop yields with plant steroids. The Plant Cell 24 (3), 842–857. Available at https://doi.org/10.1105/tpc.111.094912.

Wang, Y., Liu, C., Li, K., Sun, F., Hu, H., Li, X., et al., 2007. *Arabidopsis* EIN2 modulates stress response through abscisic acid response pathway. Plant Molecular Biology 64 (6), 633–644. Available at https://doi.org/10.1007/s11103-007-9182-7.

Wasternack, C., Strnad, M., 2018. Jasmonates: news on occurrence. International Journal of Molecular Sciences 19 (9), 2539.

Wasternack, C., Forner, S., Strnad, M., Hause, B., 2013. Jasmonates in flower and seed development. Biochimie 95 (1), 79–85. Available at https://doi.org/10.1016/j.biochi.2012.06.005.

Watkins, C.B., 2002. Ethylene synthesis, mode of action. Fruit Quality and Its Biological Basis 180–224.

Wei, Y., Liu, G., Chang, Y., He, C., Shi, H., 2018. Heat shock transcription factor 3 regulates plant immune response through modulation of salicylic acid accumulation and signalling in cassava. Molecular Plant Pathology 19 (10), 2209–2220. Available at https://doi.org/10.1111/mpp.12691.

Werner, T., Motyka, V., Strnad, M., Schmülling, T., 2001. Regulation of plant growth by cytokinin. Proceedings of the National Academy of Sciences 98 (18), 10487–10492. Available at https://doi.org/10.1073/pnas.171304098.

Werner, T., Nehnevajova, E., Köllmer, I., Novák, O., Strnad, M., Krämer, U., et al., 2010. Root-specific reduction of cytokinin causes enhanced root growth, drought tolerance, and leaf mineral enrichment in *Arabidopsis* and tobacco. The Plant Cell 22 (12), 3905–3920. Available at https://doi.org/10.1105/tpc.109.072694.

Wilen, R.W., Sacco, M., Gusta, L.V., Krishna, P., 1995. Effects of 24-epibrassinolide on freezing and thermotolerance of bromegrass (*Bromus inermis*) cell cultures. Physiologia Plantarum 95 (2), 195–202. Available at https://doi.org/10.1111/j.1399-3054.1995.tb00827.x.

Wu, Y.S., Yang, C.Y., 2019. Ethylene-mediated signaling confers thermotolerance and regulates transcript levels of heat shock factors in rice seedlings under heat stress. Botanical Studies 60 (1). Available at https://doi.org/10.1186/s40529-019-0272-z, http://link.springer.com/journal/40529.

Wu, F., et al., 2022. Biological functions of strigolactones and their crosstalk. Frontiers in Plant Science 13, 821563.

Wybouw, B., De Rybel, B., 2019. Cytokinin – a developing story. Trends in Plant Science 24 (2), 177–185. Available at https://doi.org/10.1016/j.tplants.2018.10.012.

Xing, J., Xu, Y., Tian, J., Gianfagna, T., Huang, B., 2009. Suppression of shade- or heat-induced leaf senescence in creeping bentgrass through transformation with the ipt gene for cytokinin synthesis. Journal of the American Society for Horticultural Science 134 (6), 602–609. Available at https://doi.org/10.21273/jashs.134.6.602.

Xiong, L., Zhu, J.K., 2003. Regulation of abscisic acid biosynthesis. Plant Physiology 133 (1), 29–36. Available at https://doi.org/10.1104/pp.103.025395, http://www.plantphysiol.org/.

Xu, S., Brockmöller, T., Navarro-Quezada, A., Kuhl, H., Gase, K., Ling, Z., et al., 2017. Wild tobacco genomes reveal the evolution of nicotine biosynthesis. Proceedings of the National Academy of Sciences 114 (23), 6133–6138. Available at https://doi.org/10.1073/pnas.1700073114.

Yadav, B., et al., 2021. An overview of recent advancement in phytohormones-mediated stress management. Plant Gene 25, 100264.

Yamori, W., Suzuki, K., Noguchi, K., Nakai, M., Terashima, I., 2006. Effects of Rubisco kinetics and Rubisco activation state on the temperature dependence of the photosynthetic rate in spinach leaves from contrasting growth temperatures. Plant, Cell & Environment 29 (8), 1659–1670. Available at https://doi.org/10.1111/j.1365-3040.2006.01550.x.

Yan, H., Filardo, F., Hu, X., Zhao, X., Fu, D.H., 2016. Cadmium stress alters the redox reaction and hormone balance in oilseed rape (*Brassica napus* L.) leaves. Environmental Science and Pollution Research 23 (4), 3758–3769. Available at https://doi.org/10.1007/s11356-015-5640-y.

Yang, Z., Zhang, C., Yang, X., Liu, K., Wu, Z., Zhang, X., et al., 2014. PAG1, a cotton brassinosteroid catabolism gene, modulates fiber elongation. New Phytologist 203 (2), 437–448. Available at https://doi.org/10.1111/nph.12824.

Yang, J., et al., 2019. The crosstalks between jasmonic acid and other plant hormone signalling. Frontiers in Plant Science 10, 1349.

Yao, T., Xie, R., Zhou, C., Wu, X., Li, D., 2023. Roles of brossinosteroids signaling in biotic and abiotic stresses. Journal of Agricultural and Food Chemistry 71 (21), 7947–7960. Available at https://doi.org/10.1021/acs.jafc.2c07493.

Yildirim, E., Turan, M., Guvenc, I., 2008. Effect of foliar salicylic acid applications on growth, chlorophyll, and mineral content of cucumber grown under salt stress. Journal of Plant Nutrition 31 (3), 593–612. Available at https://doi.org/10.1080/01904160801895118.

Yoon, J.Y., Hamayun, M., Lee, S.-K., Lee, I.-J., 2009. Methyl jasmonate alleviated salinity stress in soybean. Journal of Crop Science and Biotechnology 12 (2), 63–68. Available at https://doi.org/10.1007/s12892-009-0060-5.

Zaid, A., Mohammad, F., 2018. Methyl jasmonate and nitrogen interact to alleviate cadmium stress. Journal of Plant Growth Regulation 37 (4), 1331–1348.

Zentella, R., Zhang, Z.L., Park, M., Thomas, S.G., Endo, A., Murase, K., et al., 2007. Global analysis of DELLA direct targets in early gibberellin signaling in *Arabidopsis*. The Plant Cell 19 (10), 3037–3057. Available at https://doi.org/10.1105/tpc.107.054999, http://www.plantcell.org/cgi/reprint/19/10/3037.

Zhang, J., Jia, W., Yang, J., Ismail, A.M., 2006. Role of ABA in integrating plant responses to drought and salt stresses. Field Crops Research 97 (1), 111–119. Available at https://doi.org/10.1016/j.fcr.2005.08.018, www.elsevier.com/inca/publications/store/5/0/3/3/0/8.

Zhang, S., Cai, Z., Wang, X., 2009a. The primary signaling outputs of brassinosteroids are regulated by abscisic acid signaling. Proceedings of the National Academy of Sciences 106 (11), 4543–4548. Available at https://doi.org/10.1073/pnas.0900349106.

Zhang, G., Chen, M., Li, L., Xu, Z., Chen, X., Guo, J., et al., 2009b. Overexpression of the soybean GmERF3 gene, an AP2/ERF type transcription factor for increased tolerances to salt, drought, and diseases in transgenic tobacco. Journal of Experimental Botany 60 (13), 3781–3796. Available at https://doi.org/10.1093/jxb/erp214.

Zhang, C.X., Feng, B.H., Chen, T.T., Zhang, X.F., Tao, L.X., Fu, G.F., 2017. Sugars, antioxidant enzymes and IAA mediate salicylic acid to prevent rice spikelet degeneration caused by heat stress. Plant Growth Regulation 83 (2), 313–323. Available at https://doi.org/10.1007/s10725-017-0296-x.

Zhang, W., Wang, Z., Song, J., Yue, S., Yang, H., 2019. Cd^{2+} uptake inhibited by MhNCED3 from *Malus hupehensis* alleviates Cd-induced cell death. Environmental and Experimental Botany 166, 103802.
Zhang, J., et al., 2021. Transcriptome analysis of Kentucky bluegrass. PLoS One 16 (12), e0261472.
Zhang, Y., et al., 2022. Molecular mechanisms of diverse auxin responses. International Journal of Molecular Sciences 23 (20), 12495.
Zhang, Y., et al., 2024. Phytohormones-mediated strategies for mitigation of heavy metals toxicity. Plant Cell Reports 43 (4), 99.
Zhao, Y., 2010. Auxin biosynthesis and its role in plant development. Annual Review of Plant Biology 61 (1), 49–64. Available at https://doi.org/10.1146/annurev-arplant-042809-112308.
Zhou, Z., Guo, K., Elbaz, A., Yang, Z., 2009. Salicylic acid alleviates mercury toxicity by preventing oxidative stress in roots of *Medicago sativa*. Environmental and Experimental Botany 65 (1), 27–34. Available at https://doi.org/10.1016/j.envexpbot.2008.06.001.
Zhu, X.F., Jiang, T., Wang, Z.W., Lei, G.J., Shi, Y.Z., Li, G.X., et al., 2012. Gibberellic acid alleviates cadmium toxicity by reducing nitric oxide accumulation and expression of IRT1 in *Arabidopsis thaliana*. Journal of Hazardous Materials 239-240, 302–307. Available at https://doi.org/10.1016/j.jhazmat.2012.08.077.
Zolla, G., Heimer, Y.M., Barak, S., 2010. Mild salinity stimulates a stress-induced morphogenic response in *Arabidopsis thaliana* roots. Journal of Experimental Botany 61 (1), 211–224. Available at https://doi.org/10.1093/jxb/erp290.

CHAPTER

3

Regulation of auxin and cytokinin signaling in plants during salt stress adaptation

Pradeep K. Agarwal[1,2], *Anjali Chittora*[1,2], *Aayushi Gupta*[1], *and Parinita Agarwal*[1]

[1]Halophyte Biology and Biotechnology Division, CSIR-Central Salt and Marine Chemicals Research Institute (CSIR-CSMCRI), Council of Scientific and Industrial Research (CSIR), Gijubhai Badheka Marg, Bhavnagar, Gujarat, India [2]Academy of Scientific and Innovative Research (AcSIR), Ghaziabad, Uttar Pradesh, India

3.1 Introduction

Phytohormones are natural substances that regulate key physiological processes in plant growth and development, such as seed germination, seedling growth, meristem function, plant development, flowering, and the mitigation of environmental stress (Waadt et al., 2022). Different phytohormones, including cytokinin (CK), auxins, gibberellic acid (GA), abscisic acid (ABA), ethylene, jasmonic acid (JA), and salicylic acid (SA), are produced in plants and are involved in various developmental pathways and metabolic processes. CK was first identified in 1955, playing a role in cell division and cell differentiation (Miller et al., 1955). CKs can be found in various parts of the plant, including roots, stems, and flowers, and their levels can vary depending on the plant's developmental stage or environmental conditions (Osugi and Sakakibara, 2015). They also work in conjunction with other hormones, such as auxins, to ensure a balanced growth pattern in plants (Moubayidin et al., 2009). Auxins play a crucial role in promoting cell elongation, root growth, and regulating the development of flowers and fruits (Zhang et al., 2022). They are also involved in responses to light and gravity, helping plants orient themselves for optimal growth (Muday, 2001). Gibberellins are responsible for promoting stem elongation, seed germination, and the flowering process. They influence overall plant height and size, playing an important role in the early stages of plant growth (Ueguchi-Tanaka, 2023). ABA regulates plant stress responses, particularly those induced by drought and salinity stresses.

Harnessing Plant-Stimulating Molecules
DOI: https://doi.org/10.1016/B978-0-443-44994-9.00006-7

It controls seed dormancy and inhibits early germination, ensuring plant growth occurs only under favorable conditions (Sano and Marion-Poll, 2021).

Ethylene is a gaseous hormone that plays a significant role in the aging process, as it is involved in the ripening of fruits and the shedding of leaves. It also helps plants respond to stress, such as damage or environmental changes (Iqbal et al., 2017). JA is involved in the plant's defense mechanisms, particularly in response to herbivory or pathogen attacks (Zhang et al., 2017). Another important phytohormone, SA, is also involved in activating defenses against pathogens, contributing to disease resistance (Kim and Lim, 2023).

A detailed progress has enhanced our understanding of hormone biosynthesis and signaling, as well as their roles in plant growth and development. Under environmental cues, hormone signaling components interact and help plants cope with stress conditions. In this chapter, we will discuss both CKs and auxin biosynthesis, metabolism, and signaling, with an emphasis on recent findings.

3.2 Cytokinins

CK was first identified as a cell division inducer during plant tissue culture (Miller et al., 1955). Since then, extensive research has shown its involvement in a wide range of physiological processes in plants. CK primarily facilitates shoot growth in plants and is involved in several other processes, including regulating leaf development, vascular development, senescence, and others (Kyozuka, 2007; Wu et al., 2021; Zürcher and Müller, 2016). In plants, CKs interact with other plant hormones, particularly auxins and ABA (Danilova et al., 2016; Iqbal et al., 2006; Nordström et al., 2004). This chapter highlights the overview of CK metabolism and mediated metabolism under abiotic stress conditions.

3.2.1 Cytokinin metabolism

3.2.1.1 Cytokinin biosynthesis

Naturally occurring CKs are adenine derivatives, featuring isoprenoid side chains attached to the N^6 position of the adenine ring. Among higher plants, CK commonly exists in trans-zeatin (tZ), dihydrozeatin (DZ), N^6-(Δ^2-Δ2-isopentenyl) adenine (iP), and cis zeatin forms (cZ; Mok and Mok, 2001). The biosynthesis of isoprenoid CKs is initiated by N-prenylation of adenosine 5′ phosphates (AMP, ADP, or ATP) with dimethylallyl diphosphate (DMAPP), which is catalyzed by adenosine phosphates-isopentenyltransferase (IPT) and is the first and key step in this pathway (Sakakibara, 2006). The primary product of this reaction is an iP nucleotide, such as iP riboside 5′-triphosphate (iPRTP) or iP riboside 5′-diphosphate (iPRDP), as IPT exhibits a preference for ATP or ADP as substrates (Sakakibara, 2025). The cytochrome P450 monooxygenases CYP735A catalyze the conversion of iP nucleotides to tZ nucleotides. Among the iP nucleotide substrates, iP riboside 5′-monophosphate (iPRMP) is most efficiently utilized by CYP735A, as the catalytic efficiency (k_{cat}/K_m) of CYP735A for iPRDP is approximately one-third, while for iPRTP, it is nearly one-hundredth of the efficiency of iPRMP (Takei et al., 2004). The tZ riboside 5′-monophosphate (tZRMP) is further converted into DZ riboside 5′ monophosphate (DZRMP) by the zeatin reductase enzyme. The iPRMP, tZRMP, and DZRMP are converted to bioactive CK iP, tZ, and DZ by lonely guy (LOG) in a one-step reaction or in a step reaction of dephosphorylation and deribosylation by 5′-ribonucleotide phosphohydrolase grain yield 3 (GY3) and CK/purine riboside nucleosidase (CPN), respectively (Sakakibara, 2025; Frébort et al., 2011). The known pathway for the biosynthesis of cZ involves tRNA prenylation, catalyzed by tRNA-isopentenyltransferase (tRNA-IPT). Some tRNA species can undergo prenylation, and prenylated tRNA has a cis-hydroxyl group; thus, degradation of prenylated tRNA generates cZ (Kamada-Nobusada and Sakakibara, 2009) (Fig. 3.1).

3.2.1.2 *Cytokinin degradation*

In plants, CK degradation is crucial for maintaining homeostasis by regulating the accumulation and distribution of CK metabolites. This process is driven by CK oxidase/dehydrogenase (CKX), which irreversibly inactivates base and nucleoside forms (iP, tZ, iPR, and tZR) of isoprenoid CKs by cleaving their unsaturated isoprenoid side chains. Nucleotide

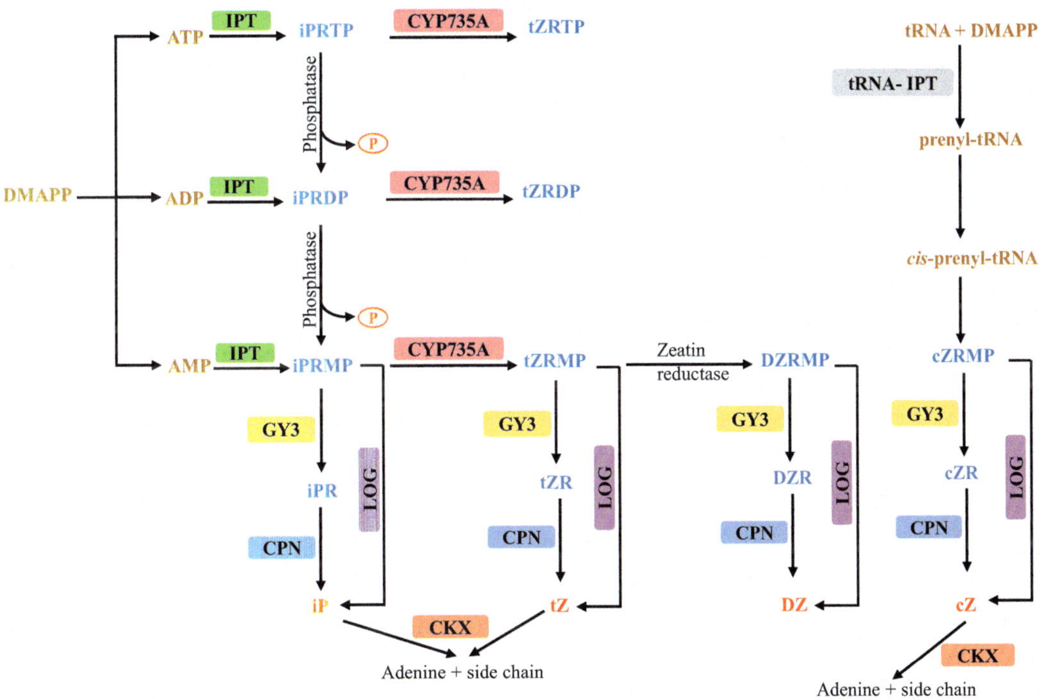

FIGURE 3.1 Cytokinin biosynthesis pathway in plants. Plant IPT utilizes ATP, ADP, or AMP as isoprenoid acceptors and DMAPP as the donor to synthesize iPRTP, iPRDP, and iPRMP, respectively. iPRTP and iPRDP are dephosphorylated by phosphatases. The iP-type nucleotides are subsequently converted to tZ-type nucleotides by the enzyme CYP735A. Nucleotide precursors such as iPRMP, tZRMP, DZRMP, and cZRMP are activated to cytokinin nucleobases either through a one-step conversion by LONELY GUY (*LOG*) or via a two-step process involving 5′-ribonucleotide phosphohydrolase GRAIN YIELD 3 (*GY3*) and cytokinin/purine riboside nucleosidase (*CPN*). Finally, active cytokinins, such as iP, tZ, and their nucleosides, are degraded by cytokinin oxidase/dehydrogenase (*CKX*) into adenine (*Ade*). *IPT*, Isopentenyltransferase.

derivatives are not targeted by CKX for degradation. The Δ^2 double bond in the isoprenoid side chain of tZ and iP plays a critical role in substrate recognition, as its reduction renders these CKs resistant to CKX activity. Consequently, DZ naturally evades CKX-mediated breakdown. Additionally, modifications to the side chain, such as O-glucosylation, further enhance resistance to CKX (Sakakibara, 2010). The glycosylation occurs mainly at the O- and N-positions and is catalyzed by UDP-glucosyltransferases (UGTs). The O-glucosylation is reversible, and the conjugated CKs can be reactivated by β-glucosidase enzymes (Li et al., 2022) (Fig. 3.1).

3.2.2 Cytokinin signaling

In plants, CK signaling operates through a multistep phosphorelay system that involves histidine kinase receptors (HKs), histidine phosphotransfer proteins (HPs), and response regulators (RRs) (Müller and Sheen, 2007). In *Arabidopsis thaliana*, the N-terminal region of CK receptors (AHK2, AHK3, and AHK4/WOL/CRE1) contains two to three transmembrane domains along with an extracellular domain. The predicted extracellular regions of these receptors include the cyclase/histidine-kinase-associated extracellular (CHASE) domain, which is believed to function as the CK-binding site (Anantharaman and Aravind, 2001). When CKs bind to the HK receptor, they dimerize, triggering autophosphorylation at conserved histidine residues within their N-terminal sensor-kinase domains. The phosphate group is then transferred to a conserved aspartate residue in the receptor's C-terminal receiver domain. From there, the phosphoryl group is relayed to HP proteins in the cytoplasm, continuing the signal transduction process (Werner and Schmülling, 2009). HP proteins function as intermediaries in the CK signaling pathway, facilitating phosphorelay between CK receptors and nuclear responses. Upon entering the nucleus, HP proteins transfer their phosphoryl group to type-B ARRs. This phosphorylation event enhances the transcriptional activity of type-B ARRs, which then activate the expression of primary CK response genes, including type-A ARRs. Type-A ARRs act as negative regulators, establishing a feedback loop that suppresses CK signaling, thereby fine-tuning the response to CK stimuli (To et al., 2004; Hwang and Sakakibara, 2006) (Fig. 3.2).

3.2.3 Regulation of salt stress tolerance by cytokinins

3.2.3.1 *Cytokinin metabolic genes*

The *LOG* gene is an activator of CK. Overexpressing *GhLOG3* from cotton in *A. thaliana* improved salt tolerance, whereas silencing *GhLOG3_At* using virus induced gene silencing (VIGS) led to increased salt sensitivity, characterized by increased hydrogen peroxide accumulation and reduced chlorophyll and proline levels, indicating positive regulation of *GhLOG3_At* toward salt stress tolerance (Wang et al., 2021). *SlIPT3* and *SlIPT4* genes in tomato play distinct roles in CK biosynthesis and salt stress response. Salt stress repressed their expression in roots, with differential expression in leaves. Overexpression of *SlIPT3* enhanced the accumulation and expression of CK metabolites and CK-related genes, thereby improving photosynthetic pigment retention and K^+/Na^+ balance under salinity stress. Transgenic *SlIPT3* tomato plants exhibited greater salt tolerance. Thus *SlIPT3* is crucial in

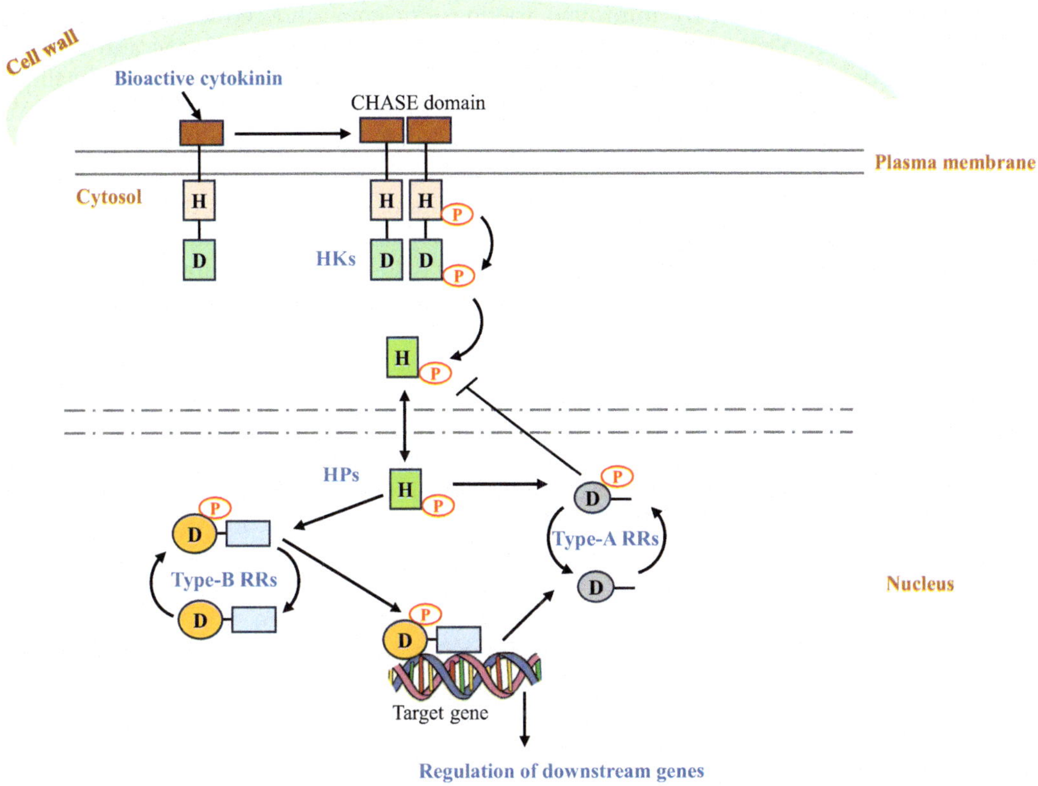

FIGURE 3.2 Cytokinin (*CK*) signaling in plants operates via a multistep phosphorelay involving histidine kinase receptors (*HK*), histidine phosphotransfer proteins (*HPs*), and response regulators (*RRs*). CK binding to receptors triggers autophosphorylation and phosphate transfer to HPs, which shuttle into the nucleus. There, type-B ARRs are activated to induce CK-responsive genes, including type-A ARRs. Type-A ARRs provide negative feedback, fine-tuning the CK response.

modulating CK metabolism to enhance salinity resilience (Žižková et al., 2015). The *CKX* gene family (*AtCKX1-AtCKX6*) plays a key role in regulating CK levels and exhibits distinct effects on plant development. In *A. thaliana*, overexpression of these genes results in decreased CK accumulation in shoots, leading to reduced activity of vegetative and floral shoot apical meristems. This reduction is associated with fewer leaf primordia and stunted shoot growth. In contrast, root development displayed an increase in the number of meristematic cells and enhanced primary root elongation, indicating that CK negatively regulates root meristem activity. These findings highlight the opposing functions of CKs in shoot and root development, underscoring their critical role in balancing growth between these organs in *A. thaliana* (Werner et al., 2003). Under salinity stress, *Arabidopsis* CK-deficient plants (CKX over-expressers; 35S:CKX1, 35S:CKX2, 35S:CKX3, 35S:CKX4, 35S:CKX5) exhibited a salt stress–resistant phenotype with slightly higher survival rates (Nishiyama et al., 2011). *Solanum tuberosum* (potato) transgenic lines with the *AtCKX2* gene exhibited reduced CK levels and enhanced tolerance to 50 mM and 100 mM NaCl concentrations, as evidenced by improved rooting efficiency and reduced salinity injury compared to control plants. However, under severe salinity stress (150 mM and 200 mM NaCl), the CK-deficient lines exhibited increased sensitivity, characterized by higher salinity injury indices and decreased shoot growth. These lines showed impaired upregulation of antioxidant enzymes, including superoxide dismutase (SOD), peroxidase (POD), and catalase (CAT), in response to salinity (Raspor et al., 2024). In *Morus indica cv. K2,* the *MiCKX4* showed significantly higher expression in response to drought and salinity stresses. The overexpression of *MiCKX4* in *A. thaliana* showed enhanced tolerance to stress, improved growth, and reduced oxidative damage compared to wild-type plants (Singhal et al., 2024). The *GhCKX6b-Dt* promoted salinity tolerance in cotton (*Gossypium hirsutum*). *GhCKX6b-Dt* expression is increased under salt stress, and a transgenic cotton plant with higher overexpression ability exhibited improved growth and biomass and lessened oxidative damage by enhancing the ROS scavenging enzymes like SOD, CAT, and POD (Liu et al., 2023). The *OsCKX2* plays an important role in regulating panicle architecture, grain yield, and drought tolerance in *Oryza sativa* indica varieties. In rice *Osckx2* mutants showed decrease in CK oxidase activity, that, lead to higher CK accumulation. These mutants exhibited enhanced panicle branching and increased grain yield compared to wild-type plants. In contrast to earlier reported functions, overexpression of the *CKX* gene provides stress tolerance; the *Osckx2* mutants demonstrated improved drought tolerance, characterized by higher survival rates and reduced membrane damage under water-limited conditions, which could lead to the development of high-yielding, stress-resilient rice cultivars suitable for climate-challenged agricultural systems (Rashid et al., 2024) (Table 3.1).

3.2.3.2 *Cytokinin signaling genes*

CK signaling core components play a significant role in modulating salt stress. In *Arabidopsis,* elevated active CK levels have been linked to negatively affecting the plant's tolerance to salt and drought stress (Nishiyama et al., 2011). The gain and loss-of-function studies of HKs revealed that *AHK1* functions as a positive regulator of drought and salt stress via both ABA-dependent and ABA-independent pathways, acting upstream of stress-related transcription factors such as AREB1, ANAC, and DREB2A. In contrast, other receptors, *AHK2, AHK3,* and *CRE1,* act as negative regulators, with their loss enhancing stress tolerance by upregulating ABA- and stress-responsive genes (Tran et al., 2007). The CK signaling components, *AHP2, AHP3,* and *AHP5,* negatively regulate drought stress responses by reducing CK levels in *Arabidopsis.* Loss-of-function mutants of these AHPs enhance drought tolerance through improved cell membrane integrity and increased ABA sensitivity. These AHPs regulate dehydration stress via CK signaling and in both ABA-dependent and ABA-independent manners (Nishiyama et al., 2013). Similarly, *OsAHP1* and *OsAHP2* act as positive regulators of the CK signaling pathway. However, knockout lines showed varied results under stress, increased sensitivity to salt, and enhanced resistance to osmotic stress (Sun et al., 2014). The CK type-B RR, *PeRR12,* acts as a negative regulator in regulating adventitious rooting and salt tolerance in poplar. Overexpression of *PeRR12* resulted in the inhibition of adventitious root formation by repressing key developmental genes (*PeWOX11, PeWOX5, PePIN1,* and *PePIN3*) and reducing salt tolerance by down-regulating *PeHK11;1.* Also, the expression of type B RR genes *PeRR2* and *PeRR10.1* was

TABLE 3.1 Cytokinin and auxin related gene regulation under salt stress.

S. no.	Plant	Gene	Transgenic plant[a]	Salinity stress	Function	References
Cytokinin related genes						
1	*Gossypium hirsutum*	*GhLOG3*	*Arabidopsis thaliana* (OE)	50 mM NaCl 100 mM NaCl	Improves physio-biochemical parameters and enhances salt tolerance	Wang et al. (2021)
2	*Solanum lycopersicum*	*SlIPT3*	*Solanum lycopersicum* (OE)	100 mM NaCl 150 mM NaCl	Enhances salt tolerance by accumulating CK metabolites, photosynthetic pigments, and maintaining K^+/Na^+ ratio	Žižková et al. (2015)
3	*Arabidopsis thaliana*	*AtCKX2*	Potato(OE)	50 and 100 mM NaCl 150 and 200 mM NaCl	Greater tolerance to low (50 mM) and moderate (100 mM) NaCl levels but were more sensitive to high salinity stress (150–200 mM NaCl)	Raspor et al. (2024)
4	*Gossypium hirsutum*	*GhCKX6b-Dt*	*Gossypium hirsutum* (KO)	100 mM NaCl	Oxidative damage and decreased salt tolerance	Liu et al. (2023)
5	*Arabidopsis thaliana*	*AHK1*	*Arabidopsis thaliana* (KO)	200 mM NaCl	Positive regulator of salt stress responses	Tran et al. (2007)
		AHK2, *AHK3*, and *CRE1*			Negative regulators of salt stress responses	
6	*Oryza sativa*	*OsAHP1* and *OsAHP2*	*Oryza sativa* (KO)	0–250 mM NaCl	Hyposensitive to exogenous cytokinin, hypersensitive to salt stress but resistant to osmotic stress	Sun et al. (2014)
7	*Populus deltoides* × *P. euramericana*, clone 'Nanlin895'	*PeRR12*	Poplar (OE)	200 mM NaCl	Acts as a repressor that mediates salt tolerance	Qi et al. (2022)
8	*Zea mays*	*ZmRR1*	*Zea mays* (OE)	100 mM NaCl	Salt tolerance via CK-mediated promotion of Cl^- exclusion from shoots	Yin et al. (2023)

(*Continued*)

TABLE 3.1 (Continued)

S. no.	Plant	Gene	Transgenic plant[a]	Salinity stress	Function	References
9	*Arabidopsis thaliana*	*CRF2*	*Arabidopsis thaliana* (KO)	150 mM NaCl	Reduced salt tolerance with decreased chlorophyll and photosynthetic efficiency	Keshishian et al. (2022)
10	*Medicago sativa*	*MsCKX*	*Arabidopsis thaliana* (OE)	100 mM NaCl 150 mM NaCl	Salt tolerance via high K^+/Na^+ ratio	Li et al. (2019)
11	*Oryza sativa*	*OsCKX2*	*Oryza sativa* (KO)	200 mM NaCl	Salt tolerance with improved growth, water content, and reduced electrolyte leakage	Joshi et al. (2018)
12	*Glycine max*	*GmAHP10*	*Glycine max* (OE)	200 mM NaCl	Salt tolerance with improved root system development, reduced H_2O_2 and MDA levels, increased antioxidant enzyme activities, and elevated proline content.	Cao et al. (2025)
13	*Oryza sativa*	*OsPHP1*	*Oryza sativa* (KO)	200 mM NaCl	Stress resilience, with better photosynthesis, increased tiller and panicle production, higher spikelet fertility and grain filling.	Yadav et al. (2025)
14	*Loliumperenne*	*LpARR10*	*Arabidopsis thaliana* (OE)	300 mM NaCl	Salt tolerance with reduced Na^+ accumulation and upregulation of *SOS1* and *SOS3* ion transport genes.	Yang et al. (2024)
15	*Oryza sativa (ZH11)*	*RST31*	*Oryza sativa* (KO)	120 mM NaCl	Salt sensitivity with increased DNA damage and higher ROS accumulation. The mutant showed of suppressed root-to-shoot transport of cytokinin.	He et al. (2024)

Auxin-related genes						
1	*Ipomoea batatas*	*IbARF5*	*Arabidopsis thaliana* (OE)	200 mM NaCl	Salt tolerance with decreased H_2O_2 accumulation	Kang et al. (2018)
2	*Solanum lycopersicum*	*SlARF4*	*Solanum lycopersicum* (KO)	100 mM NaCl 150 mM NaCl	Salt tolerance with decrease in stomatal conductance and increased relative water content	Bouzroud et al. (2020)
3	*Cucurbita pepo*	*CpARF22*	*Arabidopsis thaliana* (OE)	50 mM NaCl	Longer root phenotype and salt tolerance	Zhang et al. (2024)
4	*Arabidopsis thaliana*	*PIN1*	*Arabidopsis thaliana* (KO)	100 mM NaCl	Increased expression of cotyledon and epidermal cell development genes under salt stress	Bawa et al. (2022)
5	*Gossypium hirsutum*	*GH3.5*	*Gossypium hirsutum* (KO)	250 mM NaCl	Reduced salt tolerance with decreased chlorophyll, water content and SOD, and increased MDA and ion leakage.	Kirungu et al. (2019)
6	*Arabidopsis thaliana*	*AFB3*	*Arabidopsis thaliana* (OE)	150 mM NaCl	Salt tolerance with enhanced root development	Garrido-Vargas et al. (2020)
7	*Arabidopsis thaliana*	*TIR1 AFB2*	*Arabidopsis thaliana* (KO)	150 mM NaCl 200 mM NaCl	Enhanced salt tolerance with increased chlorophyll, germination, antioxidant enzymes	Iglesias et al. (2010)
8	*Arabidopsis thaliana*	*mTIR1*	*Arabidopsis thaliana* (OE)	100 mM NaCl 150 mM NaCl	miR393 resistant form of TIR1 enhances salt tolerance by improved physio-biochemical parameters, delayed senescence, and decreased cell death	Chen et al. (2015)
9	*Solanum tuberosum*	*ARP1*	*Arabidopsis thaliana* (OE)	200 mM NaCl	Salt tolerance with improved morphological, physio-biochemical parameters.	AlNeyadi et al. (2024)
10	*Glycine max*	*GmHXK2*	*Arabidopsis thaliana* (OE)	100 mM NaCl	Salt tolerance with increased lateral root formation, auxin and ascorbic acid accumulation, and upregulation of ion homeostasis genes.	Guo et al. (2024)

(*Continued*)

TABLE 3.1 (Continued)

S. no.	Plant	Gene	Transgenic plant[a]	Salinity stress	Function	References
11	*Trifolium repens*	*TrIAA27*	*Arabidopsis thaliana* (OE)	200 mM NaCl	Salt tolerance with improved morphological, physio-biochemical parameters.	Iqbal et al. (2024)
12	*Zea mays*	*ZmARF1*	*Arabidopsis thaliana* (OE)	200 mM NaCl	Salt tolerance with higher chlorophyll content and fluorescence, reduced water loss and ion leakage.	Liu et al. (2024)
13	*Triticum aestivum*	*TaHAK1*	*Oryza sativa* (OE)	100 mM NaCl	Salt tolerance with maintained K^+/Na^+ balance and increased antioxidant enzyme activity.	Liu et al. (2025)
14	*Populus alba* × *P. glandulosa*	*PagYUC6a*	*Populus alba* × *P. glandulosa* (OE)	200 mM NaCl	Salt tolerance with improved ROS scavenging capacity.	Zhao et al. (2025)

MDA, Malondialdehyde; *ROS*, reactive oxygen species; *SOD*, superoxide dismutase.
[a] *Overexpression lines—OE; knockout lines—KO.*

downregulated in *PeRR12* overexpressing lines, suggesting type-A RR genes play a redundant role in regulating type-B RRs (Qi et al., 2022). In *Zea mays*, type-A RR, *ZmRR1*, acts as a negative regulator of salt tolerance by inhibiting CK signaling. The ZmRR1 interacts with ZmHP2, and under salinity stress, ZmRR1 degrades, releasing ZmHP2. The ZmRR1-ZmHP2-mediated signaling activates ZmMATE29, a tonoplast-located chloride transporter that sequesters chloride into root vacuoles, reducing harmful Cl^- buildup in shoots (Yin et al., 2023). CK response factors (CRFs) are AP2/ERF transcription factors that modulate CK signaling by interacting with HPs as a side branch of the two-component pathway (Keshishian and Rashotte, 2015). The *CK Response Factor 2* (*CRF2*) plays a key role in the salt stress response in *Arabidopsis*. Salt stress upregulates *CRF2* expression in *A. thaliana*, leading to increased CK levels, particularly trans-zeatin and isopentenyl adenine. The *CRF2* overexpression improves ion homeostasis by upregulating *NHX1* and *HKT1* by maintaining potassium-sodium balance. Additionally, *CRF2* enhances reactive oxygen species (ROS) scavenging by increasing the expression of antioxidative enzymes, thereby mitigating oxidative damage. The overexpression of *CRF2* also promotes root growth under salt stress, facilitating improved water and nutrient uptake (Keshishian et al., 2022).

3.2.4 Interaction of cytokinin with other hormones

The mutant research revealed that CK receptor HKs act as negative regulators of ABA and osmotic stress signaling, while *AHK1* acts as a positive regulator (Tran et al., 2007). CK and ABA tend to play antagonistic roles in stomatal activity, with many records of CK-induced increases in stomatal conductance, transpiration, and photosynthesis in response to water stress. Under stress, CK levels decrease, accompanied by an increase in ABA content and a decrease in stomatal conductance, particularly during water shortages (Pospíšilová, 2003; Pospíšilová et al., 2005). The CK-deficient *Arabidopsis* plants exhibited enhanced tolerance to salt and drought stress, primarily through increased ABA sensitivity and improved membrane stability. Stress and ABA treatments decreased active CK levels. Additionally, CK deficiency under nonstress conditions suppressed the expression of key ABA biosynthetic genes, resulting in reduced endogenous ABA levels. These findings highlight a reciprocal regulatory mechanism between CK and ABA that modulates both stress adaptation and developmental processes in plants (Nishiyama et al., 2011). The SnRK2 kinases, central to ABA signaling, phosphorylate type-A ARR5, enhancing its stability and ABA sensitivity. Overexpression of phosphorylated ARR5 promotes drought tolerance, while type-B ARRs (ARR1/11/12) inhibit SnRK2 activity. Molecular evidence positions SnRK2s between type-B and type-A ARRs in regulating ABA response. These findings elucidate a key hormonal crosstalk between ABA and CK in balancing growth and drought stress adaptation (Huang et al., 2018).

Ethylene signaling during stress is mediated by the transcription factor EIN3, which directly represses key cold-responsive genes (*CBF1*, *CBF2*, and *CBF3*) and CK signaling repressors (*ARR5*, *ARR7*, and *ARR15*) by binding to their promoters. The *EIN3* acts as a negative regulator of both cold stress and CK signaling pathways (Shi et al., 2012).

Drought stress activates JA signaling, as evidenced by elevated transcript levels of *LOX2* and *MYC2*, while concurrently suppressing CK-related genes, such as *ARR15*. Interestingly, the expression of *AHP6*, a negative regulator of CK signaling, was upregulated under stress, reinforcing the antagonistic interaction between JA and CK pathways (Jang and Choi, 2018). This hormonal antagonism was also evident under circadian stress in *Arabidopsis*, where CK-deficient plants exhibited a marked increase in JA-responsive gene expression and accumulation of JA metabolites, resulting in enhanced cell death compared to wild-type plants (Nitschke et al., 2016).

3.3 Auxin

The word "auxin" is derived from the Greek term "auxein," signifying growth or enlargement (Went, 1934; Went and Thimann, 1937). Auxin is recognized as a crucial plant hormone involved in nearly every aspect of plant growth. Auxins are endogenous regulators whose levels can also be controlled by the environment to control the formation of roots and shoots as well as their relative growth (Sachs, 2005). Additionally, auxin contributes to various other physiological processes, such as leaf morphogenesis, flowering, and senescence (Aloni et al., 2003; Okushima et al., 2005). Multiple pathways, including biosynthesis, degradation, transport, and conjugate formation, regulate auxin homeostasis (Ludwig-Müller, 2011) (Fig. 3.3).

3.3.1 Auxin biosynthesis

Auxin availability to plants involves both de novo auxin biosynthesis and the release of auxin from conjugates. The primary endogenous auxin in plants is indole-3-acetic acid (IAA), which regulates key developmental and environmental responses. In addition to IAA, three other naturally occurring auxins include indole-3-butyric acid (IBA), 4-chloroindole-3-acetic acid (4-Cl-IAA), and phenylacetic acid (PAA), present in both free and conjugated forms (Ludwig-Müller, 2011). Tryptophan (Trp) is the well-known precursor for auxin biosynthesis. The four tryptophan (Trp)-dependent auxin biosynthesis pathways are classified according to their key intermediates: the indole-3-acetaldoxime (IAOx), indole-3-acetamide (IAM), tryptamine (TAM), and indole-3-pyruvic acid (IPA) pathways (Mano and Nemoto, 2012). The most common and highly conserved two-step auxin biosynthesis pathway involves IPA as the intermediate. The Trp aminotransferases convert precursor Trp to IPA. The *SAV3* gene, which encodes a protein with an alliinase C-terminal/aminotransferase domain, has been shown to catalyze the conversion of Trp to IPA. As a result, SAV3 is renamed TAA1 (tryptophan aminotransferase of *Arabidopsis*). The TAA1 and its close homologs (TAR1 and TAR2) play important roles in auxin biosynthesis and plant development (Tao et al., 2008; Stepanova et al., 2008). *YUCCA* genes encode flavin-containing monooxygenases that mediate the N-hydroxylation of TAM in auxin biosynthesis (Zhao et al., 2001). Although initially considered part of a separate pathway, the genetic, biochemical, and metabolic evidence indicate that the TAA and YUC families function within a single auxin biosynthetic pathway. In *Arabidopsis*, YUC catalyze the critical conversion of IPA to IAA, representing the rate-limiting step of the IPA-dependent auxin biosynthesis pathway, which is crucial for normal plant growth and development (Mashiguchi et al., 2011). In the IAM pathway, trp is converted to IAM by tryptophan

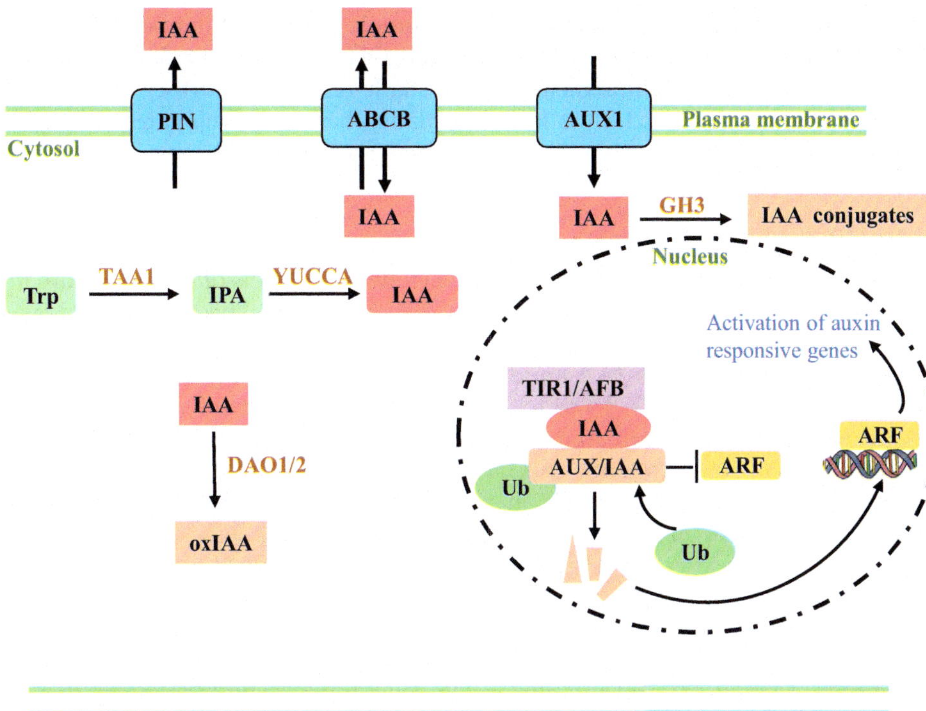

FIGURE 3.3 A schematic overview of the auxin pathway illustrating its biosynthesis, transport, conjugation, and signaling. IAA is synthesized from trp via the intermediate IPA through the action of TAA1 and YUCCA enzymes. Auxin transport is mediated by AUX1, PIN, and ABCB proteins. GH3 enzymes inactivate IAA by conjugating it to amino acids. Auxin promotes the degradation of Aux/IAA repressors through ubiquitination, thereby releasing ARFs to activate the transcription of auxin-responsive genes. *IAA*, Indole-3-acetic acid; *IPA*, indole-3-pyruvic acid; *ABCB*, ATP-binding cassette subfamily B; *ARFs*, auxin response factor.

2-monooxygenase, then to IAA by IAM hydrolase (Mano and Nemoto, 2012). The conversion of IAOx to IAA in *Arabidopsis* involves two probable intermediates: IAM and indole-3-acetonitrile (IAN). *Arabidopsis* nitrilases (NIT) can convert IAN to IAA and IAM in vitro. The study showed that IAN is predominantly derived from IAOx; it is hypothesized that IAM may serve as an intermediate in the IAOx-to-IAA conversion pathway (Sugawara et al., 2009). Moreover, there is evidence of a family of amidases that can hydrolyze IAM into IAA, implying that IAM may be a plant intermediate for IAA biosynthesis (Pollmann et al., 2009).

3.3.2 Auxin degradation

The primary route for IAA oxidation is its conversion to 2-oxindole-3-acetic acid (oxIAA), which can further undergo conjugation with a hexose moiety via its carboxyl group (Östin et al., 1998). The first step of oxidation is carried out by two dioxygenases for auxin oxidation (DAO1 and DAO2) enzymes, which are responsible for oxidizing IAA to oxIAA in *Arabidopsis* (Porco et al., 2016; Zhang et al., 2016). IAA is also metabolized into conjugates such as indole-3-acetyl aspartic acid (IAAsp) and indole-3-acetyl glutamate (IAGlu). IAAsp undergoes further oxidation to form OxIAAsp and OH-IAAsp, particularly at higher IAA levels. In contrast, IAGlu remains metabolically stable and is not further processed (Östin et al., 1998). However, more oxIAA is found in plants than IAA conjugates, indicating that the oxidation process is very active (Kowalczyk and Sandberg, 2001).

3.3.3 Auxin conjugation

Amide conjugates are synthesized in response to elevated levels of IAA exposure in plant tissues (Venis, 1972). Interestingly, multiple GH3 genes have been identified as being rapidly induced by auxin, highlighting their potential role in auxin homeostasis. In *A. thaliana*, GH3 genes encode IAA-amido synthetases, which contribute to auxin homeostasis by conjugating excess IAA to amino acids. Six recombinant GH3 enzymes synthesized IAA conjugates with amino acids such as alanine, aspartic acid, phenylalanine, and tryptophan in vitro. Mutants of GH3 genes showed increased auxin sensitivity, confirming the role of GH3 enzymes in regulating auxin activity through amino acid conjugation (Staswick et al., 2005). UGT74E2 specifically catalyzes the glucosylation of IBA to form IBA-glucose conjugates. This conjugation alters IBA and IAA homeostasis, impacting plant morphology and stress responses (Tognetti et al., 2010). 1-O-(indole-3-acetyl)-β-d-glucose: myo-inositol indoleacetyl transferase (IA-myo-inositol synthase) is a key enzyme in IAA metabolism that transfers the indole acetyl group from IAA-glucose to myo-inositol, forming IA-myo-inositol and glucose. Purified from maize endosperm, this enzyme exists in both monomeric and dimeric forms, exhibits dual transferase and hydrolase activities, and belongs to the serine carboxypeptidase-like (SCPL) acyltransferase family (Kowalczyk et al., 2003).

3.3.4 Auxin transportation

Auxin transportation is regulated by various influx/efflux carriers, which channel auxin flow and thus produce a gradient. The auxin-resistant 1/like aux1 (AUX/LAX) family of influx carriers mediates auxin influx through diffusion through the plasma membrane or active transport (Swarup and Péret, 2012). AUX1 plays a dual role in auxin transport within roots, facilitating both basipetal movement (towards the root base) and acropetal transport (towards the root tip), particularly via a phloem-associated auxin transport stream (Swarup et al., 2001). The AUX1 in *Arabidopsis* also plays a critical role in root gravitropism by facilitating auxin uptake. Disruption of AUX1 function leads to a gravitropic root growth, which can be rescued using the membrane-permeable synthetic auxin 1-naphthaleneacetic acid (1-NAA). AUX1 functions alongside the efflux carrier PIN2 to mediate localized auxin redistribution in the root apex (Marchant and Kargul, 1999). PIN proteins, belonging to the major facilitator superfamily of membrane transporters, are crucial for directional auxin transport and are oriented according to the flow of auxin within plant tissues. The auxin efflux carriers PIN1, PIN2, PIN3, PIN4, and PIN7 are found at the plasma membrane (Petrášek et al., 2006). PIN5, PIN6, and PIN8 belong to a subgroup characterized by a shorter middle hydrophilic loop, which governs auxin exchange between the endoplasmic reticulum and the cytosol (Mravec et al., 2009). PIN1 directs auxin flow through vascular tissues to the root tip, with assistance from PIN3, PIN4, and PIN7 in adjacent regions. At the tip, auxin is redistributed basipetally via PIN2 and laterally redirected by PIN2, PIN3, and PIN7 back into the central PIN1-driven stelar transport pathway, forming a continuous auxin reflux loop critical for root development (Blilou et al., 2005; Benková et al., 2003; Friml et al., 2003). Plant orthologs of the mammalian ATP-binding cassette subfamily B (ABCB)-type transporters of the multidrug resistance/phosphoglycoprotein (ABCB/MDR/PGP) protein family also play a role in auxin efflux (Verrier et al., 2008). The ABCB15, 16, 17, 18, and 22 were identified as novel auxin transporters. Knockdown of these genes reduced lateral root density and root clock amplitude. Their activity in the lateral root cap and epidermis supports shootward auxin flow, highlighting their role in modulating auxin oscillations, which are essential for root patterning (Chen et al., 2023).

3.3.5 Auxin signaling

Auxin signaling begins with the perception of auxin because of its contact with a receptor (Leyser, 2002). There are four distinct classes of auxin perception mechanisms have been identified: the auxin binding protein 1 (ABP1) in association with transmembrane kinases (TMK1–TMK4); the F-box protein family including transport inhibitor response 1 (TIR1) and auxin signaling F-box proteins (AFB1–AFB5); the cell cycle–related S-phase kinase-associated protein 2a (SKP2a); and the noncanonical transcription factor auxin response factor 3 (Yu et al., 2022) (ARF3). ABP1 is proposed to function in auxin signaling by binding IAA in the apoplast and forming a complex with TMKs, potentially acting as a coreceptor. The ABP1–IAA–TMK complex influences cellular processes, such as PIN polarity and endocytic trafficking (Yu et al., 2022). SKP2a acts as an auxin-binding F-box protein that enhances cell cycle progression by promoting the ubiquitin-mediated degradation of transcriptional repressors, thereby facilitating the G1-to-S phase transition and directly linking auxin perception to cell division control (Yu et al., 2022). Additionally, auxin binds to TIR1/AFB to initiate the signaling (Strader and Zhao, 2016). The binding of Aux/IAA repressors stabilizes the association of auxin and TIR1/AFB receptors (Dos Santos Maraschin et al., 2009). The ubiquitination and the resulting degradation of Aux/IAA repressors are caused by their binding to TIR/AFB (as part of the SCF complex). The ARF transcriptional responses are activated as a function of the absence of Aux/IAA (Leyser, 2002) (Fig. 3.3).

3.3.6 Regulation of salt stress tolerance by auxin

Under salt stress, auxin levels rapidly decline due to tight multilevel regulation (Table 3.1). Nitric oxide contributes to this reduction by downregulating *PIN* gene expression (thereby limiting auxin transport), stabilizing the repressor *IAA17* (inhibiting auxin signaling), and facilitating the suppression of root meristem growth (Liu et al., 2015). The *AtGH3.5* gene, which encodes IAA amido-conjugate synthases, was induced by high salinity. Overexpression of *AtGH3.5* resulted in a decline in free auxin and an increase in auxin conjugates such as IAA-Asp (Park et al., 2007). The *IbARF5* gene from sweet potato enhances carotenoid accumulation and improves tolerance to salt and drought. In transgenic *Arabidopsis*, elevated ABA, proline, and SOD activity were observed, accompanied by reduced H_2O_2 levels. *IbARF5* upregulates genes related to abiotic stress responses (Kang et al., 2018). In tomato, *SlARF4* acts as a negative regulator of stress tolerance. The *arf4* mutants exhibited enhanced tolerance to salt and osmotic stress by promoting root development and chlorophyll and increasing soluble sugars. The *ARF4* suppression resulted in the upregulation of antioxidant and ABA biosynthesis genes (Bouzroud et al., 2020). Overexpression of *AFB3* in *Arabidopsis* enhances salt stress tolerance by promoting primary and lateral root growth, as well as improving seed germination rates, compared to the wild type. In the auxin-dependent signaling cascade, the transcription factor SZF1 acts downstream of AFB3 and regulates NAC4 in response to this stress (Garrido-Vargas et al., 2020). In *Helianthus tuberosus*, the conserved miR390-TAS3-tasiARF pathway regulates salt stress responses. Under moderate salt (100 mM NaCl), miR390 is upregulated, promoting tasiARF (trans-acting small interfering RNAs influencing ARF) production and ARF3/4 cleavage, enhancing salt tolerance. High salt (300 mM NaCl) represses this pathway, reducing tasiARF levels and ARF cleavage. Thus ARF2 contributes positively to salinity tolerance (Wen et al., 2020). The *IAR3* acts as a novel target of miR167a in *Arabidopsis*, linking microRNA regulation to auxin homeostasis under osmotic stress. Under stress, miR167a levels decrease, leading to increased *IAR3*, which converts inactive IAA–alanine to active IAA, promoting lateral root development. The *iar3* mutants failed to exhibit root changes and stress tolerance, highlighting the functional role of IAR3 (Kinoshita et al., 2012). In *Arabidopsis*, an auxin/redox association plays a role in salinity adaptation. When compared to wild-type plants, the auxin receptor double mutant (*tir1 afb2*) showed increased resistance to salinity as measured by chlorophyll content, germination rate, root elongation, and enhanced antioxidant enzymes (Iglesias et al., 2010). Salt stress increases the expression of miR393, which in turn lowers the levels of TIR1 and AFB2, two auxin receptors. Salt stress also stabilized Aux/IAA repressors, resulting in downregulation of auxin signaling. The *mir393ab* mutant exhibited decreased inhibition of lateral root growth when exposed to NaCl, indicating that miR393 is involved in lateral root initiation, proliferation, and elongation during salinity stress (Iglesias et al., 2014).

3.3.7 Interaction of auxin with other hormones

Ethylene influences auxin biosynthesis by regulating key genes in the tryptophan-dependent pathway. The *Arabidopsis weak ethylene insensitive 8* (WEI8) mutant revealed that TAA1, which converts trp to IPA, is ethylene-responsive. Ethylene also upregulates *WEI12* and *WEI17*, enhancing trp production. Thus ethylene promotes local auxin synthesis, especially in root tips, linking the two hormonal pathways (Stepanova et al., 2008). Ethylene suppresses lateral root formation in *Arabidopsis* by enhancing auxin transport and interfering with AUX1-mediated IAA uptake. Ethylene-insensitive mutants resist this effect, exhibiting increased root branching, which highlights ethylene–auxin crosstalk in regulating lateral root development (Negi et al., 2008).

Gibberellin deficiency in *Arabidopsis* reduces auxin transport by lowering the abundance of PIN proteins via enhanced vacuolar degradation. GA treatment restores PIN levels and auxin flow, thereby supporting PIN-dependent processes such as cotyledon development and root gravitropism. This reveals a molecular basis for GA–auxin crosstalk through GA-mediated stabilization of PIN proteins (Willige et al., 2011).

ABA and auxin interact to regulate lateral root emergence via an ABI3–ERF1 module. ABI3 negatively regulates lateral root emergence by activating ERF1, which in turn upregulates ABI3, thereby forming a feedback loop that facilitates rapid signal amplification. Their physical interaction reduces their DNA-binding activity, fine-tuning the expression of lateral root-related genes such as PIN1, AUX1, ARF7, and ABI5. This module functions to balance ABA and auxin signaling during lateral root development (Zhang et al., 2023).

3.4 Auxin–cytokinin crosstalk

AZG1 (Arabidopsis Zinc-dependent Gatekeeper 1), a CK transporter, plays an important role in modulating root stress responses in *A. thaliana*. AZG1 physically interacts with the auxin efflux carrier PIN-FORMED1 (PIN1), establishing a direct link between CK and auxin transport mechanisms. Functional analysis revealed that AZG1 facilitates the cellular uptake of CKs, thereby influencing auxin distribution patterns mediated by PIN1. Under stress conditions, such as salinity and osmotic challenges, AZG1 expression is upregulated, leading to altered auxin fluxes and enhanced root adaptability. These findings highlight the importance of AZG1 in integrating CK and auxin signaling pathways to fine-tune root architecture and stress resilience (Tessi et al., 2023). CK-mediated mechanism regulates PIN1-dependent auxin transport polarity during lateral root organogenesis. CK promotes the removal of PIN1 from specific cell regions, altering its polarity and directing auxin flow. This mechanism depends on PIN1 phosphorylation, with phosphor-mimic forms and increased kinase activity reducing sensitivity (Marhavý et al., 2014). Auxin and CK interact to regulate root meristem activity in *Arabidopsis*. The CK-responsive transcription factor ARR1 activates SHY2/IAA3, a repressor of auxin signaling, which downregulates PIN auxin transporters and redistributes auxin to promote cell differentiation. In contrast, auxin triggers SHY2 degradation, maintaining PIN activity and cell division. This reciprocal regulation forms a loop that balances cell division and differentiation, ultimately controlling root meristem size and root growth (Ioio et al., 2008). Phyllotaxy in maize is regulated by a complex crosstalk between auxin and CK. The *ABPH1* gene, a CK-inducible regulator, requires auxin or its polar transport for expression and positively regulates *PIN1* in a feedback loop. In *abph1* mutants, reduced auxin levels and PIN1 expression resulted in delayed leaf initiation, an enlarged shoot apical meristem, and altered phyllotaxy, revealing the integrated role of both hormones in shoot patterning (Lee et al., 2009). In *Arabidopsis*, ABA and CK induce ABI4, which decreases polar auxin transport by downregulating PIN1, reducing auxin levels, and limiting lateral root formation. In contrast, auxin suppresses ABI4, allowing higher auxin transport and promoting lateral root development. ABI4 acts as a key integrator of hormonal signals, balancing inhibition by ABA and CK with promotion by auxin (Shkolnik-Inbar and Bar-Zvi, 2010). In cucumber, two parthenocarpic genotypes (PPC-6 and DG-8) and a nonparthenocarpic line (IMPU-1) were examined for hormonal regulation of fruit set. Parthenocarpic lines had higher auxin and gibberellin levels, while CK varied with stage. With zeatin or IAA + zeatin there was fruit induction in IMPU-1. Upregulation of CK biosynthesis genes (IPT, LOG, and CYP735A) and AUX–GA pathways confirmed their role in parthenocarpy. The study reveals a coordinated CK–auxin–gibberellin network governing fruit set in cucumber (Mandal et al., 2022).

Recent advances in plant hormone research are increasingly being translated into practical breeding and biotechnology applications. Genome editing has rapidly emerged as a transformative platform for precise crop improvement, enabling targeted manipulation of genes that regulate complex agronomic traits. By integrating functional genomics with CRISPR/Cas-based technologies, researchers can dissect and reprogram key hormonal and developmental pathways to enhance yield potential and stress resilience. A genome-wide association study of 348 maize lines identified *Zea mays ZmCKX6* as a key regulator of leaf width. CRISPR/Cas9-mediated *ckx6* knockout maize mutants exhibited narrower leaves, shorter plants, and reduced yield due to altered CK, auxin, and gibberellin levels, revealing that ZmCKX6 modulates CK homeostasis and plant architecture demonstrating the potential of genome editing to optimize yield-related traits (Wang et al., 2025). Similarly, CRISPR/Cas9-mediated editing of the CK RR gene *OsRR22* in rice generated novel allelic variants that markedly enhanced salinity tolerance by reducing Na^+ accumulation and improving ionic balance, while maintaining normal agronomic performance (Sheng et al., 2023). In tomato, CRISPR/Cas9-mediated mutagenesis of CK signaling genes *SlHP2* and *SlHP3* produced homozygous double mutants (*slhp2,3*) that displayed enhanced drought tolerance, with improved water retention, deeper roots, reduced stomatal density, and lower oxidative damage under PEG-induced stress. Downregulation of CK signaling regulators, stomatal development genes, and suppression of ABA were observed, suggesting that targeted genome editing of CK signaling can effectively improve drought resilience (Aydin et al., 2025). In *Dendrobium huoshanense*, drought stress alters various miRNAs, including miR156, miR157d, and miR160a-5p, which target auxin response factors and CK receptors (Wang et al., 2022). These miRNAs likely regulate auxin and CK signaling, shaping the plant's morphological and physiological drought responses. Understanding the miRNA hormone regulatory network thus offers practical applications for molecular breeding and genome editing to develop stress-tolerant crops, thereby sustaining productivity under unfavorable conditions.

3.5 Conclusion and future prospects

CK and auxin play pivotal roles in regulating plant growth, development, and stress responses, particularly under conditions of salt stress. CK biosynthesis, degradation, and signaling through key components modulate developmental plasticity and adaptive mechanisms. Similarly, auxin homeostasis, mediated via biosynthesis, transport, degradation, and signaling, coordinates root and shoot architecture under abiotic stress. Salt stress induces complex hormonal crosstalk. CK and auxin signaling can act antagonistically or synergistically with other hormones to modulate stomatal conductance, antioxidant defense, and ion homeostasis, and govern root patterning and lateral root emergence. Understanding auxin-CK crosstalk under abiotic stresses is crucial. Modifying key genes can optimize hormone distribution, promote developmental growth, and enhance stress resilience in crop plants.

Acknowledgments

CSIR-CSMCRI communication number 106/2025. The authors are thankful to the Council of Scientific and Industrial Research (CSIR), New Delhi, India, for financial assistance. AC is thankful to DST-INSPIRE research fellowship and AcSIR for enrollment in PhD. PA is supported by DSTWOS-A scheme for the financial assistance.

Author contributions

AC was involved in literature collection, preparation of table and figures, and writing and editing of the manuscript. **AG** is involved in preparation of table and writing of manuscript. **PKA** and **PA** were involved in the analysis of the literature and reviewing and editing of the manuscript.

Data availability

This is a book chapter and does not contain the data.

Declarations

Conflict of interest

The authors declare that they have no conflict of interest.

References

AlNeyadi, S., Kappachery, S., Tanveer, A., Khan, S., Karumannil, M., AlHosani, et al., 2024. Shi, ectopic expression of potato ARP1 encoding auxin-repressed protein confers salinity stress tolerance in *Arabidopsis thaliana*. PLoS ONE 19 (10), e0309452. Available at https://doi.org/10.1371/journal.pone.0309452.

Aloni, R., Schwalm, K., Langhans, M., Ullrich, C.I., 2003. Gradual shifts in sites of free-auxin production during leaf-primordium development and their role in vascular differentiation and leaf morphogenesis in Arabidopsis. Planta 216 (5), 841–853. Available at https://doi.org/10.1007/s00425-002-0937-8, link.springer.de/link/service/journals/00425/index.htm.

Anantharaman, V., Aravind, L., 2001. The CHASE domain: a predicted ligand-binding module in plant cytokinin receptors and other eukaryotic and bacterial receptors. Trends in Biochemical Sciences 26 (10), 579–582. Available at https://doi.org/10.1016/S0968-0004(01)01968-5, www.elsevier.com/locate/tibs.

Aydin, A., Yerlikaya, B.A., Yerlikaya, S., Yilmaz, N.N., Kavas, M., 2025. CRISPR-mediated mutation of cytokinin signaling genes (SlHP2 and SlHP3) in tomato: morphological, physiological, and molecular characterization. The Plant Genome 18 (1). Available at https://doi.org/10.1002/tpg2.20542.

Bawa, G., Liu, Z., Wu, R., Zhou, Y., Liu, H., Sun, S., et al., 2022. PIN1 regulates epidermal cells development under drought and salt stress using single-cell analysis. Frontiers in Plant Science 13. Available at https://doi.org/10.3389/fpls.2022.1043204, https://www.frontiersin.org/journals/plant-science.

Benková, E., Michniewicz, M., Sauer, M., Teichmann, T., Seifertová, D., Jürgens, G., et al., 2003. Local, efflux-dependent auxin gradients as a common module for plant organ formation. Cell 115 (5), 591–602. Available at https://doi.org/10.1016/S0092-8674(03)00924-3, https://www.sciencedirect.com/journal/cell.

Blilou, I., Xu, J., Wildwater, M., Willemsen, V., Paponov, I., Friml, J., et al., 2005. The PIN auxin efflux facilitator network controls growth and patterning in *Arabidopsis* roots. Nature 433 (7021), 39–44. Available at https://doi.org/10.1038/nature03184.

Bouzroud, S., Gasparini, K., Hu, G., Barbosa, M.A.M., Rosa, B.L., Fahr, M., et al., 2020. Down regulation and loss of auxin response factor 4 function using CRISPR/Cas9 alters plant growth, stomatal function and improves tomato tolerance to salinity and osmotic stress. Genes 11 (3). Available at https://doi.org/10.3390/genes11030272, https://www.mdpi.com/2073-4425/11/3/272/pdf.

Cao, L., Cao, J., Wang, L., Wang, J., Che, Z., 2025. Genome-wide analysis of AHP genes in soybean and the role of GmAHP10 in improving salt stress tolerance. Functional and Integrative Genomics 25 (1). Available at https://doi.org/10.1007/s10142-025-01636-8, https://www.springer.com/journal/10142.

Chen, Z., Hu, L., Han, N., Hu, J., Yang, Y., Xiang, T., et al., 2015. Overexpression of a miR393-resistant form of transport inhibitor response protein 1 (mTIR1) enhances salt tolerance by increased osmoregulation and Na+ exclusion in *Arabidopsis thaliana*. Plant and Cell Physiology 56 (1), 73–83. Available at https://doi.org/10.1093/pcp/pcu149. Available from: http://pcp.oxfordjournals.org/.

Chen, J., Hu, Y., Hao, P., Tsering, T., Xia, J., Zhang, Y., et al., 2023. ABCB-mediated shootward auxin transport feeds into the root clock. EMBO Reports 24 (4), e56271.

Danilova, M.N., Kudryakova, N.V., Doroshenko, A.S., Zabrodin, D.A., Vinogradov, N.S., Kuznetsov, V.V., 2016. Molecular and physiological responses of *Arabidopsis thaliana* plants deficient in the genes responsible for ABA and cytokinin reception and metabolism to heat shock. Russian Journal of Plant Physiology 63 (3), 308–318. Available at https://doi.org/10.1134/s1021443716030043.

Dos Santos Maraschin, F., Memelink, J., Offringa, R., 2009. Auxin-induced, SCF TIR1 -mediated poly-ubiquitination marks AUX/IAA proteins for degradation. The Plant Journal 59 (1), 100–109. Available at https://doi.org/10.1111/j.1365-313x.2009.03854.x.

Frébort, I., Kowalska, M., Hluska, T., Frébortová, J., Galuszka, P., 2011. Evolution of cytokinin biosynthesis and degradation. Journal of Experimental Botany 62 (8), 2431–2452.

Friml, J., Vieten, A., Sauer, M., Weijers, D., Schwarz, H., Hamann, T., et al., 2003. Efflux-dependent auxin gradients establish the apical-basal axis of *Arabidopsis*. Nature 426 (6963), 147–153. Available at https://doi.org/10.1038/nature02085.

Garrido-Vargas, F., Godoy, T., Tejos, R., O'Brien, J.A., 2020. Overexpression of the auxin receptor AFB3 in *Arabidopsis* results in salt stress resistance and the modulation of NAC4 and SZF1. International Journal of Molecular Sciences 21 (24), 9528. Available at https://doi.org/10.3390/ijms21249528.

Guo, Y., Liu, C., Chen, S., Tian, Z., 2024. GmHXK2 promotes the salt tolerance of soybean seedlings by mediating AsA synthesis, and auxin synthesis and distribution. BMC Plant Biology 24 (1). Available at https://doi.org/10.1186/s12870-024-05301-3. Available from: https://bmcplantbiol.biomedcentral.com/.

He, L., Chen, T., Zhao, C., Zhao, L., Zhao, Q., Yao, S., et al., 2024. RST31 controls salt tolerance in rice (*Oryza sativa*) by regulating the cytokinin pathway. Plant Physiology and Biochemistry 216, 109075. Available at https://doi.org/10.1016/j.plaphy.2024.109075.

Huang, X., Hou, L., Meng, J., You, H., Li, Z., Gong, Z., et al., 2018. The antagonistic action of abscisic acid and cytokinin signaling mediates drought stress response in *Arabidopsis*. Molecular Plant 11 (7), 970–982. Available at https://doi.org/10.1016/j.molp.2018.05.001.

Hwang, I., Sakakibara, H., 2006. Cytokinin biosynthesis and perception. Physiologia Plantarum 126 (4), 528–538. Available at https://doi.org/10.1111/j.1399-3054.2006.00665.x.

Iglesias, M.J., Terrile, M.C., Bartoli, C.G., D'Ippólito, S., Casalongué, C.A., 2010. Auxin signaling participates in the adaptative response against oxidative stress and salinity by interacting with redox metabolism in *Arabidopsis*. Plant Molecular Biology 74 (3), 215–222. Available at https://doi.org/10.1007/s11103-010-9667-7.

Iglesias, M.J., Terrile, M.C., Windels, D., Lombardo, M.C., Bartoli, C.G., Vazquez, F., et al., 2014. MiR393 regulation of auxin signaling and redox-related components during acclimation to salinity in *Arabidopsis*. PLoS ONE 9 (9). Available at https://doi.org/10.1371/journal.pone.0107678, http://www.plosone.org/article/fetchObject.action?uri=info%3Adoi%2F10.1371%2Fjournal.pone.0107678&representation=PDF.

Ioio, R.D., Nakamura, K., Moubayidin, L., Perilli, S., Taniguchi, M., Morita, M.T., et al., 2008. A Genetic framework for the control of cell division and differentiation in the root meristem. Science 322 (5906), 1380–1384. Available at https://doi.org/10.1126/science.1164147.

Iqbal, M., Ashraf, M., Jamil, A., 2006. Seed enhancement with cytokinins: changes in growth and grain yield in salt stressed wheat plants. Plant Growth Regulation 50 (1), 29–39. Available at https://doi.org/10.1007/s10725-006-9123-5.

Iqbal, N., Khan, N.A., Ferrante, A., Trivellini, A., Francini, A., Khan, M.I.R., 2017. Ethylene role in plant growth, development and senescence: interaction with other phytohormones. Frontiers in Plant Science 8. Available at https://doi.org/10.3389/fpls.2017.00475, http://journal.frontiersin.org/article/10.3389/fpls.2017.00475/full.

Iqbal, M.Z., Liang, Y., Anwar, M., Fatima, A., Hassan, M.J., Ali, A., et al., 2024. Overexpression of auxin/indole-3-acetic acid gene TrIAA27 enhances biomass, drought, and salt tolerance in *Arabidopsis thaliana*. Plants 13 (19), 2684. Available at https://doi.org/10.3390/plants13192684.

Jang, G., Choi, Y.D., 2018. Drought stress promotes xylem differentiation by modulating the interaction between cytokinin and jasmonic acid. Plant Signaling & Behavior 13 (3), e1451707. Available at https://doi.org/10.1080/15592324.2018.1451707.

Joshi, R., Sahoo, K.K., Tripathi, A.K., Kumar, R., Gupta, B.K., Pareek, A., et al., 2018. Knockdown of an inflorescence meristem-specific cytokinin oxidase – OsCKX2 in rice reduces yield penalty under salinity stress condition. Plant, Cell and Environment 41 (5), 936–946. Available at https://doi.org/10.1111/pce.12947, http://www.blackwellpublishing.com/journals/PCE.

Kamada-Nobusada, T., Sakakibara, H., 2009. Molecular basis for cytokinin biosynthesis. Phytochemistry 70 (4), 444–449. Available at https://doi.org/10.1016/j.phytochem.2009.02.007.

Kang, C., He, S., Zhai, H., Li, R., Zhao, N., Liu, Q., 2018. A sweetpotato auxin response factor gene (ibarf5) is involved in carotenoid biosynthesis and salt and drought tolerance in transgenic *Arabidopsis*. Frontiers in Plant Science 9. Available at https://doi.org/10.3389/fpls.2018.01307, https://www.frontiersin.org/journals/plant-science.

Keshishian, E.A., Cliver, B.R., McLaughlin, W.F., Hallmark, H.T., Plačková, L., Goertzen, L.R., et al., 2022. CYTOKININ RESPONSE FACTOR 2 is involved in modulating the salt stress response. Plant Journal 110 (4), 1097–1110. Available at https://doi.org/10.1111/tpj.15726, http://onlinelibrary.wiley.com/journal/10.1111/(ISSN)1365-313X.

Keshishian, E.A., Rashotte, A.M., 2015. Plant cytokinin signalling. Essays in Biochemistry 58, 13–27. Available at https://doi.org/10.1042/BSE0580013, http://essays.biochemistry.org/content/ppebio/58/13.full.pdf.

Kim, T.J., Lim, G.H., 2023. Salicylic acid and mobile regulators of systemic immunity in plants: transport and metabolism. Plants 12 (5). Available at https://doi.org/10.3390/plants12051013, www.mdpi.com/journal/plants.

Kinoshita, N., Wang, H., Kasahara, H., Liu, J., MacPherson, C., Machida, Y., et al., 2012. IAA-Ala resistant3, an evolutionarily conserved target of miR167, mediates *Arabidopsis* root architecture changes during high osmotic stress. The Plant Cell 24 (9), 3590–3602. Available at https://doi.org/10.1105/tpc.112.097006.

Kirungu, J.N., Magwanga, R.O., Lu, P., Cai, X., Zhou, Z., Wang, X., et al., 2019. Functional characterization of Gh_A08G1120 (GH3.5) gene reveal their significant role in enhancing drought and salt stress tolerance in cotton. BMC Genetics 20 (1). Available at https://doi.org/10.1186/s12863-019-0756-6.

Kowalczyk, S., Jakubowska, A., Zielińska, E., Bandurski, R.S., 2003. Bifunctional indole-3-acetyl transferase catalyses synthesis and hydrolysis of indole-3-acetyl-myo-inositol in immature endosperm of *Zea mays*. Physiologia Plantarum 119 (2), 165–174. Available at https://doi.org/10.1034/j.1399-3054.2003.00158.x.

Kowalczyk, M., Sandberg, G., 2001. Quantitative analysis of indole-3-acetic acid metabolites in *Arabidopsis*. Plant Physiology 127 (4), 1845–1853. Available at https://doi.org/10.1104/pp.010525, http://www.plantphysiol.org/.

Kyozuka, J., 2007. Control of shoot and root meristem function by cytokinin. Current Opinion in Plant Biology 10 (5), 442–446. Available at https://doi.org/10.1016/j.pbi.2007.08.010.

Lee, B.H., Johnston, R., Yang, Y., Gallavotti, A., Kojima, M., Travençolo, B.A.N., et al., 2009. Studies of aberrant phyllotaxy1 mutants of maize indicate complex interactions between auxin and cytokinin signaling in the shoot apical meristem1[W][OA]. Plant Physiology 150 (1), 205–216. Available at https://doi.org/10.1104/pp.109.137034, http://www.plantphysiol.org/cgi/reprint/150/1/205.

Leyser, O., 2002. Molecular genetics of auxin signaling. Annual Review of Plant Biology 53 (1), 377–398. Available at https://doi.org/10.1146/annurev.arplant.53.100301.135227.

Liu, M., Cui, Y., Peng, F., Wang, S., Cui, R., Liu, X., et al., 2023. Antioxidant system was triggered to alleviate salinity stress by cytokinin oxidase/dehydrogenase gene GhCKX6b-Dt in cotton. Environmental Sciences Europe 35 (1). Available at https://doi.org/10.1186/s12302-023-00788-3, https://enveurope.springeropen.com.

Liu, L., Gong, Y., Yahaya, B.S., Chen, Y., Shi, D., Liu, F., et al., 2024. Maize auxin response factor ZmARF1 confers multiple abiotic stresses resistances in transgenic *Arabidopsis*. Plant Molecular Biology 114 (4). Available at https://doi.org/10.1007/s11103-024-01470-9, https://www.springer.com/journal/11103.

Liu, W., Li, R.J., Han, T.T., Cai, W., Fu, Z.W., Lu, Y.T., 2015. Salt stress reduces root meristem size by nitric oxidemediated modulation of auxin accumulation and signaling in *Arabidopsis*. Plant Physiology 168 (1), 343–356. Available at https://doi.org/10.1104/pp.15.00030, http://www.plantphysiol.org/content/168/1/321.full.pdf.

Liu, J., Xu, X.-L., Wang, B., Xiao, Y., Zhang, M.-C., Liu, H.-T., et al., 2025. TaHAK1 promotes salt tolerance via synergistic modulation of K+/Na+ ion homeostasis and auxin signaling in rice. Plant Physiology and Biochemistry 228, 110308. Available at https://doi.org/10.1016/j.plaphy.2025.110308.

Li, S., An, Y., Hailati, S., Zhang, J., Cao, Y., Liu, Y., et al., 2019. Overexpression of the cytokinin oxidase/dehydrogenase (CKX) from medicago sativa enhanced salt stress tolerance of *Arabidopsis*. Journal of Plant Biology 62 (5), 374–386. Available at https://doi.org/10.1007/s12374-019-0141-z, http://www.springer.com/life+sci/plant+sciences/journal/12374.

Li, L., Zheng, Q., Jiang, W., Xiao, N., Zeng, F., Chen, G., et al., 2022. Molecular regulation and evolution of cytokinin signaling in plant abiotic stresses. Plant and Cell Physiology 63 (12), 1787–1805. Available at https://doi.org/10.1093/pcp/pcac071, http://pcp.oxfordjournals.org/.

Ludwig-Müller, J., 2011. Auxin conjugates: their role for plant development and in the evolution of land plants. Journal of Experimental Botany 62 (6), 1757–1773. Available at https://doi.org/10.1093/jxb/erq412.

Mandal, N.K., Kumari, K., Kundu, A., Arora, A., Bhowmick, P.K., Iquebal, M.A., et al., 2022. Cross-talk between the cytokinin, auxin, and gibberellin regulatory networks in determining parthenocarpy in cucumber. Frontiers in Genetics 13. Available at https://doi.org/10.3389/fgene.2022.957360, https://www.frontiersin.org/journals/genetics#.

Mano, Y., Nemoto, K., 2012. The pathway of auxin biosynthesis in plants. Journal of Experimental Botany 63 (8), 2853–2872. Available at https://doi.org/10.1093/jxb/ers091.

Marchant, A., Kargul, J., May, S.T., Muller, P., Delbarre, A., Perrot-Rechenmann, C., Bennett, M.J., 1999. AUX1 regulates root gravitropism in *Arabidopsis* by facilitating auxin uptake within root apical tissues. EMBO Journal 18 (8), 2066–2073. Available at http://emboj.embopress.org/, https://doi.org/10.1093/emboj/18.8.2066, http://emboj.embopress.org/.

Marhavý, P., Duclercq, J., Weller, B., Feraru, E., Bielach, A., Offringa, R., et al., 2014. Cytokinin controls polarity of PIN1-dependent auxin transport during lateral root organogenesis. Current Biology 24 (9), 1031–1037. Available at https://doi.org/10.1016/j.cub.2014.04.002, http://www.current-biology.com.

Mashiguchi, K., Tanaka, K., Sakai, T., Sugawara, S., Kawaide, H., Natsume, M., et al., 2011. The main auxin biosynthesis pathway in *Arabidopsis*. Proceedings of the National Academy of Sciences 108 (45), 18512–18517. Available at https://doi.org/10.1073/pnas.1108434108.

Miller, C.O., Skoog, F., Von Saltza, M.H., Strong, F.M., 1955. Kinetin, a cell division factor from deoxyribonucleic acid. Journal of the American Chemical Society 77 (5), 1392. Available at https://doi.org/10.1021/ja01610a105.

Mok, D.W.S., Mok, M.C., 2001. Cytokinin metabolism and action. Annual Review of Plant Biology 52, 89–118. Available at https://doi.org/10.1146/annurev.arplant.52.1.89, http://www.annualreviews.org/journal/arplant.

Moubayidin, L., Di Mambro, R., Sabatini, S., 2009. Cytokinin-auxin crosstalk. Trends in Plant Science 14 (10), 557–562. Available at https://doi.org/10.1016/j.tplants.2009.06.010.

Mravec, J., Skůpa, P., Bailly, A., Hoyerová, K., Křeček, P., Bielach, A., et al., 2009. Subcellular homeostasis of phytohormone auxin is mediated by the ER-localized PIN5 transporter. Nature 459 (7250), 1136–1140. Available at https://doi.org/10.1038/nature08066.

Muday, G.K., 2001. Auxins and tropisms. Journal of Plant Growth Regulation 20 (3), 226–243. Available at https://doi.org/10.1007/s003440010027.

Müller, B., Sheen, J., 2007. Advances in cytokinin signaling. Science 318 (5847), 68–69. Available at https://doi.org/10.1126/science.1145461.

Negi, S., Ivanchenko, M.G., Muday, G.K., 2008. Ethylene regulates lateral root formation and auxin transport in *Arabidopsis thaliana*. Plant Journal 55 (2), 175–187. Available at https://doi.org/10.1111/j.1365-313X.2008.03495.x.

Nishiyama, R., Watanabe, Y., Fujita, Y., Le, D.T., Kojima, M., Werner, T., et al., 2011. Analysis of cytokinin mutants and regulation of cytokinin metabolic genes reveals important regulatory roles of cytokinins in drought, salt and abscisic acid responses, and abscisic acid biosynthesis. The Plant Cell 23 (6), 2169–2183. Available at https://doi.org/10.1105/tpc.111.087395, http://www.plantcell.org/content/23/6/2169.full.pdf+html.

Nishiyama, R., Watanabe, Y., Leyva-Gonzalez, M.A., Van Ha, C., Fujita, Y., Tanaka, M., et al., 2013. *Arabidopsis* AHP2, AHP3, and AHP5 histidine phosphotransfer proteins function as redundant negative regulators of drought stress response. Proceedings of the National Academy of Sciences 110 (12), 4840–4845. Available at https://doi.org/10.1073/pnas.1302265110.

Nitschke, S., Cortleven, A., Iven, T., Feussner, I., Havaux, M., Riefler, M., et al., 2016. Circadian stress regimes affect the circadian clock and cause jasmonic acid-dependent cell death in cytokinin-deficient *arabidopsis* plants. The Plant Cell 28 (7), 1616–1639. Available at https://doi.org/10.1105/tpc.16.00016, http://www.plantcell.org/content/28/7/1616.full.pdf.

Nordström, A., Tarkowski, P., Tarkowska, D., Norbaek, R., Åstot, C., Dolezal, K., et al., 2004. Auxin regulation of cytokinin biosynthesis in *Arabidopsis thaliana*: a factor of potential importance for auxin–cytokinin-regulated development. Proceedings of the National Academy of Sciences 101 (21), 8039–8044. Available at https://doi.org/10.1073/pnas.0402504101.

Okushima, Y., Mitina, I., Quach, H.L., Theologis, A., 2005. AUXIN RESPONSE FACTOR 2 (ARF2): a pleiotropic developmental regulator. Plant Journal 43 (1), 29–46. Available at https://doi.org/10.1111/j.1365-313X.2005.02426.x.

Östin, A., Kowalyczk, M., Bhalerao, R.P., Sandberg, G., 1998. Metabolism of indole-3-acetic acid in *Arabidopsis*. Plant Physiology 118 (1), 285–296. Available at https://doi.org/10.1104/pp.118.1.285, http://www.plantphysiol.org/.

Osugi, A., Sakakibara, H., 2015. Q&A: How do plants respond to cytokinins and what is their importance? BMC Biology 13 (1). Available at https://doi.org/10.1186/s12915-015-0214-5.

Park, J.E., Park, J.Y., Kim, Y.S., Staswick, P.E., Jeon, J., Yun, J., et al., 2007. GH3-mediated auxin homeostasis links growth regulation with stress adaptation response in *Arabidopsis*. Journal of Biological Chemistry 282 (13), 10036–10046. Available at https://doi.org/10.1074/jbc.M610524200SouthKorea, http://www.jbc.org/cgi/reprint/282/13/10036.

Petrášek, J., Mravec, J., Bouchard, R., Blakeslee, J.J., Abas, M., Seifertová, D., et al., 2006. PIN proteins perform a rate-limiting function in cellular auxin efflux. Science 312 (5775), 914–918. Available at https://doi.org/10.1126/science.1123542.

Pollmann, S., Düchting, P., Weiler, E.W., 2009. Tryptophan-dependent indole-3-acetic acid biosynthesis by 'IAA-synthase' proceeds via indole-3-acetamide. Phytochemistry 70 (4), 523–531. Available at https://doi.org/10.1016/j.phytochem.2009.01.021.

Porco, S., Pěnčík, A., Rashed, A., Voß, U., Casanova-Sáez, R., Bishopp, A., et al., 2016. Dioxygenase-encoding AtDAO1 gene controls IAA oxidation and homeostasis in *Arabidopsis*. Proceedings of the National Academy of Sciences 113 (39), 11016–11021. Available at https://doi.org/10.1073/pnas.1604375113.

Pospíšilová, J., 2003. Interaction of cytokinins and abscisic acid during regulation of stomatal opening in bean leaves. Photosynthetica 41 (1), 49–56. Available at https://doi.org/10.1023/a:1025852210937.

Pospíšilová, J., Vágner, M., Malbeck, J., Trávníčková, A., Baťková, P., 2005. Interactions between abscisic acid and cytokinins during water stress and subsequent rehydration. Biologia Plantarum 49 (4), 533–540. Available at https://doi.org/10.1007/s10535-005-0047-0.

Qi, H., Cai, H., Liu, X., Liu, S., Ding, C., Xu, M., 2022. The cytokinin type-B response regulator PeRR12 is a negative regulator of adventitious rooting and salt tolerance in poplar. Plant Science 325, 111456. Available at https://doi.org/10.1016/j.plantsci.2022.111456.

Rashid, A., Achary, V.M.M., Abdin, M.Z., Karippadakam, S., Parmar, H., Panditi, V., et al., 2024. C ytokinin oxidase2-deficient mutants improve panicle and grain architecture through cytokinin accumulation and enhance drought tolerance in indica rice. Plant Cell Reports 43 (8), 207.

Raspor, M., Mrvaljević, M., Savić, J., Ćosić, T., Kaleri, A.R., Pokimica, N., et al., 2024. Cytokinin deficiency confers enhanced tolerance to mild, but decreased tolerance to severe salinity stress in in vitro grown potato. Frontiers in Plant Science 14. Available at https://doi.org/10.3389/fpls.2023.1296520.

Sachs, T., 2005. Auxin's role as an example of the mechanisms of shoot/root relations. Plant and Soil 268 (1), 13–19. Available at https://doi.org/10.1007/s11104-004-0173-z.

Sakakibara, H., 2010. Cytokinin biosynthesis and metabolism. Plant Hormones: Biosynthesis, Signal Transduction, Action!. Springer Netherlands, Japan, pp. 95–114. Available at http://www.springerlink.com/openurl.asp?genre=book&isbn=978-1-4020-2684-3, https://doi.org/10.1007/978-1-4020-2686-7_5.

Sakakibara, H., 2025. Five unaddressed questions about cytokinin biosynthesis. Journal of Experimental Botany 76 (7), 1941–1949. Available at https://doi.org/10.1093/jxb/erae348, http://jxb.oxfordjournals.org/.

Sakakibara, H., 2006. CYTOKININS: activity, biosynthesis, and translocation. Annual Review of Plant Biology 57 (1), 431–449. Available at https://doi.org/10.1146/annurev.arplant.57.032905.105231.

Sano, N., Marion-Poll, A., 2021. ABA metabolism and homeostasis in seed dormancy and germination. International Journal of Molecular Sciences 22 (10), 5069. Available at https://doi.org/10.3390/ijms22105069.

Sheng, X., Ai, Z., Tan, Y., Hu, Y., Guo, X., Liu, X., et al., 2023. Novel salinity-tolerant third-generation hybrid rice developed via CRISPR/Cas9-mediated gene editing. International Journal of Molecular Sciences 24 (9), 8025. Available at https://doi.org/10.3390/ijms24098025.

Shi, Y., Tian, S., Hou, L., Huang, X., Zhang, X., Guo, H., et al., 2012. Ethylene signaling negatively regulates freezing tolerance by repressing expression of CBF and Type-A ARR genes in *Arabidopsis*. The Plant Cell 24 (6), 2578–2595. Available at https://doi.org/10.1105/tpc.112.098640.

Shkolnik-Inbar, D., Bar-Zvi, D., 2010. ABI4 mediates abscisic acid and cytokinin inhibition of lateral root formation by reducing polar auxin transport in *Arabidopsis*. The Plant Cell 22 (11), 3560–3573. Available at https://doi.org/10.1105/tpc.110.074641.

Singhal, C., Singh, A., Sharma, A.K., Khurana, P., 2024. Identification of CKX gene family in morusindica cv K2 and functional characterization of MiCKX4 during abiotic stress. Stress Biology 4 (1). Available at https://doi.org/10.1007/s44154-024-00173-x, https://link.springer.com/journal/44154.

Staswick, P.E., Serban, B., Rowe, M., Tiryaki, I., Maldonado, M.T., Maldonado, M.C., et al., 2005. Characterization of an *Arabidopsis* enzyme family that conjugates amino acids to indole-3-acetic acid. The Plant Cell 17 (2), 616–627. Available at https://doi.org/10.1105/tpc.104.026690, http://www.plantcell.org/.

Stepanova, A.N., Robertson-Hoyt, J., Yun, J., Benavente, L.M., Xie, D.Y., Doležal, K., et al., 2008. TAA1-mediated auxin biosynthesis is essential for hormone crosstalk and plant development. Cell 133 (1), 177–191. Available at https://doi.org/10.1016/j.cell.2008.01.047, https://www.sciencedirect.com/journal/cell.

Strader, L.C., Zhao, Y., 2016. Auxin perception and downstream events. Current Opinion in Plant Biology 33, 8–14. Available at https://doi.org/10.1016/j.pbi.2016.04.004, www.elsevier.com/inca/publications/store/6/0/1/3/1/4/index.htt.

Sugawara, S., Hishiyama, S., Jikumaru, Y., Hanada, A., Nishimura, T., Koshiba, T., et al., 2009. Biochemical analyses of indole-3-acetaldoxime-dependent auxin biosynthesis in *Arabidopsis*. Proceedings of the National Academy of Sciences 106 (13), 5430–5435. Available at https://doi.org/10.1073/pnas.0811226106.

Sun, L., Zhang, Q., Wu, J., Zhang, L., Jiao, X., Zhang, S., et al., 2014. Two rice authentic histidine phosphotransfer proteins, OsAHP1 and OsAHP2, mediate cytokinin signaling and stress responses in rice. Plant Physiology 165 (1), 335–345. Available at https://doi.org/10.1104/pp.113.232629, http://www.plantphysiol.org/content/165/1/335.full.pdf.

Swarup, R., Friml, J:, Marchant, A., Ljung, K., Sandberg, G., Palme, K., et al., 2001. Localization of the auxin permease AUX1 suggests two functionally distinct hormone transport pathways operate in the *Arabidopsis* root apex. Genes & Development 15 (20), 2648–2653. Available at https://doi.org/10.1101/gad.210501.

Swarup, R., Péret, B., 2012. AUX/LAX family of auxin influx carriers-an overview. Frontiers in Plant Science 3. Available at https://doi.org/10.3389/fpls.2012.00225, http://journal.frontiersin.org/Journal/10.3389/fpls.2012.00225/pdf.

Takei, K., Yamaya, T., Sakakibara, H., 2004. *Arabidopsis* CYP735A1 and CYP735A2 encode cytokinin hydroxylases that catalyse the biosynthesis of trans-Zeatin. Journal of Biological Chemistry 279 (40), 41866–41872. Available at https://doi.org/10.1074/jbc.M406337200.

Tao, Y., Ferrer, J.L., Ljung, K., Pojer, F., Hong, F., Long, J.A., et al., 2008. Rapid synthesis of auxin via a new tryptophan-dependent pathway is required for shade avoidance in plants. Cell 133 (1), 164–176. Available at https://doi.org/10.1016/j.cell.2008.01.049, https://www.sciencedirect.com/journal/cell.

Tessi, T.M., Maurino, V.G., Shahriari, M., Meissner, E., Novak, O., Pasternak, T., et al., 2023. AZG1 is a cytokinin transporter that interacts with auxin transporter PIN1 and regulates the root stress response. New Phytologist 238 (5), 1924–1941. Available at https://doi.org/10.1111/nph.18879, http://onlinelibrary.wiley.com/journal/10.1111/(ISSN)1469-8137.

Tognetti, V.B., van Aken, O., Morreel, K., Vandenbroucke, K., van de Cotte, B., de Clercq, I., et al., 2010. Perturbation of indole-3-butyric acid homeostasis by the UDP-glucosyltransferase UGT74E2 modulates *Arabidopsis* architecture and water stress tolerance. The Plant Cell 22 (8), 2660–2679. Available at https://doi.org/10.1105/tpc.109.071316, http://www.plantcell.org/cgi/reprint/22/8/2660.pdf.

To, J.P.C., Haberer, G., Ferreira, F.J., Deruère, J., Mason, M.G., Schaller, G.E., et al., 2004. Type-a *Arabidopsis* response regulators are partially redundant negative regulators of cytokinin signaling. The Plant Cell 16 (3), 658–671. Available at https://doi.org/10.1105/tpc.018978, http://www.plantcell.org/.

Tran, L.-S.P., Urao, T., Qin, F., Maruyama, K., Kakimoto, T., Shinozaki, K., et al., 2007. Functional analysis of AHK1/ATHK1 and cytokinin receptor histidine kinases in response to abscisic acid, drought, and salt stress in *Arabidopsis*. Proceedings of the National Academy of Sciences 104 (51), 20623–20628. Available at https://doi.org/10.1073/pnas.0706547105.

Ueguchi-Tanaka, M., 2023. Gibberellin metabolism and signaling. Bioscience, Biotechnology, and Biochemistry 87 (10), 1093–1101. Available at https://doi.org/10.1093/bbb/zbad090.

Venis, M.A., 1972. Auxin-induced conjugation systems in peas. Plant Physiology 49 (1), 24–27. Available at https://doi.org/10.1104/pp.49.1.24.

Verrier, P.J., Bird, D., Burla, B., Dassa, E., Forestier, C., Geisler, M., et al., 2008. Plant ABC proteins - a unified nomenclature and updated inventory. Trends in Plant Science 13 (4), 151–159. Available at https://doi.org/10.1016/j.tplants.2008.02.001.

Waadt, R., Seller, C.A., Hsu, P.K., Takahashi, Y., Munemasa, S., Schroeder, J.I., 2022. Plant hormone regulation of abiotic stress responses. Nature Reviews Molecular Cell Biology 23 (10), 680–694. Available at https://doi.org/10.1038/s41580-022-00479-6, https://www.nature.com/molcellbio.

Wang, R., Liu, L., Kong, Z., Li, S., Lu, L., kabir, N., et al., 2021. Identification of GhLOG gene family revealed that GhLOG3 is involved in regulating salinity tolerance in cotton (*Gossypium hirsutum* L.). Plant Physiology and Biochemistry 166, 328–340. Available at https://doi.org/10.1016/j.plaphy.2021.06.011, http://www.journals.elsevier.com/plant-physiology-and-biochemistry/.

Wang, Y., Dai, J., Chen, R., Song, C., Wei, P., Cai, Y., Wang, Y., Han, B., 2022. miRNA-Based Drought Regulation in the Important Medicinal Plant Dendrobium huoshanense. Journal of Plant Growth Regulation 41, 1099–1108. Available at https://doi.org/10.1007/s00344-021-10366-7.

Wang, H., Yan, Y., Wu, T., Wang, Y., Sun, H., Li, W., et al., 2025. Natural variation in the cytokinin oxidase gene ZmCKX6 influences leaf morphology and yield-related traits in maize. The Plant Journal 123 (1). Available at https://doi.org/10.1111/tpj.70332.

Went, F.W., 1934. A Test Method for Rhizocaline, the Root-forming Substance.

Went, F.W., Thimann, K.V., 1937. Phytohormones. Macmillan, New York, pp. 294.

Wen, F.L., Yue, Y., He, T.F., Gao, X.M., Zhou, Z.S., Long, X.H., 2020. Identification of miR390-TAS3-ARF pathway in response to salt stress in Helianthus tuberosus L. Gene 738. Available at https://doi.org/10.1016/j.gene.2020.144460, www.elsevier.com/locate/gene.

Werner, T., Motyka, V., Laucou, V., Smets, R., Van Onckelen, H., Schmülling, T., 2003. Cytokinin-deficient transgenic *Arabidopsis* plants show multiple developmental alterations indicating opposite functions of cytokinins in the regulation of shoot and root meristem activity. The Plant Cell 15 (11), 2532–2550. Available at https://doi.org/10.1105/tpc.014928, http://www.plantcell.org/.

Werner, T., Schmülling, T., 2009. Cytokinin action in plant development. Current Opinion in Plant Biology 12 (5), 527–538. Available at https://doi.org/10.1016/j.pbi.2009.07.002.

Willige, B.C., Isono, E., Richter, R., Zourelidou, M., Schwechheimer, C., 2011. Gibberellin regulates PIN-FORMED abundance and is required for auxin transport-dependent growth and development in *Arabidopsis thaliana*. The Plant Cell 23 (6), 2184–2195. Available at https://doi.org/10.1105/tpc.111.086355, http://www.plantcell.org/content/23/6/2184.full.pdf+html.

Wu, W., Du, K., Kang, X., Wei, H., 2021. The diverse roles of cytokinins in regulating leaf development. Horticulture Research 8 (1). Available at https://doi.org/10.1038/s41438-021-00558-3, www.nature.com/hortres.

Yadav, C., Rawat, N., Singla-Pareek, S.L., Pareek, A., 2025. Knockdown of OsPHP1 leads to improved yield under salinity and drought in rice via regulating the complex set of TCS members and cytokinin signalling. Plant, Cell & Environment 48 (4), 2769–2782. Available at https://doi.org/10.1111/pce.15337.

Yang, H., Zhang, Q., Li, Z., Zhao, R., Liu, X., Chi, Y., et al., 2024. Molecular characterization of the cytokinin responsive type-B response regulators in perennial ryegrass highlights LpARR10 conferring salt tolerance via trans-activating SOS1 and SOS3. Environmental and Experimental Botany 217, 105566. Available at https://doi.org/10.1016/j.envexpbot.2023.105566.

Yin, P., Liang, X., Zhao, H., Xu, Z., Chen, L., Yang, X., et al., 2023. Cytokinin signaling promotes salt tolerance by modulating shoot chloride exclusion in maize. Molecular Plant 16 (6), 1031–1047. Available at https://doi.org/10.1016/j.molp.2023.04.011, https://www.cell.com/molecular-plant/home.

Yu, Z., Zhang, F., Friml, J., Ding, Z., 2022. Auxin signaling: research advances over the past 30 years. Journal of Integrative Plant Biology 64 (2), 371–392. Available at https://doi.org/10.1111/jipb.13225, http://onlinelibrary.wiley.com/journal/10.1111/(ISSN)1744-7909.

Zhang, Q., Gong, M., Xu, X., Li, H., Deng, W., 2022. Roles of auxin in the growth, development, and stress tolerance of horticultural plants. Cells 11 (17), 2761. Available at https://doi.org/10.3390/cells11172761.

Zhang, J., Lin, J.E., Harris, C., Campos Mastrotti Pereira, F., Wu, F., Blakeslee, J.J., et al., 2016. DAO1 catalyzes temporal and tissue-specific oxidative inactivation of auxin in *Arabidopsis thaliana*. Proceedings of the National Academy of Sciences 113 (39), 11010–11015. Available at https://doi.org/10.1073/pnas.1604769113.

Zhang, Mj, Xue, Yy, Xu, S., Jin, Xr, Man, Xc, 2024. Identification of ARF genes in cucurbita pepo L and analysis of expression patterns, and functional analysis of CpARF22 under drought, salt stress. BMC Genomics 25 (1). Available at https://doi.org/10.1186/s12864-024-09992-8, https://bmcgenomics.biomedcentral.com/.

Zhang, L., Zhang, F., Melotto, M., Yao, J., He, S.Y., 2017. Jasmonate signaling and manipulation by pathogens and insects. Journal of Experimental Botany 68 (6), 1371–1385. Available at https://doi.org/10.1093/jxb/erw478, http://jxb.oxfordjournals.org/.

Zhang, J., Zhao, P., Chen, S., Sun, L., Mao, J., Tan, S., et al., 2023. The ABI3-ERF1 module mediates ABA-auxin crosstalk to regulate lateral root emergence. Cell Reports 42 (7), 112809. Available at https://doi.org/10.1016/j.celrep.2023.112809.

Zhao, Y., Christensen, S.K., Fankhauser, C., Cashman, J.R., Cohen, J.D., Weigel, D., et al., 2001. A role for flavin monooxygenase-like enzymes in auxin biosynthesis. Science 291 (5502), 306–309. Available at https://doi.org/10.1126/science.291.5502.306.

Zhao, Y., Song, S., Guo, Y., Tian, Z., Shang, Y., Ding, Y., et al., 2025. Overexpression of auxin synthesis gene PagYUC6a in poplar (*Populus alba* × *P. glandulosa*) enhances salt tolerance. International Journal of Biological Macromolecules 311 (Pt 3), 143712. Available at https://doi.org/10.1016/j.ijbiomac.2025.143712.

Žižková, E., Dobrev, P.I., Muhovski, Y., Hošek, P., Hoyerová, K., Haisel, D., et al., 2015. Tomato (*Solanum lycopersicum* L.) SlIPT3 and SlIPT4 isopentenyltransferases mediate salt stress response in tomato. BMC Plant Biology 15 (1). Available at https://doi.org/10.1186/s12870-015-0415-7, http://www.biomedcentral.com/bmcplantbiol/.

Zürcher, E., Müller, B., 2016. Cytokinin synthesis, signaling, and function-advances and new insights. International Review of Cell and Molecular Biology 324, 1–38. Available at https://doi.org/10.1016/bs.ircmb.2016.01.001, http://www.elsevier.com.

CHAPTER

4

Crosstalk of salicylic acid with other growth regulators and signaling molecules in mediating abiotic stress tolerance in plants

Harmanjit Kaur, Km Sandhya Devi, Anshika Yadav, and Farheen Fatima

Department of Botany, University of Allahabad, Prayagraj, Uttar Pradesh, India

4.1 Introduction

Due to sessile nature, plants are often exposed to diverse abiotic stressors throughout their life span, such as water logging (WL), drought, heat, cold, salt, and metal toxicity, that impede their growth, development, yield, and seed/grain quality (Ghosh and Roychoudhury, 2024; Nazir et al., 2023). To cope with difficult environmental surroundings, plants have developed several adaptive mechanisms that help them to survive and complete their reproductive cycle (Du et al., 2024). Further, responses of plants to stressful environments are generally influenced by the synthesis of certain chemical signals (Zhang and Sonnewald, 2017).

In recent years, there has been an increasing interest in understanding the role of hormone-like plant growth regulators (PGRs) and other signaling molecules in plant adaptation and tolerance mechanisms against different abiotic stresses (Mahajan et al., 2025b). Salicylic acid (SA) is established as the sixth phytohormone that belongs to the class of phenolic compounds and contains a carboxylic group, an aromatic ring, and a hydroxyl group (Khan et al., 2015; Ding and Ding, 2020; Bagautdinova et al., 2022; Mahajan et al., 2023). Salicin, a naturally occurring SA derivative, was isolated for the first time by Johan Büchner in 1828 from willow tree (*Salix* sp.) bark (Muthulakshmi and Lingakumar, 2017). It is derived from the Latin word "salix," and initially this natural compound was identified for the first time by an Italian chemist, Raffaele Piria, in 1838 from willow bark (Sharma et al., 2020). In 1859, SA was synthesized chemically by

Harnessing Plant-Stimulating Molecules
DOI: https://doi.org/10.1016/B978-0-443-44994-9.00005-5

Kolbe and his team, thereby making it more easily accessible and inexpensive (Maruri-López et al., 2019; Ding and Ding, 2020), although its saleable synthetic production started in Germany in 1874 (Sharma et al., 2020). The identification of SA's function as a secondary metabolite in plant physiology did not happen till the late 20th century (Wani et al., 2017; Sharma et al., 2020). Different plants contain SA in amounts ranging from 0.1 to 10 $\mu g\ g^{-1}$ fresh weight, depending on the plant species, and considerable levels of SA are stored in either methylated and/or glucosylated forms (Liu et al., 2022a). SA plays a pivotal role in several aspects of plant growth, such as seed germination, stomatal opening, senescence, respiration, cell growth, flowering, seedling development, thermogenesis, biosynthesis of lignin and pigments, disease resistance, and allelopathy (Rodas-Junco et al., 2020; Talaat and Shawky, 2022). Additionally, SA is involved as an important signaling molecule (Sytar et al., 2019) that helps plants to cope with stressful conditions by influencing several plant processes including photosynthesis, antioxidant defense, plant water relations, proline nitrogen (N) metabolism, and glycine betaine (Khan et al., 2015; Wani et al., 2017; Bagautdinova et al., 2022). It is synthesized in cytoplasm and chloroplasts (Hassoon and Abduljabbar, 2019) and is transported to the region of action via the apoplast pathway (Lim et al., 2020). In plants, SA can be produced by two different and specific pathways, shikimic acid and malonic acid pathways (Lefevere et al., 2020). Shikimic acid pathway, also known as the isochorismate (IC) pathway, occurs in chloroplast and starts with chorismate that is changed to IC with the help of the enzyme isochorismate synthase (ICS) (Ding and Ding, 2020; Lefevere, Bauters, and Gheysen, 2020). Over 90% of the SA synthesized during the instigation of stress responses in plants is believed to be generated via this pathway (Khan et al., 2015). On the other hand, the phenylalanine ammonia-lyase (PAL) pathway occurs in the cytoplasm (Ding and Ding, 2020; Janda et al., 2020), in which prephenic acid is produced from chorismic acid by chorismate mutase, which is further changed into l-phenylalanine via a reductive reaction and subsequently to trans-cinnamic acid (t-CA) by the activity of PAL (Lefevere et al., 2020). Two compounds could be synthesized from t-CA: ortho-coumaric acid and benzaldehyde. SA could be directly synthesized from ortho-coumaric acid while aldehyde oxidase transforms benzaldehyde into benzoic acid, which is then converted into SA via activity of benzoic acid 2-hydroxylase (Maruri-López et al., 2019; Ding and Ding, 2020).

Last few decades have witnessed a lot of interest from the scientific community in exploiting plant hormones or PGRs and certain signaling molecules to mitigate the negative effects of stressful conditions on plants. An intricate mechanism modulated by plant hormones (auxins, gibberellins, cytokinins, abscisic acid [ABA], ethylene, jasmonic acid [JA], and brassinosteroids [BRs]), other PGRs (melatonin (MT), chitosan, polyamines (PAs), and triacontanol [TRIA]), and gasotransmitters (nitric oxide [NO], hydrogen sulfide [H_2S], hydrogen peroxide [H_2O_2]) exists in plants for averting stress at physiological, biochemical, and molecular levels. SA interacts with all these plant hormones and PGRs along with signaling molecules to maintain coordination of plant responses to abiotic stressors. Most recently, SA has drawn significant interest in research studies due to its beneficial role in protecting crops against detrimental effects of abiotic stresses that are being intensified due to global warming and constantly increasing industrial pollution. Moreover, crosstalk between various hormones and/or PGRs and signaling molecules has become a major focus of research in plant biology with an aim to discern a complete plant hormone signaling network. Although significant achievement in comprehending the effects and functions of SA in plants has been made in last few years, several knowledge gaps still exist regarding SA crosstalk with other key PGRs and signaling molecules involved in plant growth and development, especially under stressful conditions. Further, how SA reinforces plant intrinsic oxidative defense machinery under stressful cues needs to be understood. This chapter discusses the potential role of SA in boosting plant tolerance under diverse abiotic conditions. It also summarizes our current understanding of interactions of SA with other PGRs and signaling molecules. The intricacy of these interactions is highlighted, with anticipation of invigorating future research to address existing knowledge gaps in the current topic, mainly with respect to regulating plant growth and defense under challenging abiotic conditions.

4.2 Effects of salicylic acid on plants under various abiotic stresses

4.2.1 Temperature stress

4.2.1.1 Low temperature

Low-temperature stress, including freezing temperatures below 0°C and chilling temperatures below 20°C, is a significant abiotic factor that adversely impacts plant growth, development, physiology, biochemical processes, and overall productivity (Ritonga and Chen, 2020; Manasa et al., 2022). Low-temperature stress interferes with several metabolic pathways, with membrane system damage being a major factor contributing to chilling injury, especially in fruits and vegetables (Song et al., 2022). Plants detect low-temperature signals via cell membrane receptors that are influenced by both temperature and light conditions (Soliman et al., 2018; Zheng et al., 2021; Guo et al., 2018).

SA helps in protecting the plants from chilling injury by regulating several physiological processes, including photosynthesis, carbon and nitrogen metabolism, reactive oxygen species (ROS) scavenging, and proline metabolism (Esim and Atici, 2015; Soliman et al., 2018; Soleimani Aghdam et al., 2016; Shin et al., 2018). A study conducted by Shin et al. (2018) demonstrated enhancement in the freezing tolerance of spinach (*Spinacia oleracea*) leaves in response to a 4-day treatment with 0.5 mM exogenous SA, which reduced the membrane leakage and restricted ROS accumulation without negatively affecting leaf health. The authors argued that the improved tolerance in spinach leaves was linked to the increased levels of compatible solutes such as proline and the antioxidant ascorbic acid (AsA), potentially mediated by SA-induced signaling molecules like NO and H_2O_2. SA is commonly applied as a foliar spray to enhance crop productivity and improve cold tolerance, as was demonstrated in wheat (*Triticum aestivum*) (Wang et al., 2021a). It has been reported that SA aids in protecting the plants under cold stress by stabilizing chloroplast membranes, preserving the D1 protein (Fig. 4.1) and supporting overall photosynthetic efficiency (Huang et al., 2016; Saleem et al., 2021). Prestorage treatment with 1 $\mu mol\ L^{-1}$ SA for 15 minutes at 4°C effectively delayed and reduced internal browning in peach (*Prunus persica*) fruit, a common symptom of chilling injury (Zhao et al., 2021). Moreover, SA influenced sugar metabolism, leading to a significant rise in total soluble sugars (TSS), primarily due to an increase in sucrose levels. This was further supported by

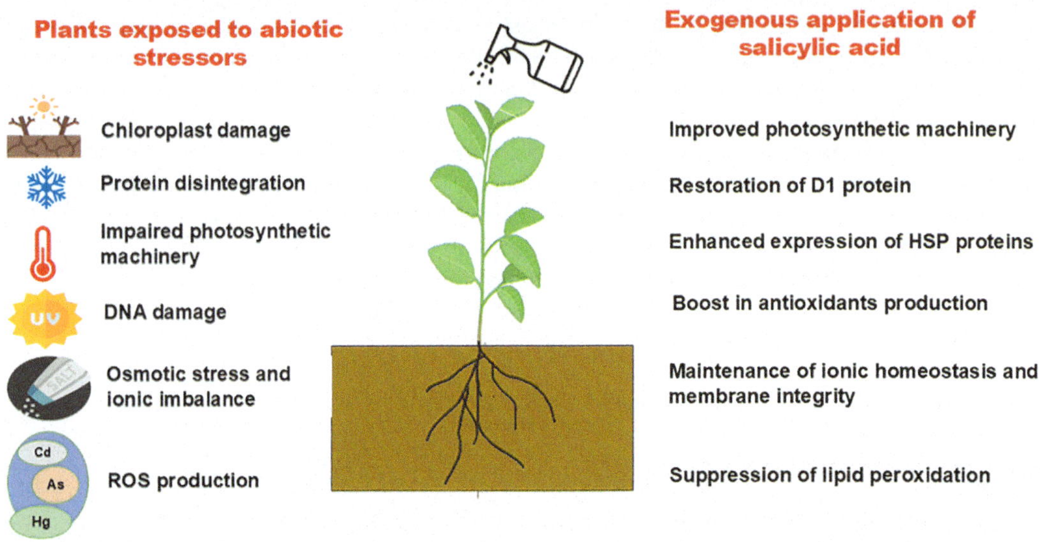

FIGURE 4.1 Role of salicylic acid in plants under different abiotic stressors. *ROS*, Reactive oxygen species; *HSP*, heat shock proteins.

marked changes in the expression of genes involved in sucrose biosynthesis and degradation. Additionally, SA treatment enhanced the expression of two dehydration-responsive element binding (DREB) transcription factors (TFs), which are key regulators of cold stress responses, thereby contributing to improved cold tolerance in peach fruit (Zhao et al., 2021).

Cold stress impairs plant proteins, disrupts membrane integrity, inhibits enzyme activity, and negatively affects the photosynthetic process (Megha et al., 2018; Manasa S et al., 2022). Small signaling molecules assist plants in managing cold stress by triggering downstream signaling pathways. These molecules include cyclic guanosine monophosphate (cGMP), calcium (Ca^{2+}), ROS, H_2S, NO, sphingolipids, and phosphatidic acid (Soliman et al., 2018; Zheng et al., 2021; Manasa S et al., 2022). Exogenous application of SA in winter jujube (*Ziziphus jujube)* effectively reduced weight loss and decay and also suppressed the respiratory rate and ethylene production (Sang et al., 2022). Moreover, SA-treated fruits maintained higher levels of nonenzymatic antioxidants such as total phenols, flavonoids, AsA, and glutathione (GSH), along with sustained activities of antioxidant enzymes including superoxide dismutase (SOD), catalase (CAT), ascorbate peroxidase (APX), and glutathione reductase (GR). Furthermore, SA activated ROS metabolic pathways, modulated hormone signaling, and induced the expression of specific TFs, thereby boosting antioxidant defenses, delaying postharvest senescence, and preserving fruit quality (Sang et al., 2022).

4.2.1.2 *High temperature*

The increasing temperatures observed in the 21st century are largely driven by climate change and global warming, resulting in heightened levels of heat stress (Chaudhry and Sidhu, 2022). Heat stress is generally defined as exposure to high temperatures beyond a critical threshold for a prolonged duration, leading to irreversible harm to a plant's growth and development (Khan et al., 2020). Elevated temperatures can lead to significant crop losses and widespread food shortages. Globally, crop production is under severe threat due to high-temperature stress. A heat shock, however, refers to a sudden, short-term rise in temperature—typically 10°C–15°C above normal levels—while heat stress is characterized by the intensity, speed, and duration of the temperature increase (Medina et al., 2021).

SA functions as a signaling molecule under heat stress, contributing to enhanced thermotolerance (Medina et al., 2021) and triggering the plant's defense mechanisms (Paolacci et al., 2017). The heat stress response is complex and involves various components, including ABA, ethylene, NADP oxidases, ultraviolet (UV) radiation, heat shock proteins (HSPs), SA (Fig. 4.1) (Choudhary et al., 2024a), and the enzyme phospholipase C (Medina et al., 2021; Hu et al., 2022). Treating wheat with 0.5 mM SA helped the plants to cope with heat stress by boosting proline levels, since SA increased the activity of γ-glutamyl kinase (GK), an enzyme involved in proline synthesis, and reduced the activity of catabolizing enzyme proline oxidase (PROX) (Khan et al., 2014). As a result, the plants maintained better water balance and continued photosynthesis even under heat stress conditions (Fig. 4.1). In addition, authors reported that SA kept ethylene levels in check by lowering the activity of 1-aminocyclopropane carboxylic acid synthase (ACS), an enzyme responsible for producing ethylene. This prevented too much ethylene buildup, which can be harmful under stress (Khan et al., 2014). Cingoz and Gurel (2016) demonstrated how SA and high temperature affected cardenolide production and heat tolerance in *Digitalis trojana* callus cultures. They found that when the callus was first treated with SA and then exposed to high temperatures for 2 hours, it produced much more cardenolide (up to 472.28 mg per gram of dry weight) and became more tolerant to heat. High temperature alone (for 2 or 4 hours) reduced the activity of two important enzymes that helped in combating stress, i.e., SOD and CAT. However, pretreating with SA boosted the activity of both the enzymes and the levels of proline, total phenolics, and flavonoids in the callus, which protected the plant cells from heat stress (Cingoz and Gurel, 2016). SA has been reported to enhance heat tolerance in rice plants during the pollen mother cell (PMC) meiosis stage by increasing pollen viability and higher seed-setting rates under heat stress (Feng et al., 2018). This protective effect was largely attributed to SA's ability to reduce excessive ROS in the anthers by inhibiting caspase-3 activation, which in turn suppressed programmed cell death (PCD) in anther tissues, thereby preventing heat-induced tapetum degradation. Furthermore, SA regulated the expression of genes critical for tapetum development, including EAT1 (eternal tapetum 1), MIL2 (microsporeless 2), and DMT1 (defective tapetum and meiocytes 1), aiding in prevention of heat-stress-induced damage to tapetum cells (Feng et al., 2018). Authors suggested that the protective effects of SA might operate, at least in part, through H_2O_2-dependent pathways. Supporting this, rice plants treated with SA or with paclobutrazol (PAC) combined with H_2O_2 under heat stress showed higher pollen viability, increased H_2O_2 levels, and lower malondialdehyde (MDA) levels compared to those treated with water, PAC alone, or SA combined with the H_2O_2 scavenger dimethylthiourea (DMTU). This study strongly indicated that H_2O_2 plays a central role in mediating SA-induced heat tolerance in plants. In another study, SA was found to enhance essential oil content, regulate antioxidant enzyme activities, and reduce both MDA production and ROS accumulation under heat stress in *Mentha* sp. (Haydari et al., 2019). In tomato (*Solanum lycopersicum*), heat stress significantly impaired biomass production, disrupted the photosynthetic system, enhanced ROS levels, and altered antioxidant enzyme activities (Shah Jahan et al., 2019). However, the application of SA mitigated these negative effects by protecting the photosynthetic machinery, stabilizing antioxidant enzyme activity, lowering MDA and H_2O_2 levels, restoring pigment concentrations, and minimizing oxidative damage by preserving the integrity of cellular membranes in tomato seedlings (Shah Jahan et al., 2019). Similarly, Wassie et al. (2020) reported SA-induced mitigation of heat-induced membrane damage in alfalfa (*Medicago sativa*) through lowered electrolyte leakage (EL) and MDA levels along with regulation of the activities of key antioxidant enzymes such as CAT, SOD, and peroxidase [POD]). Likewise, Ghani et al. (2021) also reported SA-mediated enhancement in both leaf and root biomass in oilseed rape (*Brassica napus*) coupled with increased activity of antioxidant enzymes under normal and under heat stress conditions. More recently, Saeed et al. (2022) investigated the effects of SA (1 mM and 2 mM) on heat-stressed seedlings of three cultivars of chili (*Capsicum frutescens*), namely, Moro, Tilhari, and Ren-02, and reported that foliar application of SA, particularly at the 2 mM concentration, significantly enhanced heat tolerance in the Moro and Ren-02 cultivars, which was manifested as increased biomass production, higher chlorophyll and carotenoid content, improved relative water content (RWC), and elevated activities of CAT and POD. The study conducted by Choudhary et al. (2024b) reported that pretreatment of fenugreek with 0.5 mM SA improved plant growth, reduced leaf EL, enhanced membrane thermostability, boosted antioxidant enzyme activities, and upregulated several stress-responsive proteins under heat stress conditions. Furthermore, differential protein profiling identified 12 proteins specifically in the SA plus heat treatment, out of which four protein spots (11a, 11b, 11c, and 11d) were closely associated with stress defense and were predominantly upregulated, highlighting their involvement in the plant's protective mechanisms facilitated by SA treatment (Choudhary et al., 2024b). In another study, application of SA extended the grain-filling period and increased the number of grains per panicle, which led to heavier grains and higher yields in heat-stressed maize (*Zea mays*) plants (Guo et al., 2024). SA also increased the number of endosperm cells, their DNA content, and production and storage of sugars and starch in the endosperm, ultimately boosting better development of grains under heat stress. In heat-tolerant maize, SA also helped in keeping the plant hormone levels balanced, activated important hormone signaling pathways, and triggered the metabolism of alpha-linolenic acid and JA, which are linked to stress responses. In heat-sensitive maize, SA influenced the production of protective compounds like phenylpropanoids and flavonoids (Guo et al., 2024). Overall, in this study, SA treatment led to changes in the plant's physical traits, chemical processes, gene activity, and metabolites—helping the maize plants to cope better with high temperatures. Heat stress has also been reported to impair kernel development, starch accumulation, and starch quality in plants. In this context, Wang et al. (2024) explored the effect of SA on grain yield and starch quality of two waxy maize varieties (*Zea mays* L. *var. ceratina* Kulesh), JKN2000 (heat-tolerant) and SYN5 (heat-sensitive), treated with SA after pollination and then subjected to HT during the grain-filling stage (Wang et al., 2024). SA ensured the integrity of starch granules, enlarged the granule size, and reduced the starch crystallinity by increasing amylopectin branching, thereby improving pasting and thermal properties and in vitro digestibility of starch under heat stress (Wang et al., 2024).

4.2.2 Water stress

4.2.2.1 *Waterlogging*

Waterlogging (WL), caused by factors such as excessive rainfall, poor drainage, uneven land surfaces, and soil erosion, is one of the major environmental challenges that hinders sustainable agriculture and global crop productivity (Garcia et al., 2020; Kaur et al., 2020; Zhang et al., 2021). It affects roughly 16% of the world's cultivated land and leads to estimated annual economic losses of around $74 billion (Kaur et al., 2020). WL significantly impairs various physiological and biochemical traits of plants. For instance, it has been reported that WL causes a rise in lipid peroxidation, H_2O_2, and methylglyoxal (MG) levels in plants, indicating heightened oxidative stress (Fig. 4.1). Prolonged exposure to WL also leads to reduced leaf RWC, proline concentration, decline in photosynthetic pigments, and disruption in the cellular redox equilibrium (Anee et al., 2019). WL also inflicts damage on the roots, diminishing their permeability and hydraulic conductivity, which in turn triggers stomatal closure. Furthermore, the impairment in photosynthesis leads to an accumulation of ROS in plant tissues under WL conditions (Yeung et al., 2018; Anee et al., 2019; Zhou et al., 2020). Plants have evolved a range of adaptive responses to cope with the oxygen deficiency induced by WL stress (Islam et al., 2020). Morphological responses include the development of adventitious roots and enhanced stem elongation (Yamauchi et al., 2018), while short-term physiological adjustments involve reduced stomatal conductance and a consequent decline in net photosynthetic rate (Jacobsen et al., 2007; Bhusal et al., 2020). Enzymatic activities of APX, monodehydroascorbate reductase (MDHAR), and glutathione peroxidase (GPX) generally increase under WL conditions, whereas activities of dehydroascorbate reductase (DHAR), GR, and CAT decrease. WL also affects the glyoxalase system, increasing Gly II activity and, to a lesser extent, Gly I activity (Anee et al., 2019).

The application of exogenous SA has been reported to reduce the accumulation of ROS, affect antioxidative enzymes, and improve photosynthesis, thereby promoting the growth of plants under WL conditions. In this context, application of exogenous SA alleviated WL-induced damages, improved WL tolerance in waxy corn through the activation of antioxidative defense systems, elimination of ROS, and improved photosynthesis efficiency in leaves (Wang et al., 2021c). Exogenous application of SA promoted the formation of axile roots and surface adventitious roots that originated from basal stem nodes, but inhibited their elongation, leading to the formation of a shallow root system in WL- wheat plants (Koramutla et al., 2022). Further, SA enhanced the formation of root aerenchyma, a key anatomical adaptive response of plants to WL. The authors suggested that the role of SA in promoting the development of axile and surface adventitious roots in waterlogged wheat plants is ethylene independent, while the induction of aerenchyma formation by SA requires the presence of ethylene. A recent investigation highlighted the vital role of SA in enhancing the defense mechanisms of mung bean (*Vigna radiata*) plants under WL stress. SA pretreatment, applied before the onset of WL, improved various morphological, physiological, and biochemical traits. Mung bean plants pretreated with SA, especially the BU mung bean-6 variety, showed marked improvements in shoot dry weight, leaf succulence, leaf area, and levels of photosynthetic pigments compared to plants subjected to WL alone. SA-treated plants also exhibited reduced ROS accumulation, indicated by lower H_2O_2, MDA, and EL levels—particularly in BU mung bean-6—suggesting SA's role in mitigating oxidative damage. Additionally, enhanced enzymatic activities of CAT, APX, POD, and glutathione S-transferase (GST), along with increased levels of nonenzymatic antioxidants such as total flavonoids and carotenoids, were linked to decreased ROS levels (Fig. 4.1). SA-pretreated plants also maintained higher levels of proline, soluble sugars, and free amino acids under WL, contributing to better water retention and osmotic regulation, especially in BU mung bean-6. These SA-induced morphological and physiological enhancements significantly improved yield attributes in BU mung bean-6 compared to BU mung bean-4 under WL conditions. Collectively, these findings underscored the effectiveness of SA in enhancing mung bean yield under WL environments (Sultana et al., 2024).

4.2.2.2 Drought

Drought is an important abiotic stress factor limiting crop productivity worldwide and its impact is increasing with climate change (Oguz et al., 2022; Mahajan et al., 2025c). Drought aggravates other stresses like salinity, pathogen attack, and heat, which damages the plants and causes major setbacks to agricultural productivity every year (Ahluwalia et al., 2021). It is the key important environmental stress that occurs due to temperature dynamics, light intensity, and low rainfall. Despite this, its cumulative, not obvious, impact and multidimensional nature severely affects the plant's morphological, physiological, biochemical, and molecular attributes with an adverse impact on photosynthetic capacity (Seleiman et al., 2021), thereby decreasing transpiration and increasing the plants' efficiency to use the stored water. Constant water loss through transpiration results in leaf water deficits (Bijalwan et al., 2022), which decrease relative water and chlorophyll content and increase the accumulation of osmolytes, epicuticular wax content, antioxidant enzymatic activities, ROS, secondary metabolites, membrane lipid peroxidation, and ABA (Ahmad et al., 2022). Plants are more sensitive to drought stress during certain critical stages like germination, seedling formation, flowering, fertilization, and grain formation periods (Oguz et al., 2022). Plants cope with drought stress regimes through changes in cellular osmotic potential, water potential, and activation of natural defense systems in the form of antioxidant enzymes and accumulation of osmolytes including proteins, proline, glycine betaine, phenolic compounds, and soluble sugars (Wahab et al., 2022).

SA is a potent phenolic signaling biomolecule and a multifaceted PGR, which participates in a wide range of growth, metabolism, and defense systems (Arif et al., 2020). It is a promising compound that can increase plant tolerance to drought stress and can affect many aspects of physiological and biochemical processes. Studies using *Arabidopsis* (*Arabidopsis thaliana)* mutants revealed that plants with high SA levels (*snc1*) showed significantly less impact from PEG-simulated drought stress, experiencing reduced growth inhibition, less leaf water loss, and improved photosynthesis compared to wild-type plants (He et al., 2014). Conversely, mutants with SA deficiency (*nahG, sid2*) or blocked SA signaling (*npr1-1*) suffered more severely under drought conditions. High SA also led to increased accumulation of H_2O_2 and a more pronounced increase in antioxidant enzyme activities (SOD and POD) with a greater decrease in CAT activity, along with a significant increase in proline levels. This drought tolerance in *snc1* was directly dependent on SA, as reducing SA levels in these plants (*snc1/nahG*) eliminated their drought resistance. These findings strongly suggest that endogenous SA levels and signaling are vital for drought tolerance in *Arabidopsis* (He et al., 2014). SA treatments successfully improved the performance of water-stressed *Musa acuminata* cv. 'Berangan' shoot tips cultured in vitro on MS media by boosting proliferation rate, fresh weight, RWC, and chlorophyll concentrations and proline content (Bidabadi et al., 2012). Moreover, SA enhanced the plant's resilience to oxidative stress, which was evidenced by a marked reduction in H_2O_2 and MDA in SA-treated shoot tips during water stress (Bidabadi et al., 2012). Similarly, application of SA at 0.5 mM effectively ameliorated the negative effects of drought stress on Sardari ecotypes of winter wheat by improving the photosynthetic performance, keeping membrane permeability, inducing stress proteins, and enhancing the activity of antioxidant enzymes (Khalvandi et al., 2021). In another study, exogenous application of SA at 100 mM significantly increased leaf N, phosphorus (P), and potassium (K^+) contents in wheat cultivars under drought stress in hydroponic culture, with Barani-17 demonstrating significantly higher values than Anaj-17 (Ahmad et al., 2021). SA application increased the enzymatic activities of SOD (85%), APX (60%), total phenolic content (60%), and 30% of antioxidant capacity concomitant with decreased oxidative stress and improved photosynthetic performance and growth of *Aristotelia chilensis* plants subjected to moderate drought stress (González-Villagra et al., 2022). Similarly, foliar spraying of SA at 2.00–4.00 mgL^{-1} on drought-stressed sweet potato varieties (Zheshu 77 and Zheshu 13) enhanced the activity levels of SOD and POD and protected the photosynthetic apparatus of leaves (Huang et al., 2022). Further, SA downregulated the expression of ABA-related genes, such as NECD-like3, and decreased the ABA content in leaves. In another study, foliar spray of SA at 100 mg L^{-1} enhanced the antioxidant activity (CAT, APX, POD, and SOD), AsA, total soluble protein, TSS, total phenolics, proline, anthocyanin, endogenous SA, and amylase activity under drought stress in two rice (*Oryza sativa*) cultivars (Asma et al., 2023). SA not only increased plant growth, dry weight, and metabolism or metabolic activities but also increased the nutritional status of the plant by decreasing the concentration of ROS in the membranous bounded organelles under drought stress (Asma et al., 2023). Similar results in which SA-induced drought tolerance has been reported in black cumin (*Nigella sativa*) (Kabiri et al., 2014), *Artemisia aucheri* (Abbaspour and Ehsanpour, 2016), balsam (*Impatiens walleriana*) (Antonić et al., 2016), barley (*Hordeum vulgare*) (Sharafizadeh, 2017), and wheat (Loutfy et al., 2022). Henschel et al. (2022) demonstrated that SA at a concentration of 100 μM increased shoot biomass at W40% irrigation levels and enhanced storage root biomass at both W60% and W40% in radish (*Raphanus sativus*) plants. Additionally, SA reversed the reduction in storage root volume observed at W60%. Foliar application of 2 mM SA in *Impatiens walleriana* reduced the adverse effect of water deficit stress and improved growth and ornamental plant attributes (Safari et al., 2022). Researchers observed that expression of the gene encoding for Δ1-pyrroline-5-carboxylate synthetase (P5CS) was slightly increased under control treatment (95% FC + SA 0 mM) and then significantly increased at 75% FC and upon SA treatments. Further application of SA reduced the amount of EL and MDA through increased antioxidant activities and water balance (Safari et al., 2022). SA supplementation (0.5 mM) improved RWC and proline levels in drought-stressed barley genotypes (BARI barley 5 and BARI barley 7), indicating the osmotic adjustment functions of SA under water-deficit conditions (Islam et al., 2023). Recently, Sperdouli et al. (2024) provided insights into the mechanisms of foliar sprayed SA action on PSII function in basil plants under nonstress and mild drought-stress conditions. Authors reported that SA application enhanced PSII function of basil plants by inducing the nonphotochemical quenching (NPQ) mechanism, thus dissipating the excess excitation energy under nonstressed conditions while under mild drought stress, the redox status of the plastoquinone pool (qp) in SA-sprayed leaves initiated the acclimation response to mild drought stress by retaining the same electron transport rate (ETR) as that of control plants.

4.2.3 Salinity stress

Salt stress has emerged as a predominant abiotic factor that jeopardizes global crop growth and yield (Song et al., 2025). Over 9% of global arable land is under saline-alkali stress, with a particular impact on arid and semiarid areas of the world (FAO, 2024). Improper use of water resources in irrigation that contain a significant amount of salt, faulty agronomic practices such as improper fertilization, climate change, etc., are gradually increasing soil salinity of arable lands across the globe (Atta et al., 2023; Mahajan et al., 2025a). Sulfate and chlorides of calcium, sodium (Na^+), and magnesium are commonly associated with the development of soil salinity. However, the detrimental effects of Na^+ salts on plants have been reported to be more pronounced than that of Ca^{2+} (Bryla et al., 2021; Wu et al., 2023b). For instance, excess uptake of Na^+ and Cl^- disrupts Na^+/K^+ ratio in plant tissues, leading to osmotic stress, ionic toxicity, and nutritional imbalances (Fig. 4.1) (Isayenkov and Maathuis, 2019). Salinity decreases plant production by impairing various physiological, biochemical, and molecular functions. It hampers germination, growth, photosynthesis, transpiration, and stomatal conductance by decreasing leaf water potential and turgor pressure and generates osmotic stress (Arif et al., 2020). Salinity also affects nutritional balance, membrane stability by increasing lipid peroxidation, ion toxicity, cell death, EL, and ROS accumulation, which in turn triggers oxidative stress in plants (Khan et al., 2014; Islam et al., 2024; Sultana et al., 2025a).

SA efficiently prevented the adverse effects of salt stress on the growth and physiology of *Dianthus superbus* through increased thickness of the mesophyll layer and induction of the expression of salt-induced genes to resist higher salinity levels and for improving the metabolic level of the plants (Zheng et al., 2018). Specifically, SA application leads to a marked improvement in growth parameters, which are typically compromised under saline conditions. Barwal et al. (2024) reported the role of SA in bolstering the antioxidant defense system in salt-stressed maize through enhancement in the activities of antioxidant enzymes, including SOD, CAT, POX, APX, GR, and AsA, which collectively mitigated oxidative stress by scavenging ROS (Fig. 4.1). Further, the reduction in ROS accumulation was followed by decreased MDA levels, indicating reduced lipid peroxidation and cellular membrane damage, which are compromised under saline conditions (Fig. 4.1). The authors observed that SA application led to improvement in growth parameters and chlorophyll content, indicating better photosynthetic efficiency and overall plant vigor, thereby conferring greater resilience to salt-induced oxidative damage. Additionally, SA induced the accumulation of proline, soluble protein, and phenols, modulated the uptake and distribution of essential ions, such as Na^+ and K^+, maintaining ionic homeostasis under salt stress (Fig. 4.1). Future research on SA-mediated salt stress tolerance in maize plants, as well as in other crop plants, needs a focus on elucidating the molecular signaling pathways and gene networks involved. Advanced genomic and proteomic techniques can be employed to identify key regulatory genes and proteins (Barwal et al., 2024). SA at 1 mM concentration increased the morphological traits such as shoot fresh/dry weight, total dry/fresh weight, leaf area, and physiological properties including photosynthesis pigments, proline and flavonoid content, and antioxidant enzyme activities (CAT and POX) and lessen the harmful consequences of salt stress in soybean (*Glycine max* L.) (Alizadeh et al., 2024). According to Roshdy et al. (2021), foliar application of SA at a concentration of 90 ppm was a beneficial treatment for enhancing the growth, yield, and quality of fruits in strawberry plants (*Fragari* × *ananassa* Duch. cv. Camarosa) under normal or saline-stressed conditions. Study conducted by Jini and Joseph (2017) demonstrated enhancement in growth, yield, nutrient values, and activities of SOD, CAT, and POD of rice varieties (ASD16 and BR26) after SA treatment to high and low saline soils. Authors traced the effects of SA on Na^+, K^+, and Cl^- ionic accumulation under salt stress conditions by inductively coupled plasma optical emission spectrometry and ion chromatography and observed that the increased accumulation of Na^+ and Cl^- ions in saline-stressed rice plants was reduced by SA application. Further, an increased concentration of endogenous SA level was detected by the authors from the SA-treated salt-stressed rice varieties through liquid chromatography-electrospray Ionization-tandem mass spectrometry (Jini and Joseph, 2017). This study proved that the application of SA alleviates the adverse effects of salt stress through regulation of physiological processes in plants.

SA has been reported to reduce the deleterious impact of salt stress on the leaf anatomical structure. In this context, foliar application of 100 ppm of SA on cowpea (*Vigna unguiculata*) plants exposed to 6000 ppm of salinity stress resulted in positive variations in the anatomical structure of the median portion of the main stems and blades of mature leaves (El-Taher et al., 2022). The results revealed that SA treatment substantially increased the plant height, number of compound leaves, number of internodes per plant, fresh weights of leaves and stems, productivity, photosynthetic pigment content, and concentrations of N, P, and K^+ of the cowpea plants compared to the controls, clearly evidencing the positive roles of SA in increasing the salt stress tolerance of cowpea plants through improvement in the morphological and physiological attributes (El-Taher et al., 2022). Layachi and Kechrid (2025) observed reduced photosynthetic pigments in faba bean (*Vicia faba*) plants under elevated salt levels, which was attributed to breakdown of the ultrastructure of chloroplast, increase in activity of chlorophyllase enzyme and inhibition of chlorophyll synthesis (Shams et al., 2019). However, application of SA to salt-stressed faba bean increased the chlorophyll contents as SA plays a crucial role in chloroplast biosynthesis, maintaining chlorophyll stability, and regulating photosynthesis by protecting chloroplast against toxic effects of ROS (Layachi and Kechrid, 2025). Similar results have been reported in French bean (*Phaseolus vulgaris*) (Youssef et al., 2023), wheat (Loutfy et al., 2020), and mustard (*Brassica juncea*) (Husen et al., 2018; Sultana et al., 2025a) compared the efficacy of different application methods of SA for alleviating the salinity stress in tomato and found root SA pretreatment + foliar SA treatment to be superior to other methods in exhibiting protective effects against salt-induced toxicity by enhancing growth parameters, increasing the accumulation of K^+, Ca^{2+}, Mg^{2+}, S with minimum Na^+ and maximum proline accumulation, higher CAT and APX activities, and lowered H_2O_2 generation, which resulted in ionic equilibrium and a decrease in osmotic stress (Fig. 4.1). Authors suggested that applications of SA in the form of a combination of root pretreatment and foliar spray could be used for tomato cultivation under salt stress conditions. In another study, Omidi et al. (2022) investigated the interaction of several levels of NaCl salinity (0, 4, 8, and 12 ds m^{-1}) with SA (0, 0.5, 1, and 2 mM) in Damask rose (*Rosa damascena* Mill.) and found that even at low concentrations (0.5 mM), SA managed to neutralize the negative effects of salinity through increased antioxidant enzymes and their gene expressions. Similarly, SA treatment (10 mM and 20 mM) played a crucial role in maintaining osmotic balance, facilitating mineral ion uptake, and aiding in ROS detoxification by promoting the production of enzymatic and nonenzymatic antioxidants and osmolytes in salt-stressed canola plants (Ilyas et al., 2024). Pirasteh-Anosheh et al. (2022) analyzed the pragmatic role of SA seed priming in imparting salinity tolerance to barley and wheat plants in terms of morphophysiological parameters, harvest index, and yield attributes and observed elevation in all these traits. The researchers concluded that SA-induced enhancement in plant traits was primarily due to enhanced K^+ concentrations and reduced Na^+ and Na^+/K^+ ratios, indicating SA's role in balancing these ionic ratios to maintain the ion homeostasis in plants under adverse conditions of salinity. Recently, Hanif et al. (2024) also demonstrated that SA protected barley from salinity stress by improving plant growth and enhancing morphophysiological and biochemical attributes.

In a nutshell, application of SA provides a practical basis for wide cultivation of crops and constitutes a sustainable and effective approach to mitigate the adverse effects of salinity. Use of SA could prove to be an ecological and economical solution to deal with salt-affected soils, especially in arid and marginal regions. However, more studies are direly needed to optimize the SA concentration under different salinity levels before making its use on a large scale. Future molecular studies with different levels of SA application could provide a deeper understanding of their role in modulating salinity tolerance mechanisms in plants so that the best optimal SA concentration could be determined.

4.2.4 Heavy metal stress

Heavy metal toxicity has emerged as one of the greatest risks to crop productivity that could get significantly worse over the next few decades (Nazir et al., 2024; Riyazuddin et al., 2022). Depending on their geological origin in the soil or through man-made activities, the concentration of heavy metals in the soil can range from 1 to 100,000 mg kg^{-1} (Aslam et al., 2021). Soils contaminated with heavy metals exert toxic effects on plants mainly through suppression of absorption and utilization of nutrients, disruption of cellular structures and functions, and accumulation of heavy metal ions in plants, which in turn leads to disturbance in physiological processes such as photosynthesis, respiration, and transpiration, and ultimately inhibiting growth and development (Rahman et al., 2023). Heavy metal ions disrupt plant mineral nutrition by directly competing with the elements essentially needed by plants, thereby making it tough for the plants to absorb and utilize nutrients, which limit their normal growth (Song et al., 2023).

SA is a natural phytohormone and a signaling molecule that is crucial for plant growth and development and participates in several abiotic stress responses, such as those induced by heavy metals (Subašić et al., 2022). The main functions of SA in imparting plant tolerance to stresses include activation of plant defense mechanisms (Alam et al., 2023; Chávez-Arias et al., 2022), improved photosynthesis (Li et al., 2023; Sherin et al., 2022), regulation of normal growth and development (Gondor et al., 2022; Hundare et al., 2022; Liu et al., 2022b; Lone et al., 2023), and enhancement of tolerance to several abiotic stresses, including heavy metals (Rahman et al., 2023). The exogenous application of SA to plants induces acclimatization impacts by improving resistance and leading to tolerance against heavy metal stress, which is achieved through the activation of stress signaling hormonal pathways and overexpression of stress-related genes and enzymes (Rahman et al., 2023). Exogenous application of SA has been reported to restore heavy metal-induced damage to various photosynthetic traits in plants, thereby improving their stress tolerance. Another major mechanism involved in SA-regulated heavy metal stress tolerance in plants is the modulation of ROS detoxification machinery, which protects the plants against oxidative stress. SA has also been documented to modulate the levels of nonenzymatic antioxidant metabolites, including proline, upon exposure to heavy metal stress (Saini et al., 2021). Impaired development of chlorophyll has been reported under heavy metal stress, which may be the result of interference of heavy metals with protein synthesis (Fig. 4.1), the integral structural parts of a chloroplast (Gillani et al., 2021). Exogenous SA treatment in *Alyssum inflatum* led to increased growth and plant biomass by reducing nickel (Ni) uptake, improving photosynthetic pigment levels, and increasing N metabolism and mitotic activities (Najafi-Kakavand et al., 2023). SA has been found to reduce cadmium (Cd) accumulation in tomato plants and alter its distribution by enhancing cell wall polysaccharide synthesis and related gene expression, leading to cell wall thickening and blocking Cd penetration (Jia et al., 2021). Additionally, SA decreased pectin methyl esterase activity, reduced cell wall Cd accumulation, and altered the Cd partition ratio in Cd-stressed tomato plants (Jia et al., 2021). In another study, SA application increased the percent distribution of Cd in cell wall fractions (from 35.7% to 40.6%) and decreased Cd distribution in the other fractions in rice roots, indicating that SA application increases Cd deposition in the root cell wall, thus reducing Cd toxicity to plant cells (Pan et al., 2021). Moreover, SA promoted lignin biosynthesis for thickening of the cell wall in rice roots, thereby enhancing defense against Cd toxicity. This could be attributed to the fact that SA application is capable of activating the PAL, an enzyme shared between SA and lignin biosynthesis pathway, and consequently promoting lignin deposition in the root cell wall (Pan et al., 2021). In another study, Cd induced an increase in pectin, HC1, HC2, and cellulose, as well as in carboxyl (–OH) and hydroxyl groups (–COOH) of cell wall components, which played an important role in Cd absorption in black nightshade (*Solanum nigrum* L.) (Wang et al., 2021b). Guo et al. (2021) argued that cell wall thickening is a key mechanism for defense against Cd toxicity in plants. SA has a great potential in controlling ROS production in plants exposed to metal stresses. For instance, SA treatment increased the antioxidant enzyme activities in potato under Cd stress, which resulted from the dramatic upregulation of genes, viz., StAPX and StSOD (Li et al., 2019). Similarly, seed priming with SA minimized the oxidative stress in terms of MDA content in roots and shoots of wheat seedlings under chromium (Cr^{6+}) and zinc (Zn^{2+}) stresses (Mazumder et al., 2022). In another study, SA pretreatment minimized the accumulation of vanadium (V) in root and shoot, enhanced the antioxidant and ascorbate-glutathione (AsA-GSH) pathway enzyme activities, decreased the ROS and MG contents, upregulated the antioxidant and glyoxalase-related genes, and improved the photosynthetic efficiency of the rice plants under V stress (Altaf et al., 2022). Khalid et al. (2023) demonstrated that sweet pepper (*Capsicum annuum*) seedlings exposed to very high mercury (Hg) toxicity exhibited great H_2O_2, whereas SA significantly decreased the H_2O_2 content by stimulating the activities of SOD, POD, and CAT. Authors further reported that SA-treated plants showed a lesser decrease in photosynthetic system II (PSII), plant length (shoot and root), fresh and dry biomass (shoot and root), and number of leaves as compared to non-SA plants. Moreover, SA restricted the accumulation of Hg in the roots and leaves of sweet pepper plants and hindered the translocation of Hg from roots to fruits (Khalid et al., 2023). Similarly in canola, increased SA supply lowered arsenic (As) uptake, reduced oxidative stress by reducing ROS accumulation, and increased proline accumulation, which ultimately resulted in alleviation of As toxicity by osmotic adjustment and protection of enzyme structure. SA supplementation significantly reduced the harmful effects of As on plant growth, development, and photosynthesis by increasing enzymatic and nonenzymatic antioxidant defense, S-assimilation, production of carbohydrates, and, in addition, the concentration of phytochelatins (PCs) and nonprotein thiols (NPTs), which simultaneously reduced the As-induced stress by increasing the functionality of Rubisco and photosynthesis in canola (Bano et al., 2022). SA also confers tolerance against the harmful effects of cadmium and copper stress through improved antioxidant systems, proline buildup, reduced ROS, and also stabilizes membrane stability (Zahra et al., 2024).

In conclusion, since human health is seriously threatened by the absorption of heavy metals from polluted soil and their accumulation, particularly in food crops, exogenous application of SA can help to alleviate the toxicity of heavy metals in crop plants, leading to improved yields and food security.

4.2.5 Ultraviolet stress

UV radiation is a part of the nonionizing radiation region of the electromagnetic spectrum and comprises about 9% of the emitted solar radiation (Correa et al., 2023). Among different abiotic stresses, ultraviolet-B (UV-B) radiation coming to the Earth's surface due to depletion of the stratospheric ozone layer is of serious concern to all living organisms. It has long been considered a stressor for plants, negatively influencing their growth, DNA, proteins, membranes, and yield, ultimately posing a threat to food security (Choudhary et al., 2021; Mareri et al., 2022). Few reports have demonstrated the beneficial effects of SA on plants under UV stress. For instance, exogenous application of SA reduced the negative effects of UV-B in peas through an increase in CO_2 assimilation, which in turn increased the growth and biomass accumulation (Martel and Qaderi, 2016). SA-mediated activation of antioxidant enzymes was considered as a major factor for SA-mediated regulation of UV-induced oxidative stress in pepper (*Capsicum annuum*) leaves (Mahdavian et al., 2007). Similarly, increased photosynthetic rate, pollen viability, leaf phenolic concentration, and yield in UV-B-stressed rice was also reported upon exogenous application of SA (Mohammed and Tarpley, 2013). Therefore considering these reports, it can be concluded that not much work has been done to investigate the role of SA in alleviating UV stress in plants; hence, there is an imperative need for further research in this direction. Understanding the mechanism by which SA enhances UV stress tolerance could lead to the development of crop varieties with increased resistance to UV radiation and would be crucial for ensuring food security and maintaining plant productivity in a changing environment.

4.3 Crosstalk of salicylic acid with other plant growth regulators

4.3.1 Salicylic acid and auxins

Auxin (IAA, indole-3-acetic acid) is one of the most important plant hormones that regulate plant growth and development, including morphogenesis of shoots and roots and their elongation (Naser and Shani, 2016). Besides regulating the plant growth and development, auxin also aids the plants to survive under diverse abiotic stresses (Ma et al., 2022). For example, in faba beans exposed to salt stress, IAA enhanced the accumulation of osmolytes (soluble sugars and proteins), regulated ionic homeostasis, improved antioxidant activity, as well as increased the number of nodules for better biological N fixation (Latef et al., 2021). Pretreatment of Picual olive (*Olea europaea cv.* Picual) plants with combined application of IAA and SA significantly improved biomass production under saline conditions (del Pilar Cordovilla et al., 2023). IAA and SA favored the Na^+ exclusion mechanism by reducing Na^+ transport from roots to leaves and improved the K^+/Na^+ ratio. Further, IAA and SA pretreatments promoted the accumulation of Spd and Spm, respectively, increased the antioxidant system, the stabilization of membranes, and greater pigment protection (Fig. 4.2) (del Pilar Cordovilla et al., 2023). In okra plants exposed to salinity, IAA and SA individually or in combination induced salt tolerance via induction of antioxidant enzymes (Esan et al., 2017). With respect to heavy metal toxicity, Zulfiqar et al. (2024) argued that a decrease in IAA and cytokinin concentration was responsible for the reduction in morphological characteristics of wheat under Cd stress. IAA and SA inhibited Cd transport from roots to shoots of wheat seedlings and were involved in inhibition of oxidative burst through stimulation of nonenzymatic (carotenoids, proline) antioxidant

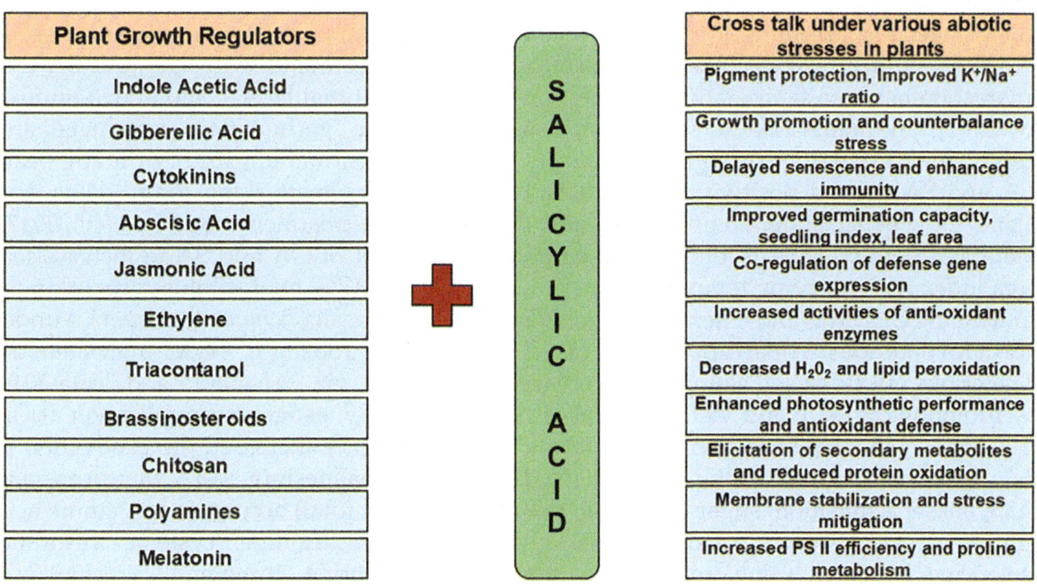

FIGURE 4.2 Crosstalk of salicylic acid with different plant growth regulators (*PGRs*) under various abiotic stresses.

machinery responsible for a tight regulation of ROS homeostasis during Cd stress (Agami and Mohamed, 2013). Likewise, SA and IAA alleviated the salt and heavy metal stress in tomato in terms of improved growth and physiological parameters (Fig. 4.2) (Malik and Lal, 2018). Hence, research on SA, including its interaction with other plant hormones and its role in various abiotic stress responses, is crucial for optimizing its use in crop protection and improvement.

4.3.2 Salicylic acid and gibberellic acid

Gibberellic acid (GA_3), an endogenous tetracyclic diterpenoid plant hormone, regulates many growth and developmental aspects of crop plants. GA_3 plays a pivotal role in mitigating abiotic stress–induced perturbations in plants by modulating various physio-biochemical and molecular processes (Shah et al., 2023). In general, SA and GA, as two important plant growth phytohormones, are the key regulators for increasing plant tolerance to abiotic stress, which is achieved by some pathways involved in plant defense mechanisms including stimulated expression of the genes associated with antioxidant activity, modulation of cellular redox homeostasis, and alteration in transcription element activities. In this context, the crosstalk between SA and GA has been reported in *Arabidopsis* involving the regulatory role of GAs in SA biosynthesis, where expression of responsive genes of GA increased the endogenous levels of SA under heavy metal stress (Emamverdian et al., 2020). Bahrani and Pourreza (2012) reported a significant increase in germination percentage and germination rate in wheat under saline conditions by SA and GA. Seed priming of Sorghum (*Sorghum bicolor* L.) seeds with SA and GA under drought stress improved the germination percentage, germination index, weight of utilized (mobilized) seed, seed reserve utilization efficiency, seedling dry weight, and seed reserve depletion percentage. Also, priming improved enzyme activity as compared to the unprimed seeds (Roghayyeh et al., 2014). Similarly, hormonal priming treatments with GA and SA and interaction between temperatures (suboptimal [< 25°C], optimal [25°C], and above-optimal [> 25°C]) in maize had a significant positive effect on germination percentage and germination rate under drought and salinity stress (Bahrabadi et al., 2022). In mung bean, GA and SA exerted positive and significant impact on dry matter and grain yields, and plants grew better under 50 ppm of SA and 100 ppm of GA treatments (Keikha et al., 2017). Mahmoudi et al. (2022) recommended combined treatment of GA and SA to increase leaf area index (LAI), canopy temperature depression (CTD), RWC, total soluble carbohydrates content (SCC), DMY, and essential oil yield (EOY) in sweet basil (*Ocimum basilicum* L.) under normal conditions. Foliar application of 1.0 mM SA and 150 $mg{\cdot}L^{-1}$ GA3 alleviated the deleterious effects of soil salinity in two varieties of wheat, viz., Anaj-2017 and Ujala-2016, by promoting growth and yield (Iqbal et al., 2022). Similarly, exogenous application of GA (100 mg L^{-1}) in conjunction with SA (100 mg L^{-1}) significantly increased the production of date fruits via regulating the nutrient (N, P, K^+, Ca^{2+}, magnesium, Na^+, zinc, iron, and manganese) acquisition, sugar (reducing, nonreducing, and total) accumulation, amino acid (glutamic acid, aspartic acid, proline, glycine, alanine, arginine, cysteine, histidine, isoleucine, leucine, lysine, methionine, phenylalanine, serine, threonine, tyrosine, and valine) profile, antioxidant (total phenols and tannins content as well as POD and CAT activities) response, and phytohormone (IAA, cytokinins, gibberellic acid, and SA) performance. Authors supported the potential of the combined application of GA and SA as a strategy to enhance the growth and development of fruit plants (Talaat et al., 2023). Few studies have indicated distinct adaptive responses of plants to exogenous GA3 and SA. In this context Ciobanu and Ionescu (2021) observed that SA increased the levels of POD activity, acid phosphatase activity, soluble protein, and glucose concentration in the roots of wheat plants, while GA heightened the levels of lipid peroxidation products (TBARS) and glucose compared to control (Fig. 4.2). On the same lines, Haider Abbasi et al. (2020) demonstrated stimulatory effects of various concentrations of GA in terms of maximum biomass accumulation in *Ajuga integrifolia* by stimulating shoot growth compared with SA and control. SA was effective in increasing total phenolic and flavonoid contents, whereas phytochemical production was significantly high in case of GA treatment (Haider Abbasi et al., 2020). On the same lines, Goldani et al. (2021) observed that SA-treated water-deficit wild petunia (*Persian petunia*) plants exhibited reduced negative effects of water stress on most of the qualitative and quantitative attributes compared to the control, while GA_3 application had little impact on the studied traits. Importantly, stomatal conductance and RWC were improved in water-stressed plants treated with SA. Moreover, SA-treated plants produced more flowers at all water-deficit levels compared to GA_3-sprayed plants in wild petunia (Goldani et al., 2021).

4.3.3 Salicylic acid and cytokinins

Cytokinin is essential for the development and structural maintenance of cellular components, playing a significant role in sustaining cellular integrity (Hudeček et al., 2023). Moreover, it regulates protein synthesis processes vital for the assembly of the photosynthetic machinery, which is fundamental to energy generation and overall plant vitality (Kieber and Schaller, 2018). In addition, kinetin is instrumental in orchestrating key physiological processes, including cell division, reproductive development, apical dominance, nutrient acquisition, and the mobilization of assimilates to sink tissues. It effectively retards senescence and stimulates metabolic and transport activities within plant systems (Fig. 4.2) (Ahanger et al., 2020). Under suboptimal conditions, kinetin also enhances growth dynamics, crop yield, quality parameters, and tolerance to abiotic stress factors such as salinity and heavy metal exposure (Ahanger et al., 2018; Kaya et al., 2018). Furthermore, phytohormones such as SA and cytokinin analogs like kinetin (KN) exhibit considerable potential in alleviating oxidative stress (Abdelaal et al., 2020; Kapoor and Hasanuzzaman, 2020).

A study on sage (*Salvia splendens)* demonstrated that application of SA at 20 ppm significantly increased flower spike count, total flower production, seed yield per plant, and the test weight of 1000 seeds. A concentration of 30 ppm promoted early bud initiation and flower anthesis, while 40 ppm delayed flower senescence and enhanced pod development. Additionally, foliar spraying with 15 ppm kinetin improved plant height and canopy spread at 90 days after transplanting (DAT) under nonstress conditions (Kumari et al., 2025b). The findings also revealed that the use of KN and SA significantly improved seed yield, oil content, oil yield, and the composition of key fatty acids in safflower (*Carthamus tinctorius* L.). Notably, the oleic acid content—a crucial parameter for oilseed crops—was positively influenced by seed priming with kinetin under unstressed conditions. Furthermore, seed priming proved to be more effective than foliar application (Culpan and Gürsoy, 2025). Khalil et al. (2021) reported that presowing seed soaking treatments with SA, either applied alone or in alternating combinations with KN and Ca^{2+}, conferred enhanced tolerance to nickel and/or lead stress in plants. These treatments significantly boosted antioxidant defense systems, promoted proline accumulation, and reduced ROS, thereby maintaining membrane stability under heavy metal stress conditions.

Numerous studies demonstrated that foliar application of synthetic cytokinin (BAP, 6-Benzylaminopurine) and SA mitigated the adverse effects of abiotic stresses across various crop species. In faba bean, BAP and SA applications significantly reduced Na^+ accumulation and lipid peroxidation while enhancing K^+ content, K^+/Na^+ ratio, soluble sugar levels, antioxidant enzyme activities, and membrane stability index under variable salinity conditions (Fig. 4.2) (Samea-Andabjadid et al., 2018). In soybean, foliar treatment with SA and KN alleviated WL stress by decreasing ROS accumulation and increasing antioxidant enzyme activities, particularly glyoxalase I (Gly I) and glyoxalase II (Gly II), which facilitated MG detoxification (Hasanuzzaman et al., 2022).

Foliar application of cytokinin and SA over a 16-day period significantly enhanced drought stress tolerance in sunflower, resulting in a marked improvement in seed and oil yield. Specifically, seed yield increased by 69%, 72%, 37%, and 64% following treatment with cytokinin at 25 and 50 $\mu M \cdot L^{-1}$ and SA at 200 and 400 $mM \cdot L^{-1}$, respectively. Under irrigation intervals of 8, 12, and 16 days, cytokinin at 50 μM led to seed yield gains of 33%, 16.5%, and 72%, along with oil yield increases of 33.5%, 19.5%, and 50% (Hossini et al., 2024). Similarly, Kardan et al. (2024) reported that foliar application of KN and SA in soybean under salt stress led to reduced levels of Na^+, H_2O_2, MDA, and proline, while increasing uptake of K^+, Ca^{2+}, and magnesium ions, chlorophyll concentrations, LAI, antioxidant activity, membrane stability index, soluble sugars, and leaf RWC (Fig. 4.2).

The above reports suggest that the integrated use of phytohormones like cytokinin, kinetin, and SA offers promising agronomic strategies for optimizing growth, resilience, and yield. Their targeted application—especially in seed priming or foliar regimes—can significantly fortify plants against environmental stressors while enhancing economic traits like seed quality and oil yield.

4.3.4 Salicylic acid and ethylene

Ethylene is a gaseous phytohormone with an important role in plant development and multiple stress responses. It has been reported that in many plant species, various abiotic stresses affect ethylene biosynthesis and signaling pathways. Application of exogenous SA could protect plants from many environmental stresses, which are caused by higher ethylene emanation and ethylene-caused oxidative stress damage. For instance, in canola plants, foliar spray of ethylene lowered chlorophyll, carotenoid contents, and anthocyanin accumulation, while SA alleviated these effects. Further, SA induced an increase in AsA content in canola plants after the ethylene treatments and played an important role in the alleviation of damages caused by stress conditions in canola plants (Tirani et al., 2013). In another study, foliar application of SA at concentrations of 0.2 and 0.3 mM effectively suppressed ethylene evolution from germinating sweet pepper seeds by up to 30% under 60 mM NaCl stress (Ahmed et al., 2020). Further, SA elevated the levels of TSS and SOD activity, leading to improved nutrient use efficiency (NUE), particularly K^+ use efficiency, under salinity stress in sweet pepper (Ahmed et al., 2020). Similarly, SA pretreatment enhanced salt tolerance of barley seedlings by modulating the accumulation of other PGRs in both roots and shoots. Specifically, SA exhibited antagonistic effects on ethylene in barley plants subjected to high salt stress (Torun et al., 2022). On the contrary, ethylene intervened in the effect of SA in the presence of SO_4^{2-} to induce S assimilation, upregulated the enzymatic and nonenzymatic antioxidants, and imparted NaCl tolerance in mustard plants (Rasheed et al., 2021).

4.3.5 Salicylic acid and abscisic acid

Abscisic acid (ABA) was first isolated from immature cotton boll in 1963. Since its role in promoting shedding, it is called abscisin II. The compound was officially named ABA in 1965. As an endogenous hormone, ABA is widely used in agricultural production for its ability to regulate plant growth, such as participation in seed dormancy and germination, flowering and fruiting, storage protein and lipid synthesis, grain filling process, and plant–microbial interactions to regulate root structure, leaf senescence, and stomatal opening and closing (Sah et al., 2016). Application of SA (1 mM) and ABA (0.1 mM) improved the morphological and anatomical characters, such as stem diameter, number of vessels/bundle in stem as well as leaf, lamina thickness, and average diameter of xylem vessel, and also increased the chlorophyll concentrations and yield components of faba bean under drought stress (Abdelaal, 2015). Likewise, foliar application of exogenous SA and ABA was reported to improve growth quality of rice at the reproductive stage under unstressed conditions, which was mainly related to the enhancement of photosynthesis, increased seedling index and leaf area, chlorophyll content, and enhanced protective enzyme activity (Chen et al., 2022). In another study, exogenous application of ABA and SA affected carotenoid biosynthesis when applied on low-temperature-stressed maize seedlings (Xiang et al., 2024). The applied ABA interacted with the ABA-responsive *cis*-acting elements (ABREs) in the promoter regions of *PSY3*, *ZDS*, and *CHYB* and activated their expressions in maize seedlings under chilling stress (Xiang et al., 2024). Application of SA did not yield positive results in alleviating chilling stress; however, neoxanthin content was notably boosted by SA (Xiang et al., 2024). In another study, foliar spray of SA (2.00–4.00 $mg \cdot L^{-1}$) downregulated the expression of ABA-related genes, such as *NECD-like3* and decreased the ABA content in leaves of sweet potato under drought conditions (Huang et al., 2022). Seo et al. (2014) observed that application of 0.5 mM SA and 0.1 mM ABA did not significantly reduce the drought injuries on waxy corn (*Zea mays* var. ceratina L.). Researchers suggested that pretreatment of the PGRs before water deficit stress or divided application at low concentration may be required to obtain the reduction effect on drought stress of waxy corn.

4.3.6 Salicylic acid and jasmonic acid

Jasmonic acid (JA) is an endogenous growth-regulating substance, initially identified as a stress-related hormone in higher plants. It has been reported to play a significant role in plant signaling pathways in response to abiotic stresses, including cold, drought, salinity, heavy metals, and light (Wang et al., 2020). Exogenous SA and JA applications effectively mitigated the negative impacts of drought stress on garden thyme (Ghaderi, 2015). Results of Mahmoud et al. (2021) provided perspectives on the possible function of SA and methyl jasmonate (MeJA) in ameliorating salt stress in citrus. Authors reported that SA and MeJA enhanced the expression of enzymatic antioxidants: POD, CSD, CAT, PAL, APX1, GSTs, aquaporin proteins, and Na cotransporters and also altered the expression of PR1, PR3, PR4, and PR5, while only MeJA altered the expression of PR2 in salt-stressed citrus plants. JA and SA can enhance the resistance of pulses to abiotic stresses such as drought, temperature extremes, metal toxicity, and salt stress, which could be achieved through the regulation of specific gene expression. The WRKY70 TF plays a significant role in the posttranscriptional regulation and modulation of genes, such as NPR1, VSP1, VSP2, PR2, and PR10, which are associated with stimulating the plant defense response in the pulse crops. The regulation of JA and SA signaling pathways in pulse crops under stress conditions is complex, requiring a deeper understanding of the underlying mechanisms. Therefore developing well-organized strategies for the exogenous application of JA/SA and its derivatives in pulses becomes crucial in mitigating the impact of these stresses (Sultana et al., 2025b). Some reports have indicated a greater positive impact of SA than JA in terms of mitigating abiotic stresses in plants. For example, combined foliar sprays of JA and SA increased the content of osmolytes (e.g., proline, glycine betaine, and soluble sugars) while levels of oxidative markers in date palm leaves under NaCl stress decreased, although individually SA had a greater effect than JA in alleviating the salinity effect (Suhim et al., 2023). Similarly, foliar sprays of JA and SA decreased Na^+ entry into leaf cells and alleviated oxidative and osmotic damage by increasing antioxidant enzyme activities and boosting levels of soluble sugars, proteins, and glycine betaine in salt-stressed soybean plants (Fig. 4.2) (Farhangi-Abriz and Ghassemi-Golezani, 2018). While exogenous JA had minimal impact on K^+ and Ca^+ contents, plant biomass, or seed yield under salinity, SA proved to be superior than JA. SA enhanced chlorophyll content by reducing proline synthesis, and, crucially, increased K^+ and Ca^+ levels, along with chlorophyll content and stability index. By concurrently reducing ionic, oxidative, and osmotic stresses, SA treatment considerably improved both plant biomass and seed yield better than JA in salt-stressed soybean (Farhangi-Abriz and Ghassemi-Golezani, 2018). On the contrary, Ilyas et al. (2017) reported JA (100 μM) to be more effective than SA (10 Mm) in mitigating the drought effects on wheat. Some researchers have highlighted greater beneficial effects of combined applications of SA and JA than their individual applications under abiotic stresses. For instance, combined MeJA+SA pretreatment effectively mitigated drought-induced oxidative stress in maize seedlings by lowering MDA, LOX, and H_2O_2 levels, modulating osmolytes and antioxidant enzyme activities (Tayyab et al., 2020). Similar results have been reported in *Matricaria chamomilla* cultivars (Bodgold and Hungary breed seeds) (Nazarli et al., 2014), *Dracocephalum moldavica* L. (Abbaspour and Rezaei, 2014), French bean (Mohi-Ud-Din et al., 2021), and *Thymus vulgaris* (Nasri et al., 2022) under drought conditions. Few researchers have reported that appropriate concentrations of SA and JA could act as a biodegradable, fast, and cost-effective solution for increasing secondary metabolites (Fig. 4.2) in certain plants that could be used as a raw material for pharmaceutical, cosmetic, and food industries. In this context, a study conducted by Tafreshi et al. (2025) revealed that the application of 3 mM SA and 1.5 mM JA led to the highest (11.68%) seed oil content in rosehip (*Rosa canina* L.) plants under unstressed conditions. Eicosanoic acid and palmitic acid production increased under this treatment, while the amount of linoleic acid decreased as the concentrations of SA and JA increased. Further, spraying with 3 mM SA and 1.5 mM JA increased vitamin C content and total phenol content. Additionally, the treatment of 2 mM SA and 1.5 mM JA resulted in the highest total flavonoid content, antioxidant activity, and total carotenoids (Tafreshi et al., 2025). JA and SA have been reported to induce certain TFs that play a positive regulatory role in inducing salt stress tolerance in plants. For example, Wen et al. (2023) demonstrated the role of MeJA- and SA-induced *ZjWRKY18* TF in triterpenoid accumulation, leading to increased tolerance to salt stress in tobacco (*Nicotiana tabacum*) and *Arabidopsis*. Authors observed that silencing and overexpression of *ZjWRKY18* decreased and increased triterpenoid accumulation, respectively, and that *ZjWRKY18* promoted the activity of *ZjHMGR* and *ZjFPS* by directly binding to their promoters (Wen et al., 2023). Farhangi-Abriz et al. (2019) reported differential effects of foliar sprays of JA and SA in canola (*Brassica napus*, Cultivar: Ocapi) on root growth under varying salt stress levels (0, 40, 70, and 100 mM NaCl). Authors observed that JA promoted lateral root growth but decreased primary root growth and shoot dry weight, without significantly altering root water content under salt stress. In contrast, SA had no effect on lateral root growth but positively enhanced primary root length, weight, diameter, and root density. These improvements induced by SA subsequently led to increased water content in root tissues and overall significant enhancement in canola growth under saline conditions (Farhangi-Abriz et al., 2019).

4.3.7 Salicylic acid and brassinosteroids

Salicylic acid (SA) and Brassinosteroids (BRs) are known to interact and jointly mitigate plant stress (Banerjee and Roychoudhury, 2018). Studies have shown that BAK1, a core receptor in BR signaling, enhances SA production, as was evidenced through higher accumulation of SA and H_2O_2 in BAK1-overexpressing transgenic plants (Kim et al., 2017). The combined application of SA and 24-epibrassinolide (EBL), a synthetic BR, was found to be effective in reducing lead toxicity in mustard by limiting lead uptake, boosting antioxidant enzyme activity, and improving water content (Kohli et al., 2018). Additionally, seed pretreatment with SA and EBL in mustard alleviated the harmful impacts of heavy metals while enhancing photosynthetic performance, phenolic compounds, and pigment concentrations (Fig. 4.2) (Kohli et al., 2017a,b). In wheat, drought conditions were mitigated through foliar applications of SA and BR, demonstrating their ability to improve plant tolerance to water deficit (Emam and Jahani Doghezloo, 2017). Similarly, in chickpea, treatments with SA and EBL—both individually and in combination—significantly enhanced antioxidant enzyme activities (SOD, CAT, POD, APX, and GR) under both drought and nonstress conditions (Sujatha Edupuganti and Anuradha, 2019). The coapplication notably elevated ascorbate (AsA) levels more effectively than individual treatments (Sujatha Edupuganti and Anuradha, 2019). Recently, a study highlighted the role of BR biosynthetic regulator Oryza sativa Small Grain (OsSLG) in promoting Cd tolerance in rice (Zhang et al., 2025). OsSLG activated the SA pathway during Cd exposure, which led to reduced Cd accumulation and ROS levels along with enhanced seedling growth. This study clearly indicated synergistic interaction between BR and SA signaling in enhancing rice defense mechanisms against Cd toxicity.

Brassinosteroid signaling kinases (BSKs), critical components of the BR pathway, are regulated through de-S-acylation influenced by SA. Most BSKs in *Arabidopsis* undergo reversible S-acylation, which is crucial for their membrane localization and function. SA disrupts this localization by reducing their S-acylation, primarily through the induction of the enzyme ABAPT11 (alpha/beta hydrolase domain-containing protein 17-like acyl protein thioesterase 11), which de-S-acylates BSKs. This mechanism reveals how SA integrates with BR signaling to regulate plant development (Liu et al., 2023). SA signaling is primarily mediated by nonexpressor of pathogenesis-related genes (NPRs) and TGACG motif-binding transcription factors (TGAs). These components are potentially influenced by other hormones and environmental factors. However, the crosstalk between SA and other signaling networks remains poorly understood. Recent studies have shown that BR enhances SA responses by suppressing BR-INSENSITIVE 2 (BIN2), a negative regulator that inhibits clade I TGAs—key redox-sensitive TFs in *Arabidopsis*. Both BR and bikinin (a BIN2 inhibitor) synergistically amplified SA-mediated immune responses, including resistance to *Pseudomonas syringae* pv. tomato DC3000 (Pst DC3000). Biochemical and genetic analyses revealed that BIN2 interacts with TGA1 and TGA4, but not other TGAs. BIN2 phosphorylates TGA4 at serine 202, which hinders its redox-dependent interaction with NPR1 and reduces TGA4 stability. Supporting this, transgenic *Arabidopsis* plants overexpressing a TGA4-YFP mutant with a serine-to-alanine substitution at position 202 (S202A) exhibited stronger SA responses than wild-type TGA4-YFP lines (Kim et al., 2022).

4.3.8 Salicylic acid and melatonin

Melatonin (MT), also known as N-acetyl-5-methoxytryptamine, is a widespread indoleamine signaling compound in plants. It plays a vital role in maintaining postharvest fruit quality by regulating diverse metabolic processes (Wei et al., 2018; Yang et al., 2024; Madebo et al., 2022). For instance, MT application in postharvest yellow-flesh peaches activated antioxidant defenses, enhanced the activity of antioxidant enzymes, reduced ROS and MDA levels, and significantly extended shelf life (Wu et al., 2023a).

Recent study revealed that treatments with SA, MT, or their combination effectively lowered browning index, reduced weight loss, and preserved the levels of AsA, soluble sugars, and soluble proteins in grape *(Vitis vinifera* cv.'Cuibao seedless') berries under unstressed conditions, with MT as the most effective treatment, followed by SA and then SA+MT combination (Wang et al., 2025). These treatments also increased the levels of total phenols, flavonoids, proline, and activities of key antioxidant enzymes such as SOD, CAT, and PAL. Meanwhile, they decreased MDA levels and the activity of browning-related enzymes like polyphenol oxidase and POD. These treatments also upregulated genes like *VvPAL*, *VvCHI*, *VvF3H2*, and *VvANR*, promoting phenolic biosynthesis and reducing enzymatic browning (Fig. 4.2) (Wang et al., 2025). In lettuce, coapplication of SA and MT enhanced tolerance to salinity and drought stress by boosting yield and growth parameters, chlorophyll content, carotenoids, ionic balance (K^+/Na^+, Ca^{2+}/Na^+), mineral uptake, and antioxidant activity. Irrigating lettuce at 75% field capacity with both the hormones increased water use efficiency by 3.5% and resulted in only a 1.9% yield reduction compared to the control, making it a practical strategy for water-limited regions (Kiremit et al., 2024). In sword lily (*Gladiolus grandiflorus*), MT and SA alleviated As-induced stress, improved plant tolerance by reducing oxidative damage, with combined application outperforming individual treatments (Zulfiqar et al., 2023). A similar synergistic effect was noted in drought-stressed tomato plants, where MT and SA improved drought resistance by increasing dry weight, chlorophyll content, water status, and the photochemical efficiency of photosystem II (*Fv/Fm*) (Kaya et al., 2023). This improvement was linked to enhanced nitrosative processes and upregulation of the glyoxalase system, helping detoxify MG. In pepper plants, the joint application of SA and MT boosted growth and physiological traits, enhanced N metabolism enzyme activity, and lowered As accumulation, suggesting cooperative action in reducing heavy metal toxicity (Kaya et al., 2022). In pear fruits, exogenous applications of 0.1–0.2 mM MT and 0.2 mM SA delayed weight loss, maintained fruit quality attributes such as firmness, soluble solids content, and titratable acidity, and decreased ethylene emission and cell wall–degrading enzyme activity viz. Pectin Methylesterase (PME), Polygalacturonase (PG) and Lipoxygenase (LOX), while boosting antioxidant enzymes (SOD, POD, CAT, APX) (Cheng et al., 2025). Although combined treatments did not show additive effects, both SA and MT increased endogenous levels of each other and delayed ripening and senescence during storage (Cheng et al., 2025). Further, gene expression analysis revealed enhancement in Gene Ontology (GO) and Kyoto Encyclopedia of Genes and Genomes (KEGG) pathways related to hormone signaling, transcription, and cell wall metabolism. Additionally, MT promoted SA signaling by upregulating *PR1-like* genes and downregulating *NPR-like* genes, and both the hormones jointly suppressed ethylene signaling via modulation of genes like *ETR1*, *EBF1*, *EIN3-like1*, and *CTR1* (Cheng et al., 2025). In wheat, the coapplication of SA and MT restored salt tolerance by preserving photosynthetic efficiency, which in turn enhanced carbon assimilation and growth of wheat plants under saline conditions (Talaat, 2021). SA and MT combination also improved root H^+-pump activity, detoxified ROS, restricted Na^+ absorption, boosted nutrient uptake (N, P, K^+, iron [Fe], zinc [Zn], copper [Cu]), and stabilized cell membranes, enhancing overall plant resilience under salt stress (Talaat and Shawky, 2022). Similar results were reported in heat-stressed *Mentha arvensis* and *Mentha piperita* (Haydari et al., 2019). Enhancing plant productivity and stress resilience depends on the activation of metabolic and defense responses mediated by SA and MT, especially under drought conditions. If restrictions on genetically modified organisms are lifted, producing transgenic plants capable of overexpressing genes responsible for SA and MT biosynthesis could open new avenues for improving crop performance and resilience.

4.3.9 Salicylic acid and polyamines

SA and polyamines (PAs) act together in plants and impact various physiological processes, including growth, development, and responses to several abiotic stresses. SA can affect PA metabolism, and PAs can, in turn, regulate SA synthesis, establishing a complex relationship between these two compounds (Szepesi et al., 2022). On these lines, a study on oats under drought stress highlighted the interaction between SA and PA biosynthesis, revealing that SA modulates PA biosynthesis by altering the expression of related genes (Canales et al., 2019). Notably, SA reduced putrescine levels and increased spermine levels during drought, which corresponded to downregulation of the *ADC* gene and upregulation of *AdoMetDC* (Canales et al., 2019). In another study, Takács et al. (2016)

revealed that genes involved in PA biosynthesis, such as those encoding arginine decarboxylase, ornithine decarboxylase, spermidine synthase, and spermine synthase, were strongly upregulated by 1 mM SA in tomato leaves under illuminated or dark environments. Conversely, enzymes involved in PA catabolism, including diamine oxidases and PA oxidases, exhibited increased transcript levels and enzyme activity in response to 0.1 mM SA in tomato irrespective of light or dark conditions (Takács et al., 2016). The combined application of SA and spermine enhanced seed germination, antioxidant enzymes, proline, chlorophyll, carotenoid contents, and shoot biomass in *Echinacea purpurea* under drought conditions (Darvizheh et al., 2019). Similarly, treatment with SA and putrescine improved fruit quality and elevated the levels of bioactive compounds in drought-stressed *Echinacea purpurea* (Darvizheh et al., 2019). In tomato subjected to salt stress, Fariduddin et al. (2018) demonstrated that treatment with spermidine (1.0 mM) and/or SA (10^{-5} M) enhanced antioxidant enzyme activity and proline accumulation in leaves, which effectively counteracted the negative impacts of salt stress.

4.3.10 Salicylic acid and tricontanol

Tricontanol (TRIA) is an important growth regulator and is considered to influence micropropagation, morphogenesis, and promotion of metabolism in many plant species. Exogenous application of TRIA has been proved to be effective in boosting plant growth under both normal and stressed environmental conditions. Under abiotic stress conditions, such as salinity, drought, and heavy metal toxicity, TRIA mitigates the negative impacts by promoting stress tolerance mechanisms (El-Beltagi et al., 2025). Very few reports are available in literature regarding the combined application of SA and TRIA in plants under abiotic stressful conditions. Presowing seed treatment with SA (1 mM) and TRIA (25 μM) in maize under lead stress upregulated the antioxidant defense system, reduced ROS generation, and increased the production of osmoprotectants (Iqbal et al., 2023). Similarly, application of TRIA at 25 mg L^{-1} provided the highest plant growth, flowering characterizations, and chemical composition, followed by SA at 50 mg L^{-1} in pansy plants under salt stress (El-Kinany, 2020).

4.3.11 Salicylic acid and chitosan

Chitosan is a naturally occurring biopolymer, amino polysaccharide, and biodegradable substance that is extracted from the exoskeletons of marine crustaceans for commercial use (Román-Doval et al., 2023). It has been reported that chitosan and its oligosaccharides enhance crop growth by promoting photosynthesis, regulating primary photochemistry, and preventing pathogen proliferation, offering a practical option for sustainable crop production (Mukhtar Ahmed et al., 2020). Chitosan enhanced the ability of wheatgrass (*Agropyron repens*) cultivar cells to tolerate drought by increasing the activity of cytoprotective enzymes that scavenged O_2 and H_2O_2 and reduced the formation of organic materials (Pongprayoon et al., 2022). In *Aconitum napellus*, combined application of chitosan and SA alleviated Cr stress through enhancement in various morphological, physiological, and biochemical parameters (Ramzan et al., 2024). Lo'Ay and Taher (2018) used chitosan/PVP blending in SA at different concentrations as a coating for guava fruits to alleviate chilling injury during cold storage. Authors reported that it minimized oxidative reactions by scavenging O^{2}– and H_2O_2 generation, which maintained plasma cell membrane compartments by reducing lipid peroxidation (MDA) and protein oxidation (PCG) accumulation. Similarly, the combination of SA and chitosan improved the physiological parameters and cucurbitacin contents in *C. colocynthis* which proved to be effective in protecting the plants against adverse effects of drought stress (Azin et al., 2021). In another study, foliar application of chitosan-SA nanocomposite improved plant growth and increased plant height under salinity conditions in spearmint (*Mentha spicata)* under hydroponic conditions (Hassanpouraghdam et al., 2022). The positive effects of chitosan and SA on elicitation of secondary metabolites and antioxidant activity in safflower under in vitro salinity stress has also been reported (Golkar et al., 2019). Similar results were reported by Seraj et al. (2024) in milk thistle *(Silybum marianum*) under drought stress. On the contrary, Morovvat et al. (2021) observed that foliar application of chitosan and SA did not lead to significant differences in the yield, yield components, and physiological characteristics of potato plants under drought and nondrought conditions. The researchers suggested that the best way to deal with reduced potato yield in dry areas is to irrigate the field or change the planting date to avoid extreme heat and the higher water requirements of the plants. Hence, considering the progressive abiotic stress incidence in many parts of the world, it seems that the combined application of SA and chitosan can be exploited in the agricultural extension sections, and the resulting mixture could be used in plant breeding programs in abiotic-stressed areas to possibly reduce the adverse effects on plants.

4.4 Crosstalk of salicylic acid with signaling molecules

4.4.1 Salicylic acid and nitric oxide

Nitric oxide (NO) is an essential gas transmitter, which acts as a signaling molecule during plant stress (Khatoon et al., 2023; Singhal et al., 2021). NO deserves special attention due to its participation in plant defense mechanisms in response to biotic and abiotic stress factors, in particular, drought, salinity, unfavorable temperatures, and excess of metal ions (Asgher et al., 2017). SA and NO are key signaling molecules required to activate the plant's innate immunity against abiotic stresses and biotrophic attackers (Fig. 4.3) (Rai et al., 2021). Nitric oxide provokes stress defense responses in plants either by acting as an antioxidant or as an elicitor of ROS scavenging system, thus stimulating the activities of various antioxidative enzymes and metabolites (Corpas and Palma, 2020). The exogenous application of SA in combination with NO ameliorated arsenate-induced oxidative stress in maize plants by a simultaneous stimulation of the components of the antioxidant and glyoxalase systems (Kaya et al., 2020). In another study, SA and NO supplementation in the presence of sulfur minimized heat stress-caused oxidative stress in wheat by maintaining a fine-tuning among the antioxidant defense system components and ethylene production, which increased the chlorophyll content and eventually protected PS II activity under heat stress (Rasheed et al., 2022). Similarly, SA+NO addition successfully mitigated the adverse impacts of Ag on *Hypericum perforatum* seedlings (Ghasemifar et al., 2023). Likewise, foliar

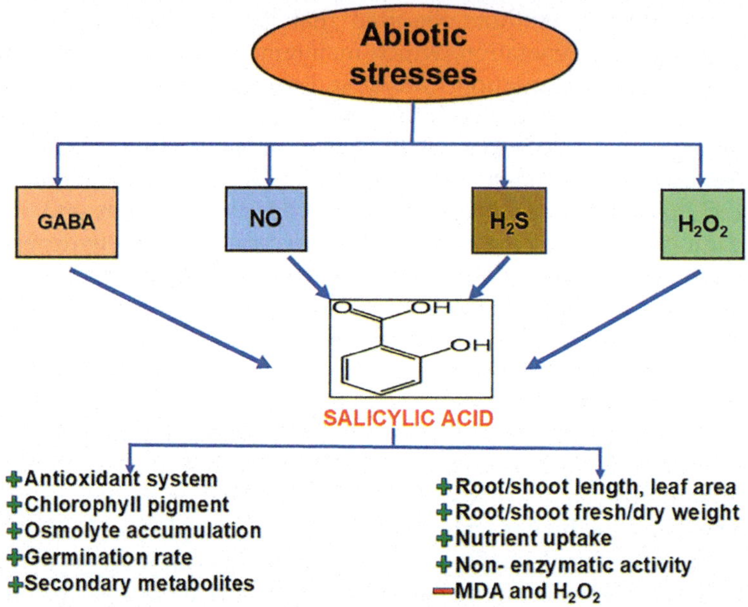

FIGURE 4.3 Interaction of salicylic acid with signaling molecules, i.e., gamma-aminobutyric acid (*GABA*), nitric oxide (*NO*), hydrogen sulfide (H_2S), and hydrogen peroxide (H_2O_2) in plants under abiotic stresses.

application of combined SA and NO was effective in ameliorating high-temperature-induced oxidative stress in *Lablab purpureus* plants by regulating pollen viability and germination, accumulation of proline, scavenging of ROS by stimulating the activities of antioxidant enzymes, and by modulating the expression of stress-responsive genes (Rai et al., 2020). In another study, foliar application of SA and NO alleviated the adverse effects of alkaline stress on *Rosa hybrida* 'Beverly Watson' by increasing growth parameters and essential elements such as K, Ca, Zn, Fe, Mn, and Cu in leaf and root and decreasing Na content in leaf and root (Babasheikhali et al., 2020). Similarly, SA and sodium nitroprusside (SNP) as NO donors, effectively combated salinity stress through improvement in the redox status of quinoa (*Chenopodium quinoa)* plants (Hajihashemi et al., 2022). Similar positive results of SA and SNP were reported in safflower plants under Zn toxicity (Namdjoyan et al., 2017). It was suggested that NO might be involved in SA-induced pectin synthesis, pectin dimethyl esterification, and lignin biosynthesis as a downstream signaling molecule, contributing to reduced Cd accumulation in Cd-stressed rice seedlings (Pan et al., 2021). On the contrary, NO–SA combination increased the Cd concentration in the cell wall of *Arachis hypogaea* to prevent organelles from toxic effects (Xu et al., 2015). NO and SA exhibited differential effects on saffron (*Crocus sativus)* salinity tolerance, with NO showing a more positive impact in terms of increased growth, accumulation of compatible solutes (proline and sugars), enhanced antioxidant enzyme activities (SOD and POX), and increased production of secondary metabolites (phenol, flavonoid, anthocyanin, and carotenoids) (Fig. 4.3) (Babaei et al., 2021). Conversely, SA treatment did not significantly improve plant growth under salinity stress in saffron (Babaei et al., 2021).

4.4.2 Salicylic acid and hydrogen sulfide

Hydrogen sulfide (H_2S) has been considered as the third gaseous signaling molecule after NO and carbon monoxide (Guo et al., 2016). H_2S, a small flammable colorless gas with a characteristic odor of rotten eggs, has long been known as a phytotoxin, which may be connected with the mass extinction of species in the Permian period (Lisjak et al., 2013). In plants, L-cysteine desulfhydrase (L-DES, E.C. 4.4.1.1.) is considered to be a key enzyme in H_2S biosynthesis (Lisjak et al., 2013). SA and H_2S enhanced heat tolerance in maize seedlings by strengthening the activities of antioxidant enzymes and increasing osmolyte content (Fig. 4.3) (Li, 2015). It has been suggested that H_2S might act as a downstream signal molecule of SA in the acquisition of stress tolerance in plants, for example, heat tolerance in *Zea mays* (Li et al., 2015) and Cd tolerance in *Arabidopsis* (Qiao et al., 2015). Thus the positive role of SA under the stress condition is partially dependent on H_2S. Further, SA has been reported to induce endogenous H_2S generation by enhancing the activity of H_2S synthesis enzyme, DES (Li, 2015). In pepper plants under Pb stress, SA improved enzyme activities of the AsA-GSH cycle system (Kaya et al., 2020). In addition, exogenous SA enhanced the H_2S content, which was further reinforced by H_2S donor NaHS. Authors argued that SA triggers endogenous H_2S accumulation, which further regulates the AsA-GSH cycle to resist Pb toxicity (Kaya, 2021). In another study, the combined application of H_2S and SA indicated a synergistic relationship in mitigating the detrimental effects of Cd on nutrient homeostasis and overall health of mustard, primarily by boosting the antioxidant pathway (Kaur et al., 2022). Similarly, under combined application of SA and H_2S, an increase in SOD and CAT activity was witnessed in wheat plantlets under heat stress (Karpets et al., 2020).

4.4.3 Salicylic acid and hydrogen peroxide

H_2O_2 being a versatile molecule acts as an important signaling molecule at normal levels, whereas under abiotic or biotic stress conditions it induces oxidative stress (Saxena et al., 2016). Both SA and H_2O_2 are signaling molecules participating in plant defense responses; however, they play different roles. It is known that SA induces plant resistance and regulates expression of genes related to stress responses. On the other hand, H_2O_2 acts as a signaling molecule in diverse plant processes, including stress responses, and can be either produced by or involved in SA-mediated pathways (Sohag et al., 2020). Under unstressed conditions, seed priming with SA and H_2O_2 alleviated Fe deficiency through the modulation of growth, root acidification capacity, and photosynthetic performance in *Sulla carnosa* (Jelali et al., 2021). Presoaking seeds with SA + H_2O_2 induced the transport of the nutrient solution into plants, allowing the enhancement of Fe and K contents. Authors argued that the stimulation of nutrient uptake in *S. carnosa* plants grown from priming seeds could be due to the enhancement of H^+ ATPase activity. *Sid2* seeds of *Arabidopsis* (defective in SA biosynthesis) under high salinity were hypersensitive to H_2O_2 for germination but the physiological concentrations of SA modulated antioxidant activity to prevent oxidative damage, thereby reducing the inhibitory effects of high salinity on germination (Fig. 4.3) (Lee et al., 2010). It has been reported that H_2O_2 and ABA form a positive feedback loop to mediate SA-induced freezing tolerance in wheat plants (Wang et al., 2018). In another study, pretreatments of seeds with a combination of SA (0.5 mM) and H_2O_2 (10 mM) increased the accumulation of compatible osmolytes, elevated antioxidant response, and improved photosynthetic pigments during water deficit stress (Fig. 4.3) (Singh et al., 2021). Sohag et al. (2020) demonstrated that exogenous application of 5 $mmol{\cdot}L^{-1}$ H_2O_2 or 0.5 and 1 $mmol{\cdot}L^{-1}$ SA was equally effective in improving the water stress tolerance of rice seedlings and attributed it to H_2O_2 or SA-mediated alleviation of drought-induced overaccumulation of ROS, possibly by enhancing the activities of antioxidant enzymes. Several possible mechanisms have been proposed by Sohag et al. (2020) through which SA and H_2O_2 in combination induce and maintain homeostasis and exert synergistic effect on plants, ultimately inducing tolerance to different stresses: (1) changes in ROS levels via increased antioxidant defense machinery of plants; (2) metabolism and energy supply through enhanced α-amylase activity leading to rapid mobilization of starch in seeds, which ultimately supplies metabolites and energy for seed germination and seedling growth under stressful conditions; (3) hormone metabolism and regulation: for instance, upregulation of GA biosynthesis genes *ZmGA20ox1* and *ZmGA3ox2* and downregulation of GA catabolism gene *ZmGA2ox1* and ABA biosynthesis gene *ZmNCED1*. In addition, the expression of *ZmCPK11* and *ZmSnRK2.1* encoding response receptors in the ABA signaling pathway has also been reported to be upregulated by SA and H_2O_2 (Li et al., 2017a). Taken together, these reports indicate that application of H_2O_2 or SA could be recommended to farmers as an effective measure for the better establishment of plants under diverse abiotic environments.

4.4.4 Salicylic acid and gamma-aminobutyric acid

Gamma-aminobutyric acid (GABA), a four-carbon amino acid present in animals, plants, and microorganisms, acts as a signaling molecule to improve plant stress tolerance and defense responses (Hasan et al., 2021; Heli et al., 2022; Kumari et al., 2025a). SA and GABA are both important signaling molecules involved in plant stress responses. On one hand, SA regulates plant growth and development and has a defensive role against diverse environmental cues, while on the other hand, GABA, a nonprotein amino acid, is also involved in mitigating the effects of abiotic stresses like drought, heavy metals, and salinity. In general, both SA and GABA have been reported to act synergistically in augmenting plant tolerance to different stresses. For example, application of SA and GABA mitigated chlorophyll degradation in As-stressed sunflower (*Helianthus annuus)* plants and improved plant health by enhancing photosynthetic pigments (Fig. 4.3), nutrient uptake, reducing As accumulation, regulating osmotic pressure, boosting the accumulation of secondary metabolites, and minimizing oxidative damage, resulting in decreased As translocation and toxicity (Nawaz et al., 2024). In another study, Nawaz et al. (2025) showed that the combination of GABA and SA exhibited the most significant impact on As translocation compared to individual treatments, with the lowest As removal efficiency observed in sunflower. In a study involving creeping bent grass (*Agrostis stolonifera*) subjected to drought stress, the extent of damage was found to be successfully reduced, with the exogenous application of GABA (0.5 mM) and SA (10 μM). Despite having comparable physiological effects on reducing drought stress, metabolic profiling revealed that GABA and SA had effects on shared metabolic pathways and also induced distinct alterations in metabolic accumulation under stress. SA significantly accelerated the buildup of carbohydrates and amino acids while GABA primarily improved the metabolism of organic acids, amino acids, and secondary metabolites for stress defense and energy production (Li et al., 2017b). Thus GABA and SA could serve as valuable stress mitigation tools for crops facing abiotic stresses in the future. Understanding the intricate mechanisms by which SA, along with different signaling molecules, mediates plants responses to abiotic stress can be used to improve crop yields and quality under stress conditions, contributing to food security and sustainable agriculture.

4.5 Conclusions and future prospects

Plants have evolved complex signaling mechanisms to withstand numerous environmental challenges, such as salinity, drought, heavy metals, extreme temperatures, and UV exposure. SA plays a pivotal role in these pathways, impacting plant growth, development, and stress responses. SA exercises its influence over these processes by regulating protein synthesis, gene expression, and enzymatic actions. Outstanding progress has been made in recent years in terms of understanding the role of SA in plant defense. Similarly, phytohormones and/or PGRs along with other signaling molecules also perform significant regulatory roles in the complex processes that plants use for sustaining optimum growth and well-being. SA is involved in complex crosstalk with other phytohormones and signaling molecules, particularly during stressful environmental conditions, which could be manipulated in genetic engineering and to increase crop growth. Nevertheless, how auxin restrains SA signaling and how SA synchronizes with ethylene and JA in plant defense machinery needs to be investigated in the future. Addressing these issues in the background of new evidence on SA and other plant hormones/signaling molecules and incorporating the knowledge into hormonal crosstalk networks is crucial to gain in-depth perspectives on modulating plant growth and defense under challenging conditions. In recent times, with the help of sophisticated technologies, mutants impacting hormone biosynthesis can be generated in several plants, which could be exploited to investigate SA crosstalk with other hormones in diverse plant species. It would not interesting to know how different plants may have distinct ways to control SA signaling and coordinate its influence with other PGRs.

References

Abbaspour, H., Rezaei, H., 2014. Effects of salicylic acid and jasmonic acid on Hill reaction and photosynthetic pigment (*Dracocephalum moldavica* L.) in different levels of drought stress.

Abbaspour, J., Ehsanpour, A., 2016. The impact of salicylic acid on some physiological responses of Artemisia aucheri Boiss. under in vitro drought stress. Acta agriculturae Slovenica 107 (2), 287–298. Available at https://doi.org/10.14720/aas.2016.107.2.03.

Abdelaal, K., 2015. Effect of salicylic acid and abscisic acid on morpho-physiological and anatomical characters of faba bean plants (*Vicia faba* L.) under drought stress. Journal of Plant Production 6 (11), 1771–1788. Available at https://doi.org/10.21608/jpp.2015.52096.

Abdelaal, K.A.A., Attia, K.A., Alamery, S.F., El-Afry, M.M., Ghazy, A.I., Tantawy, D.S., et al., 2020. Exogenous application of proline and salicylic acid can mitigate the injurious impacts of drought stress on barley plants associated with physiological and histological characters. Sustainability (Switzerland) 12 (5). Available at https://doi.org/10.3390/su12051736, https://res.mdpi.com/d_attachment/sustainability/sustainability-12-01736/article_deploy/sustainability-12-01736-v2.pdf.

Agami, R.A., Mohamed, G.F., 2013. Exogenous treatment with indole-3-acetic acid and salicylic acid alleviates cadmium toxicity in wheat seedlings. Ecotoxicology and Environmental Safety 94, 164–171. Available at https://doi.org/10.1016/j.ecoenv.2013.04.013.

Ahanger, M.A., Alyemeni, M.N., Wijaya, L., Alamri, S.A., Alam, P., Ashraf, M., et al., 2018. Potential of exogenously sourced kinetin in protecting *Solanum lycopersicum* from NaCl-induced oxidative stress through up-regulation of the antioxidant system, ascorbate-glutathione cycle and glyoxalase system. PLoS One 13 (9). Available at https://doi.org/10.1371/journal.pone.0202175, https://journals.plos.org/plosone/article/file?id=10.1371/journal.pone.0202175&type=printable.

Ahanger, M.A., Mir, R.A., Alyemeni, M.N., Ahmad, P., 2020. Combined effects of brassinosteroid and kinetin mitigates salinity stress in tomato through the modulation of antioxidant and osmolyte metabolism. Plant Physiology and Biochemistry 147, 31–42. Available at https://doi.org/10.1016/j.plaphy.2019.12.007, http://www.journals.elsevier.com/plant-physiology-and-biochemistry/.

Ahluwalia, O., Singh, P.C., Bhatia, R., 2021. A review on drought stress in plants: implications, mitigation and the role of plant growth promoting rhizobacteria. Resources, Environment and Sustainability 5. Available at https://doi.org/10.1016/j.resenv.2021.100032, https://www.sciencedirect.com/journal/Resources-Environment-and-Sustainability.

Ahmad, A., Aslam, Z., Naz, M., Hussain, S., Javed, T., Aslam, S., et al., 2021. Exogenous salicylic acid-induced drought stress tolerance in wheat (*Triticum aestivum* L.) grown under hydroponic culture. PLoS One 16 (12). Available at https://doi.org/10.1371/journal.pone.0260556, https://journals.plos.org/plosone/article?id=10.1371/journal.pone.0260556.

Ahmad, H.M., Fiaz, S., Hafeez, S., Zahra, S., Shah, A.N., Gul, B., et al., 2022. Plant growth-promoting rhizobacteria eliminate the effect of drought stress in plants: a review. Frontiers in Plant Science 13. Available at https://doi.org/10.3389/fpls.2022.875774, https://www.frontiersin.org/journals/plant-science.

Ahmed, W., Imran, M., Yaseen, M., Haq, T.U., Jamshaid, M.U., Rukh, S., et al., 2020. Role of salicylic acid in regulating ethylene and physiological characteristics for alleviating salinity stress on germination, growth and yield of sweet pepper. PeerJ. 2020 (3). Available at https://doi.org/10.7717/peerj.8475, https://peerj.com/.

Alam, P., Balawi, T.A., Faizan, M., 2023. Salicylic acid's impact on growth, photosynthesis, and antioxidant enzyme activity of triticum aestivum when exposed to salt. Molecules (Basel, Switzerland) 28 (1). Available at https://doi.org/10.3390/molecules28010100, http://www.mdpi.com/journal/molecules.

Alizadeh, M.M., Gerami, M., Majidian, P., Ghorbani, H.R., 2024. The potential application of biochar and salicylic acid to alleviate salt stress in soybean (*Glycine max* L). Heliyon 10 (4). Available at https://doi.org/10.1016/j.heliyon.2024.e26677, https://www.sciencedirect.com/science/journal/24058440.

Altaf, M.M., Diao, X.P., Wang, H., Khan, L.U., Rehman, A.U., Shakoor, A., et al., 2022. Salicylic acid induces vanadium stress tolerance in rice by regulating the AsA-GSH cycle and glyoxalase system. Journal of Soil Science and Plant Nutrition 22 (2), 1983–1999. Available at https://doi.org/10.1007/s42729-022-00788-x, https://www.springer.com/life+sciences/plant+sciences/journal/42729.

Anee, T.I., Nahar, K., Rahman, A., Mahmud, J.A., Bhuiyan, T.F., Alam, M.U., et al., 2019. Oxidative damage and antioxidant defense in sesamum indicum after different waterlogging durations. Plants 8 (7), 196. Available at https://doi.org/10.3390/plants8070196.

Antonić, D., Milošević, S., Cingel, A., Lojić, M., Trifunović-Momčilov, M., Petrić, M., et al., 2016. Effects of exogenous salicylic acid on *Impatiens walleriana* L. grown in vitro under polyethylene glycol-imposed drought. South African Journal of Botany 105, 226–233. Available at https://doi.org/10.1016/j.sajb.2016.04.002.

Arif, Y., Sami, F., Siddiqui, H., Bajguz, A., Hayat, S., 2020. Salicylic acid in relation to other phytohormones in plant: a study towards physiology and signal transduction under challenging environment. Environmental and Experimental Botany 175, 104040. Available at https://doi.org/10.1016/j.envexpbot.2020.104040.

Asgher, M., Per, T.S., Masood, A., Fatma, M., Freschi, L., Corpas, F.J., et al., 2017. Nitric oxide signaling and its crosstalk with other plant growth regulators in plant responses to abiotic stress. Environmental Science and Pollution Research 24 (3), 2273–2285. Available at https://doi.org/10.1007/s11356-016-7947-8, http://www.springerlink.com/content/0944-1344.

Aslam, M., Aslam, A., Sheraz, M., Ali, B., Ulhassan, Z., Najeeb, U., et al., 2021. Lead toxicity in cereals: mechanistic insight into toxicity, mode of action, and management. Frontiers in Plant Science 11. Available at https://doi.org/10.3389/fpls.2020.587785, https://www.frontiersin.org/journals/plant-science.

Asma, Hussain, I., Ashraf, M.Y., Saleem, M.H., Ashraf, M.A., Ali, B., et al., 2023. Alleviating effects of salicylic acid spray on stage-based growth and antioxidative defense system in two drought-stressed rice (*Oryza sativa* L.) cultivars. Turkish Journal of Agriculture and Forestry 47 (1), 79–99. Available at https://doi.org/10.55730/1300-011X.3066, https://journals.tubitak.gov.tr/cgi/viewcontent.cgi?article=3066&context=agriculture.

Atta, K., Mondal, S., Gorai, S., Singh, A.P., Kumari, A., Ghosh, T., et al., 2023. Impacts of salinity stress on crop plants: improving salt tolerance through genetic and molecular dissection. Frontiers in Plant Science 14. Available at https://doi.org/10.3389/fpls.2023.1241736, https://www.frontiersin.org/journals/plant-science.

Azin, Z., Emamjomeh, A., Bahabadi, S.E., Hasanein, P., 2021. Effect of exogenous chitosan, salicylic acid, and their combination on some physiological parameters of *Citrullus colocynthis* (L.) under drought stress. Iranian Journal of Plant Physiology 11 (5), 3885–3897. Available at https://doi.org/10.30495/ijpp.2021.687854, http://ijpp.iau-saveh.ac.ir/.

Babaei, S., Niknam, V., Behmanesh, M., 2021. Comparative effects of nitric oxide and salicylic acid on salinity tolerance in saffron (*Crocus sativus*). Plant Biosystems - An International Journal Dealing with all Aspects of Plant Biology 155 (1), 73–82. Available at https://doi.org/10.1080/11263504.2020.1727975.

Babasheikhali, M.M., Jabbarzadeh, Z., Amiri, J., Barin, M., 2020. Impact of salicylic acid and nitric oxide on improving growth and nutrients uptake of rose in alkaline soil conditions. Journal of Plant Nutrition 43 (5), 667–681.

Bagautdinova, Z.Z., Omelyanchuk, N., Tyapkin, A.V., Kovrizhnykh, V.V., Lavrekha, V.V., Zemlyanskaya, E.V., 2022. Salicylic acid in root growth and development. International Journal of Molecular Sciences 23 (4). Available at https://doi.org/10.3390/ijms23042228, https://www.mdpi.com/1422-0067/23/4/2228/pdf.

Bahrabadi, E., Tavakkol Afshari, R., Mahallati, M.N., Seyyedi, S.M., 2022. Abscisic, gibberellic, and salicylic acids effects on germination indices of corn under salinity and drought stresses. TJournal of Crop Improvement 36 (1), 73–89. Available at https://doi.org/10.1080/15427528.2021.1908474, http://www.tandfonline.com/toc/wcim20/current.

Bahrani, A., Pourreza, J., 2012. Gibberlic acid and salicylic acid effects on seed germination and seedlings growth of wheat (*Triticum aestivum* L.) under salt stress condition. World Applied Sciences Journal 18 (5), 633–641. Available at https://doi.org/10.5829/idosi.wasj.2012.18.05.1372Iran, http://idosi.org/wasj/wasj18(5)12/6.pdf.

Banerjee, A., Roychoudhury, A., 2018. Interactions of brassinosteroids with major phytohormones: antagonistic effects. Journal of Plant Growth Regulation 37 (4), 1025–1032. Available at https://doi.org/10.1007/s00344-018-9828-5.

Bano, K., Kumar, B., Alyemeni, M.N., Ahmad, P., 2022. Exogenously-sourced salicylic acid imparts resilience towards arsenic stress by modulating photosynthesis, antioxidant potential and arsenic sequestration in brassica napus plants. Antioxidants 11 (10). Available at https://doi.org/10.3390/antiox11102010, www.mdpi.com/journal/antioxidants.

Barwal, S.K., Shah, S.H., Pawar, A., Siddiqui, M.H., Agnihotri, R.K., Vimala, Y., et al., 2024. Mechanistic insights of salicylic acid-mediated salt stress tolerance in *Zea mays* L. seedlings. Heliyon 10 (14). Available at https://doi.org/10.1016/j.heliyon.2024.e34486, https://www.sciencedirect.com/science/journal/24058440.

Bhusal, N., Kim, H.S., Han, S.G., Yoon, T.M., 2020. Photosynthetic traits and plant–water relations of two apple cultivars grown as bi-leader trees under long-term waterlogging conditions. Environmental and Experimental Botany 176. Available at https://doi.org/10.1016/j.envexpbot.2020.104111, www.elsevier.com/inca/publications/store/2/6/7.

Bidabadi, S.S., Mahmood, M., Baninasab, B., Ghobadi, C., 2012. Influence of salicylic acid on morphological and physiological responses of banana (*Musa acuminata* cv. 'Berangan', AAA) shoot tips to in vitro water stress induced by polyethylene glycol. Plant OMICS 5 (1), 33–39. Available at http://www.pomics.com/bidabadi_5_1_2012_33_39.pdf.

Bijalwan, P., Sharma, M., Kaushik, P., 2022. Review of the effects of drought stress on plants: a systematic approach. Preprints, 2022020014.

Bryla, D.R., Scagel, C.F., Lukas, S.B., Sullivan, D.M., 2021. Ion-specific limitations of sodium chloride and calcium chloride on growth, nutrient uptake, and mycorrhizal colonization in northern and southern highbush blueberry. Journal of the American Society for Horticultural Science 146 (6), 399–410. Available at https://doi.org/10.21273/JASHS05084-21, https://journals.ashs.org/downloadpdf/journals/jashs/146/6/article-p399.pdf.

Canales, F.J., Montilla-Bascón, G., Rispail, N., Prats, E., 2019. Salicylic acid regulates polyamine biosynthesis during drought responses in oat. Plant Signaling & Behavior 14 (10), e1651183. Available at https://doi.org/10.1080/15592324.2019.1651183.

Chaudhry, S., Sidhu, G.P.S., 2022. Climate change regulated abiotic stress mechanisms in plants: a comprehensive review. Plant Cell Reports 41 (1). Available at https://doi.org/10.1007/s00299-021-02759-5, link.springer.de/link/service/journals/00299/index.htm.

Chávez-Arias, C.C., Ramírez-Godoy, A., Restrepo-Díaz, H., 2022. Influence of drought, high temperatures, and/or defense against arthropod herbivory on the production of secondary metabolites in maize plants. a review. Current Plant Biology 32, 100268. Available at https://doi.org/10.1016/j.cpb.2022.100268.

Chen, G., Zheng, D., Feng, N., Zhou, H., Mu, D., Liu, L., et al., 2022. Effects of exogenous salicylic acid and abscisic acid on growth, photosynthesis and antioxidant system of rice. Chilean Journal of Agricultural Research 82 (1), 21–32. Available at https://doi.org/10.4067/s0718-58392022000100021.

Cheng, A., Zhao, P., Wang, X., Luo, S., Xu, P., Ye, Y., et al., 2025. Relationships between melatonin and salicylic acid treatments in delaying the senescence of postharvest pear fruit. Postharvest Biology and Technology 219, 113288. Available at https://doi.org/10.1016/j.postharvbio.2024.113288.

Choudhary, K.K., Singh, S., Agrawal, M., Agrawal, S.B., 2021. Role of jasmonic and salicylic acid signaling in plants under UV-B stress. Jasmonates and salicylates signaling in plants. Springer International Publishing, Cham, pp. 45–63.

Choudhary, R., Rajput, V.D., Ghodake, G., Ahmad, F., Meena, M., Rehman, R.U., et al., 2024a. Comprehensive journey from past to present to future about seed priming with hydrogen peroxide and hydrogen sulfide concerning drought, temperature, UV and ozone stresses- a review. Plant and Soil 500 (1-2), 351–373. Available at https://doi.org/10.1007/s11104-024-06499-9, https://www.springer.com/journal/11104.

Choudhary, S., Bhat, T.M., Alwutayd, K.M., Abd El-Moneim, D., Naaz, N., 2024b. Salicylic acid enhances thermotolerance and antioxidant defense in *Trigonella foenum* graecum L. under heat stress. Heliyon 10 (6).

Cingoz, G.S., Gurel, E., 2016. Effects of salicylic acid on thermotolerance and cardenolide accumulation under high temperature stress in *Digitalis trojana* Ivanina. Plant Physiology and Biochemistry 105, 145–149. Available at https://doi.org/10.1016/j.plaphy.2016.04.023, http://www.journals.elsevier.com/plant-physiology-and-biochemistry/.

Ciobanu, G., Ionescu, C., 2021. Comparative study on the effects of wheat plants exposure to gibberellic acid and salicylic acid. Annals of the University of Craiova, Series Chemistry 27 (2), 60–74. Available at https://doi.org/10.52846/auc.chem.2021.2.07.

Corpas, F., Palma, J., 2020. Assessing nitric oxide (no) in higher plants: an outline. Nitrogen 1 (1), 12–20. Available at https://doi.org/10.3390/nitrogen1010003.

Correa, M.D.S.S., Saavedra, M.E.R.R., Parra, E.A.E., Ontiveros, E.N., Flores, J.D.C.B., Montiel, J.G.O., et al., 2023. Ultraviolet radiation and its effects on plants. Abiotic stress in plants-adaptations to climate change. IntechOpen.

Culpan, E., Gürsoy, M., 2025. Foliar and seed application of salicylic acid and kinetin impacts of the seed yield, oil content and fatty acid composition of safflower (*Carthamus tinctorius* L). Journal of the American Oil Chemists' Society 102 (7), 1061–1071. Available at https://doi.org/10.1002/aocs.12949.

Darvizheh, H., Zahedi, M., Abbaszadeh, B., Razmjoo, J., 2019. Changes in some antioxidant enzymes and physiological indices of purple coneflower (*Echinacea purpurea* L.) in response to water deficit and foliar application of salicylic acid and spermine under field condition. Scientia Horticulturae 247, 390–399. Available at https://doi.org/10.1016/j.scienta.2018.12.037.

del Pilar Cordovilla, M., Aparicio, C., Melendo, M., Bueno, M., 2023. Exogenous application of indol-3-acetic acid and salicylic acid improves tolerance to salt stress in olive plantlets (*Olea europaea* L. Cultivar Picual) in growth chamber environments. Agronomy 13 (3), 647. Available at https://doi.org/10.3390/agronomy13030647.

Ding, P., Ding, Y., 2020. Stories of salicylic acid: a plant defense hormone. Trends in Plant Science 25 (6), 549–565. Available at https://doi.org/10.1016/j.tplants.2020.01.004.

Du, B., Haensch, R., Alfarraj, S., Rennenberg, H., 2024. Strategies of plants to overcome abiotic and biotic stresses. Biological Reviews 99 (4), 1524–1536. Available at https://doi.org/10.1111/brv.13079.

El-Beltagi, H.S., Gad, M., Abdel-Haleem, M., Shalaby, T.A., Rezk, A.A., Hamdy Khedr, E., 2025. Triacontanol: a multifunctional growth regulator for enhancing stress tolerance. Journal of Crop Health 77 (2). Available at https://doi.org/10.1007/s10343-025-01144-6, https://link.springer.com/journal/10343.

El-Kinany, R., 2020. The beneficial role of salicylic acid, triacontanol and δ-aminolevulinic acid on the growth, flowering and chemical composition of pansy (*Viola wittrockiana* Gams) under salt stress conditions. Hortscience Journal of Suez Canal University 9 (1), 13–30. Available at https://doi.org/10.21608/hjsc.2020.119462.

El-Taher, A.M., Abd El-Raouf, H.S., Osman, N.A., Azoz, S.N., Omar, M.A., Elkelish, A., et al., 2022. Effect of salt stress and foliar application of salicylic acid on morphological, biochemical, anatomical, and productivity characteristics of cowpea (*Vigna unguiculata* L.) plants. Plants 11 (1). Available at https://doi.org/10.3390/plants11010115, https://www.mdpi.com/2223-7747/11/1/115/pdf.

Emam, Y., Jahani Doghezloo, M., 2017. Effect of salicylic acid and brassinosteroid on alleviation of water stress in two wheat cultivars. Journal of Crop Production and Processing 7 (1), 127–139. Available at https://doi.org/10.18869/acadpub.jcpp.7.1.127.

Emamverdian, A., Ding, Y., Mokhberdoran, F., 2020. The role of salicylic acid and gibberellin signaling in plant responses to abiotic stress with an emphasis on heavy metals. Plant Signaling & Behavior 15 (7), 1777372. Available at https://doi.org/10.1080/15592324.2020.1777372.

Esan, A.M., Masisi, K., Dada, F.A., Olaiya, C.O., 2017. Comparative effects of indole acetic acid and salicylic acid on oxidative stress marker and antioxidant potential of okra (*Abelmoschus esculentus*) fruit under salinity stress. Scientia Horticulturae 216, 278–283. Available at https://doi.org/10.1016/j.scienta.2017.01.007, www.elsevier.com/inca/publications/store/5/0/3/3/1/6.

Esim, N., Atici, Ö., 2015. Effects of exogenous nitric oxide and salicylic acid on chilling-induced oxidative stress in wheat (*Triticum aestivum*). Frontiers in Life Science 8 (2), 124–130. Available at https://doi.org/10.1080/21553769.2014.998299.

FAO, 2024. Available from: https://www.fao.org/soils-portal/data-hub/soil-maps-and-databases/global-map-of-salt-affected-soils/en/ [accessed 20.07.25].

Farhangi-Abriz, S., Ghassemi-Golezani, K., 2018. How can salicylic acid and jasmonic acid mitigate salt toxicity in soybean plants? Ecotoxicology and Environmental Safety 147, 1010–1016. Available at https://doi.org/10.1016/j.ecoenv.2017.09.070.

Farhangi-Abriz, S., Alaee, T., Tavasolee, A., 2019. Salicylic acid but not jasmonic acid improved canola root response to salinity stress. Rhizosphere 9, 69–71. Available at https://doi.org/10.1016/j.rhisph.2018.11.009.

Fariduddin, Q., Khan, T.A., Yusuf, M., Aafaqee, S.T., Khalil, R.R.A.E., 2018. Ameliorative role of salicylic acid and spermidine in the presence of excess salt in lycopersicon esculentum. Photosynthetica 56 (3), 750–762. Available at https://doi.org/10.1007/s11099-017-0727-y.

Feng, B., Zhang, C., Chen, T., Zhang, X., Tao, L., Fu, G., 2018. Salicylic acid reverses pollen abortion of rice caused by heat stress. BMC Plant Biology 18 (1). Available at https://doi.org/10.1186/s12870-018-1472-5.

Garcia, N., da-Silva, C.J., Cocco, K.L.T., Pomagualli, D., de Oliveira, F.K., da Silva, J.V.L., et al., 2020. Waterlogging tolerance of five soybean genotypes through different physiological and biochemical mechanisms. Environmental and Experimental Botany 172. Available at https://doi.org/10.1016/j.envexpbot.2020.103975, www.elsevier.com/inca/publications/store/2/6/7.

Ghaderi, A.A., 2015. Effects of salicylic acid and jasmonic acid on morphological and physiological traits of medicinal plant *Thymus vulgaris* under drought stress (*Thymus vulgaris* L.) (Doctoral dissertation, University of Zabol).

Ghani, M.A., Abbas, M.M., Ali, B., Ziaf, K., Azam, M., Anjum, R., et al., 2021. Role of salicylic acid in heat stress tolerance in tri-genomic brassica napus L. Bioagro 33 (1), 13–20. Available at https://doi.org/10.51372/BIOAGRO331.2, https://revistas.uclave.org/index.php/bioagro/article/view/3018/1877.

Ghasemifar, E., Habibi, G., Bakhshi-Khaniki, G., 2023. Pretreatment with salicylic acid and nitric oxide alleviates silver nanoparticles toxicity in medicinal *Hypericum perforatum* plants. Journal of Plant Process and Function 11 (52), 83–93.

Ghosh, P., Roychoudhury, A., 2024. Molecular basis of salicylic acid–phytohormone crosstalk in regulating stress tolerance in plants. Brazilian Journal of Botany 47 (3), 735–750. Available at https://doi.org/10.1007/s40415-024-00983-3.

Gillani, S.R., Murtaza, G., Mehmood, A., 2021. Mitigation of lead stress in *Triticum aestivum* L. seedlings by treating with salicylic acid. Pakistan Journal of Botany 53 (1), 39–44. Available at https://doi.org/10.30848/PJB2021-1(44), http://www.pakbs.org/pjbot/papers/1610610993.pdf.

Goldani, M., Dolatkhahi, A., Parsa, M., Vahdati, N., Rasouli, Z., 2021. Investigation of improving the drought tolerance in *Persian petunia* (*Petunia* sp.) by exogenous application of salicylic acid and gibberellic acid. Acta Scientiarum Polonorum Hortorum Cultus 20 (1), 37–48. Available at https://doi.org/10.24326/asphc.2021.1.4.

Golkar, P., Taghizadeh, M., Yousefian, Z., 2019. The effects of chitosan and salicylic acid on elicitation of secondary metabolites and antioxidant activity of safflower under in vitro salinity stress. Plant Cell, Tissue and Organ Culture (PCTOC 137 (3), 575–585. Available at https://doi.org/10.1007/s11240-019-01592-9.

Gondor, O.K., Pál, M., Janda, T., Szalai, G., 2022. The role of methyl salicylate in plant growth under stress conditions. Journal of Plant Physiology 277, 153809. Available at https://doi.org/10.1016/j.jplph.2022.153809.

González-Villagra, J., Reyes-Díaz, M.M., Tighe-Neira, R., Inostroza-Blancheteau, C., Escobar, A.L., Bravo, L.A., 2022. Salicylic acid improves antioxidant defense system and photosynthetic performance in *Aristotelia chilensis* plants subjected to moderate drought stress. MDPI, Chile Plants 11 (5). Available at https://doi.org/10.3390/plants11050639, https://www.mdpi.com/2223-7747/11/5/639/pdf.

Guo, H., Xiao, T., Zhou, H., Xie, Y., Shen, W., 2016. Hydrogen sulfide: a versatile regulator of environmental stress in plants. Acta Physiologiae Plantarum 38 (1), 1–13. Available at https://doi.org/10.1007/s11738-015-2038-x, http://www.springerlink.com/content/1861-1664.

Guo, X., Liu, D., Chong, K., 2018. Cold signaling in plants: insights into mechanisms and regulation. Journal of Integrative Plant Biology 60 (9), 745–756. Available at https://doi.org/10.1111/jipb.12706.

Guo, X., Luo, J., Du, Y., Li, J., Liu, Y., Liang, Y., et al., 2021. Coordination between root cell wall thickening and pectin modification is involved in cadmium accumulation in *Sedum alfredii*. Environmental Pollution 268 (Pt A), 115665. Available at https://doi.org/10.1016/j.envpol.2020.115665.

Guo, J., Wang, Z., Li, J., Qu, L., Chen, Y., Li, G., et al., 2024. Salicylic acid promotes endosperm development and heat-tolerance of waxy maize (*Zea mays* L. var. ceratina Kulesh) under heat stress. Plant Stress 14, 100684. Available at https://doi.org/10.1016/j.stress.2024.100684.

Haider Abbasi, B., Ullah, M.A., Nadeem, M., Tungmunnithum, D., Hano, C., 2020. Exogenous application of salicylic acid and gibberellic acid on biomass accumulation, antioxidant and anti-inflammatory secondary metabolites production in multiple shoot culture of *Ajuga integrifolia* Buch. Ham. ex D. Don. Industrial Crops and Products 145, 112098. Available at https://doi.org/10.1016/j.indcrop.2020.112098.

Hajihashemi, S., Jahantigh, O., Alboghobeish, S., 2022. The redox status of salinity-stressed *Chenopodium quinoa* under salicylic acid and sodium nitroprusside treatments. Frontiers in Plant Science 13. Available at https://doi.org/10.3389/fpls.2022.1030938.

Hanif, S., Mahmood, A., Javed, T., Bibi, S., Zia, M.A., Asghar, S., et al., 2024. Exogenous application of salicylic acid ameliorates salinity stress in barley (*Hordeum vulgare* L). BMC Plant Biology 24 (1). Available at https://doi.org/10.1186/s12870-024-04968-y, https://bmcplantbiol.biomedcentral.com/.

Hasan, M.M., Alabdallah, N.M., Alharbi, B.M., Waseem, M., Yao, G., Liu, X.D., et al., 2021. Gaba: a key player in drought stress resistance in plants. International Journal of Molecular Sciences 22 (18). Available at https://doi.org/10.3390/ijms221810136, https://www.mdpi.com/1422-0067/22/18/10136/pdf.

Hasanuzzaman, M., Ahmed, N., Saha, T., Rahman, M., Rahman, K., Alam, M.M., et al., 2022. Exogenous salicylic acid and kinetin modulate reactive oxygen species metabolism and glyoxalase system to confer waterlogging stress tolerance in soybean (*Glycine max* L). Plant Stress 3. Available at https://doi.org/10.1016/j.stress.2022.100057, www.journals.elsevier.com/plant-stress.

Hassanpouraghdam, M.B., Mohammadi, L., Gohari, G., Vojodi Mehrabani, L., 2022. The effects of foliar application of chitosan-salicylic acid nanocomposite on *Mentha spicata* L. under salinity stress in hydroponic conditions. Journal of Vegetables Sciences 5 (10), 35–51.

Hassoon, A.S., Abduljabbar, I.A., 2019. Review on the role of salicylic acid in plants. In: Hasanuzzaman, I., Teixeira Filho, M.C.M., Fujita, M., Nogueira, T.A.R. (Eds.), Sustainable crop production. Intech Open Book Series, London, pp. 352.

Haydari, M., Maresca, V., Rigano, D., Taleei, A., Shahnejat-Bushehri, A.A., Hadian, J., et al., 2019. Salicylic acid and melatonin alleviate the effects of heat stress on essential oil composition and antioxidant enzyme activity in *Mentha* × *Piperita* and *Mentha arvensis* L. Antioxidants 8 (11), 547. Available at https://doi.org/10.3390/antiox8110547.

He, Q., Zhao, S., Ma, Q., Zhang, Y., Huang, L., Li, G., et al., 2014. Endogenous salicylic acid levels and signaling positively regulate *Arabidopsis* response to polyethylene glycol-simulated drought stress. Journal of Plant Growth Regulation 33 (4), 871–880. Available at https://doi.org/10.1007/s00344-014-9438-9.

Heli, Z., Hongyu, C., Dapeng, B., Yee Shin, T., Yejun, Z., Xi, Z., et al., 2022. Recent advances of γ-aminobutyric acid: physiological and immunity function, enrichment, and metabolic pathway. Frontiers in Nutrition 9. Available at https://doi.org/10.3389/fnut.2022.1076223.

Henschel, J.M.I., Dantas, E.F.O., Soares, V.D.A., Santos, S.K.D., Santos, L.W.O.D., Dias, T.J., et al., 2022. Salicylic acid mitigates the effects of mild drought stress on radish (*Raphanus sativus*) growth. Functional Plant Biology 49 (9), 822–831. Available at https://doi.org/10.1071/FP22040, http://www.publish.csiro.au/nid/103.htm.

Hossini, F., Meskarbashee, M., Rahnama, A., 2024. The effect of foliar application of cytokinin and salicylic acid on yield and physiological characteristics of sunflower under drought stress conditions. Journal of Plant Production Research.

Hu, Y.R., Wang, Y., Chen, Y.J., Chai, Q.Q., Dong, H.Z., Shen, J.W., et al., 2022. Salicylic acid enhances heat stress resistance of *Pleurotus ostreatus* (Jacq.) P. Kumm through metabolic rearrangement. Antioxidants 11 (5). Available at https://doi.org/10.3390/antiox11050968, https://www.mdpi.com/2076-3921/11/5/968/pdf?version=1652450908.

Huang, C., Wang, D., Sun, L., Wei, L., 2016. Effects of exogenous salicylic acid on the physiological characteristics of *Dendrobium officinale* under chilling stress. Plant Growth Regulation 79 (2), 199–208. Available at https://doi.org/10.1007/s10725-015-0125-z.

Huang, C., Liao, J., Huang, W., Qin, N., 2022. Salicylic acid protects sweet potato seedlings from drought stress by mediating abscisic acid-related gene expression and enhancing the antioxidant defense system. International Journal of Molecular Sciences 23 (23), 14819. Available at https://doi.org/10.3390/ijms232314819.

Hudeček, M., Nožková, V., Plíhalová, L., Plíhal, O., 2023. Plant hormone cytokinin at the crossroads of stress priming and control of photosynthesis. Frontiers in Plant Science 13. Available at https://doi.org/10.3389/fpls.2022.1103088.

Hundare, A., Joshi, V., Joshi, N., 2022. Salicylic acid attenuates salinity-induced growth inhibition in in vitro raised ginger (*Zingiber officinale* Roscoe) plantlets by regulating ionic balance and antioxidative system. Plant Stress 4, 100070. Available at https://doi.org/10.1016/j.stress.2022.100070.

Husen, A., Iqbal, M., Sohrab, S.S., Ansari, M.K.A., 2018. Salicylic acid alleviates salinity-caused damage to foliar functions, plant growth and antioxidant system in Ethiopian mustard (*Brassica carinata* A. Br). Agriculture & Food Security 7 (1). Available at https://doi.org/10.1186/s40066-018-0194-0.

Ilyas, N., Gull, R., Mazhar, R., Saeed, M., Kanwal, S., Shabir, S., et al., 2017. Influence of salicylic acid and jasmonic acid on wheat under drought stress. Communications in Soil Science and Plant Analysis 48 (22), 2715–2723. Available at https://doi.org/10.1080/00103624.2017.1418370, http://www.tandfonline.com/loi/lcss20.

Ilyas, M., Maqsood, M.F., Shahbaz, M., Zulfiqar, U., Ahmad, K., Naz, N., et al., 2024. Alleviating salinity stress in canola (*Brassica napus* L.) through exogenous application of salicylic acid. BMC Plant Biology 24 (1). Available at https://doi.org/10.1186/s12870-024-05314-y, https://bmcplantbiol.biomedcentral.com/.

Iqbal, M.S., Zahoor, M., Akbar, M., Ahmad, K.S., Hussain, S.A., Munir, S., et al., 2022. Alleviating the deleterious effects of salt stress on wheat (*Triticum aestivum* L.) by foliar application of gibberellic acid and salicylic acid. Applied Ecology and Environmental Research 20 (1), 119–134. Available at https://doi.org/10.15666/aeer/2001_119134.

Iqbal, J., Perveen, S., Parveen, A., Saeed, M., Zafar, S., Iqbal, N., 2023. Interactive effect of salicylic acid and triacontanol to improve lead stress tolerance in maize (*Zea mays* L). Arabian Journal of Geosciences 16 (5). Available at https://doi.org/10.1007/s12517-023-11317-0.

Isayenkov, S.V., Maathuis, F.J., 2019. Plant salinity stress: many unanswered questions remain. Frontiers in Plant Science 10, 80.

Islam, K.N., Khan, M.M.H., Islam, M.M., Uddin, M.M., Latif, M.A., 2020. Performance of different cultivars of mungbean in coastal region of Bangladesh. SAARC Journal of Agriculture 18 (1), 161–172. Available at https://doi.org/10.3329/sja.v18i1.48390.

Islam, S.M.N., Paul, N., Rahman, M.M., Haque, M.A., Rohman, M.M., Mostofa, M.G., 2023. Salicylic acid application mitigates oxidative damage and improves the growth performance of barley under drought stress. Phyton-International Journal of Experimental Botany 92 (5), 1513–1537. Available at https://doi.org/10.32604/phyton.2023.025175, https://www.techscience.com/phyton/v92n5/51851/pdf.

Islam, S., Mohammad, F., Shakeel, A., Corpas, F.J., 2024. Glycine betaine: a multifaceted protectant against salt stress in Indian mustard through ionic homeostasis, ROS scavenging and osmotic regulation. Physiologia Plantarum 176 (5). Available at https://doi.org/10.1111/ppl.14530, http://onlinelibrary.wiley.com/journal/10.1111/(ISSN)1399-3054.

Jacobsen, A.L., Agenbag, L., Esler, K.J., Pratt, R.B., Ewers, F.W., Davis, S.D., 2007. Xylem density, biomechanics and anatomical traits correlate with water stress in 17 evergreen shrub species of the mediterranean-type climate region of South Africa. Journal of Ecology 95 (1), 171–183. Available at https://doi.org/10.1111/j.1365-2745.2006.01186.x.

Janda, T., Szalai, G., Pál, M., 2020. Salicylic acid signalling in plants. International Journal of Molecular Sciences 21 (7), 2655. Available at https://doi.org/10.3390/ijms21072655.

Jelali, N., Ben Youssef, R., Boukari, N., Zorrig, W., Dhifi, W., Abdelly, C., 2021. Salicylic acid and H_2O_2 seed priming alleviates Fe deficiency through the modulation of growth, root acidification capacity and photosynthetic performance in Sulla carnosa. Plant Physiology and Biochemistry 159, 392–399. Available at https://doi.org/10.1016/j.plaphy.2020.11.039, http://www.journals.elsevier.com/plant-physiology-and-biochemistry/.

Jia, H.L., Wang, X.H., Wei, T., Wang, M., Liu, X., Hua, L., et al., 2021. Exogenous salicylic acid regulates cell wall polysaccharides synthesis and pectin methylation to reduce Cd accumulation of tomato. Ecotoxicology and Environmental Safety 207, 111550. Available at https://doi.org/10.1016/j.ecoenv.2020.111550.

Jini, D., Joseph, B., 2017. Physiological mechanism of salicylic acid for alleviation of salt stress in rice. Rice Science 24 (2), 97–108. Available at https://doi.org/10.1016/j.rsci.2016.07.007.

Kabiri, R., Nasibi, F., Farahbakhsh, H., 2014. Effect of exogenous salicylic acid on some physiological parameters and alleviation of drought stress in *Nigella sativa* plant under hydroponic culture. Plant Protection Science 50 (1), 43–51. Available at https://doi.org/10.17221/56/2012-pps.

Kapoor, R.T., Hasanuzzaman, M., 2020. Exogenous kinetin and putrescine synergistically mitigate salt stress in *Luffa acutangula* by modulating physiology and antioxidant defense. Physiology and Molecular Biology of Plants 26 (11), 2125–2137. Available at https://doi.org/10.1007/s12298-020-00894-z, http://www.springerlink.com/content/120962/.

Kardan, J., Amirnia, R., Ghiyasi, M., Sharifi, L., 2024. Effects of foliar application of salicylic acid and kinetin on soybean physiological responses at different levels of salt stress. Journal of Agricultural Science and Sustainable Production 34 (2), 335–353.

Karpets, Y.V., Shkliarevskyi, M.A., Horielova, E.I., Kolupaev, Y.E., 2020. Participation of hydrogen sulfide in induction of antioxidant system in roots of wheat plantlets and their heat resistance by salicylic acid. Applied Biochemistry and Microbiology 56 (4), 467–472. Available at https://doi.org/10.1134/S0003683820040079, http://www.kluweronline.com/issn/0003-6838.

Kaur, G., Singh, G., Motavalli, P.P., Nelson, K.A., Orlowski, J.M., Golden, B.R., 2020. Impacts and management strategies for crop production in waterlogged or flooded soils: a review. Agronomy Journal 112 (3), 1475–1501. Available at https://doi.org/10.1002/agj2.20093, https://acsess.onlinelibrary.wiley.com/journal/14350645.

Kaur, H., Hussain, S.J., Al-Huqail, A.A., Siddiqui, M.H., Al-Huqail, A.A., Khan, M.I.R., et al., 2022. Hydrogen sulphide and salicylic acid regulate antioxidant pathway and nutrient balance in mustard plants under cadmium stress. Plant Biology 24 (4), 660–669. Available at https://doi.org/10.1111/plb.13322.

Kaya, C., Akram, N.A., Ashraf, M., 2018. Kinetin and indole acetic acid promote antioxidant defense system and reduce oxidative stress in maize (*Zea mays* L.) plants grown at boron toxicity. SJournal of Plant Growth Regulation 37 (4), 1258–1266. Available at https://doi.org/10.1007/s00344-018-9827-6, http://www.springer.com/life+sci/plant+sciences/journal/344.

Kaya, C., Ashraf, M., Alyemeni, M.N., Corpas, F.J., Ahmad, P., 2020. Salicylic acid-induced nitric oxide enhances arsenic toxicity tolerance in maize plants by upregulating the ascorbate-glutathione cycle and glyoxalase system. Journal of Hazardous Materials 399. Available at https://doi.org/10.1016/j.jhazmat.2020.123020, www.elsevier.com/locate/jhazmat.

Kaya, C., 2021. Salicylic acid-induced hydrogen sulphide improves lead stress tolerance in pepper plants by upraising the ascorbate-glutathione cycle. Physiologia Plantarum 173 (1), 8–19. Available at https://doi.org/10.1111/ppl.13159, http://onlinelibrary.wiley.com/journal/10.1111/(ISSN)1399-3054.

Kaya, C., Sarıoglu, A., Ashraf, M., Alyemeni, M.N., Ahmad, P., 2022. The combined supplementation of melatonin and salicylic acid effectively detoxifies arsenic toxicity by modulating phytochelatins and nitrogen metabolism in pepper plants. Environmental Pollution 297. Available at https://doi.org/10.1016/j.envpol.2021.118727, https://www.journals.elsevier.com/environmental-pollution.

Kaya, C., Ugurlar, F., Ashraf, M., Alyemeni, M.N., Ahmad, P., 2023. Exploring the synergistic effects of melatonin and salicylic acid in enhancing drought stress tolerance in tomato plants through fine-tuning oxidative-nitrosative processes and methylglyoxal metabolism. Scientia Horticulturae 321. Available at https://doi.org/10.1016/j.scienta.2023.112368, https://www.sciencedirect.com/science/journal/03044238.

Keikha, M., Noori, M., Keshtehgar, A., 2017. Effect of salicylic acid and gibberellin on yield and yield components of Mungbean (*Vigna radiata*). Iranian Journal Pulses Research 7 (2), 138–151.

Khalid, M.F., Elezz, A.A., Jawaid, M.Z., Ahmed, T., 2023. Salicylic acid restricts mercury translocation by activating strong antioxidant defense mechanisms in sweet pepper (*Capsicum annum* L). Environmental Technology and Innovation 32. Available at https://doi.org/10.1016/j.eti.2023.103283, http://www.journals.elsevier.com/environmental-technology-and-innovation/.

Khalil, R., Haroun, S., Bassyoini, F., Nagah, A., Yusuf, M., 2021. Salicylic acid in combination with kinetin or calcium ameliorates heavy metal stress in *Phaseolus vulgaris* plant. Journal of Agriculture and Food Research 5, 100182. Available at https://doi.org/10.1016/j.jafr.2021.100182.

Khalvandi, M., Siosemardeh, A., Roohi, E., Keramati, S., 2021. Salicylic acid alleviated the effect of drought stress on photosynthetic characteristics and leaf protein pattern in winter wheat. Heliyon 7 (1), e05908. Available at https://doi.org/10.1016/j.heliyon.2021.e05908.

Khan, M.I.R., Asgher, M., Khan, N.A., 2014. Alleviation of salt-induced photosynthesis and growth inhibition by salicylic acid involves glycinebetaine and ethylene in mungbean (*Vigna radiata* L). Plant Physiology and Biochemistry 80, 67–74. Available at https://doi.org/10.1016/j.plaphy.2014.03.026.

Khan, M.I.R., Fatma, M., Per, T.S., Anjum, N.A., Khan, N.A., 2015. Salicylic acid-induced abiotic stress tolerance and underlying mechanisms in plants. Frontiers in Plant Science 6 JUNE. Available at https://doi.org/10.3389/fpls.2015.00462, http://journal.frontiersin.org/article/10.3389/fpls.2015.00462/pdf.

Khan, M.A., Asaf, S., Khan, A.L., Jan, R., Kang, S.M., Kim, K.M., et al., 2020. Thermotolerance effect of plant growth-promoting *Bacillus cereus* SA1 on soybean during heat stress. BMC Microbiology 20 (1). Available at https://doi.org/10.1186/s12866-020-01822-7, http://www.biomedcentral.com/bmcmicrobiol/.

Khatoon, S., Jain, K., Mahajan, M., Khan, M.I.R., 2023. Nitric oxide biosynthesis under stressful environments. Nitric Oxide in Developing Plant Stress Resilience. Elsevier, India, pp. 17–30. Available at http://doi.org/10.1016/B978-0-323-91209-9.00013-0, https://www.sciencedirect.com/book/9780323912099.

Kieber, J.J., Schaller, G.E., 2018. Cytokinin signaling in plant development. Company of Biologists Ltd, United States Development (Cambridge) 145 (4). Available at https://doi.org/10.1242/dev.149344, https://journals.biologists.com/dev/article/145/4/dev155275/48583/Connexin-30-controls-astroglial-polarization.

Kim, S.Y., Shang, Y., Joo, S.H., Kim, S.K., Nam, K.H., 2017. Overexpression of BAK1 causes salicylic acid accumulation and deregulation of cell death control genes. Biochemical and Biophysical Research Communications 484 (4), 781–786. Available at https://doi.org/10.1016/j.bbrc.2017.01.166, http://www.sciencedirect.com/science/journal/0006291X.

Kim, Y.W., Youn, J.H., Roh, J., Kim, J.M., Kim, S.K., Kim, T.W., 2022. Brassinosteroids enhance salicylic acid-mediated immune responses by inhibiting BIN2 phosphorylation of clade I TGA transcription factors in *Arabidopsis*. Molecular Plant 15 (6), 991–1007. Available at https://doi.org/10.1016/j.molp.2022.05.002, https://www.cell.com/molecular-plant/home.

Kiremit, M.S., Akınoğlu, G., Mitrovica, B., Rakıcıoğlu, S., 2024. Enhancing drought-salinity stress tolerance in lettuce: synergistic effects of salicylic acid and melatonin. South African Journal of Botany 172, 212–226. Available at https://doi.org/10.1016/j.sajb.2024.07.021, https://www.sciencedirect.com/science/journal/02546299.

Kohli, S.K., Handa, N., Kaur, R., Kumar, V., Khanna, K., Bakshi, P., et al., 2017a. Role of salicylic acid in heavy metal stress tolerance: insight into underlying mechanism. Salicylic Acid: A Multifaceted Hormone. Springer Singapore, India, pp. 123–144. Available at http://doi.org/10.1007/978-981-10-6068-7_7, http://www.springer.com/in/book/9789811060670.

Kohli, S.K., Handa, N., Sharma, A., Kumar, V., Kaur, P., Bhardwaj, R., 2017b. Synergistic effect of 24-epibrassinolide and salicylic acid on photosynthetic efficiency and gene expression in *Brassica juncea* L. Under Pb stress. Turkish Journal of Biology 41 (6), 943–953. Available at https://doi.org/10.3906/biy-1707-15, http://journals.tubitak.gov.tr/biology/issues/biy-17-41-6/biy-41-6-10-1707-15.pdf.

Kohli, S.K., Handa, N., Sharma, A., Gautam, V., Arora, S., Bhardwaj, R., et al., 2018. Combined effect of 24-epibrassinolide and salicylic acid mitigates lead (Pb) toxicity by modulating various metabolites in *Brassica juncea* L. seedlings. Protoplasma 255 (1), 11–24. Available at https://doi.org/10.1007/s00709-017-1124-x, www.springer.co.at/springer.py?Page=40&Key=456&cat=9&id_journal=4.

Koramutla, M.K., Tuan, P.A., Ayele, B.T., 2022. Salicylic acid enhances adventitious root and aerenchyma formation in wheat under waterlogged conditions. International Journal of Molecular Sciences 23 (3). Available at https://doi.org/10.3390/ijms23031243, https://www.mdpi.com/1422-0067/23/3/1243/pdf.

Kumari, P., Tsomu, T., Singh, A.K., Singh, R.P., Maying, B., Tasung, A., 2025a. Impact of salicylic acid and kinetin on morpho-physiological traits, flowering patterns and seed yield in salvia (*Salvia splendens*). Journal of Applied Horticulture 27 (1), 113–117. Available at https://doi.org/10.37855/jah.2025.v27i01.22, http://horticultureresearch.net/index.php.

Kumari, S., Kaur, P., Mahajan, M., Nayak, S.R., Khanna, R.R., Rehman, M.T., et al., 2025b. γ-aminobutyric acid (GABA) supplementation modulates phosphorus retention, production of carbon metabolites and defense metabolism under arsenic toxicity in wheat. Plant Science 356. Available at https://doi.org/10.1016/j.plantsci.2025.112504, https://www.sciencedirect.com/science/journal/01689452.

Latef, A.A.H.A., Tahjib-Ul-Arif, M., Rhaman, M.S., 2021. Exogenous auxin-mediated salt stress alleviation in faba bean (*Vicia faba* L.). Agronomy 11 (3), 547. Available at https://doi.org/10.3390/agronomy11030547.

Layachi, N., Kechrid, Z., 2025. Improvement of salt tolerance in faba bean (*Vicia faba* L.) by salicylic acid application. Applied Ecology and Environmental Research 23 (1), 1239–1255. Available at https://doi.org/10.15666/aeer/2301_12391255.

Lee, S., Kim, S.G., Park, C.M., 2010. Salicylic acid promotes seed germination under high salinity by modulating antioxidant activity in *Arabidopsis*. New Phytologist 188 (2), 626–637. Available at https://doi.org/10.1111/j.1469-8137.2010.03378.x.

Lefevere, H., Bauters, L., Gheysen, G., 2020. Salicylic acid biosynthesis in plants. Frontiers in Plant Science 11. Available at https://doi.org/10.3389/fpls.2020.00338.

Li, Z.-G., 2015. Synergistic effect of antioxidant system and osmolyte in hydrogen sulfide and salicylic acid crosstalk-induced heat tolerance in maize (*Zea mays* L.) seedlings. Plant Signaling & Behavior 10 (9), e1051278. Available at https://doi.org/10.1080/15592324.2015.1051278.

Li, Z.G., Xie, L.R., Li, X.J., 2015. Hydrogen sulfide acts as a downstream signal molecule in salicylic acid-induced heat tolerance in maize (*Zea mays* L.) seedlings. Journal of Plant Physiology 177, 121–127. Available at https://doi.org/10.1016/j.jplph.2014.12.018, www.urbanfischer.de/journals/jpp/p_physio.htm.

Li, Z., Xu, J., Gao, Y., Wang, C., Guo, G., Luo, Y., et al., 2017a. The synergistic priming effect of exogenous salicylic acid and H_2O_2 on chilling tolerance enhancement during maize (*Zea mays* L.) seed germination. Frontiers in Plant Science 8. Available at https://doi.org/10.3389/fpls.2017.01153, http://journal.frontiersin.org/article/10.3389/fpls.2017.01153/full.

Li, Z., Yu, J., Peng, Y., Huang, B., 2017b. Metabolic pathways regulated by abscisic acid, salicylic acid and γ-aminobutyric acid in association with improved drought tolerance in creeping bentgrass (*Agrostis stolonifera*). Physiologia Plantarum 159 (1), 42–58. Available at https://doi.org/10.1111/ppl.12483.

Li, Q., Wang, G., Wang, Y., Yang, D., Guan, C., Ji, J., 2019. Foliar application of salicylic acid alleviate the cadmium toxicity by modulation the reactive oxygen species in potato. Ecotoxicology and Environmental Safety 172, 317–325. Available at https://doi.org/10.1016/j.ecoenv.2019.01.078.

Li, Y., Han, X., Ren, H., Zhao, B., Zhang, J., Ren, B., et al., 2023. Exogenous SA or 6-BA maintains photosynthetic activity in maize leaves under high temperature stress. Crop Journal 11 (2), 605–617. Available at https://doi.org/10.1016/j.cj.2022.08.006, http://www.elsevier.com/journals/the-crop-journal/2214-5141.

Lim, G.H., Liu, H., Yu, K., Liu, R., Shine, M.B., Fernandez, J., et al., 2020. The plant cuticle regulates apoplastic transport of salicylic acid during systemic acquired resistance. Science Advances 6 (19). Available at https://doi.org/10.1126/sciadv.aaz0478, https://advances.sciencemag.org/content/advances/6/19/EAAZ0478.full.pdf.

Lisjak, M., Teklic, T., Wilson, I.D., Whiteman, M., Hancock, J.T., 2013. Hydrogen sulfide: environmental factor or signalling molecule? Plant, Cell and Environment 36 (9), 1607–1616. Available at https://doi.org/10.1111/pce.12073, http://www.blackwellpublishing.com/journals/PCE.

Liu, J., Qiu, G., Liu, C., Li, H., Chen, X., Fu, Q., et al., 2022a. Salicylic acid, a multifaceted hormone combats abiotic stresses in plants. Life (Chicago, Ill.: 1978) 12, 886.

Liu, N., Xu, Y., Li, Q., Cao, Y., Yang, D., Liu, S., et al., 2022b. A lncRNA fine-tunes salicylic acid biosynthesis to balance plant immunity and growth. Cell Host & Microbe 30 (8), 1124. Available at https://doi.org/10.1016/j.chom.2022.07.001.

Liu, X., Chen, Z., Huang, L., Ouyang, Y., Wang, Z., Wu, S., et al., 2023. Salicylic acid attenuates brassinosteroid signaling via protein de-S-acylation. The EMBO Journal 42 (13). Available at https://doi.org/10.15252/embj.2022112998.

Lo'Ay, A.A., Taher, M.A., 2018. Effectiveness salicylic acid blending in chitosan/PVP biopolymer coating on antioxidant enzyme activities under low storage temperature stress of 'Banati'guava fruit. Scientia Horticulturae 238, 343–349.

Lone, R., Hassan, N., Bashir, B., Rohela, G.K., Malla, N.A., 2023. Role of growth elicitors and microbes in stress management and sustainable production of Sorghum. Plant Stress 9. Available at https://doi.org/10.1016/j.stress.2023.100179, www.journals.elsevier.com/plant-stress.

Loutfy, N., Sakuma, Y., Gupta, D.K., Inouhe, M., 2020. Modifications of water status, growth rate and antioxidant system in two wheat cultivars as affected by salinity stress and salicylic acid. Journal of Plant Research 133 (4), 549–570. Available at https://doi.org/10.1007/s10265-020-01196-x, www.springer.com/journal/10265.

Loutfy, N., Hassanein, A.M., Inouhe, M., Salem, J.M., 2022. Biological aspects and proline metabolism genes influenced by polyethylene glycol and salicylic acid in two wheat cultivars. Egyptian Journal of Botany 62 (3), 671–685. Available at https://doi.org/10.21608/ejbo.2022.124280.1921, https://ejbo.journals.ekb.eg/article_237300_5029107360caf0e3435acc625146265d.pdf.

Ma, C., Yuan, S., Xie, B., Li, Q., Wang, Q., Shao, M., 2022. IAA plays an important role in alkaline stress tolerance by modulating root development and ros detoxifying systems in rice plants. International Journal of Molecular Sciences 23 (23), 14817. Available at https://doi.org/10.3390/ijms232314817.

Madebo, M.P., Zheng, Y., Jin, P., 2022. Melatonin-mediated postharvest quality and antioxidant properties of fresh fruits: a comprehensive meta-analysis. Comprehensive Reviews in Food Science and Food Safety 21 (4), 3205–3226. Available at https://doi.org/10.1111/1541-4337.12961, http://onlinelibrary.wiley.com/journal/10.1111/(ISSN)1541-4337.

Mahajan, M., Nazir, F., Jahan, B., Siddiqui, M.H., Iqbal, N., Khan, M.I.R., 2023. Salicylic acid mitigates arsenic stress in rice (*Oryza sativa*) via modulation of nitrogen–sulfur assimilation, ethylene biosynthesis, and defense systems. Agriculture 13 (7), 1293. Available at https://doi.org/10.3390/agriculture13071293.

Mahajan, M., Poor, P., Kaur, H., Aher, R.R., Palakolanu, S.R., Khan, M.I.R., 2025a. Salt stress tolerance and abscisic acid in plants: associating role of plant growth regulators and transcription factors. Plant Physiology and Biochemistry 228. Available at https://doi.org/10.1016/j.plaphy.2025.110303, https://www.sciencedirect.com/science/journal/09819428.

Mahajan, M., Sahoo, R.N., Khan, M.I.R., 2025b. Lanthanum supplementation with abscisic acid modulates rhizosphere dynamics through changes in nitric oxide synthesis in wheat. Physiologia Plantarum 177 (1). Available at https://doi.org/10.1111/ppl.70126, http://onlinelibrary.wiley.com/journal/10.1111/(ISSN)1399-3054.

Mahajan, M., Khatoon, S., Prakash, S., Khan, M.I.R., 2025c. Plant stress hormones: fostering drought stress resilience. Optimizing Plant Health under Abiotic-Stress Environments. CRC Press, pp. 123–142.

Mahdavian, K., Kalantari, K.M., Ghorbanli, M., 2007. The effect of different concentrations of salicylic acid on protective enzyme activities of pepper (*Capsicum annuum* L.) plants. Pakistan Journal of Biological Sciences 10 (18), 3162–3165. Available at https://doi.org/10.3923/pjbs.2007.3162.3165, http://scialert.net/previous.php?issn=1028-8880.

Mahmoud, L.M., Vincent, C.I., Grosser, J.W., Dutt, M., 2021. The response of salt-stressed Valencia sweet orange (*Citrus sinensis*) to salicylic acid and methyl jasmonate treatments. Plant Physiology Reports 26 (1), 137–151. Available at https://doi.org/10.1007/s40502-020-00563-z, https://link.springer.com/journal/volumesAndIssues/40502.

Mahmoudi, D., Shekari, F., Afsahi, K., Maleki, A., 2022. Evaluation of the responses of basil to the application of salicylic acid and gibberellic acid. Zemdirbyste 109 (2), 131–138. Available at https://doi.org/10.13080/z-a.2022.109.017, http://www.zemdirbyste-agriculture.lt/wp-content/uploads/2022/06/109_2_str17.pdf.

Malik, Z.A., Lal, E.P., 2018. Effect of salicylic acid and indole acetic acid on growth, physiological fluctuations and proline content of varieties of tomato (*Lycopersicum esculentum*) under induced cadmium and NaCl stress. Annals of Plant and Soil Research 20 (3), 276–280.

Manasa S, L., Panigrahy, M., Panigrahi, K.C.S., Rout, G.R., 2022. Overview of cold stress regulation in plants. Botanical Review 88 (3), 359–387. Available at https://doi.org/10.1007/s12229-021-09267-x, https://www.springer.com/journal/12229.

Mareri, L., Parrotta, L., Cai, G., 2022. Environmental stress and plants. International Journal of Molecular Sciences 23 (10), 5416. Available at https://doi.org/10.3390/ijms23105416.

Martel, A.B., Qaderi, M.M., 2016. Does salicylic acid mitigate the adverse effects of temperature and ultraviolet-B radiation on pea (*Pisum sativum*) plants? Environmental and Experimental Botany 122, 39–48. Available at https://doi.org/10.1016/j.envexpbot.2015.09.002, www.elsevier.com/inca/publications/store/2/6/7.

Maruri-López, I., Aviles-Baltazar, N.Y., Buchala, A., Serrano, M., 2019. Intra and extracellular journey of the phytohormone salicylic acid. Frontiers in Plant Science 10. Available at https://doi.org/10.3389/fpls.2019.00423, https://www.frontiersin.org/articles/10.3389/fpls.2019.00423/pdf.

Mazumder, M.K., Sharma, P., Moulick, D., Tata, S.K., Choudhury, S., 2022. Salicylic acid ameliorates zinc and chromium-induced stress responses in wheat seedlings: a biochemical and computational analysis. Cereal Research Communications 50 (3), 407–418. Available at https://doi.org/10.1007/s42976-021-00201-w, https://link.springer.com/journal/42976.

Medina, E., Kim, S.H., Yun, M., Choi, W.G., 2021. Recapitulation of the function and role of ros generated in response to heat stress in plants. Plants 10 (2), 1–13. Available at https://doi.org/10.3390/plants10020371, https://www.mdpi.com/2223-7747/10/2/371.

Megha, S., Basu, U., Kav, N.N.V., 2018. Regulation of low temperature stress in plants by microRNAs. Plant, Cell & Environment 41 (1), 1–15. Available at https://doi.org/10.1111/pce.12956.

Mohammed, A.R., Tarpley, L., 2013. Effects of enhanced ultraviolet-B (UV-B) radiation and antioxidative-type plant growth regulators on rice (*Oryza sativa* L.) leaf photosynthetic rate, photochemistry and physiology. Journal of Agricultural Science 5 (5). Available at https://doi.org/10.5539/jas.v5n5p115.

Mohi-Ud-Din, M., Talukder, D., Rohman, M., Ahmed, J.U., Jagadish, S.V.K., Islam, T., et al., 2021. Exogenous application of methyl jasmonate and salicylic acid mitigates drought-induced oxidative damages in French bean (*Phaseolus vulgaris* L). Plants 10 (10), 2066. Available at https://doi.org/10.3390/plants10102066.

Morovvat, S.A., Sadrabadi, R., Noferest, K.S., Darban, A.S., Salati, M., 2021. Effects of foliar application chitosan and salicylic acid on physiological characteristics and yield under deficit irrigation condition. Agrivita 43 (1), 101–113. Available at https://doi.org/10.17503/agrivita.v43i1.2796, https://agrivita.ub.ac.id/index.php/agrivita/article/download/2274/1357.

Mukhtar Ahmed, K.B., Khan, M.M.A., Siddiqui, H., Jahan, A., 2020. Chitosan and its oligosaccharides, a promising option for sustainable crop production- a review. Carbohydrate Polymers 227. Available at https://doi.org/10.1016/j.carbpol.2019.115331, http://www.elsevier.com/wps/find/journaldescription.cws_home/405871/description#description.

Muthulakshmi, S., Lingakumar, K., 2017. Role of salicylic acid (SA) in plants—a review. International Journal of Applied Research 3, 33–37.

Najafi-Kakavand, S., Karimi, N., Ghasempour, H.R., Raza, A., Chaichi, M., Modarresi, M., 2023. Role of jasmonic and salicylic acid on enzymatic changes in the root of two alyssum inflatum náyr. populations exposed to nickel toxicity. Journal of Plant Growth Regulation 42 (3), 1647–1664. Available at https://doi.org/10.1007/s00344-022-10648-8, https://www.springer.com/journal/344.

Namdjoyan, S., Kermanian, H., Abolhasani Soorki, A., Modarres Tabatabaei, S., Elyasi, N., 2017. Interactive effects of Salicylic acid and nitric oxide in alleviating zinc toxicity of Safflower (*Carthamus tinctorius* L). Ecotoxicology (London, England) 26 (6), 752–761. Available at https://doi.org/10.1007/s10646-017-1806-3.

Naser, V., Shani, E., 2016. Auxin response under osmotic stress. Plant Molecular Biology 91 (6), 661–672. Available at https://doi.org/10.1007/s11103-016-0476-5.

Nasri, M., Karimi, Z., Oveisi, M., Kasraie, P., Larijani, H., 2022. Impact of foliar application of salicylic acid and jasmonic acid on biochemical and agronomic properties of *Thymus vulgaris* L. under drought stress. Journal of Crops Improvement 24 (4), 1359–1372.

Nawaz, M., Hussain, I., Mahmood-ur-Rehman, Ashraf, M.A., Rasheed, R., 2024. Salicylic acid and gemma-aminobutyric acid mediated regulation of growth, metabolites, antioxidant defense system and nutrient uptake in sunflower (*Helianthus annuus* L.) under arsenic stress. Dose-Response. 22 (2). Available at https://doi.org/10.1177/15593258241258407, https://journals.sagepub.com/home/DOS.

Nawaz, M., Hussain, I., Rasheed, R., Ashraf, M.A., Mahmood-ur-Rahman, 2025. Gamma-aminobutyric acid and salicylic acid regulated osmolyte accumulation, nutrient acquisition, and ROS metabolism to decrease arsenic effects on sunflower (*Helianthus annuus* L.) at vegetative stage. Journal of Soil Science and Plant Nutrition 25 (2), 2605–2626. Available at https://doi.org/10.1007/s42729-025-02287-1, https://www.springer.com/journal/42729.

Nazarli, H., Ahmadi, A., Hadian, J., 2014. Salicylic acid and methyl jasmonate enhance drought tolerance in chamomile plants. Journal of HerbMed Pharmacology 3 (2), 87–92.

Nazir, F., Kumari, S., Mahajan, M., Khan, M.I.R., 2023. An explicit story of plant abiotic stress resilience: overtone of selenium, plant hormones and other signaling molecules. Plant and Soil 486 (1-2), 135–163. Available at https://doi.org/10.1007/s11104-022-05826-2, https://www.springer.com/journal/11104.

Nazir, F., Khatoon, S., Mahajan, M., Kumari, S., AlAjmi, M.F., Rehman, M.T., et al., 2024. Enhanced arsenic stress tolerance in landrace and improved rice cultivars through modulation of gibberellic acid (GA3) synthesis and antioxidant metabolism via phosphorus and silicon supplementation. Plant Stress 13. Available at https://doi.org/10.1016/j.stress.2024.100511, https://www.sciencedirect.com/science/journal/2667064X.

Oguz, M.C., Aycan, M., Oguz, E., Poyraz, I., Yildiz, M., 2022. Drought stress tolerance in plants: interplay of molecular, biochemical and physiological responses in important development stages. Physiologia 2 (4), 180–197. Available at https://doi.org/10.3390/physiologia2040015.

Omidi, M., Khandan-Mirkohi, A., Kafi, M., Zamani, Z., Ajdanian, L., Babaei, M., 2022. Biochemical and molecular responses of *Rosa damascena* mill. cv. Kashan to salicylic acid under salinity stress. BMC Plant Biology 22 (1). Available at https://doi.org/10.1186/s12870-022-03754-y.

Pan, J., Guan, M., Xu, P., Chen, M., Cao, Z., 2021. Salicylic acid reduces cadmium (Cd) accumulation in rice (*Oryza sativa* L.) by regulating root cell wall composition via nitric oxide signaling. Science of the Total Environment 797, 149202. Available at https://doi.org/10.1016/j.scitotenv.2021.149202.

Paolacci, A.R., Catarcione, G., Ederli, L., Zadra, C., Pasqualini, S., Badiani, M., et al., 2017. Jasmonate-mediated defence responses, unlike salicylate-mediated responses, are involved in the recovery of grapevine from bois noir disease. BMC Plant Biology 17 (1). Available at https://doi.org/10.1186/s12870-017-1069-4, http://www.biomedcentral.com/bmcplantbiol/.

Pirasteh-Anosheh, H., Ranjbar, G., Hasanuzzaman, M., Khanna, K., Bhardwaj, R., Ahmad, P., 2022. Salicylic acid-mediated regulation of morpho-physiological and yield attributes of wheat and barley plants in deferring salinity stress. Journal of Plant Growth Regulation 41 (3), 1291–1303. Available at https://doi.org/10.1007/s00344-021-10358-7.

Pongprayoon, W., Panya, A., Jaresitthikunchai, J., Phaonakrop, N., Roytrakul, S., 2022. Phosphoprotein profile of rice (*Oryza sativa* L.) seedlings under osmotic stress after pretreatment with chitosan. Plants 11 (20), 2729. Available at https://doi.org/10.3390/plants11202729.

Qiao, Z., Jing, T., Liu, Z., Zhang, L., Jin, Z., Liu, D., et al., 2015. H2S acting as a downstream signaling molecule of SA regulates Cd tolerance in *Arabidopsis*. Plant and Soil 393 (1-2), 137–146. Available at https://doi.org/10.1007/s11104-015-2475-8.

Rahman, S.U., Han, J.C., Ahmad, M., Gao, S., Khan, K.A., Li, B., et al., 2023. Toxic effects of lead (Pb), cadmium (Cd) and tetracycline (TC) on the growth and development of *Triticum aestivum*: a meta-analysis. Science of the Total Environment 904. Available at https://doi.org/10.1016/j.scitotenv.2023.166677, www.elsevier.com/locate/scitotenv.

Rai, K.K., Rai, N., Aamir, M., Tripathi, D., Rai, S.P., 2020. Interactive role of salicylic acid and nitric oxide on transcriptional reprogramming for high temperature tolerance in *Lablab purpureus* L.: structural and functional insights using computational approaches. Journal of Biotechnology 309, 113–130. Available at https://doi.org/10.1016/j.jbiotec.2020.01.001, www.elsevier.com/locate/jbiotec.

Rai, K.K., Pandey, N., Rai, N., Rai, S.K., Pandey-Rai, S., 2021. Salicylic acid and nitric oxide: insight into the transcriptional regulation of their metabolism and regulatory functions in plants. Frontiers in Agronomy 3. Available at https://doi.org/10.3389/fagro.2021.781027, www.frontiersin.org/journals/agronomy#.

Ramzan, M., Javed, T., Hassan, A., Ahmed, M.Z., Ashraf, H., Shah, A.A., et al., 2024. Protective effects of the exogenous application of salicylic acid and chitosan on chromium-induced photosynthetic capacity and osmotic adjustment in *Aconitum napellus*. BMC Plant Biology 24 (1). Available at https://doi.org/10.1186/s12870-024-05634-z, https://bmcplantbiol.biomedcentral.com/.

Rasheed, F., Sehar, Z., Fatma, M., Iqbal, N., Masood, A., Anjum, N., et al., 2021. Involvement of ethylene in reversal of salt stress by salicylic acid in presence of sulfur in mustard (*Brassica juncea* L.). Research Square. Available at https://doi.org/10.21203/rs.3.rs-668051/v1, https://www.researchsquare.com/browse.

Rasheed, F., Mir, I.R., Sehar, Z., Fatma, M., Gautam, H., Khan, S., et al., 2022. Nitric oxide and salicylic acid regulate glutathione and ethylene production to enhance heat stress acclimation in wheat involving sulfur assimilation. Plants 11 (22). Available at https://doi.org/10.3390/plants11223131, www.mdpi.com/journal/plants.

Ritonga, F.N., Chen, S., 2020. Physiological and molecular mechanism involved in cold stress tolerance in plants. Plants 9 (5), 560. Available at https://doi.org/10.3390/plants9050560.

Riyazuddin, R., Nisha, N., Ejaz, B., Khan, M.I.R., Kumar, M., Ramteke, P.W., et al., 2022. A comprehensive review on the heavy metal toxicity and sequestration in plants. Biomolecules 12 (1). Available at https://doi.org/10.3390/biom12010043, https://www.mdpi.com/2218-273X/12/1/43/pdf.

Rodas-Junco, B.A., Nic-Can, G.I., Muñoz-Sánchez, A., Hernández-Sotomayor, S.M.T., 2020. Phospholipid signaling is a component of the salicylic acid response in plant cell suspension cultures. International Journal of Molecular Sciences 21 (15), 5285. Available at https://doi.org/10.3390/ijms21155285.

Roghayyeh, S., Saeede, R., Omid, A., Mohammad, S., 2014. The effect of salicylic acid and gibberellin on seed reserve utilization, germination and enzyme activity of sorghum (*Sorghum bicolor* L.) seeds under drought stress. Journal of Stress Physiology & Biochemistry 10 (1), 5–13.

Román-Doval, R., Torres-Arellanes, S.P., Tenorio-Barajas, A.Y., Gómez-Sánchez, A., Valencia-Lazcano, A.A., 2023. Chitosan: properties and its application in agriculture in context of molecular weight. Polymers 15 (13). Available at https://doi.org/10.3390/polym15132867, http://www.mdpi.com/journal/polymers.

Roshdy, A.E.D., Alebidi, A., Almutairi, K., Al-Obeed, R., Elsabagh, A., 2021. The effect of salicylic acid on the performances of salt stressed strawberry plants, enzymes activity, and salt tolerance index. Agronomy 11 (4). Available at https://doi.org/10.3390/agronomy11040775, https://www.mdpi.com/2073-4395/11/4/775/pdf.

Saeed, F., Rasul, S., Batool, S., Zafar, U., Manzoor, H., 2022. Exogenous applications of salicylic acid alleviate the damaging effects of heat stress in chili (*Capsicum frutescens* L.) through improved antioxidant defense system. International Journal of Applied and Experimental Biology 2 (1), 59–68. Available at https://doi.org/10.56612/ijaeb.v2i1.51.

Safari, M., Mousavi-Fard, S., Rezaei Nejad, A., Sorkheh, K., Sofo, A., 2022. Exogenous salicylic acid positively affects morpho-physiological and molecular responses of *Impatiens walleriana* plants grown under drought stress. International Journal of Environmental Science and Technology 19 (2), 969–984. Available at https://doi.org/10.1007/s13762-020-03092-2.

Sah, S.K., Reddy, K.R., Li, J., 2016. Abscisic acid and abiotic stress tolerance in crop plants. Frontiers in Plant Science 7 2016. Available at https://doi.org/10.3389/fpls.2016.00571, http://journal.frontiersin.org/article/10.3389/fpls.2016.00571/full.

Saini, S., Kaur, N., Pati, P.K., 2021. Phytohormones: key players in the modulation of heavy metal stress tolerance in plants. Ecotoxicology and Environmental Safety 223, 112578. Available at https://doi.org/10.1016/j.ecoenv.2021.112578.

Saleem, M., Fariduddin, Q., Janda, T., 2021. Multifaceted role of salicylic acid in combating cold stress in plants: a review. Journal of Plant Growth Regulation 40 (2), 464–485. Available at https://doi.org/10.1007/s00344-020-10152-x.

Samea-Andabjadid, S., Ghassemi-Golezani, K., Nasrollahzadeh, S., Najafi, N., 2018. Exogenous salicylic acid and cytokinin alter sugar accumulation, antioxidants and membrane stability of faba bean. Acta Biologica Hungarica 69 (1), 86–96. Available at https://doi.org/10.1556/018.68.2018.1.7.

Sang, Y., Liu, Y., Tang, Y., Yang, W., Guo, M., Chen, G., 2022. Transcriptome sequencing reveals mechanism of improved antioxidant capacity and maintained postharvest quality of winter jujube during cold storage after salicylic acid treatment. Postharvest Biology and Technology 189, 111929. Available at https://doi.org/10.1016/j.postharvbio.2022.111929.

Saxena, I., Srikanth, S., Chen, Z., 2016. Cross talk between H_2O_2 and interacting signal molecules under plant stress response. Frontiers in Plant Science 7. Available at https://doi.org/10.3389/fpls.2016.00570 2016.

Seleiman, M.F., Al-Suhaibani, N., Ali, N., Akmal, M., Alotaibi, M., Refay, Y., et al., 2021. Drought stress impacts on plants and different approaches to alleviate its adverse effects. Plants 10 (2), 1–25. Available at https://doi.org/10.3390/plants10020259, https://www.mdpi.com/2223-7747/10/2/259/pdf.

Seo, Y., Park, K., Chang, E., Ryu, S., Park, J., Kim, K., 2014. Effect of salicylic acid and abscisic acid on drought stress of Waxy Corn. Korean Journal of Crop Science 59 (1), 54–58. Available at https://doi.org/10.7740/kjcs.2014.59.1.054.

Seraj, R.G.M., Behnamian, M., Ahmadikhah, A., Esmaeilzadeh-Salestani, K., 2024. Foliar application of chitosan and salicylic acid improved milk thistle's tolerance to drought stress. Russian Journal of Plant Physiology 71 (3), 62.

Shah Jahan, M., Wang, Y., Shu, S., Zhong, M., Chen, Z., Wu, J., et al., 2019. Exogenous salicylic acid increases the heat tolerance in tomato (*Solanum lycopersicum* L) by enhancing photosynthesis efficiency and improving antioxidant defense system through scavenging of reactive oxygen species. Scientia Horticulturae 247, 421–429. Available at https://doi.org/10.1016/j.scienta.2018.12.047, www.elsevier.com/inca/publications/store/5/0/3/3/1/6.

Shah, S.H., Islam, S., Mohammad, F., Siddiqui, M.H., 2023. Gibberellic acid: a versatile regulator of plant growth, development and stress responses. Journal of Plant Growth Regulation 42 (12), 7352–7373. Available at https://doi.org/10.1007/s00344-023-11035-7, https://www.springer.com/journal/344.

Shams, M., Ekinci, M., Ors, S., Turan, M., Agar, G., Kul, R., et al., 2019. Nitric oxide mitigates salt stress effects of pepper seedlings by altering nutrient uptake, enzyme activity and osmolyte accumulation. Physiology and Molecular Biology of Plants 25 (5), 1149–1161. Available at https://doi.org/10.1007/s12298-019-00692-2.

Sharafizadeh, M., 2017. Effect of salicylic acid and drought stress on germination and activity of antioxidant enzymes of barely. Iranian Journal of Seed Science and Technology 6 (2), 161–169.

Sharma, A., Sidhu, G.P.S., Araniti, F., Bali, A.S., Shahzad, B., Tripathi, D.K., et al., 2020. The role of salicylic acid in plants exposed to heavy metals. Molecules (Basel, Switzerland) 25 (3). Available at https://doi.org/10.3390/molecules25030540, https://www.mdpi.com/1420-3049/25/3/540/pdf.

Sherin, G., Aswathi, K.P.R., Puthur, J.T., 2022. Photosynthetic functions in plants subjected to stresses are positively influenced by priming. Plant Stress 4, 100079. Available at https://doi.org/10.1016/j.stress.2022.100079.

Shin, H., Min, K., Arora, R., 2018. Exogenous salicylic acid improves freezing tolerance of spinach (*Spinacia oleracea* L.) leaves. Cryobiology 81, 192–200. Available at https://doi.org/10.1016/j.cryobiol.2017.10.006.

Singh, S., Prakash, P., Singh, A.K., 2021. Salicylic acid and hydrogen peroxide improve antioxidant response and compatible osmolytes in wheat (*Triticum aestivum* L.) under water deficit. Agricultural Research 10 (2), 175–186. Available at https://doi.org/10.1007/s40003-020-00490-3, http://www.springer.com/life+sciences/cell+biology/journal/40003.

Singhal, R.K., Jatav, H.S., Aftab, T., Pandey, S., Mishra, U.N., Chauhan, J., et al., 2021. Roles of nitric oxide in conferring multiple abiotic stress tolerance in plants and crosstalk with other plant growth regulators. Journal of Plant Growth Regulation 40 (6), 2303–2328. Available at https://doi.org/10.1007/s00344-021-10446-8, https://www.springer.com/journal/344.

Sohag, A.A.M., Tahjib-Ul-Arif, M., Brestič, M., Afrin, S., Sakil, M.A., Hossain, M.T., et al., 2020. Exogenous salicylic acid and hydrogen peroxide attenuate drought stress in rice. Plant, Soil and Environment 66 (1), 7–13. Available at https://doi.org/10.17221/472/2019-PSE, https://www.agriculturejournals.cz/publicFiles/472_2019-PSE.pdf.

Soleimani Aghdam, M., Jannatizadeh, A., Sheikh-Assadi, M., Malekzadeh, P., 2016. Alleviation of postharvest chilling injury in anthurium cut flowers by salicylic acid treatment. Scientia Horticulturae 202, 70–76. Available at https://doi.org/10.1016/j.scienta.2016.02.025, www.elsevier.com/inca/publications/store/5/0/3/3/1/6.

Soliman, M.H., Alayafi, A.A.M., El Kelish, A.A., Abu-Elsaoud, A.M., 2018. Acetylsalicylic acid enhance tolerance of *Phaseolus vulgaris* L. to chilling stress, improving photosynthesis, antioxidants and expression of cold stress responsive genes. Botanical Studies 59 (1). Available at https://doi.org/10.1186/s40529-018-0222-1, http://link.springer.com/journal/40529.

Song, C., Wang, K., Xiao, X., Liu, Q., Yang, M., Li, X., et al., 2022. Membrane lipid metabolism influences chilling injury during cold storage of peach fruit. Food Research International 157, 111249. Available at https://doi.org/10.1016/j.foodres.2022.111249.

Song, J., Song, Q., Wang, D., Liu, Y., 2023. Mitigation strategies for excessive cadmium in rice. Comprehensive Reviews in Food Science and Food Safety 22 (5), 3847–3869. Available at https://doi.org/10.1111/1541-4337.13210.

Song, X., Chen, J., Xu, C., Cai, X., Song, W., Chang, A., et al., 2025. Physiological and molecular mechanisms of exogenous salicylic acid in enhancing salt tolerance in tobacco seedlings by regulating antioxidant defence system and gene expression. Frontiers in Plant Science 16. Available at https://doi.org/10.3389/fpls.2025.1545865.

Sperdouli, I., Panteris, E., Moustaka, J., Aydın, T., Bayçu, G., Moustakas, M., 2024. Mechanistic insights on salicylic acid-induced enhancement of photosystem ii function in basil plants under non-stress or mild drought stress. International Journal of Molecular Sciences 25 (11), 5728. Available at https://doi.org/10.3390/ijms25115728.

Subašić, M., Šamec, D., Selović, A., Karalija, E., 2022. Phytoremediation of cadmium polluted soils: current status and approaches for enhancing. Soil Systems 6 (1), 3. Available at https://doi.org/10.3390/soilsystems6010003.

Suhim, A.A., Awad, K.M., Jaffer, O.N., Abass, M.H., 2023. The impact of salicylic and jasmonic acid in mitigating salinity stress on date palm *Phoenix dactylifera* L. Barhi Cv. Basrah Journal of Agricultural Sciences 36 (1), 120–130. Available at https://doi.org/10.37077/25200860.2023.36.1.10, https://bjas.bajas.edu.iq/index.php/bjas/article/view/911/308.

Sujatha Edupuganti, L.R., Anuradha, S., 2019. Morphological and physio-biochemical changes in response to exogenous application of 24-epibrassinolide and salicylic acid under water stress in chickpea. Journal of Pharmacognosy and Phytochemistry 8 (4), 2443–2452.

Sultana, S., Rahman, M.M., Das, A.K., Haque, M.A., Rahman, M.A., Islam, S.M.N., et al., 2024. Role of salicylic acid in improving the yield of two mung bean genotypes under waterlogging stress through the modulation of antioxidant defense and osmoprotectant levels. Plant Physiology and Biochemistry 206. Available at https://doi.org/10.1016/j.plaphy.2023.108230, https://www.sciencedirect.com/science/journal/09819428.

Sultana, R., Imam, Z., Kumar, R.R., Banu, V.S., Nahakpam, S., Bharti, R., et al., 2025a. Signaling and defence mechanism of jasmonic and salicylic acid response in pulse crops: role of WRKY transcription factors in stress response. Journal of Plant Growth Regulation 44 (1), 5–21. Available at https://doi.org/10.1007/s00344-023-11203-9, https://www.springer.com/journal/344.

Sultana, S., Altaf-Un-Nahar, M., Islam, M.R., Roy, M., Rahman, F., Azam, M.G., et al., 2025b. Foliar and root applications of salicylic acid alleviate salinity stress by modulating morpho-physiological and biochemical aspects in tomato (*Solanum lycopersicum*). Discover Plants 2 (1). Available at https://doi.org/10.1007/s44372-025-00117-3.

Sytar, O., Kumari, P., Yadav, S., Brestic, M., Rastogi, A., 2019. Phytohormone priming: regulator for heavy metal stress in plants. Journal of Plant Growth Regulation 38 (2), 739–752. Available at https://doi.org/10.1007/s00344-018-9886-8.

Szepesi, Á., Poór, P., Bakacsy, L., 2022. Short-term salicylic acid treatment affects polyamine metabolism causing ROS–NO imbalance in tomato roots. Plants 11 (13), 1670. Available at https://doi.org/10.3390/plants11131670.

Tafreshi, Y.M., Eghlima, G., Hatami, M., Vafadar, M., 2025. Exploring the potential impact of salicylic acid and jasmonic acid in promoting seed oil content, vitamin C and antioxidant activity in rosehip (*Rosa canina* L). BMC Plant Biology 25 (1). Available at https://doi.org/10.1186/s12870-025-06251-0, https://bmcplantbiol.biomedcentral.com/.

Takács, Z., Poór, P., Tari, I., 2016. Comparison of polyamine metabolism in tomato plants exposed to different concentrations of salicylic acid under light or dark conditions. Plant Physiology and Biochemistry 108, 266–278. Available at https://doi.org/10.1016/j.plaphy.2016.07.020.

Talaat, N.B., 2021. Co-application of melatonin and salicylic acid counteracts salt stress-induced damage in wheat (*Triticum aestivum* L.) photosynthetic machinery. Journal of Soil Science and Plant Nutrition 21 (4), 2893–2906. Available at https://doi.org/10.1007/s42729-021-00576-z, https://www.springer.com/life+sciences/plant+sciences/journal/42729.

Talaat, N.B., Shawky, B.T., 2022. Synergistic effects of salicylic acid and melatonin on modulating Ion homeostasis in salt-stressed wheat (*Triticum aestivum* L.) plants by enhancing root H+-Pump activity. Plants 11 (3). Available at https://doi.org/10.3390/plants11030416, https://www.mdpi.com/2223-7747/11/3/416/pdf.

Talaat, N.B., Nesiem, M.R.A., Gadalla, E.G., Ali, S.F., 2023. Gibberellic acid and salicylic acid dual application improves date palm fruit growth by regulating the nutrient acquisition, amino acid profile, and phytohormone performance. Journal of Soil Science and Plant Nutrition 23 (4), 6216–6231. Available at https://doi.org/10.1007/s42729-023-01479-x, https://www.springer.com/journal/42729.

Tayyab, N., Naz, R., Yasmin, H., Nosheen, A., Keyani, R., Sajjad, M., et al., 2020. Combined seed and foliar pre-treatments with exogenous methyl jasmonate and salicylic acid mitigate drought-induced stress in maize. PLoS One 15 (5). Available at https://doi.org/10.1371/journal.pone.0232269, https://journals.plos.org/plosone/article/file?id=10.1371/journal.pone.0232269&type=printable.

Tirani, M.M., Nasibi, F., Kalantari, K.M., 2013. Interaction of salicylic acid and ethylene and their effects on some physiological and biochemical parameters in canola plants (*Brassica napus* L.). Photosynthetica 51 (3), 411–418. Available at https://doi.org/10.1007/s11099-013-0041-2.

Torun, H., Novák, O., Mikulík, J., Strnad, M., Ayaz, F.A., 2022. The effects of exogenous salicylic acid on endogenous phytohormone status in *Hordeum vulgare* L. under salt stress. Plants 11 (5). Available at https://doi.org/10.3390/plants11050618, https://www.mdpi.com/2223-7747/11/5/618/pdf.

Wahab, A., Abdi, G., Saleem, M.H., Ali, B., Ullah, S., Shah, W., et al., 2022. Plants' physio-biochemical and phytohormonal responses to alleviate the adverse effects of drought stress: a comprehensive review. Plants 11 (13). Available at https://doi.org/10.3390/plants11131620, https://www.mdpi.com/2223-7747/11/13/1620/pdf?version=1655777519.

Wang, W., Wang, X., Huang, M., Cai, J., Zhou, Q., Dai, T., et al., 2018. Hydrogen peroxide and abscisic acid mediate salicylic acid-induced freezing tolerance in wheat. Frontiers in Plant Science 9. Available at https://doi.org/10.3389/fpls.2018.01137.

Wang, J., Song, L., Gong, X., Xu, J., Li, M., 2020. Functions of jasmonic acid in plant regulation and response to abiotic stress. International Journal of Molecular Sciences 21 (4), 1446. Available at https://doi.org/10.3390/ijms21041446.

Wang, J., Shi, S.H., Wang, D.Y., Sun, Y., Zhu, M., Li, F.H., 2021a. Exogenous salicylic acid ameliorates waterlogging stress damages and improves photosynthetic efficiency and antioxidative defense system in waxy corn. Photosynthetica 59 (1), 84–94. Available at https://doi.org/10.32615/ps.2021.005.

Wang, J., Chen, X., Chu, S., Hayat, K., Chi, Y., Zhi, Y., et al., 2021b. Influence of Cd toxicity on subcellular distribution, chemical forms, and physiological responses of cell wall components towards short-term Cd stress in *Solanum nigrum*. Environmental Science and Pollution Research 28 (11), 13955–13969. Available at https://doi.org/10.1007/s11356-020-11505-5.

Wang, W., Wang, X., Huang, M., Cai, J., Zhou, Q., Dai, T., et al., 2021c. Alleviation of field low-temperature stress in winter wheat by exogenous application of salicylic acid. Journal of Plant Growth Regulation 40 (2), 811–823. Available at https://doi.org/10.1007/s00344-020-10144-x.

Wang, Z., Qu, L., Li, J., Niu, S., Guo, J., Lu, D., 2024. Effects of exogenous salicylic acid on starch physicochemical properties and in vitro digestion under heat stress during the grain-filling stage in waxy maize. International Journal of Biological Macromolecules 254 (Pt 1), 127765. Available at https://doi.org/10.1016/j.ijbiomac.2023.127765.

Wang, X.F., Dong, F.X., Jiang, Y.S., Wei, J.F., Zhao, Q.F., Ji, W., 2025. Exogenous melatonin and salicylic acid enhance postharvest quality and boost antioxidant capacity in *Vitis vinifera* cv. '*Cuibao seedless*' grape berries. Plant Physiology and Biochemistry 223. Available at https://doi.org/10.1016/j.plaphy.2025.109850, https://www.sciencedirect.com/science/journal/09819428.

Wani, A.B., Chadar, H., Wani, A.H., Singh, S., Upadhyay, N., 2017. Salicylic acid to decrease plant stress. Springer Verlag, India Environmental Chemistry Letters 15 (1), 101–123. Available at https://doi.org/10.1007/s10311-016-0584-0, https://www.springer.com/journal/10311.

Wassie, M., Zhang, W., Zhang, Q., Ji, K., Cao, L., Chen, L., 2020. Exogenous salicylic acid ameliorates heat stress-induced damages and improves growth and photosynthetic efficiency in Alfalfa (*Medicago sativa* L.). Ecotoxicology and Environmental Safety 191, 110206. Available at https://doi.org/10.1016/j.ecoenv.2020.110206.

Wei, Y., Chang, Y., Zeng, H., Liu, G., He, C., Shi, H., 2018. RAV transcription factors are essential for disease resistance against cassava bacterial blight via activation of melatonin biosynthesis genes. Journal of Pineal Research 64 (1). Available at https://doi.org/10.1111/jpi.12454.

Wen, C., Zhang, Z., Shi, Q., Duan, X., Du, J., Wu, C., et al., 2023. Methyl jasmonate-and salicylic acid-induced transcription factor ZjWRKY18 regulates triterpenoid accumulation and salt stress tolerance in jujube. International Journal of Molecular Sciences 24 (4), 3899.

Wu, C., Hao, W., Yan, L., Zhang, H., Zhang, J., Liu, C., et al., 2023a. Postharvest melatonin treatment enhanced antioxidant activity and promoted GABA biosynthesis in yellow-flesh peach. Food Chemistry 419, 136088. Available at https://doi.org/10.1016/j.foodchem.2023.136088.

Wu, D., Chen, C., Liu, Y., Yang, L., Yong, J.W.H., 2023b. Iso-osmotic calcium nitrate and sodium chloride stresses have differential effects on growth and photosynthetic capacity in tomato. Scientia Horticulturae 312. Available at https://doi.org/10.1016/j.scienta.2023.111883, www.elsevier.com/inca/publications/store/5/0/3/3/1/6.

Xiang, N., Zhang, B., Hu, J., Li, K., Guo, X., 2024. Modulation of carotenoid biosynthesis in maize (*Zea mays* L.) seedlings by exogenous abscisic acid and salicylic acid under low temperature. Plant Cell Reports 43 (1). Available at https://doi.org/10.1007/s00299-023-03106-6.

Xu, L.L., Fan, Z.Y., Dong, Y.J., Kong, J., Bai, X.Y., 2015. Effects of exogenous salicylic acid and nitric oxide on physiological characteristics of two peanut cultivars under cadmium stress. Biologia plantarum 59 (1), 171–182. Available at https://doi.org/10.1007/s10535-014-0475-9.

Yamauchi, T., Colmer, T.D., Pedersen, O., Nakazono, M., 2018. Regulation of root traits for internal aeration and tolerance to soil waterlogging-flooding stress. Plant Physiology 176 (2), 1118–1130. Available at https://doi.org/10.1104/pp.17.01157, http://www.plantphysiol.org/content/176/2/1118.full.pdf.

Yang, Z., Yang, X., Wei, S., Shen, F., Ji, W., 2024. Exogenous melatonin delays leaves senescence and enhances saline and alkaline stress tolerance in grape seedlings. Plant Signaling & Behavior 19 (1). Available at https://doi.org/10.1080/15592324.2024.2334511.

Yeung, E., van Veen, H., Vashisht, D., Sobral Paiva, A.L., Hummel, M., Rankenberg, T., et al., 2018. A stress recovery signaling network for enhanced flooding tolerance in *Arabidopsis thaliana*. Proceedings of the National Academy of Sciences 115 (26). Available at https://doi.org/10.1073/pnas.1803841115.

Youssef, S.M., López-Orenes, A., Ferrer, M.A., Calderón, A.A., 2023. Foliar application of salicylic acid enhances the endogenous antioxidant and hormone systems and attenuates the adverse effects of salt stress on growth and yield of french bean plants. Horticulturae 9 (1). Available at https://doi.org/10.3390/horticulturae9010075, www.mdpi.com/journal/horticulturae.

Zahra, A., Zafar, M., Younas, A., Hassan, T., Nazeer, A., Arshad, N., et al., 2024. Interaction of heavy metals stress (Cadmium, Copper) and salicylic acid on growth, yield and physiological parameters in two varieties of pea. Plantation Health 03 (01), 87–101.

Zhang, H., Sonnewald, U., 2017. Differences and commonalities of plant responses to single and combined stresses. The Plant Journal 90 (5), 839–855. Available at https://doi.org/10.1111/tpj.13557.

Zhang, Y., Liu, G., Dong, H., Li, C., 2021. Waterlogging stress in cotton: damage, adaptability, alleviation strategies, and mechanisms. The Crop Journal 9 (2), 257–270. Available at https://doi.org/10.1016/j.cj.2020.08.005.

Zhang, Z., Li, L., Qiu, S., Sun, Y., Zhang, R., Chen, D., et al., 2025. BAHD acyltransferase OsSLG mediates rice cadmium tolerance by integrating the brassinosteroid and salicylic acid pathway. Plant Science 356, 112503. Available at https://doi.org/10.1016/j.plantsci.2025.112503.

Zhao, Y., Song, C., Brummell, D.A., Qi, S., Lin, Q., Bi, J., et al., 2021. Salicylic acid treatment mitigates chilling injury in peach fruit by regulation of sucrose metabolism and soluble sugar content. Food Chemistry 358, 129867. Available at https://doi.org/10.1016/j.foodchem.2021.129867.

Zheng, J., Ma, X., Zhang, X., Hu, Q., Qian, R., 2018. Salicylic acid promotes plant growth and salt-related gene expression in *Dianthus superbus* L. (Caryophyllaceae) grown under different salt stress conditions. Physiology and Molecular Biology of Plants 24 (2), 231–238. Available at https://doi.org/10.1007/s12298-017-0496-x.

Zheng, S., Su, M., Wang, L., Zhang, T., Wang, J., Xie, H., et al., 2021. Small signaling molecules in plant response to cold stress. Journal of Plant Physiology 266. Available at https://doi.org/10.1016/j.jplph.2021.153534, www.urbanfischer.de/journals/jpp/p_physio.htm.

Zhou, W., Chen, F., Meng, Y., Chandrasekaran, U., Luo, X., Yang, W., et al., 2020. Plant waterlogging/flooding stress responses: from seed germination to maturation. Plant Physiology and Biochemistry 148, 228–236. Available at https://doi.org/10.1016/j.plaphy.2020.01.020.

Zulfiqar, A., Gul, B., Saleem, A., Islam, A., Zulfiqar, U., Ali, M.F., et al., 2024. Salicylic acid mitigates cadmium stress in wheat: experimental insights into growth and biochemical parameters. Scientifica 2024 (1). Available at https://doi.org/10.1155/sci5/6887694, https://onlinelibrary.wiley.com/journal/6168.

Zulfiqar, F., Moosa, A., Ferrante, A., Nafees, M., Darras, A., Nazir, M.M., et al., 2023. Melatonin and salicylic acid synergistically improve arsenic induced oxidative stress tolerance in ornamental sword lily. Scientia Horticulturae 322. Available at https://doi.org/10.1016/j.scienta.2023.112389, https://www.sciencedirect.com/science/journal/03044238.

CHAPTER

5

Exploring the protective effects of abscisic acid against salt stress in rice

Moksh Mahajan[1], *Gurleen Kaur*[1], *Iqra Aleem*[1], *Tabassum Jehan*[2], *and M. Iqbal R. Khan*[1]

[1]Department of Botany, Jamia Hamdard, New Delhi, Delhi, India [2]Department of Botany, Zakir Husain Delhi College, University of Delhi, New Delhi, Delhi, India

5.1 Introduction

Plants, being sessile organisms, remain constantly exposed to a wide range of environmental fluctuations that vary in intensity, duration, and frequency (Khatoon et al., 2024; Sharma et al., 2022). Among these fluctuations, abiotic stresses such as salt stress, drought, extreme temperatures, heavy metals, and water imbalances are major limiting factors that adversely affect plant growth, development, and productivity (Mahajan et al., 2025c; Zhang et al., 2023). Salt stress, in particular, is one of the most damaging abiotic stresses globally (Mahajan et al., 2025b). It affects more than 20% of irrigated agricultural land and continues to increase due to climate change, seawater intrusion, and poor irrigation practices (Mukhopadhyay et al., 2021). High salt levels disrupt multiple physiological and biochemical processes by inducing osmotic stress, causing ion toxicity, and generating excessive reactive oxygen species (ROS), ultimately leading to reduced nutrient uptake, membrane instability, metabolic imbalance, and significant yield losses (Khatoon et al., 2025; Shelar et al., 2024) (Fig. 5.1).

Rice (*Oryza sativa*), one of the world's most important staple crops feeding more than half of the global population (Mahajan et al., 2023), is especially sensitive to salt stress (Chen et al., 2024). This sensitivity stems from its poor ability to regulate sodium ions (Na^+) and chloride (Cl^-) influx, low efficiency in maintaining a favorable Na^+/K^+ (potassium ions) ratio, and limited potential to accumulate osmoprotectants under stress conditions (Liu et al., 2019). As a semiaquatic crop common in lowland irrigated ecosystems, rice readily absorbs salt-containing water, making it particularly vulnerable to salt accumulation in the soil. Even moderate salt levels can severely inhibit seedling establishment, reduce leaf area, suppress photosynthetic activity, accelerate leaf senescence, and negatively impact grain filling and final yield (Zuo et al., 2024). The rapid

Harnessing Plant-Stimulating Molecules
DOI: https://doi.org/10.1016/B978-0-443-44994-9.00004-3

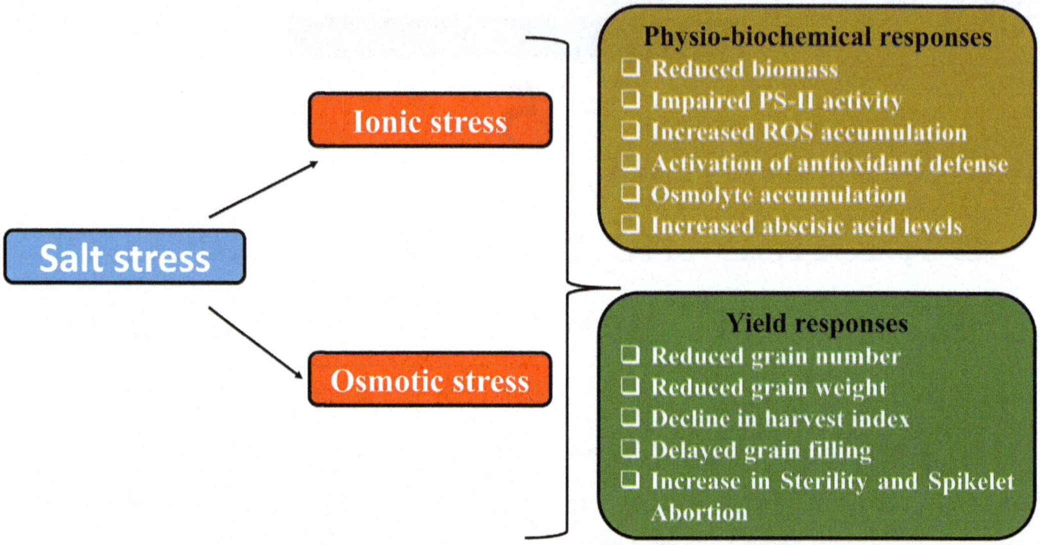

FIGURE 5.1 Physiological, biochemical and yield responses in rice under salt stress. Salt stress induces a wide range of adverse responses in plants leading to crucial physio-biochemical and yield alterations including stomatal closure, reactive oxygen species (ROS) accumulation, enhanced defense system, increased osmolyte accumulation leading to spikelet sterility, reduced grain number and weight, and decreased overall biomass and grain yield.

expansion of salt-affected soils, thus, represents a critical challenge for global rice production and long-term food security (Saleem et al., 2025).

To fight against abiotic stresses, particularly salt stress, plants employ a wide range of molecules that help them in fighting against these stresses. One of them is phytohormones that not only regulate plant growth and development but also help them to endure and ameliorate abiotic stress (Nazir et al., 2023a). Abscisic acid (ABA) is a key phytohormone that orchestrates plant adaptive responses to abiotic stresses, functioning as a central component of stress-signaling networks (Mahajan et al., 2025a; Mahajan et al., 2025d). Under salt stress conditions, endogenous ABA levels rise rapidly in both roots and shoots, triggering a series of physiological and molecular defense responses (Mahajan et al., 2025b). These include stomatal closure to minimize water loss, regulation of ion transporters to sustain Na^+/K^+ homeostasis, enhanced synthesis of osmoprotectants such as proline and soluble sugars, and activation of antioxidant defense enzymes to mitigate oxidative damage by ROS (Jiang et al., 2024). Additionally, ABA-mediated signaling influences the expression of stress-responsive genes, enhances cellular stability, and supports adaptive adjustments that enable plants to withstand salt environments (Huang et al., 2021). Through its multifaceted functions, ABA acts as a key hormone coordinating the plant's defense network during salt stress (Jiang et al., 2024). This chapter focuses on the impact of salt stress on plants, with particular emphasis on the heightened sensitivity of rice to salt conditions. It further highlights the central role of ABA as a major stress hormone involved in regulating physiological, biochemical, and molecular responses that enhance salt tolerance. Overall, this chapter aims to provide a comprehensive understanding of ABA-mediated mechanisms contributing to improved resilience in rice exposed to salt stress.

5.2 Abscisic acid and salt stress

Salt stress is one of the most pervasive abiotic constraints limiting plant growth, productivity, and geographic distribution (Nazir et al., 2023b). It imposes two interconnected forms of stress: an immediate osmotic stress and a delayed ionic stress (Mahajan et al., 2025b). The osmotic phase arises when high external salt concentrations reduce soil water potential, restricting water uptake and causing rapid cellular dehydration. Prolonged exposure leads to the ionic phase, characterized by excessive Na^+ and Cl^- accumulation in plant tissues, which interferes with enzyme function, disrupts membrane stability, and compromises metabolic homeostasis (Balasubramaniam et al., 2023). To counter these challenges, plants rely heavily on ABA, a central phytohormone that integrates environmental sensing with adaptive physiological and molecular responses. Under salt stress conditions, ABA functions as a long-distance stress signal linking root perception of salt with adaptive responses in shoots and leaves, coordinating whole-plant tolerance (Khan, 2025). Salt stress rapidly induces ABA biosynthesis, primarily through the upregulation of 9-cis-epoxycarotenoid dioxygenase (NCED), the rate-limiting enzyme in ABA production. *NCED* genes are activated in both roots and shoots under salt stress, with roots responding earlier due to their direct exposure to salt conditions. Once synthesized, ABA is transported via the xylem to aerial tissues, where it initiates multiple adaptive mechanisms, including stomatal closure to reduce transpirational water loss, osmotic adjustment through compatible solute accumulation, enhancement of antioxidant defenses, and regulation of ion transporters to limit Na^+ toxicity, collectively maintaining water balance, preserving cellular integrity, and supporting metabolic stability under prolonged salt exposure (Huang et al., 2018). Further, ABA plays a multifaceted role in enhancing salt tolerance in rice by integrating molecular and physiological responses. Also, it regulates stress-responsive gene expression through crosstalk with other phytohormones, modulates osmotic balance via compatible solute accumulation, and promotes the expression of stress-protective proteins (Fahad et al., 2015). ABA also mitigates Na^+ and Cl^- accumulation, reduces transpiration, and regulates stomatal aperture to conserve water (Xu et al., 2024) while stimulating the biosynthesis of osmoprotectants like proline to help counter oxidative damage. ABA has also been proved to improve the growth and productivity of the plants under salt stress. Reports also demonstrated that exogenous ABA (25 μM) application improved growth characteristics and the development of spikelets in rice plants under salt stress (1.878 dS m^{-1}; Hussain et al., 2023).

Recent studies demonstrate that ABA improves seedling morphology by increasing shoot and root biomass, leaf area, and root surface development under salt stress (Chen et al., 2022). It enhances ion balance, restores hormonal equilibrium, and boosts antioxidant defenses (Chen et al., 2022). ABA also safeguards photosynthesis by maintaining chlorophyll content and photosynthetic efficiency and by activating ROS-scavenging enzymes such as superoxide dismutase (SOD), peroxidase (POD), and ascorbate peroxidase (APX), thereby, reducing oxidative stress markers including malondialdehyde (MDA) and hydrogen peroxide (H_2O_2) (Mahajan et al., 2025b). Salt-induced ABA accumulation in roots leads to morphological adaptations, such as shortened and thickened primary roots, accompanied by increased expression of ABA-responsive genes (Huang et al., 2021).

5.2.1 Abscisic acid–regulated ion homeostasis and cellular ion transport

Maintaining ion homeostasis under salt stress is critical, as excess cytosolic Na^+ disrupts enzyme activity, displaces essential K^+, and destabilizes cellular membranes. ABA plays a central role in preserving ion balance by promoting Na^+ exclusion, enhancing vacuolar sequestration, and maintaining a favorable cytosolic K^+/Na^+ ratio.

A key mechanism underlying this regulation is the SOS (salt overly sensitive) signaling pathway. Cytosolic calcium ion (Ca^{2+}) spikes induced by salt stress are detected by SOS3, which activates the kinase SOS2. SOS2 then phosphorylates and activates the plasma membrane Na^+/H^+ antiporter SOS1, driving Na^+ efflux from root cells and reducing cytotoxicity (Zhao et al., 2021). Recent studies indicate that phosphatidic acid (PA) interacts with SOS2, enhancing its activity and simultaneously coordinating Na^+ efflux via SOS1 with K^+ uptake through the SOS2–SOS3-like calcium-binding protein 8 (SCaBP8)–Arabidopsis K^+ transporter 1 (AKT1) module, thereby maintaining Na^+/K^+ homeostasis (Li et al., 2023). ABA modulates this pathway through crosstalk with SOS signaling, fine-tuning ion fluxes under salt stress. In addition to regulating plasma membrane transporters, ABA controls vacuolar NHX-type Na^+/H^+ exchangers, which compartmentalize Na^+ into vacuoles, lowering cytosolic Na^+ toxicity (Bui et al., 2025). ABA contributes to maintaining a favorable Na^+/K^+ balance under salt stress by increasing K^+ accumulation in rice leaves (Jiang et al., 2024). Secondary messengers including Ca^{2+}, ROS, and nitric oxide (NO) further refine these ABA-mediated ion homeostasis responses, enabling dynamic and precise regulation of cellular ion fluxes during salt stress (Xu et al., 2024).

5.2.2 Abscisic acid–driven osmotic adjustment and cellular protection

Salt stress reduces water availability, lowering soil water potential, and causing osmotic imbalance. ABA promotes osmotic adjustment by inducing the accumulation of compatible solutes such as proline, glycine betaine, trehalose, raffinose-family oligosaccharides, and sugar alcohols like mannitol (Balasubramaniam et al., 2023). Proline synthesis, in particular, is regulated by ABA through activation of Δ^1-pyrroline-5-carboxylate synthetase (P5CS), as seen in salt-stress responses (Szabados and Savouré, 2010). In addition to osmolyte accumulation, ABA enhances the plant's antioxidant defense system, mitigating oxidative damage caused by salt-induced ROS. It increases the activities of key ROS-scavenging enzymes, including SOD, catalase (CAT), and APX, thereby maintaining cellular redox homeostasis under salt stress. Furthermore, ABA regulates aquaporin expression and activity, supporting efficient water transport across membranes and helping to prevent cellular dehydration, which is critical for maintaining turgor and metabolic activity during osmotic stress (Aizaz et al., 2024). Together, these ABA-mediated mechanisms provide a coordinated response that enhances cellular protection, water retention, and overall stress tolerance in rice.

5.2.3 Abscisic acid–mediated stomatal regulation under salt stress

ABA is a central regulator of plant adaptation to salt stress, coordinating stomatal and root responses to maintain water and ion balance. Under salt stress, ABA rapidly accumulates in guard cells, triggering a signaling cascade in which pyrabactin resistance/pyrabactin resistance-like/regulatory component of ABA receptors (PYR/PYL/RCAR) inhibit protein phosphatase 2 C (PP2C), releasing SNF1-related protein kinase 2 (SnRK2), particularly SnRK2.6/OST1 (open stomata 1), to phosphorylate downstream targets such as slow anion channel-associated 1 (SLAC1) and guard cell outward-rectifying K^+ (GORK) channels. This leads to ion efflux, reduced guard-cell turgor, and stomatal closure, minimizing transpirational water loss (Khan, 2025). ABA-induced ROS production via NADPH oxidases (respiratory burst oxidase homologs [RBOHs]) amplifies this signal, coordinating Ca dynamics and ion-channel activity. Mitochondrial proteins like voltage-dependent anion channel 3 (VDAC3) and aquaporins such as plasma membrane intrinsic protein (PIP2;1) regulate ROS homeostasis and water flux, ensuring efficient and reversible stomatal closure under stress (Qin et al., 2024). In roots, ABA gates Na-specific transcriptional responses, limiting excessive gene expression and cellular damage while maintaining growth under salt stress. This regulation prevents ionic toxicity and supports adaptive root function, contributing to overall plant resilience (Lamers et al., 2025).

ABA's dual role in controlling stomatal aperture in leaves and regulating root stress responses underscores its pivotal function in salt-stress tolerance and its potential for enhancing crop resilience in salt-stressed environments.

5.2.4 Abscisic acid–signaling pathway and its integration in salt stress tolerance

ABA signaling pathway functions as a central integrator of salt stress, converting elevated ABA levels into coordinated physiological and molecular responses (Mahajan et al., 2025c). The core signaling mechanism involves PYR/PYL/RCAR receptors, clade A PP2C phosphatases, and SnRK2 kinases: when ABA binds its receptors, PP2Cs are inhibited, freeing SnRK2s (such as SnRK2.2, 2.3, and 2.6) to autophosphorylate and activate downstream targets, including ABF/AREB (ABA-responsive element-binding proteins) transcription factors, ion transporters, and stress-responsive genes (Singh and Roychoudhury, 2023).

Beyond this canonical module, ABA signaling exhibits extensive crosstalk with other hormone pathways. Sah et al. (2016) described how ABA antagonizes gibberellins (GA) to restrict growth under stress and cooperates with jasmonic acid (JA) to amplify stress responses. Additionally, secondary signaling molecules such as Ca^{2+}, mitogen-activated protein kinase (MAPK) cascades, and ROS integrate with ABA signaling to form a multitiered regulatory network, facilitating both immediate and long-term adaptive responses to salinity (Sah et al., 2016).

This network allows plants to translate salt-induced hormonal cues into finely tuned physiological adjustments that preserve water status, maintain ionic homeostasis, and protect cellular structures under salt stress conditions.

5.3 Role of abscisic acid supplementation in rice during salt stress

Exogenous application of ABA has proven to be an effective strategy to enhance salt tolerance in rice by reinforcing the plant's natural defense responses. When externally supplied, ABA supplements the endogenous hormonal pool that becomes rapidly depleted under severe salt stress, thereby, sustaining protective mechanisms such as stomatal regulation, improved photosynthetic performance, and osmotic adjustment. Exogenous ABA also strengthens the antioxidant machinery, limiting oxidative cell damage caused by excess ROS during salt exposure. By stabilizing cellular homeostasis and supporting continued growth and yield formation, exogenously applied ABA serves as a practical physiological intervention for mitigating salt stress in rice (Table 5.1).

TABLE 5.1 Exogenous application of abscisic acid (ABA) ameliorates the salt stress in rice plants.

Salt concentration	Abscisic acid (ABA) concentration	Responses	References
100 mM NaCl	5 μM	Increase in chlorophyll retention, upregulated OsTPP3 transcript levels	Ye et al. (2023)
0.4% NaCl	0.03% S-ABA	Increase in plant height, dry-matter weight, and total chlorophyll content by 13.54%, 52.49%, and 75.34%; enhanced SOD, CAT, APX, and POD; increased number of spikelets per panicle by 24.77%, percentage of filled grains by 26.62%, 1000-grain weight by 21.48%, and yield	Wang et al. (2025)
100 mmol L^{-1}	0.03% S-ABA	Increase in plant height by 13.13% and dry-matter weight by 34.32%, reduced Na^+ content	Chen et al. (2022)
100 mM	10 μM	Prominent root apical swelling in rice seedlings	Huang et al. (2021)
100 mmol L^{-1}	0.03% S-ABA	Reduced Na^+/K^+ content by 73.68%; improved spikelet fertility and promoted effective grain filling	Jiang et al. (2024)
50 mM	1′,4′-trans-diol-ABA (2 mg L^{-1})	Increase in root length; improved K^+ content	Jie et al. (2025)
0.5% w/v	100 μM	Increase in proline content; upregulated OsP5CS1 transcript levels	Sripinyowanich et al. (2013)

ABA, Abscisic acid; *APX*, ascorbate peroxidase; *CAT*, catalase; *OsP5CS1*, *Oryza sativa* Δ^1-pyrroline-5-carboxylate synthetase 1; *OsTPP3*, *Oryza sativa* trehalose-6-phosphate phosphatase 3; *POD*, peroxidase; *S-ABA*, (+)-S-abscisic acid (bioactive stereoisomer of abscisic acid); *SOD*, superoxide dismutase.

5.3.1 Effect on growth and photosynthesis during abscisic acid supplementation in salt-stressed rice plants

Exogenous ABA application has been proved to reverse the salt-induced reduction in photosynthetic efficiency in rice by inducing stomatal closure, accelerating chlorophyll degradation, and generating ROS that damage thylakoid membranes. Reports demonstrated that 5 μM ABA treatment increased chlorophyll retention by nearly 30% in rice seedlings exposed to 100 mM NaCl (Ye et al., 2023). This improvement occurs because ABA fine-tunes guard-cell signaling, enabling controlled stomatal regulation rather than complete closure, thereby helping maintain carbon-dioxide (CO_2) assimilation during moderate stress. ABA simultaneously enhances ROS-scavenging enzymes that protect chloroplast membranes from oxidative damage. Additionally, reports suggested that the exogenous application of trans-abscisic acid (S-ABA; 0.03% S-ABA) through spraying in the 3-leaf/1-heart stage and twice in the break stage significantly increased the plant height, dry-matter weight, and total chlorophyll content by 13.54%, 52.49%, and 75.34%, respectively, compared with salt-stressed (0.4% sodium chloride [NaCl]) rice plants (Wang et al., 2025). In salt-stressed (100 mmol L^{-1} NaCl) Yuxiang youzhan (YXYZ) rice variety, S-ABA (0.03% S-ABA) sprayed at the stage of young panicle differentiation increased the plant height by 13.13% and dry-matter weight by 34.32%, respectively (Jiang et al., 2024). Also, exogenous supplementation of 5 mg·L^{-1} ABA alleviated the salt stress (50 mM) by restoring the decreased plant height caused by salt stress, which resulted in a slight increase in Q3 (on average by 1.69%) and Z3 (on average by 1.53%), respectively, from 6 to 10 days (Chen et al., 2022) (Table 5.1).

5.3.2 Effect on root growth during abscisic acid supplementation in salt-stressed rice plants

Salt stress typically reduces shoot growth, root elongation, and biomass accumulation in rice. However, ABA plays an important role in modifying root structure to improve stress resilience. Huang et al. (2021) demonstrated that exogenous application of 10 μM ABA under salt stress (100 mM) induced prominent root apical swelling in rice seedlings, primarily due to the expansion of cortical cells. This ABA-driven modification of the root apex reflects a physiological adjustment aimed at improving root function under salt stress. Such structural changes may enhance the ability of the roots to access water and maintain nutrient uptake in salt-prone soils. Although this study did not report increases in overall biomass, the observed cellular expansion highlights ABA's critical role in remodeling root architecture to support growth under stressful environments. Treatment with NaCl (50 mM) and 1′,4′-trans-diol-ABA (2 mg L^{-1}) led to a significant increase in root length by 6.48%–85.09% for HHZ and 10.47%–89.90% for XLY900 as compared to NaCl treatment alone (Jie et al., 2025) (Table 5.1).

5.3.3 Effect on antioxidant metabolism during abscisic acid supplementation in salt-stressed rice plants

Under high salt levels, rice accumulates excessive ROS, which can lead to lipid peroxidation, membrane damage, and oxidative stress. Exogenous ABA enhances the activity of key ROS-scavenging enzymes, including SOD, CAT, APX, and POD, thereby reducing ROS levels and improving membrane stability (Jiang et al., 2024). Further, spraying with 0.03% S-ABA to 0.4% NaCl salt-stressed rice plants significantly increased the SOD, APX, and POD activities by 65.28% in wax stages, 55.51% in milk stages, and 55.42% in wax stages, respectively (Wang et al., 2025) (Table 5.1).

5.3.4 Effect on osmoregulation during abscisic acid supplementation in salt-stressed rice plants

ABA plays a central role in maintaining ionic balance in rice under salt stress. Exogenous ABA reduces Na^+ accumulation and increases K^+ retention, thereby improving the Na^+/K^+ ratio and protecting cellular functions (Jiang et al., 2024). ABA also regulates transcription factors such as OsNAC3, which control the expression of Na^+/K^+ transporter genes including *OsHKT1;4* and *OsHKT1;5*, limiting Na^+ translocation to shoots (Zhang et al., 2021a). Additionally, the WD40-repeat protein OsABT enhances ABA-mediated salt tolerance by supporting the activity of ion transporters such as OsSOS1 and OsHAK5 (Wen et al., 2022). Together, these mechanisms allow ABA to orchestrate ion transport and homeostasis, mitigating ionic toxicity and improving rice tolerance to salt stress. Spraying with 0.03% S-ABA to 0.4% NaCl salt-stressed rice plants significantly increased the soluble protein content by 2.60% in the filling stage (Wang et al., 2025). In salt-stressed (100 mmol L^{-1} NaCl) Yuxiang youzhan (YXYZ) rice variety, spraying S-ABA (0.03% S-ABA) reduced the Na^+/K^+ content by 73.68% (Jiang et al., 2024). Treatment with NaCl (50 mM) and 1′,4′-trans-diol-ABA (2 mg L^{-1}) led to increased K^+ content by 0.75% (leaf) and 15.84% (root) in XLY900 (Jie et al., 2025).

Ye et al. (2023) reported that exogenous ABA at 5 μM, applied under 100 mM NaCl stress, significantly upregulated OsTPP3, leading to increased trehalose content and improved salt tolerance in rice seedlings. These results indicate that ABA-mediated trehalose accumulation is a key mechanism for maintaining osmotic balance and supporting rice growth under salt stress. Also, ABA supplementation (5 mg·L^{-1}) to salt stress (50 mM) significantly reduced the Na^+ content (by 8.31% in leaves and 23.54% in roots, respectively) (Chen et al., 2022). 100 μM ABA was topically applied to the salt-treated (0.5% w/v NaCl) seedlings, which showed increased proline content being detected within 3 days after treatment in both rice lines. Also, within 6 days of exogenous ABA application together with salt stress, the OsP5CS1 transcript levels appeared to be upregulated in the salt-tolerant LPT123-TC171 line (Sripinyowanich et al., 2013) (Table 5.1).

5.3.5 Effect on yield attributes during abscisic acid supplementation in salt-stressed rice plants

ABA, particularly in its trans-synthetic form, S-ABA, plays a vital role in protecting rice reproductive development under salt stress. By maintaining ionic balance, enhancing antioxidant defense, and modulating endogenous hormones, ABA stabilizes reproductive organs, improves spikelet fertility, and promotes effective grain filling (Jiang et al., 2024). Exogenous application of S-ABA (0.03% S-ABA) through spraying in the 3-leaf/1-heart stage and twice in the break stage significantly increased the number of spikelets per panicle by 24.77%, percentage of filled grains by 26.62%, 1000-grain weight by 21.48%, and yield by 29.29%, respectively (Wang et al., 2025; Zhang et al., 2021b; Table 5.1).

5.4 Conclusion and future prospects

Salt stress remains a major limitation to rice productivity, disrupting osmotic balance, ion homeostasis, and redox stability. ABA plays a pivotal regulatory role in coordinating rice responses to salt stress through integrated molecular, biochemical, and physiological mechanisms. Under salt stress, ABA biosynthesis and transport are rapidly enhanced, activating the PYR/PYL–PP2C–SnRK2 signaling cascade, which regulates stress-responsive genes involved in ion transport, osmotic adjustment, and antioxidant defense. ABA improves osmotic balance by promoting accumulation of compatible solutes such as proline, glycine betaine, trehalose, and soluble sugars that maintain cell hydration and protect proteins and membranes. It also stabilizes ion homeostasis by modulating Na^+/H^+ antiporters, K^+ uptake transporters, and components of the SOS pathway to maintain a favorable Na^+/K^+ ratio. Furthermore, ABA enhances oxidative stress tolerance by stimulating antioxidant enzymes, such as SOD, CAT, and APX, and activating transcription factors including NAC (NAM, ATAF, and CUC), dehydration-responsive element binding (DREB), and AREB/ABF that regulate detoxification pathways. At the physiological level, ABA improves stomatal control, root system development, photosynthetic efficiency, and overall plant vigor under saline conditions. Exogenous ABA application further alleviates salt-induced damage by enhancing antioxidant capacity, supporting nutrient uptake, and sustaining biomass accumulation. Collectively, ABA functions as a central integrator of multiple signaling networks, interacting with calcium signaling, MAPK cascades, ROS, and other hormones to orchestrate a robust adaptive response to salinity in rice. A deeper understanding of ABA-mediated tolerance mechanisms offers significant potential for developing salt-tolerant rice varieties. Integration of genomics, transcriptomics, molecular breeding, and CRISPR-based gene-editing technologies will be essential for precise manipulation of ABA-regulated pathways to strengthen rice resilience against rising soil salt levels, thereby contributing to global food security.

References

Aizaz, M., Lubna, Jan, R., Asaf, S., Bilal, S., Kim, K.M., et al., 2024. Regulatory dynamics of plant hormones and transcription factors under salt stress. Biology 13 (9), 673. Available at https://doi.org/10.3390/biology13090673, http://www.mdpi.com/journal/biology.

Balasubramaniam, T., Shen, G., Esmaeili, N., Zhang, H., 2023. Plants' response mechanisms to salinity stress. Plants 12 (12), 2253. Available at https://doi.org/10.3390/plants12122253.

Bui, H.H.T., Nguyen, D.H., Dinh, L.T.T., Trinh, H.T.T., Vu, T.K., Bui, V.N., 2025. Comprehensive in silico analysis of the NHX (Na^+/H^+ antiporter) gene in rice (*Oryza sativa* L.). International Journal of Plant Biology 16 (1), 6. Available at https://doi.org/10.3390/ijpb16010006, https://www.mdpi.com/journal/ijpb.

Chen, G., Zheng, D., Feng, N., Zhou, H., Mu, D., Zhao, L., et al., 2022. Physiological mechanisms of ABA-induced salinity tolerance in leaves and roots of rice. Scientific Reports 12 (1), 8228. Available at https://doi.org/10.1038/s41598-022-11408-0.

Chen, T., Niu, Y., Yang, C., Liang, Y., Xu, J., 2024. Screening of rice (*Oryza sativa* L.) genotypes for salinity tolerance and dissecting determinants of tolerance mechanism. Plants 13 (7), 1036. Available at https://doi.org/10.3390/plants13071036.

Fahad, S., Hussain, S., Matloob, A., Khan, F.A., Khaliq, A., Saud, S., et al., 2015. Phytohormones and plant responses to salinity stress: a review. Plant Growth Regulation 75 (2), 391–404. Available at https://doi.org/10.1007/s10725-014-0013-y, http://www.springeronline.com/sgw/cda/frontpage/0,11855,1-40109-70-35547532-0,00.html.

Huang, Y., Guo, Y., Liu, Y., Zhang, F., Wang, Z., Wang, H., et al., 2018. 9-cis-Epoxycarotenoid dioxygenase 3 regulates plant growth and enhances multi-abiotic stress tolerance in rice. Frontiers in Plant Science 9, 162. Available at https://doi.org/10.3389/fpls.2018.00162.

Huang, Y., Zhou, J., Li, Y., Quan, R., Wang, J., Huang, R., et al., 2021. Salt stress promotes abscisic acid accumulation to affect cell proliferation and expansion of primary roots in rice. International Journal of Molecular Sciences 22 (19), 10892. Available at https://doi.org/10.3390/ijms221910892.

Hussain, S., Nanda, S., Ashraf, M., Siddiqui, A.R., Masood, S., Khaskheli, M.A., et al., 2023. Interplay impact of exogenous application of abscisic acid (ABA) and brassinosteroids (BRs) in rice growth, physiology, and resistance under sodium chloride stress. Life (Chicago, Ill.: 1978) 13 (2), 498. Available at https://doi.org/10.3390/life13020498, http://www.mdpi.com/journal/life/.

Jiang, W., Wang, X., Wang, Y., Du, Y., Zhang, S., Zhou, H., et al., 2024. S-ABA enhances rice salt tolerance by regulating Na^+/K^+ balance and hormone homeostasis. Metabolites 14 (4), 181. Available at https://doi.org/10.3390/metabo14040181.

Jie, Y., Feng, S., Zheng, D., Feng, N., Zhou, H., Huang, X., et al., 2025. 1′, 4′-trans-diol-ABA enhances the growth and physiological characteristics of rice (*Oryza sativa* L.) seedlings under salt stress. PLoS One 20 (6), e0284734.

Khan, N., 2025. Molecular insights into ABA-mediated regulation of stress tolerance and development in plants. International Journal of Molecular Sciences 26 (16), 7872. Available at https://doi.org/10.3390/ijms26167872.

Khatoon, S., Mahajan, M., Kumari, S., Iqbal, N., Wahid, I., Khan, M.I.R., 2024. Green synthesized gold nanoparticles induce adaptation in photosynthetic responses, sugar and nitrogen metabolism, and seed yield of salt-stressed mustard plants. Clean Technologies and Environmental Policy 1–17.

Khatoon, S., Mahajan, M., Khan, M.I.R., 2025. Ethylene–potassium interactions safeguard ionic balance, chlorophyll biosynthesis, and source–sink metabolism to sustain wheat growth and yield responses under salt stress. Plant Stress 18, 101049. Available at https://doi.org/10.1016/j.stress.2025.101049, https://www.sciencedirect.com/science/journal/2667064X.

Lamers, J., Zhang, Y., van Zelm, E., Leong, C.K., Meyer, A.J., de Zeeuw, T., et al., 2025. Abscisic acid signaling gates salt-induced responses of plant roots. Proceedings of the National Academy of Sciences of the United States of America 122 (6), e2406373122. Available at https://doi.org/10.1073/pnas.2406373122, https://www.pnas.org/doi/10.1073/pnas.2406373122.

Li, J., Shen, L., Han, X., He, G., Fan, W., Li, Y., et al., 2023. Phosphatidic acid–regulated SOS2 controls sodium and potassium homeostasis in Arabidopsis under salt stress. The EMBO Journal 42 (8), EMBJ2022112401. Available at https://doi.org/10.15252/embj.2022112401.

Liu, J., Shabala, S., Shabala, L., Zhou, M., Meinke, H., Venkataraman, G., et al., 2019. Tissue-specific regulation of Na^+ and K^+ transporters explains genotypic differences in salinity stress tolerance in rice. Frontiers in Plant Science 10, 1361. Available at https://doi.org/10.3389/fpls.2019.01361.

Mahajan, M., Nazir, F., Jahan, B., Siddiqui, M.H., Iqbal, N., Iqbal, M., et al., 2023. Salicylic acid mitigates arsenic stress in rice (*Oryza sativa*) via modulation of nitrogen–sulfur assimilation, ethylene biosynthesis, and defense systems. Agriculture 13 (7), 1293. Available at https://doi.org/10.3390/agriculture13071293.

Mahajan, M., Poor, P., Kaur, H., Aher, R.R., Palakolanu, S.R., Khan, M.I.R., 2025b. Salt stress tolerance and abscisic acid in plants: associating role of plant growth regulators and transcription factors. Plant Physiology and Biochemistry 228, 110303. Available at https://doi.org/10.1016/j.plaphy.2025.110303, https://www.sciencedirect.com/science/journal/09819428.

Mahajan, M., Chaudhary, S., Khan, M.I.R., 2025d. Lanthanum (a rare earth element)-induced abscisic acid responses to modulate stomatal dynamics, nitrogen-carbon assimilation, metabolomic profile and yield traits in wheat. Journal of Hazardous Materials 497, 139763. Available at https://doi.org/10.1016/j.jhazmat.2025.139763, https://www.sciencedirect.com/science/journal/03043894.

Mahajan, M., Khatoon, S., Prakash, S., Khan, M.I.R., Mahajan, M., Khatoon, S., 2025c. Plant stress hormones: fostering drought stress resilience. In optimizing plant health under abiotic-stress environments. (2025c) Optimizing Plant Health under Abiotic-Stress Environments: Ensuring Food and Nutritional Security. CRC Press, India, pp. 123–142. (2025c), https://doi.org/10.1201/9781032630380-7, https://www.taylorfrancis.com/books/edit/10.1201/9781032630380/.

Mahajan, M., Sahoo, R.N., Khan, M.I.R., 2025a. Lanthanum supplementation with abscisic acid modulates rhizosphere dynamics through changes in nitric oxide synthesis in wheat. Physiologia Plantarum 177 (1), e70126. Available at https://doi.org/10.1111/ppl.70126, http://onlinelibrary.wiley.com/journal/10.1111/(ISSN)1399-3054.

Mukhopadhyay, R., Sarkar, B., Jat, H.S., Sharma, P.C., Bolan, N.S., 2021. Soil salinity under climate change: challenges for sustainable agriculture and food security. Journal of Environmental Management 280, 111736. Available at https://doi.org/10.1016/j.jenvman.2020.111736, https://www.sciencedirect.com/journal/journal-of-environmental-management.

Nazir, F., Kumari, S., Mahajan, M., Khan, M.I.R., 2023a. An explicit story of plant abiotic stress resilience: overtone of selenium, plant hormones and other signaling molecules. Plant and Soil 486 (1-2), 135–163. Available at https://doi.org/10.1007/s11104-022-05826-2, https://www.springer.com/journal/11104.

Nazir, F., Mahajan, M., Khatoon, S., Albaqami, M., Ashfaque, F., Chhillar, H., et al., 2023b. Sustaining nitrogen dynamics: a critical aspect for improving salt tolerance in plants. Frontiers in Plant Science 14, 1087946. Available at https://doi.org/10.3389/fpls.2023.1087946, https://www.frontiersin.org/journals/plant-science.

Qin, H., Yang, W., Liu, Z., Ouyang, Y., Wang, X., Duan, H., et al., 2024. Mitochondrial VOLTAGE-DEPENDENT ANION CHANNEL 3 regulates stomatal closure by abscisic acid signaling. Plant Physiology 194 (2), 1041–1058. Available at https://doi.org/10.1093/plphys/kiad516.

Sah, S.K., Reddy, K.R., Li, J., 2016. Abscisic acid and abiotic stress tolerance in crop plants. Frontiers in Plant Science 7, 571. Available at https://doi.org/10.3389/fpls.2016.00571, http://journal.frontiersin.org/article/10.3389/fpls.2016.00571/full.

Saleem, M.A., Khan, A., Tu, J., Huang, W., Liu, Y., Feng, N., et al., 2025. Salinity stress in rice: multilayered approaches for sustainable tolerance. International Journal of Molecular Sciences 26 (13), 6025. Available at https://doi.org/10.3390/ijms26136025.

Sharma, M., Kumar, P., Verma, V., Sharma, R., Bhargava, B., Irfan, M., 2022. Understanding plant stress memory response for abiotic stress resilience: molecular insights and prospects. Plant Physiology and Biochemistry 179, 10–24. Available at https://doi.org/10.1016/j.plaphy.2022.03.004.

Shelar, P.V., Mankar, G.D., Sontakke, O.P., Wayase, U.R., Bhosale, K.S., Nikalje, G.C., et al., 2024. B. A review on physio-biochemical and molecular mechanisms of salt tolerance in crops. Current Agriculture Research 12 (2).

Singh, A., Roychoudhury, A., 2023. Abscisic acid in plants under abiotic stress: crosstalk with major phytohormones. Plant Cell Reports 42 (6), 961–974. Available at https://doi.org/10.1007/s00299-023-03013-w.

Sripinyowanich, S., Klomsakul, P., Boonburapong, B., Bangyeekhun, T., Asami, T., Gu, H., et al., 2013. Exogenous ABA induces salt tolerance in indica rice (*Oryza sativa* L.): The role of OsP5CS1 and OsP5CR gene expression during salt stress. Environmental and Experimental Botany 86, 94–105. Available at https://doi.org/10.1016/j.envexpbot.2010.01.009.

Szabados, L., Savouré, A., 2010. Proline: a multifunctional amino acid. Trends in Plant Science 15 (2), 89–97. Available at https://doi.org/10.1016/j.tplants.2009.11.009.

Wang, X., Ma, G., Feng, N., Zheng, D., Zhou, H., Li, J., et al., 2025. Abscisic acid can improve the salt tolerance and yield of rice by improving its physiological characteristics. Agronomy 15 (2), 309. Available at https://doi.org/10.3390/agronomy15020309.

Wen, D., Bao, L., Huang, X., Qian, X., Chen, E., Shen, B., 2022. OsABT is involved in abscisic acid signaling pathway and salt tolerance of roots at the rice seedling stage. International Journal of Molecular Sciences 23 (18), 10656. Available at https://doi.org/10.3390/ijms231810656.

Xu, J., Lu, X., Liu, Y., Lan, W., Wei, Z., Yu, W., et al., 2024. Interaction between ABA and NO in plants under abiotic stresses and its regulatory mechanisms. Frontiers in Plant Science 15, 1330948. Available at https://doi.org/10.3389/fpls.2024.1330948.

Ye, N., Wang, Y., Yu, H., Qin, Z., Zhang, J., Duan, M., et al., 2023. Abscisic acid enhances trehalose content via OsTPP3 to improve salt tolerance in rice seedlings. Plants 12 (14), 2665. Available at https://doi.org/10.3390/plants12142665.

Zhang, Y., Kilambi, H.V., Liu, J., Bar, H., Lazary, S., Egbaria, A., et al., 2021a. ABA homeostasis and long-distance translocation are redundantly regulated by ABCG ABA importers. Science Advances 7 (43), eabf6069. Available at https://doi.org/10.1126/sciadv.abf6069, https://www.science.org/doi/10.1126/sciadv.abf6069.

Zhang, X., Long, Y., Chen, X., Zhang, B., Xin, Y., Li, L., et al., 2021b. A NAC transcription factor OsNAC3 positively regulates ABA response and salt tolerance in rice. BMC Plant Biology 21 (1), 546.

Zhang, Y., Xu, J., Li, R., Ge, Y., Li, Y., Li, R., 2023. Plants' response to abiotic stress: mechanisms and strategies. International Journal of Molecular Sciences 24 (13), 10915. Available at https://doi.org/10.3390/ijms241310915.

Zhao, S., Zhang, Q., Liu, M., Zhou, H., Ma, C., Wang, P., 2021. Regulation of plant responses to salt stress. International Journal of Molecular Sciences 22 (9), 4609. Available at https://doi.org/10.3390/ijms22094609.

Zuo, G., Huo, J., Yang, X., Mei, W., Zhang, R., Khan, A., et al., 2024. Photosynthetic mechanisms underlying NaCl-induced salinity tolerance in rice (*Oryza sativa*). BMC Plant Biology 24 (1), 41. Available at https://doi.org/10.1186/s12870-024-04723-3.

CHAPTER

6

Endophyte-driven strategies for abiotic stress tolerance in plants

Nagarathnamma Yammanuri, Vishnu Babu Vallepu, Krishna Kumar Guduru, Siva Bolupalli, Shajidabi Shaik, Chandra Sekhar Akila, and Chandra Obul Reddy Puli

Department of Botany, Yogi Vemana University, Kadapa, Andhra Pradesh, India

6.1 Introduction

As sessile organisms, plants are constantly subjected to biotic and abiotic stresses that affect their growth, development, and ultimately yield. Among the abiotic stressors, drought, salinity, heat, cold, waterlogging, and heavy metals are significant causes of yield losses worldwide, as they impair plants' physiological processes (Anand et al., 2023; Morales-Vargas et al., 2024). These constraints are exacerbated by climate change, which increases the frequency of droughts and expands salt-affected soils and increases the frequency of droughts (Verma et al., 2021a; Morales-Vargas et al., 2024). As the global population is expected to reach nearly 10 billion people by 2050, one of the major issues the world will face is how to feed the rising population; thus the need to make crops more resilient to stress and increase their yield becomes imperative (van Dijk et al., 2021).

Conventional approaches, such as crop rotation and breeding, and advanced technologies, including marker-assisted breeding and transgenic plants, have paved the way to a partial solution to the problem. However, in most cases, they face limitations such as cost, time, environmental safety, and public acceptance. In the meantime, endophytes, the microorganisms that establish symbiotic associations with plants and live inside their tissues without causing any harm, have become the plant's new, powerful, green, and sustainable allies, offering a straightforward solution to the problem of abiotic stress tolerance to a great extent (Anand et al., 2023; Ameen et al., 2024; Morales-Vargas et al., 2024). Microorganisms, including bacteria and

Harnessing Plant-Stimulating Molecules
DOI: https://doi.org/10.1016/B978-0-443-44994-9.00013-4

fungi, form deep, close, and complex relationships with their host plants (Morales-Vargas et al., 2024). Microorganisms improve host plants' stress resilience by installing a full range of concerted actions, which includes the accumulation of osmolytes and antioxidants to resist reactive oxygen species (ROS) attack, managing plant water uptake and keeping balance, and the control of primary phytohormones and genes activated by stress (Ameen et al., 2024; Pandey et al., 2023; Morales-Vargas et al., 2024).

This chapter will examine the details of signaling and regulatory processes that allow endophytes to support plants during stressful conditions. This, in turn, significantly reduces the impact of adverse conditions on agriculture and helps secure the future of sustainable farming.

6.2 Endophytes as hidden allies

6.2.1 Definition and historical context

"Endophytic" is a term derived from Greek, where "phyton" means plant and "endon" means inside. Therefore, an endophyte is a microorganism, primarily fungi and bacteria, that lives in a symbiotic relationship within a plant without causing disease symptoms, at least for most of its life cycle (Hardoim et al., 2015; Liao et al., 2025). These microorganisms are integral to the plant microbiome, having been found in every plant species studied (Watts et al., 2023; Liao et al., 2025).

The German botanist Heinrich Friedrich link first referred to 'Entophytae' in 1809; later, Anton de Bary formally established the term in 1866 to refer to "organisms that grow within plant tissues" (De Bary, 1866; Petrini, 1991; Schulz et al., 2002; Hardoim et al., 2015; Fagorzi and Mengoni, 2022; Kumar et al., 2025a). For most of the 19th century, the prevailing opinion was that healthy plants lacked bacteria and were therefore internally sterile (Hardoim et al., 2015). Victor Galippe's work in 1887, which involved discovering bacteria inside plant tissues and thereby laying the groundwork for our current understanding, was the first to challenge this belief (Compant et al., 2012). Initially, the studies focused on fungal endophytes. However, the importance of bacterial endophytes has become increasingly evident over the last few decades (Fagorzi and Mengoni, 2022).

Endophytes, which survive inside the host and derive benefits without causing disease, are considered one of the main ecological and agricultural wonders with which we can make an alliance and reap the fruits of their labor in return. Host plants provide carbon and a safe home for endophytes in mutualistic relationships, while endophytes significantly contribute to the plant's resistance to various abiotic stresses, including heavy metals, drought, and salinity. Also, they improve nutrient acquisition and growth (Watts et al., 2023; Fadiji and Babalola, 2020). This natural association demonstrates that endophyte-based practices are a powerful, sustainable approach to enhance crop productivity and manage marginal lands, which are becoming increasingly vital due to climate change (Watts et al., 2023).

6.2.2 Diversity and distribution

Endophytes are natural components of the plant microbiome, as they are present in every plant species examined to date, including trees, crops, mosses, and ferns (Gouda et al., 2016; Harrison and Griffin, 2020; Gakuubi et al., 2021; Mametja et al., 2025). Without showing the typical symptoms of the disease, they colonize the internal tissues of their hosts, such as roots, stems, leaves, seeds, and flowers (Gouda et al., 2016; Kumar et al., 2025a; Fagorzi and Mengoni, 2022; Semenzato and Fani, 2024). The variety and composition of these endophytic communities are not random; they result from the combined influence of factors such as host plant species and genotype, tissue type, environmental conditions, and geographical location (Aishwarya et al., 2025; Harrison and Griffin, 2020; Gakuubi et al., 2021).

The variety of endophytes is incredible. Based on a plant-fungus ratio, the global count of fungal endophyte species is projected to exceed 1 million, with recent analyses suggesting this number could be significantly higher, possibly reaching over 3 million species (Liao et al., 2025). However, only a minuscule fraction of the diversity has been documented, with one survey identifying about 2224 species from 877 genera (Liao et al., 2025). Research effort has not been evenly distributed. At the same time, fungal endophytes have been characterized in hosts from approximately 30% of embryophyte families; their bacterial counterparts have been surveyed in only about 10.5% of families, highlighting a significant knowledge gap (Harrison and Griffin, 2020; Liao et al., 2025). Considering that endophytes are taxonomically diverse, they incorporate a wide range of bacteria and fungi. Table 6.1 gives a brief description of some of the most commonly documented genera and their general functional roles.

The distribution of endophytes throughout a plant is a complex and dispersed process. In woody plants, the fungal communities with the most significant diversity are found in the wood, *spe*cifically in the stems of these plants. In contrast, in nonwoody plants, the roots are the most diverse tissues (Harrison and Griffin, 2020). A different division of labor between tissues ensures the wide provision of necessary functions directly from the endophyte partners, thereby maintaining the health and survival of the host plant. Their involvement is varied, from nutrient uptake through the rhizophagy cycle—a process whereby the plant

TABLE 6.1 Common genera of endophytes.

Microbial group	Common genera	Functions	References
Fungal endophytes	*Colletotrichum, Fusarium, Epichloë, Curvularia, Phyllosticta, Diaporthe*	Producers of diverse bioactive compounds; some can switch between endophytic and pathogenic lifestyles	Harrison and Griffin (2020); Liao et al. (2025)
Bacterial endophytes	*Bacillus, Pseudomonas, Herbaspirillum, Burkholderia, Enterobacter, Azoarcus*	Nitrogen fixation, phosphate solubilization, production of phytohormones (e.g., IAA), and biocontrol against pathogens	White et al. (2019); Hnamte et al. (2024); Aishwarya et al. (2025)

IAA, Indole-3-acetic acid.

gets the nutrients from the bacteria living inside it (White et al., 2019)—to the alleviation of both biotic and abiotic stresses via the induction of systemic resistance and the production of protective metabolites (Aishwarya et al., 2025; Hnamte et al., 2024; Fagorzi and Mengoni, 2022).

6.2.3 Colonization and entry mechanisms

Endophyte tissue invasion is a complex, multistep process, indicating that they have achieved a safe coexistence with their symbiotic partner. The process of selecting a host is followed by chemotaxis, where endophytes are attracted to specific compounds in the root exudates of the plant (Zhou et al., 2024; Parvez et al., 2025; Lengrand et al., 2025). For example, root exudates from tomato plants can initiate the spore germination of beneficial endophytic fungi, such as *Aspergillus terreus*, while simultaneously inhibiting the germination of harmful ones (Parvez et al., 2025). The formation of the plant's microbiome relies on such preliminary communication.

Endophytes enter plants in two ways, and Tariq et al. (2025) described the methods used by endophytic organisms to access the plant.

6.2.3.1 *Passive entry*

Plants have natural natural openings on root hairs, stomata, and lenticels, which are especially used by bacterial endophytes. They also utilize the wounds caused by the emergence of lateral roots (Tariq et al., 2025).

6.2.3.2 *Active entry*

In such an operation, endophytic bacteria produce cell-wall-degrading enzymes that degrade the cell wall, namely cutinases, cellulases, and pectinases (Mushtaq et al., 2023; Tariq et al., 2025). The enzymes loosen the plant cell wall structure, allowing the endophyte to enter without directly injuring the plant. The plant's immune system is adapted to allow the endophyte to pass through at this stage. After that, systemic colonization begins within (Kandel et al., 2017; Verma et al., 2021b).

As filamentous organisms, fungi primarily use an active penetration method that combines mechanical breaking of the target material with enzymatic degradation. They develop a specialized infection structure known as appressoria, which generates high turgor pressure to physically rupture the plant cuticle and cell wall. At the same time, they release a mixture of cell wall–degrading enzymes (CWDEs), such as cutinases and cellulases, to facilitate breaking (Giraldo and Valent, 2013; Mendgen et al., 1996).

6.2.4 Signal transduction mechanism of endophyte colonization

Mutually secreted chemical signals facilitate the molecular interaction between plants and microbes in the rhizosphere. These chemical signals ultimately determine whether a relationship is symbiotic or pathogenic. Research studies indicate that beneficial bacteria and fungi use a common chemical language to establish symbiosis; in contrast, pathogens diversify and expand their molecular weapons to weaken the host's immune defenses and gain access and establish colonization.

Lipochitooligosaccharides (LCOs), which are also referred to as Nod factors for rhizobia and Myc factors for mycorrhizal fungi, are the central signaling molecules for beneficial interactions. These LCOs, which consist of a chitin-derived oligosaccharide backbone with different substitutions, are recognized by plant LysM receptor kinases (Oldroyd, 2013). The shared recognition mechanism leaves open the possibility that the ancient mycorrhizal signaling pathway has been evolutionarily co-opted for the newer rhizobial symbiosis. Nevertheless, the most recent study has gone beyond this model to show that in addition to symbiotic bacteria, fungi can also produce a broader range of bioactive chitooligosaccharides (COs) that have the potential to modulate plant immunity and facilitate colonization (Liang et al., 2014). In addition, the first plant signals, usually flavonoids and strigolactones, have been found to influence the root microbiome in a way much more advanced than previously assumed, as they not only selectively activate symbiotic pathways in certain partners but may also repel others. One of the most important revelations from recent research is the indistinguishability of the functional boundaries between symbiotic and pathogenic signals. As an example, it was discovered that some mycorrhizal fungi and beneficial endophytes not only produce but also secrete effector-like proteins that help them to evade plant immunity, a trick to which the scientists were previously attributing pathogens only (Snelders et al., 2022).

The colonization of endophytes in plants requires complex signal transduction pathways that begin with the mutual recognition of chemical signals and proceed through the establishment of beneficial symbiotic relationships in plant tissues. For instance, one of the most essential characteristics of the endophyte's successful colonization is its capacity to breach the plant's highly protected immune system (Biswas et al., 2025; Wang et al., 2023).

The subsequent stage involves the endophyte's ability to evade or overcome the host plant's robust immune system, facilitating colonization. These are the main stages of this process (Fig. 6.1).

6.2.4.1 *Immune system triggering in plants*

Pattern recognition receptors (PRRs) on plant cell membranes enable plants to detect microbial structures called microbe-associated molecular patterns (MAMPs), which are highly conserved across a wide range of microorganisms (Mengistu, 2020; Wang et al., 2023). This detection triggers the first layer of plant defense known as pattern-triggered immunity (PTI), which may also involve an oxidative burst (Mengistu, 2020).

6.2.4.2 *Immune evasion leading to symbiotic relationship*

Endophytes with beneficial traits can evade and suppress the host's powerful immune response. For instance, the flagellin of the endophyte *Paraburkholderia phytofirmans* PsJN only elicits a very weak and short oxidative burst compared to that of pathogens (Mengistu, 2020). Some endophytes, such as *Bacillus subtilis*, produce a protein called subtilomycin that interferes with the plant's ability to recognize the binding of the protein to flagellin as a threat (Mengistu, 2020). Moreover, endophytes generally influence the most vital phyto-hormone pathways by selectively altering jasmonic acid (JA) signaling, which not only helps

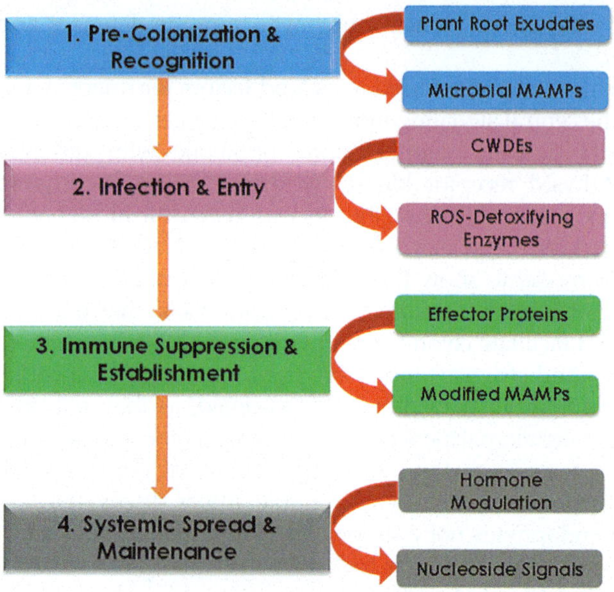

FIGURE 6.1 Key stages of signal transduction during endophyte colonization. The successful establishment of endophytes is a multistep process that involves a complex molecular conversation with the host plant. The process moves from initial recognition and infection to the suppression of plant immune responses and the long-term maintenance of the symbiosis, supported by specific signaling molecules at each stage. *CWDEs*, Cell wall–degrading enzymes; *MAMPs*, microbe-associated molecular patterns; *ROS*, reactive oxygen species; *JA*, jasmonic acid; *SA*, salicylic acid; *ET*, ethylene; *dAdo*, deoxyadenosine.

maintain the balance of the endophyte population within the host but also facilitates interaction rather than the strong defense response triggered by pathogens that commonly occur (Mengistu, 2020). Once the endophytes enter the plant, they activate the host's intracellular signaling networks to establish symbiosis and facilitate systemic benefits.

6.2.4.3 *Instigation of phytohormone signaling*

The presence of an endophyte significantly affects the plant's hormone signaling pathway. One of the major interplays is the modulation of the JA and salicylic acid (SA) pathways (Mengistu, 2020; Pandey et al., 2023). It was found that beneficial endophytes exhibit sustained activation of JA-related genes in rice roots, which consequently increases endophyte density (Mengistu, 2020). This hormonal dialogue between the two is pivotal for activating the plant's own resistance mechanisms, a phenomenon known as induced systemic resistance (ISR), which prepares the plant to better withstand subsequent pathogen attacks (Mengistu, 2020; Pandey et al., 2023).

6.2.4.4 *Secondary messengers as protagonists*

The crosstalk between calcium (Ca^{2+}) and ROS is a crucial signaling node (Wang et al., 2023). Although a severe oxidative burst is a defense response, controlled, nuanced changes in ROS can also play a crucial role as important second messengers, broadening intracellular signaling and, thus, affecting downstream metabolic pathways, ultimately leading to the establishment of the symbiotic state (Wang et al., 2023).

6.2.4.5 *Unique nucleoside signals*

Breakthrough research has identified the most precise and novel signaling pathways in the fungal endophyte *Serendipita indica* (Dunken et al., 2024). The fungus secretes two enzymes, namely, the ecto-5′-nucleotidase *Si*E5NT and the nuclease *Si* NucA, into the apoplast, which together make the nucleoside deoxyadenosine (dAdo). After being taken up by plant cells via the ENT3 transporter, this molecule primarily activates the specific TIR-NLR receptor, which then triggers the controlled death of nearby host cells, enabling the fungus to successfully colonize the root (Dunken et al., 2024).

Endophytes can colonize the intercellular spaces or vascular tissues and travel along roots, shoots, flowers, and seeds (Tariq et al., 2025). Several factors, including the plant's age, species, genetics, and microbial strain, influence the success of this colonization process. For example, *Pseudomonas fluorescens* is a bacterium that specifically colonizes olive roots, while *Firmicutes and Bacillus* may colonize the seed endosperm (Cabanás et al., 2014; Truyens et al., 2015; Tariq et al., 2025; Prieto and Mercado-Blanco, 2008).

One of the primary reasons for retaining endophytic microorganisms within plants for an extended period is the formation of biofilms. These protective microbial communities enhance the survival and function of the endophyte within plant tissues. In a biofilm, microorganisms are protected against various environmental stressors, including drought, temperature fluctuations, and heavy metals (Tariq et al., 2025). The endophyte–host interaction is also deepened by the endophyte's capacity to influence the major plant hormone groups, namely, auxins, cytokinins, and JA. Hence, this change facilitates root growth and maintains a balanced relationship, thereby preventing a pathogenic response (van Dijk et al., 2021; Tariq et al., 2025).

6.2.5 Functional roles of endophytes

After colonization, endophytes function as a hidden "second genome" that enables the host plant to perform many additional functions (Eid et al., 2021). They not only quietly survive in the plants but also have a very intricate mutualistic relationship with them, in which, through this partnership, they directly increase the plant's fitness and productivity (Li et al., 2023a). Their functions, on the one hand, solubilize nutrients needed by the plant that are naturally insoluble, such as phosphorus and iron; on the other hand, they produce a wide range of phytohormones for growth, mainly auxins and cytokinins. Finally, they achieve this by synthesizing two groups of chemical compounds: antimicrobial compounds and induced systemic resistance (ISR), which protect the host against disease (Tariq et al., 2025; Rather, 2025; Biswas et al., 2025).

However, Endophytes have attracted particular attention for their role in stress mitigation that the most research emerges, challenging the conventional view of endophytes as a primary resource for sustainable agriculture. Endophytes have demonstrated their ability to significantly support agricultural development amid the environmental challenges posed by a hotter, drier climate, increased soil salinity, and land contamination by heavy metals. They are driving the change in the host's entire physiological and molecular networks, which in turn result in the synthesis of osmoprotectants, improvement of antioxidant defense, which in turn scavenges the ROS, and regulation of the expression of genes involved in the stress response pathways

(Godara and Ramakrishna, 2023; Kumar et al., 2025b; Cui et al., 2024). By making the plant "primed" for adversity and making the plant's response stronger, endophytes are bringing about a situation where the plant is under less stress. Therefore, it can continue growing in a usual manner, while the yield can be maintained at a high level in a nonoptimal environment (Godara and Ramakrishna, 2023; Tariq et al., 2025). This feature of enabling plant robustness against abiotic stress without genetic engineering or excessive chemical inputs makes endophyte-based methods a front-runner in the development of climate-smart and sustainable agricultural systems (Godara and Ramakrishna, 2023; Biswas et al., 2025; Cui et al., 2024).

6.3 Endophyte-mediated mechanisms of abiotic stress tolerance

The fundamental ways by which endophytes strengthen plant resistance may be generally classified as follows:

6.3.1 Osmoprotection and ion homeostasis

To protect themselves against hostile environmental conditions, endophytic microbes deploy typical "compatible solutes" used as osmoprotectants: for instance, the endophytic bacteria frequently produce or accumulate glycine betaine, ectoine/hydroxyectoine, proline, trehalose, and K^+ to maintain turgor and protect proteins and membranes under drought or salinity; these characteristics are reported in many of the plant-associated *Bacilli* and *Pseudomonads* and represent one manner in which endophytes support the host stress tolerance (e.g., glycine betaine/ectoine/proline/trehalose pathways) (Ameen et al., 2024; Eswaran et al., 2024; Bruger et al., 2024). On the other hand, endophytic fungi mainly produce polyols (mannitol, arabitol, and glycerol) and trehalose as compatible solutes, which not only compensate for osmotic pressure but also serve as ROS scavengers and stabilizers during dehydration, heat, cold, and salt stresses; these functions are confirmed for filamentous fungi and yeasts and go far beyond to the endophytes, which are the source of plant abiotic-stress relief (Dijksterhuis and de Vries, 2006; Meena et al., 2015; Morales-Vargas et al., 2024).

Along with osmoprotection, endophytes play a critical role not only as a salt barrier but also in maintaining ion homeostasis. Salt stress typically results in Na^+ toxicity, leading to disturbances in the uptake of nutrients such as potassium (K^+) (Bilal et al., 2023). Endophytes are most effective at accomplishing this by blocking harmful ions (Na^+ and Cl^-) from entering the plant. At the same time, they facilitate the plant's uptake and translocation of essential ions, such as K^+ and Ca^{2+} (Bilal et al., 2023; Ameen et al., 2024). For example, the research highlights that the interaction between endophytic SB10 and glycine betaine in soybeans helps regulate sodium transporters, maintaining a stable Na^+/K^+ ratio and promoting nutrient uptake under mixed stress conditions (Bilal et al., 2023). This ionic balance is essential for maintaining enzymatic activity and metabolic processes.

Table 6.2 provides an overview of the main osmoprotectants and their respective roles in endophyte-mediated processes:

TABLE 6.2 Summary of the primary osmoprotectants and their specific roles mediated by endophytes.

Osmoprotectant	Role in stress tolerance	Endophyte-mediated mechanism and example	References
Proline	Proline serves as a strong antioxidant and osmoprotectant; it helps stabilize proteins and other cellular materials and scavenges reactive oxygen species.	Endophytes not only enhance the expression of plant biosynthesis genes (e.g., P5CS1) but they can also supply proline directly. It was found that proline induced by drought in watermelon was mainly produced in leaves and then transported to the roots.	Vujanovic et al. (2022); Wang et al. (2022)
Glycine betaine (GB)	Help to maintain the function of the photosynthetic apparatus, keep protein integrity, and also regulate ion channel function so as to avoid salinity-induced potassium (K+) loss.	Endophytic bacteria, such as *Bacillus safensis*BTL5, produce metabolites that promote GB production. The combined use of the endophyte *Trichoderma virens* (SB10) with exogenous GB improved ion homeostasis in soybeans subjected to combined heat and salinity stress.	Bilal et al. (2023); Singh et al. (2025)
Bacterial osmoprotectants	Overall, they maintain intracellular osmotic balance, stabilize soil biofilms, and improve soil conditions, providing indirect support for plants in saline soils.	Halotolerant endophytic bacteria produce and release a wide variety of compatible solutes. The main bacterial osmoprotectants are Ectoine and hydroxyectoine: extremely effective stabilizers that help enzymes and cell structures withstand drought stress and confer heat tolerance. **Trehalose:** A nonreducing disaccharide that serves as a protector of biomolecules and membranes under abiotic stress. **Proline and glycine betaine**: Bacteria, besides plants, also produce these compounds for their own osmoprotection. These compounds recruit water molecules to form a hydration shell around themselves and bind soil particles into aggregates, which makes the soil a better growing medium for the plant.	Goszcz et al. (2025)

6.3.2 Antioxidant defense and reactive oxygen species scavenging

Abiotic stresses disrupt cellular redox homeostasis, leading to the overproduction of ROS. ROS produced during redox imbalance include hydrogen peroxide (H_2O_2), superoxide radicals ($O_2\bullet^-$), and hydroxyl radicals (•OH) (Verma et al., 2021a; Sahu et al., 2022). The oxidative burst triggers the complete oxidation of lipids, proteins, and DNA, severely compromising the final outcome of plant growth and survival. To alleviate such harm, both bacterial and fungal endophytes act together to implement a well-planned dual antioxidant strategy to support the plant's defense systems (Verma et al., 2021a, 2022).

This stronger defense does not rely solely on promoting the antioxidant enzymes characteristic of the plant, but also on antioxidant biosynthesis to produce protective, low-molecular-weight metabolites. Specifically, the gene expression leading to the initiation of enzymatic antioxidants that are the mainstays of the catalytic reaction, including superoxide dismutase (SOD), catalase (CAT), and ascorbate peroxidase (APX), combines their functions into one and neutralizes different ROS types simultaneously (Sahu et al., 2022; Sadeghi et al., 2020). At the same time, they activate the generation of nonenzymatic antioxidants such as glutathione, ascorbic acid (AsA), and flavonoids, which not only scavenge ROS directly but also maintain cellular redox balance (Asomadu et al., 2024; Akram et al., 2017; Sahu et al., 2022).

6.3.3 Endophyte-mediated hormonal crosstalk to stress regulation

Endophytes delicately change the plant's hormonal balance to increase stress resistance. They act as main controllers not only because they produce hormones themselves but also because they impact the host's hormonal signaling pathways.

6.3.3.1 Abscisic acid signaling and the stress–growth trade-off

Endophytes are likely to contribute to drought tolerance in plants by modulating abscisic acid (ABA), a key hormone in stress response. The *Epichloë endophyte* significantly improved drought resistance in drunken horse grass by altering ABA levels, thereby boosting photosynthetic efficiency and growth during water shortages (Cui et al., 2022). Recently, a review suggests that several endophytes might use ABA-independent pathways; these strategies could help plants overcome the typical growth restrictions caused by ABA-dependent defense mechanisms (Ullah et al., 2025).

6.3.3.2 Jasmonic acid signaling in defense and metabolism

Endophytes could activate the JA signaling pathway, enabling plant defenses to respond quickly. Drawing a comparison with the Atractylodes lancea plantlets study, the result showed that the fungal endophyte In *Atractylodes lancea* plantlets, the fungal endophyte *Gilmaniella* sp. AL12 increased JA accumulation, which acted as a downstream signal for antimicrobial volatile oil production of volatile oil known for its antimicrobial properties (Ren and Dai, 2012). At the same time, a study done on fungal endophytes *Epichloë gansuensis* in *Achnatherum sibiricum* suggests that the endophytes that are not dependent on the JA pathway can improve host resistance to herbivores like locusts via mechanisms that are coexistent alongside the JA pathway but are different, pointing toward the plurality of strategies of stress relief (Qin et al., 2019).

When the endophytic bacterium *Pseudomonas fluorescens* SS101 was introduced into rice, a strong and sustained increase in JA levels was observed. The defense mechanism along the JA pathway was very tightly correlated to the ISR, which was experimentally confirmed as rice's resistance to the leaf blight pathogen *Xanthomonas oryzae* pv. *Oryzae* (Tran et al., 2007).

Research on the endophytic Bacillus cereus AR156 in Arabidopsis revealed that it confers ISR against the pathogen *Pseudomonas syringae* pv. tomato DC3000. The study demonstrated that the bacterium is a priming agent of the JA/ethylene-dependent defense pathways, but it also causes a very brief and low-level accumulation of SA in the early stages of the interaction, which is considered to be the factor that actually contributes to the priming without a full-blown systemic acquired resistance (SAR) response being engaged (Wang et al., 2018).

6.3.3.3 Direct hormone production and ethylene management

The research reports that the fungal endophytes *Phoma glomerata* and *Penicillium* sp. are capable of directly producing the plant growth hormones auxins (Indole-3-acetic acid [IAA]) and gibberellins (GA1, GA3, GA4, and GA7) without intermediates. Inoculating these fungi significantly enhanced plant growth and tolerance to salt and water stress, as well as root and shoot development (Waqas et al., 2012). Additionally, most beneficial bacteria, such as *Pseudomonas putida*, *Bacillus* spp., and *Variovorax paradoxus*, produce the 1-aminocyclopropane-1-carboxylate (ACC) deaminase enzyme. This enzyme degrades ACC, the compound that is the immediate precursor of ethylene in plants, reduces the accumulation of ethylene that is caused by stress and the negative impacts of root growth and senescence caused by it (Gamalero and Glick, 2015; Orozco-Mosqueda et al., 2020).

6.3.3.4 Auxin-mediated growth and stress resilience

Auxin (mostly IAA) is a key molecule that integrates plant growth and stress responses. Research on endophytes suggests that many bacterial and fungal endophytes can produce, metabolize, or modify IAA, thereby altering root system architecture (RSA), advancing resource foraging, and regulating hormone crosstalk with ABA/JA/SA under salinity, drought, and temperature stress. Recent studies have clarified (1) tryptophan-dependent microbial IAA biosynthetic pathways (e.g., indole-3-pyruvate, IPyA) in bacteria and fungi; (2) endophyte auxin control through auxin degradation; and (3) host transcriptional rewiring of TIR1/AFB–Aux/IAA–ARF signaling, resulting in lateral root (LR) and root hair (RH) proliferation conjoined with stress resistance (Chen et al., 2023; Etesami and Glick, 2024).

Both classic and recent experiments have demonstrated that auxin-producing *Azospirillum* and *Bacillus* strains modulate RSA by decreasing primary root extension while increasing lateral roots and root hairs, a typical auxin-driven RSA plasticity that enables better water and nutrient acquisition under stress. In fact, transcriptomic changes in *Arabidopsis* caused by *Azospirillum brasilense* Sp245 are very much in line with auxin response and LR/RH induction; new reviews also strongly support that auxin-dependent root branching accounts for the majority of the growth-promoting effect (Spaepen et al., 2014; Pelagio-Flores et al., 2025). The multistress-tolerant (heat and salt) endophytic bacteria *Enterobacter* sp. SA187 can completely alter the host transcriptional pattern; although most SA187 studies focus on ethylene/epigenetic signaling, its growth-promoting phenotype is accompanied by auxin-related developmental changes, implicating auxin as part of a larger network of SA187-induced resilience (Shekhawat et al., 2021; Andres-Barrao et al., 2021).

The endophytic fungus *Penicillium citrinum* secretes auxin (probably via the IPyA pathway), which directly affects the growth of *Arabidopsis* roots and shoots, as shown by DR5 reporters. Now, genomic and functional evidence reveals the capacity of fungal phytohormone biosynthesis (auxin and gibberellins) in *Penicillium* spp. (Chen et al., 2023; Lau et al., 2024) (Fig. 6.2).

In addition to biomass increases, *Piriformospora (Serendipita) indica*, a model root endophyte, reprograms host ion homeostasis and enhances drought/heat/salt resistance; auxin is involved in these developmental and signaling changes. A variety of recent experiments with different hosts (e.g., *Rhododendron* and *Paulownia*) have reached the same conclusion: stress relief through endophytes accompanied by auxin-related growth responses (Ji et al., 2024; Li et al., 2023a).

6.3.4 Molecular and genetic regulation by endophytes

Endophytes, both fungal and bacterial, play a significant role in improving plant stress tolerance. They do this by altering the host's molecular and genetic machinery. Thus, these microorganisms act as epigenetic and transcriptional reprogrammers, equipping the plant

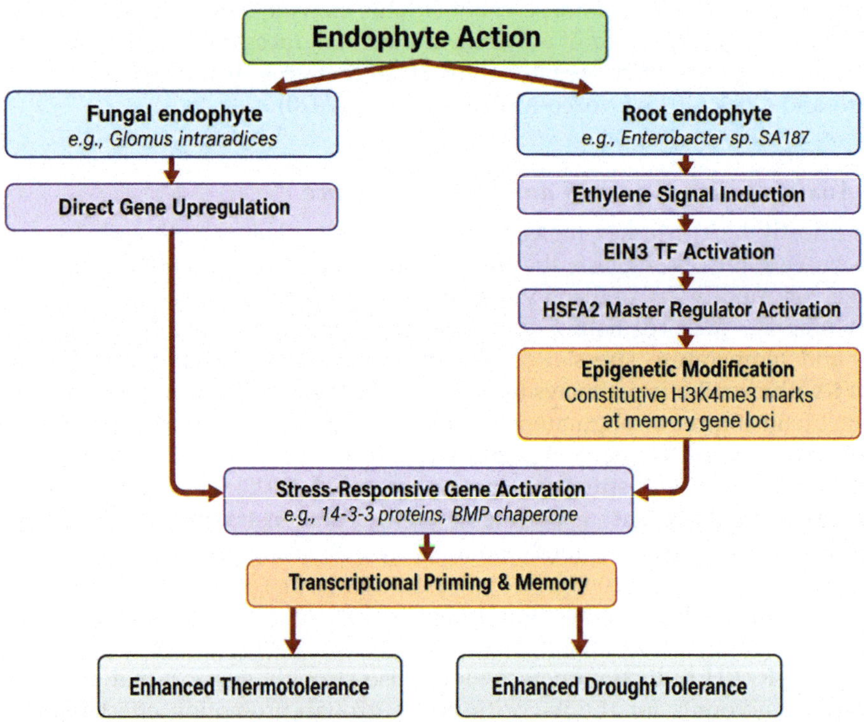

FIGURE 6.2 Molecular mechanisms of plant stress tolerance by bacterial and fungal endophytes. This diagram illustrates the distinct yet interconnected signaling pathways through which endophytes reprogram the host plant's gene expression to provide increased stress tolerance.

with a more potent and efficient response to abiotic stresses (Chen et al., 2022). Moreover, this regulation activates the key transcription factors, thereby inducing stress-memory genes that maintain the current tolerance phenotype (Fig. 6.2). Recently, extensive data on the molecular mechanisms of plant–endophyte interactions have been collected. Here are some of the most important mechanisms, with examples from this type of research, summarized in (Fig. 6.2).

6.3.4.1 Beyond single genes

The findings indicate that the gene expression patterns in the host regulated by endophytes are not limited to single genes but encompass complex gene networks. The 14-3-3 proteins regulated by *Glomus intraradices* exemplify this, as they act as central hubs that coordinate various signaling pathways and effector proteins, driving the plant's final response to drought (Porcel et al., 2006).

6.3.4.2 Epigenetic priming for long-term memory

One of the most significant discoveries is the elucidation of an epigenetic mechanism by which the bacterial endophyte *Enterobacter* sp. SA187 establishes long-lasting heat-stress memory through stable trimethylation of histone H3 lysine 4 (H3K4me3) at heat stress-memory genes. Unlike the brief change induced by thermopriming, this endophyte-induced alteration creates a stable chromatin state, which can be described as "primed," enabling rapid and robust gene activation upon secondary stress and is associated with increased thermotolerance in field conditions (Shekhawat et al., 2021).

6.3.4.3 A coordinated temporal response

The plant's transcriptional response to beneficial bacteria is highly dynamic. The concept of an initial defense response, an intermediate phase of downregulation, and a final phase of sustained TF activation suggests that the timing and pattern of gene expression are crucial in enabling long-term tolerance to be effective and not metabolically expensive for the plant (Kaleh et al., 2025).

6.3.4.4 MicroRNA-mediated regulation of endophyte colonization and host interaction

MicroRNAs (miRNAs) have been shown to significantly regulate the plant–endophyte interaction. These molecules act as gene expression regulators, balancing the suppression of defense responses and the acceptance of symbiosis. miRNAs exhibit distinct expression patterns in plants during colonization by bacterial and fungal endophytes. Such changes in miRNAs modulate signaling networks related to immunity, phytohormones, and stress tolerance (Li et al., 2023b). For instance, in the roots of raspberry, colonization by the dark septate endophyte *Phialocephala fortinii*(F5) has resulted in differential expression of 37 miRNAs targeting genes involved in hormone signaling, redox metabolism, and plant—pathogen interaction pathways. This, in turn, paves the way for fungal colonization to proceed in a controlled manner without triggering the plant immune system (Yang et al., 2022). Moreover, in switchgrass (*Panicum virgatum*), 15 miRNAs were recognized as responders to the colonization of *Rhizophagus irregularis*, and these miRNAs regulated GRAS transcription factors, SODs, and laccases that control root remodeling and redox equilibrium during fungal entry (Johnson et al., 2022).

In bacterial endophyte interactions, miRNAs regulate hormone signaling, particularly auxin, cytokinin, and JA pathways, leading to reprogramming of root architecture and maintaining immune balance (Li et al., 2023b). The host genome is not the only place where this happens, and there is evidence for the possible two-way small RNA (sRNA) exchange. In this scenario, microbial sRNA effectors might induce host gene silencing, a process that has been well described in pathogenic and symbiotic fungi (Liu et al., 2025). Altogether, miRNAs serve as one of the posttranscriptional checkpoints that can adjust the host's genetic response, thus ensuring the maintenance of a balance between defense and symbiosis and the integration of molecular signals from bacterial and fungal partners. Besides providing a deeper understanding of the endophyte colonization process, knowledge of this regulatory network also suggests a new field of research: miRNA-guided modulation of beneficial plant–microbe interactions to engineer stress-resilient crops.

6.3.5 Multiomics and systems-level insights

One of the main changes that improved our understanding of endophyte-plant interactions, as well as the shift from reporting to comprehending the system, was the adoption of genomics and metagenomics. These two -omics fields have played a crucial role in uncovering the immense genetic potential of endophytes. For example, the pan-genome research of beneficial *Flavobacterium* strains has not only allowed for the identification of a core genome rich in genes related to phosphorus solubilization, a critical feature for the growth of plants under nutrient-deficient conditions, but also made it possible to reveal that the pan-genome is open, thus enabling the appearance of new gene sets through recombination events (Kim et al., 2023). Moreover, metagenomic analysis of the desert wild plant *Senna italica* revealed that the bacterial endophytes that sustain the plant belong to the phyla *Cyanobacteria*, *Proteobacteria*, and *Actinobacteria*, indicating a taxonomic feature for survival in a harsh habitat (Samsampour et al., 2022).

Transcriptomics studies are those that concretize these genomic designs by showing how these genetic capabilities are implemented in real-time. An experiment on barley inoculated with the fungal endophyte *Piriformospora indica* demonstrated that it alters the expression of host genes involved in the drought resistance mechanism, thereby increasing the plant's water-use efficiency (Li et al., 2023a). Metabolomics and proteomics have gone beyond the transcriptome to capture the functional aspect of the interaction. The study examining the association between *Epichloë endophytes*, in perennial ryegrass concluded that rising CO_2 levels triggered significant changes in the plant's metabolome and proteome, which were defense responses to the broader environment (Geddes-McAlister, 2022). Equally, a combined metabolome-microbiome study of *Dendrobium* has concluded that the presence of *Actinobacteria* is tightly linked to flavonoid synthesis. These most widely recognized medicinal compounds mean that a direct connection between a microbial taxon and a metabolic outcome has been established (Yuan et al., 2024).

Multi-omics data enable researchers to combine datasets to build models that predict interactions (interactomes), representing the full set of molecular interactions within the system. Such a comprehensive perspective is crucial for transitioning from correlation to causality. Research at the systems level, for instance, has revealed that posttranslational modifications (PTMs) of proteins, such as phosphorylation and acetylation, are among the crucial regulatory mechanisms of stress tolerance and environmental adaptation in fungi (Geddes-McAlister, 2022; Retanal et al., 2021). As a result, these combined findings are accelerating the design of synthetic microbial communities, or SynComs, deliberately engineered to confer specific, improved resilience traits to crop plants.

6.4 Comparative advantages of endophyte-based approaches

Using bacterial and fungal endophytes represents a revolutionary shift in sustainable agriculture, offering numerous benefits over traditional methods.

One of the major additional benefits is their potential interaction in integrated management. The study indicates that using endophytes in combination is a practical strategy. A triple combination of seed treatment, seedling dip, and foliar spray was the most effective method for delivering endophytes to tomato plants and stimulating their growth (Kamran et al., 2022). Such a result not only demonstrates the possibility of their compatibility and the enhanced effects they can achieve when used together, but also opens the way for their employment as key elements in sustainable and climate-smart agriculture (Ameen et al., 2024).

6.5 Future perspectives and research directions

The next generation of sustainable agriculture, powered by bacterial and fungal endophytes, is a bright prospect. Yet, transferring such a promise from laboratory conditions to broad, reliable field applications is a challenging mission that demands a comprehensive, simultaneous strategy. The future will be about moving beyond the current simple exploitation of natural endophytes toward the deliberate design and deployment of these organisms with precision.

6.5.1 Engineering elite endophytes through synthetic biology and CRISPR

The primary area for endophyte study is progressively focusing on the development of precise microbial engineering to enhance their beneficial traits to the maximum extent. One groundbreaking technology in this regard is clustered regularly interspaced short palindromic repeats (CRISPR)-Cas9, which enables scientists to perform targeted gene disruptions, additions, and regulatory changes in both endophytic fungi and bacteria. Consequently, it becomes feasible to fabricate "designer endophytes" equipped with improved potentials. In fact, work has been done to introduce CRISPR-Cas9 into fungal endophytes for gene function research, and it can be modified to overexpress genes encoding the biosynthesis of osmoprotectants (such as proline and glycine betaine), antioxidants (such as flavonoids and AsA), and novel antibiotics. In addition, the application of synthetic biology offers the possibility of rewiring endophytes' metabolic pathways, allowing them to produce not only bioplastics, pharmaceuticals, and high-value chemicals directly on crops but also to serve as living factories. The main difficulties are developing efficient transformation protocols for diverse endophytic species and navigating the complex regulatory and public acceptance landscape for genetically modified microorganisms.

6.5.2 Designing and deploying synthetic microbial consortia

Although single-strain inoculants have shown promise, synthetic microbial consortia (SynComs) are gaining attention. However, it is time before scientists can develop the concept of combining the advantages of the bacterial and fungal endophytes. A community composed of numerous and complementary endophyte strains is carefully designed to facilitate the most robust possible interactions and provide reliable, multilayered stress tolerance. The positioning is supported by ecological reasoning: a community can implement more complex functions than a single organism. To illustrate, a SynCom might comprise: (1) a bacterium specializing in phosphate solubilization; (2) a fungus that produces antioxidant metabolites to scavenge ROS; (3) another bacterium expressing high levels of ACC deaminase to mitigate ethylene stress.

Studying desert plants led to the creation of commercial groups like "EcoBiomes," which combine multiple stress-resistant bacteria to help agriculture become drought-, heat-, and salt-resistant simultaneously. The main goal of the research is to understand how microbes interact, verify the community's stability, and ensure that the more competitive members do not take over, which is essential for maintaining reliable field performance.

6.5.3 Integration with digital and precision agriculture

Only when a precision agriculture toolkit is used alongside endophyte-based solutions can we fully realize the benefits of the latter. This involves shifting from a one-size-fits-all approach to a location-specific, data-driven method. Some features of the vision are:

6.5.3.1 Understanding the microbial population

Metagenomic profiling of soils can characterize indigenous microbial communities and identify beneficial endophytes that may be lacking.

6.5.3.2 Accuracy in endophyte infection

GPS-guided machinery and drone technology can be used to apply custom-formulated SynComs only where needed at a particular location, as determined by real-time soil sensor data on moisture, salinity, and nutrient levels.

6.5.3.3 Innovative preparations for formulations

Their viability and shelf life are maintained through developing novel encapsulation technologies that utilize biopolymers or nanomaterials to shield bacterial endophytes from desiccation and UV radiation during storage and application. This focuses on delivering the right microbes to the right place at the right time, enhancing efficacy and cost-effectiveness.

6.5.3.4 *Navigating the path to commercialization: stability, scalability, and regulation*

Challenges related to scaling up the commercial use of endophytic technology must be resolved for the technology to be used worldwide. A 2024 meta-analysis of experimental data highlighted that one of the major problems inherent in this technology was inconsistent results. Most of these inconsistencies were associated with the nonperformance of the inoculated strains in complex, competitive soil environments, leading to poor persistence and colonization failure. Therefore, as the primary research focus, scientists should direct their efforts toward enhancing microbial fitness and ecological competence. The challenge of transferring from lab-scale fermenters to industrial-scale bioreactors without compromising microbial viability and functionality is a significant hurdle for engineers.

A significant factor that limits the translation of endophyte-based research from laboratories to field trials is the issue of their long-term persistence and functionality during in vitro cultivation. The key problem in laboratory-scale growth of endophytes is their dependence on a host. Most strains are metabolically highly tuned to the complex biochemical environment of plant tissues and rely on subtle, host-specific stimuli, such as exudate metabolites, phytohormones, cell wall fragments, and local redox or oxygen microgradients, which are rarely present in standard synthetic media. Frequently, the lack of these ecological signals leads to situations in which culturability fails, and isolates enter viable but nonculturable (VBNC) states or exhibit extremely slow growth. In many cases, the strains also experience phenotypic drift or laboratory domestication, which is the first sign of decreased rapid colonization efficiency, secondary metabolite production, plasmid stability, and biofilm-forming behavior, even in the case of culturability. Moreover, some endophytes may have syntrophic growth requirements or be quorum-dependent, and their behavior might be disrupted by axenic conditions. As a result, laboratory selection is inclined to fast-growing, generalist taxa while leaving out the most functionally potent but physiologically fastidious endophytes. These factors, in combination, are the reasons for the discrepancy between laboratory and field performance that has been frequently reported in studies. To address these problems, researchers are increasingly turning to plant-mimetic culture systems, microfluidic coculture platforms, and synthetic communities (SynComs) that provide a better spatial, chemical, and ecological context for the host. The use of such advanced strategies is indispensable for conserving the native phenotype and ensuring that strains maintained in the laboratory are ecologically fit and able to alleviate stress when applied on a large scale.

Additionally, a clear and predictable regulatory system that is both clear and predictable is needed. Although microbial consortia are increasingly used as biopesticides and biofertilizers, international harmonization of regulations governing engineered endophytes is the most crucial element to facilitate innovation and enable their safe large-scale adoption in climate-smart agriculture.

References

Aishwarya, P., Sabarinathan, K.G., Gomathy, M., Meenakshisundaram, P., Manonmani, K., Kar, A.A., 2025. Endophytes as bioenhancers of plant growth: an overview. Fitoterapia 181, 106355. Available at https://doi.org/10.1016/j.fitote.2024.106355.

Akram, N.A., Shafiq, F., Ashraf, M., 2017. Ascorbic acid-a potential oxidant scavenger and its role in plant development and abiotic stress tolerance. Frontiers in Plant Science 8. Available at https://doi.org/10.3389/fpls.2017.00613.

Ameen, M., Mahmood, A., Sahkoor, A., Zia, M.A., Ullah, M.S., 2024. The role of endophytes to combat abiotic stress in plants. Plant Stress 12. Available at https://doi.org/10.1016/j.stress.2024.100435, https://www.sciencedirect.com/science/journal/2667064X.

Anand, U., Pal, T., Yadav, N., Singh, V.K., Tripathi, V., Choudhary, K.K., et al., 2023. Current scenario and future prospects of endophytic microbes: promising candidates for abiotic and biotic stress management for agricultural and environmental sustainability. Microbial Ecology 86 (3), 1455–1486. Available at https://doi.org/10.1007/s00248-023-02190-1, https://www.springer.com/journal/248.

Andres-Barrao, C., Alzubaidy, H., Jalal, R., Mariappan, K.G., De Zelicourt, A., Bokhari, A., et al., 2021. Coordinated bacterial and plant sulfur metabolism in *Enterobacter* sp. SA187-induced plant salt stress tolerance. Proceedings of the National Academy of Sciences of the United States of America 118 (46). Available at https://doi.org/10.1073/pnas.2107417118, https://www.pnas.org/content/118/46/e2107417118.

Asomadu, R.O., Ezeorba, T.P.C., Ezike, T.C., Uzoechina, J.O., 2024. Exploring the antioxidant potential of endophytic fungi: a review on methods for extraction and quantification of total antioxidant capacity (TAC). 3 Biotech 14 (5). Available at https://doi.org/10.1007/s13205-024-03970-3, https://www.springer.com/journal/13205.

Bilal, S., Shahzad, R., Asaf, S., Imran, M., Al-Harrasi, A., Lee, I.-J., 2023. Efficacy of endophytic SB10 and glycine betaine duo in alleviating phytotoxic impact of combined heat and salinity in *Glycine max* L. via regulation of redox homeostasis and physiological and molecular responses. Environmental Pollution 316 (Pt 2), 120658. Available at https://doi.org/10.1016/j.envpol.2022.120658.

Biswas, P.R., Roy, N., Ghosh, S., Tayung, K., Dutta, A.K., 2025. A review of molecular signaling and communication between root tissues and their endophytes. Rhizosphere 36. Available at https://doi.org/10.1016/j.rhisph.2025.101188, https://www.sciencedirect.com/science/journal/24522198.

Bruger, E.L., Hying, Z.T., Singla, D., Reyes, N.L.M., Pandey, S.K., Patel, J.S., et al., 2024. Enhanced catabolism of glycine betaine and derivatives provides improved osmotic stress protection in *Methylorubrum extorquens* PA1. Applied and Environmental Microbiology 90 (7). Available at https://doi.org/10.1128/aem.00310-24, https://journals.asm.org/doi/10.1128/aem.00310-24.

Cabanás, C.G.L., Schilirò, E., Valverde-Corredor, A., Mercado-Blanco, J., 2014. The biocontrol endophytic bacterium *Pseudomonas fluorescens* PICF7 induces systemic defense responses in aerial tissues upon colonization of olive roots. Frontiers in Microbiology 5. Available at https://doi.org/10.3389/fmicb.2014.00427, http://journal.frontiersin.org/Journal/10.3389/fmicb.2014.00427/full.

Chen, C.Y., Selvaraj, P., Naqvi, N.I., 2023. Functional analysis of auxin derived from a symbiotic mycobiont. Frontiers in Plant Science 14. Available at https://doi.org/10.3389/fpls.2023.1216680, https://www.frontiersin.org/journals/plant-science.

Chen, X.L., Sun, M.C., Chong, S.L., Si, J.P., Wu, L.S., 2022. Transcriptomic and metabolomic approaches deepen our knowledge of plant–endophyte interactions. Frontiers in Plant Science 12. Available at https://doi.org/10.3389/fpls.2021.700200, https://www.frontiersin.org/journals/plant-science.

Compant, S., Sessitsch, A., Mathieu, F., 2012. The 125th anniversary of the first postulation of the soil origin of endophytic bacteria – a tribute to M.L.V. Galippe. Plant and Soil 356 (1-2), 299–301. Available at https://doi.org/10.1007/s11104-012-1204-9.

Cui, X., He, W., Christensen, M.J., Yue, J., Zeng, F., Zhang, X., et al., 2022. Abscisic acid may play a critical role in the moderating effect of epichloë endophyte on *Achnatherum inebrians* under drought stress. Journal of Fungi 8 (11), 1140. Available at https://doi.org/10.3390/jof8111140.

Cui, J., Nie, F., Zhao, Y., Zhang, D., Zhou, D., Wu, J., et al., 2024. A review on plant endophytes in response to abiotic stress. Environmental Pollutants and Bioavailability 36 (1). Available at https://doi.org/10.1080/26395940.2024.2323123.

De Bary, A., 1866. Morphologie und Physiologie der Pilze, Flechten und Myxomyceten. Hofmeister's Handbook of Physiological Botany.

Dijksterhuis, J., de Vries, R.P., 2006. Compatible solutes and fungal development. The Biochemical Journal 399 (2). Available at https://doi.org/10.1042/BJ20061229.

van Dijk, M., Morley, T., Rau, M.L., Saghai, Y., 2021. A meta-analysis of projected global food demand and population at risk of hunger for the period 2010–2050. Nature Food 2 (7), 494–501. Available at https://doi.org/10.1038/s43016-021-00322-9, https://www.nature.com/natfood/.

Dunken, N., Widmer, H., Balcke, G.U., Straube, H., Langen, G., Charura, N.M., et al., 2024. A nucleoside signal generated by a fungal endophyte regulates host cell death and promotes root colonization. Cell Host and Microbe 32 (12), 2161. Available at https://doi.org/10.1016/j.chom.2024.10.020, https://www.sciencedirect.com/science/journal/19313128.

Eid, A.M., Fouda, A., Abdel-rahman, M.A., Salem, S.S., Elsaied, A., Oelmüller, R., et al., 2021. Harnessing bacterial endophytes for promotion of plant growth and biotechnological applications: an overview. Plants 10 (5). Available at https://doi.org/10.3390/plants10050935, https://www.mdpi.com/2223-7747/10/5/935/pdf.

Eswaran, S.U.D., Sundaram, L., Perveen, K., Bukhari, N.A., Sayyed, R.Z., 2024. Osmolyte-producing microbial biostimulants regulate the growth of *Arachis hypogaea* L. under drought stress. BMC microbiology 24 (1). Available at https://doi.org/10.1186/s12866-024-03320-6, https://bmcmicrobiol.biomedcentral.com/.

Etesami, H., Glick, B.R., 2024. Bacterial indole-3-acetic acid: a key regulator for plant growth, plant-microbe interactions, and agricultural adaptive resilience. Microbiological research 281, 127602. Available at https://doi.org/10.1016/j.micres.2024.127602.

Fadiji, A.E., Babalola, O.O., 2020. Exploring the potentialities of beneficial endophytes for improved plant growth. Saudi Journal of Biological Sciences 27 (12), 3622–3633. Available at https://doi.org/10.1016/j.sjbs.2020.08.002, http://www.ksu.edu.sa/sites/Colleges/CollegeofScinces/BotanyDepartment/SJBS/Default.aspx.

Fagorzi, C., Mengoni, A., 2022. Endophytes: improving plant performance. Microorganisms 10 (9), 1777. Available at https://doi.org/10.3390/microorganisms10091777.

Gakuubi, M.M., Munusamy, M., Liang, Z.X., Ng, S.B., 2021. Fungal endophytes: a promising frontier for discovery of novel bioactive compounds. Journal of Fungi 7 (10). Available at https://doi.org/10.3390/jof7100786, https://www.mdpi.com/2309-608X/7/10/786/pdf.

Gamalero, E., Glick, B.R., 2015. Bacterial modulation of plant ethylene levels. Plant Physiology 169 (1), 13–22. Available at https://doi.org/10.1104/pp.15.00284, http://www.plantphysiol.org/content/169/1/13.full.pdf.

Geddes-McAlister, J., 2022. Systems biology in fungal research. Journal of Fungi 8 (5), 478. Available at https://doi.org/10.3390/jof8050478.

Giraldo, M.C., Valent, B., 2013. Filamentous plant pathogen effectors in action. Nature Reviews. Microbiology 11 (11), 800–814. Available at https://doi.org/10.1038/nrmicro3119.

Godara, H., Ramakrishna, W., 2023. Endophytes as nature's gift to plants to combat abiotic stresses. Letters in Applied Microbiology 76 (2). Available at https://doi.org/10.1093/lambio/ovac067.

Goszcz, A., Furtak, K., Stasiuk, R., Wójtowicz, J., Musiałowski, M., Schiavon, M., et al., 2025. Bacterial osmoprotectants-a way to survive in saline conditions and potential crop allies. FEMS Microbiology Reviews 49, fuaf020. Available at https://doi.org/10.1093/femsre/fuaf020.

Gouda, S., Das, G., Sen, S.K., Shin, H.S., Patra, J.K., 2016. Endophytes: a treasure house of bioactive compounds of medicinal importance. Frontiers in Microbiology 7. Available at https://doi.org/10.3389/fmicb.2016.01538, http://journal.frontiersin.org/article/10.3389/fmicb.2016.01538/full.

Hardoim, P.R., Van Overbeek, L.S., Berg, G., Pirttilä, A.M., Compant, S., Campisano, A., et al., 2015. The hidden world within plants: ecological and evolutionary considerations for defining functioning of microbial endophytes. Microbiology and Molecular Biology Reviews 79 (3), 293–320. Available at https://doi.org/10.1128/MMBR.00050-14, http://mmbr.asm.org/content/79/3/293.full.pdf+html.

Harrison, J.G., Griffin, E.A., 2020. The diversity and distribution of endophytes across biomes, plant phylogeny and host tissues: how far have we come and where do we go from here? Environmental Microbiology 22 (6), 2107–2123. Available at https://doi.org/10.1111/1462-2920.14968, http://onlinelibrary.wiley.com/journal/10.1111/(ISSN)1462-2920.

Hnamte, L., Vanlallawmzuali, Kumar, A., Yadav, M.K., Zothanpuia, Singh, P.K., 2024. An updated view of bacterial endophytes as antimicrobial agents against plant and human pathogens. Current Research in Microbial Sciences 7. Available at https://doi.org/10.1016/j.crmicr.2024.100241, https://www.sciencedirect.com/science/journal/26665174.

Ji, H., Zhang, M., Huang, C., Lin, W., Lu, Y., Wang, P., et al., 2024. The impact of *Piriformospora indica* on plant heat and drought tolerance. Frontiers in Plant Science 15. Available at https://doi.org/10.3389/fpls.2024.1479561.

Johnson, A.C., Pendergast, T.H., Chaluvadi, S., Bennetzen, J.L., Devos, K.M., 2022. Identification of microRNAs responsive to arbuscular mycorrhizal fungi in *Panicum virgatum* (switchgrass). BMC Genomics 23 (1). Available at https://doi.org/10.1186/s12864-022-08797-x, https://bmcgenomics.biomedcentral.com/.

Kaleh, A.M., Singh, P., Ooi Chua, K., Harikrishna, J.A., 2025. Modulation of plant transcription factors and priming of stress tolerance by plant growth-promoting bacteria: a systematic review. Annals of Botany 135 (3), 387–402. Available at https://doi.org/10.1093/aob/mcae166, http://aob.oxfordjournals.org/.

Kamran, M., Imran, Q.M., Ahmed, M.B., Falak, N., Khatoon, A., Yun, B.W., 2022. Endophyte-mediated stress tolerance in plants: a sustainable strategy to enhance resilience and assist crop improvement. Cells 11 (20). Available at https://doi.org/10.3390/cells11203292, https://www.mdpi.com/journal/cells.

Kandel, S.L., Joubert, P.M., Doty, S.L., 2017. Bacterial endophyte colonization and distribution within plants. Microorganisms 5 (4). Available at https://doi.org/10.3390/microorganisms5040077, https://www.mdpi.com/2076-2607/5/4/77/pdf.

Kim, M., Cha, I.T., Lee, K.E., Li, M., Park, S.J., 2023. Pangenome analysis provides insights into the genetic diversity, metabolic versatility, and evolution of the genus *Flavobacterium*. Microbiology Spectrum 11 (5). Available at https://doi.org/10.1128/spectrum.01003-23, https://journals.asm.org/doi/pdf/10.1128/spectrum.01003-23.

Kumar, A., Chauhan, P., Kumar, A., Pooja, Mishra, T., Padiyal, A., et al., 2025a. Latest progress (2020–2024) in bacterial endophyte research with special reference to plant disease management: achievements and challenges. Discover Plants 2 (1). Available at https://doi.org/10.1007/s44372-025-00303-3.

Kumar, H., Lata, R., Khan, U., White, J.F., Gond, S.K., 2025b. Potential application of endophytic bacteria for induction of abiotic stress tolerance in plants. Symbiosis 95 (2), 153–181. Available at https://doi.org/10.1007/s13199-025-01037-2, https://www.springer.com/journal/13199.

Lau, K.J.X., Selvaraj, P., Muralishankar, V., Chen, C.Y., Muruganantham, S., Naqvi, N.I., 2024. Draft genome sequence of *Penicillium citrinum* B9, a plant growth-promoting symbiont from barley rhizosphere. Microbiology Resource Announcements 13 (6). Available at https://doi.org/10.1128/mra.00354-24, https://journals.asm.org/doi/pdf/10.1128/mra.00354-24?download=true.

Lengrand, S., Dubois, B., Renoz, F., Lebbe, O., Debode, F., Legrève, A., 2025. Seed-borne endophytic bacteria of tomato: dynamic colonization and implications in plant health. PhytoFrontiers™ 5 (3), 328–344. Available at https://doi.org/10.1094/phytofr-01-25-0004-r.

Liang, X.J., Zhang, S.H., Zhang, P., Peng, F., Ke, J., Mi, Y.N., et al., 2014. Comparative study on distribution of endophytic fungi in Eucommia barks from different habitats. Zhongguo Zhong Yao Za Zhi 39, 204–208. Available at 10.4268/cjcmm20140209.

Liao, C., Doilom, M., Jeewon, R., Hyde, K.D., Manawasinghe, I.S., Chethana, K.W.T., et al., 2025. Challenges and update on fungal endophytes: classification, definition, diversity, ecology, evolution and functions. Fungal Diversity 131 (1), 301–367. Available at https://doi.org/10.1007/s13225-025-00550-5, https://www.springer.com/journal/13225.

Liu, J., Lu, Y., Chen, X., Liu, X., Gu, Y., Li, F., 2025. The silent conversation: how small RNAs shape plant–microbe relationships. International Journal of Molecular Sciences 26 (6), 2631. Available at https://doi.org/10.3390/ijms26062631.

Li, L., Feng, Y., Qi, F., Hao, R., 2023a. Research progress of *Piriformospora indica* in improving plant growth and stress resistance to plant. Journal of Fungi 9 (10), 965. Available at https://doi.org/10.3390/jof9100965.

Li, L., Guo, N., Zhang, Y., Yuan, Z., Lu, A., Li, S., et al., 2023b. Reprogramming of fundamental miRNA and gene expression during the barley-*Piriformospora indica* Interaction. Journal of Fungi 9 (1), 24. Available at https://doi.org/10.3390/jof9010024.

Mametja, N.M., Ramadwa, T.E., Managa, M., Masebe, T.M., 2025. Recent advances and developments in bacterial endophyte identification and application: a 20-year landscape review. Plants 14 (16). Available at https://doi.org/10.3390/plants14162506, www.mdpi.com/journal/plants.

Meena, M., Prasad, V., Zehra, A., Gupta, V.K., Upadhyay, R.S., 2015. Mannitol metabolism during pathogenic fungal-host interactions under stressed conditions. Frontiers in Microbiology 6. Available at https://doi.org/10.3389/fmicb.2015.01019, http://journal.frontiersin.org/article/10.3389/fmicb.2015.01019/full.

Mendgen, K., Hahn, M., Deising, H., 1996. Morphogenesis and mechanisms of penetration by plant pathogenic fungi. Annual Review of Phytopathology 34 (1), 367–386. Available at https://doi.org/10.1146/annurev.phyto.34.1.367.

Mengistu, A.A., 2020. Endophytes: colonization, behaviour, and their role in defense mechanism. International Journal of Microbiology 2020. Available at https://doi.org/10.1155/2020/6927219, http://www.hindawi.com/journals/ijmb/.

Morales-Vargas, A.T., López-Ramírez, V., Álvarez-Mejía, C., Vázquez-Martínez, J., 2024. Endophytic fungi for crops adaptation to abiotic stresses. Microorganisms 12 (7), 1357. Available at https://doi.org/10.3390/microorganisms12071357.

Mushtaq, S., Shafiq, M., Tariq, M.R., Sami, A., Nawaz-ul-Rehman, M.S., Bhatti, M.H.T., et al., 2023. Interaction between bacterial endophytes and host plants. Frontiers in Plant Science 13. Available at https://doi.org/10.3389/fpls.2022.1092105, https://www.frontiersin.org/journals/plant-science.

Oldroyd, G.E.D., 2013. Speak, friend, and enter: signalling systems that promote beneficial symbiotic associations in plants. Nature reviews. Microbiology 11 (4), 252–263. Available at https://doi.org/10.1038/nrmicro2990.

Orozco-Mosqueda, Mdel C., Glick, B.R., Santoyo, G., 2020. ACC deaminase in plant growth-promoting bacteria (PGPB): an efficient mechanism to counter salt stress in crops. Microbiological research 235, 126439. Available at https://doi.org/10.1016/j.micres.2020.126439.

Pandey, P., Tripathi, A., Dwivedi, S., Lal, K., Jhang, T., 2023. Deciphering the mechanisms, hormonal signaling, and potential applications of endophytic microbes to mediate stress tolerance in medicinal plants. Frontiers in Plant Science 14. Available at https://doi.org/10.3389/fpls.2023.1250020.

Parvez, R., Hamayun, M., Mehmood, A., Iqbal, A., Lee, B., Hussain, A., 2025. Differential effects of tomato root exudates on the development of endophytic and pathogenic fungi depend on the foliar application of fungi. Symbiosis 96 (3), 271–290. Available at https://doi.org/10.1007/s13199-025-01077-8, https://www.springer.com/journal/13199.

Pelagio-Flores, R., Ravelo-Ortega, G., García-Pineda, E., López-Bucio, J., 2025. A century of *Azospirillum*: plant growth promotion and agricultural promise. Plant Signaling & Behavior 20 (1). Available at https://doi.org/10.1080/15592324.2025.2551609.

Petrini, O., 1991. Fungal endophytes of tree leaves. In: Andrews, J., Hirano, S. (Eds.), Microbial Ecology of Leaves. Springer, pp. 179–197. Available at https://doi.org/10.1007/978-1-4612-3168-4_9.

Porcel, R., Aroca, R., Cano, C., Bago, A., Ruiz-Lozano, J.M., 2006. Identification of a gene from the arbuscular mycorrhizal fungus *Glomus intraradices* encoding for a 14-3-3 protein that is up-regulated by drought stress during the AM symbiosis. Microbial Ecology 52 (3), 575–582. Available at https://doi.org/10.1007/s00248-006-9015-2.

Prieto, P., Mercado-Blanco, J.Ã.°s, 2008. Endophytic colonization of olive roots by the biocontrol strain *Pseudomonas fluorescens* PICF7. FEMS Microbiology Ecology 64 (2), 297–306. Available at https://doi.org/10.1111/j.1574-6941.2008.00450.x.

Qin, J., Wu, M., Liu, H., Gao, Y., Ren, A., 2019. Endophyte infection and methyl jasmonate treatment increased the resistance of *Achnatherum sibiricum* to insect herbivores independently. Toxins 11 (1), 7. Available at https://doi.org/10.3390/toxins11010007.

Rather, R.A., 2025. Shaping plant growth beneath the soil: a theoretical exploration of fungal endophyte's role as plant growth-promoting agents. MicrobiologyOpen 14 (4). Available at https://doi.org/10.1002/mbo3.70026, http://onlinelibrary.wiley.com/journal/10.1002/(ISSN)2045-8827.

Ren, C.G., Dai, C.C., 2012. Jasmonic acid is involved in the signaling pathway for fungal endophyte-induced volatile oil accumulation of *Atractylodes lancea* plantlets. BMC Plant Biology 12. Available at https://doi.org/10.1186/1471-2229-12-128China, http://www.biomedcentral.com/1471-2229/12/128.

Retanal, C., Ball, B., Geddes-McAlister, J., 2021. Post-translational modifications drive success and failure of fungal–host interactions. Journal of Fungi 7 (2), 124. Available at https://doi.org/10.3390/jof7020124.

Sadeghi, F., Samsampour, D., Askari Seyahooei, M., Bagheri, A., Soltani, J., 2020. Fungal endophytes alleviate drought-induced oxidative stress in mandarin (*Citrus reticulata* L.): toward regulating the ascorbate–glutathione cycle. Scientia Horticulturae 261, 108991. Available at https://doi.org/10.1016/j.scienta.2019.108991.

Sahu, P.K., Jayalakshmi, K., Tilgam, J., Gupta, A., Nagaraju, Y., Kumar, A., et al., 2022. ROS generated from biotic stress: effects on plants and alleviation by endophytic microbes. Frontiers in Plant Science 13. Available at https://doi.org/10.3389/fpls.2022.1042936, https://www.frontiersin.org/journals/plant-science.

Samsampour, D., Sadeghi, F., Seyahooei, M.A., Bagheri, A., Soltani, J., 2022. Metagenomic study of the communities of bacterial endophytes in the desert plant *Senna italica* and their role in abiotic stress resistance. Brazilian Journal of Biology 82, e267584.. Available at https://doi.org/10.1590/1519-6984.267584.

Schulz, B., Boyle, C., Draeger, S., Römmert, A.K., Krohn, K., 2002. Endophytic fungi: a source of novel biologically active secondary metabolites. Mycological Research 106 (9), 996–1004. Available at https://doi.org/10.1017/S0953756202006342.

Semenzato, G., Fani, R., 2024. Endophytic bacteria: a sustainable strategy for enhancing medicinal plant cultivation and preserving microbial diversity. Frontiers in Microbiology 15. Available at https://doi.org/10.3389/fmicb.2024.1477465.

Shekhawat, K., Saad, M.M., Sheikh, A., Mariappan, K., Al-Mahmoudi, H., Abdulhakim, F., et al., 2021. Root endophyte induced plant thermotolerance by constitutive chromatin modification at heat stress memory gene loci. EMBO Reports 22 (3). Available at https://doi.org/10.15252/embr.202051049, http://embor.embopress.org/.

Singh, K.M., Baksi, S., Rani, S., Jha, A.B., Dubey, R.S., Sharma, P., 2025. Glycine betaine-loaded nanoparticles: a novel approach for enhancing crop tolerance to salinity and drought stress. Plant Physiology and Biochemistry 200, 110193. Available at https://doi.org/10.1016/j.plaphy.2025.110193.

Snelders, N.C., Rovenich, H., Thomma, B.P.H.J., 2022. Microbiota manipulation through the secretion of effector proteins is fundamental to the wealth of lifestyles in the fungal kingdom. FEMS Microbiology Reviews 46 (5). Available at https://doi.org/10.1093/femsre/fuac022, https://academic.oup.com/femsre/pages/About.

Spaepen, S., Bossuyt, S., Engelen, K., Marchal, K., Vanderleyden, J., 2014. Phenotypical and molecular responses of *Arabidopsis thaliana* roots as a result of inoculation with the auxin-producing bacterium *Azospirillum brasilense*. New Phytologist 201 (3), 850–861. Available at https://doi.org/10.1111/nph.12590.

Tariq, A., Tanvir, A., Barasarathi, J., Alsohim, A.S., Mastinu, A., Sayyed, R., et al., 2025. Endophytes: key role players for sustainable agriculture: mechanisms, omics insights and future prospects. Emirates Plant Growth Regulation 105 (6), 1969–1990. Available at https://doi.org/10.1007/s10725-025-01370-y, https://www.springer.com/journal/10725.

Tran, H., Ficke, A., Asiimwe, T., Höfte, M., Raaijmakers, J.M., 2007. Role of the cyclic lipopeptide massetolide a in biological control of *Phytophthora infestans* and in colonization of tomato plants by *Pseudomonas fluorescens*. New Phytologist 175 (4), 731–742. Available at https://doi.org/10.1111/j.1469-8137.2007.02138.x.

Truyens, S., Weyens, N., Cuypers, A., Vangronsveld, J., 2015. Bacterial seed endophytes: genera, vertical transmission and interaction with plants. Environmental Microbiology Reports 7 (1), 40–50. Available at https://doi.org/10.1111/1758-2229.12181.

Ullah, A., Ullah, S., Arafat, Y., Shao, H., 2025. Endophytes-mediated mitigation of sodium stress in plants via ABA-independent signaling pathways. Plant Physiology and Biochemistry 229 (Pt B), 110476. Available at https://doi.org/10.1016/j.plaphy.2025.110476.

Verma, H., Kumar, D., Kumar, V., Kumari, M., Singh, S.K., Sharma, V.K., et al., 2021a. The potential application of endophytes in management of stress from drought and salinity in crop plants. Microorganisms 9 (8). Available at https://doi.org/10.3390/microorganisms9081729, https://www.mdpi.com/2076-2607/9/8/1729/pdf.

Verma, S.K., Sahu, P.K., Kumar, K., Pal, G., Gond, S.K., Kharwar, R.N., et al., 2021b. Endophyte roles in nutrient acquisition, root system architecture development and oxidative stress tolerance. Journal of Applied Microbiology 131 (5), 2161–2177. Available at https://doi.org/10.1111/jam.15111.

Verma, A., Shameem, N., Jatav, H.S., Sathyanarayana, E., Parray, J.A., Poczai, P., et al., 2022. Fungal endophytes to combat biotic and abiotic stresses for climate-smart and sustainable agriculture. Frontiers in Plant Science 13. Available at https://doi.org/10.3389/fpls.2022.953836, https://www.frontiersin.org/journals/plant-science.

Vujanovic, S., Vujanovic, J., Vujanovic, V., 2022. Microbiome-driven proline biogenesis in plants under stress: perspectives for balanced diet to minimize depression disorders in humans. Microorganisms 10 (11), 2264. Available at https://doi.org/10.3390/microorganisms10112264

Wang, Y., Li, Z., Xu, M., Xiao, Z., Liu, C., Du, B., et al., 2023. Signal molecules regulate the synthesis of secondary metabolites in the interaction between endophytes and medicinal plants. Processes 11 (3), 849. Available at https://doi.org/10.3390/pr11030849.

Wang, Z., Yang, Y., Yadav, V., Zhao, W., He, Y., Zhang, X., et al., 2022. Drought-induced proline is mainly synthesized in leaves and transported to roots in watermelon under water deficit. Horticultural Plant Journal 8 (4), 379–388. Available at https://doi.org/10.1016/j.hpj.2022.06.009.

Wang, S., Zheng, Y., Gu, C., He, C., Yang, M., Zhang, X., et al., 2018. *Bacillus cereus* AR156 activates defense responses to *Pseudomonas syringae* pv. tomato in *Arabidopsis thaliana* similarly to flg22. Molecular Plant-Microbe Interactions® 31 (3), 311–322. Available at https://doi.org/10.1094/mpmi-10-17-0240-r.

Waqas, M., Khan, A.L., Kamran, M., Hamayun, M., Kang, S.M., Kim, Y.H., et al., 2012. Endophytic fungi produce gibberellins and indoleacetic acid and promotes host-plant growth during stress. Molecules 17 (9), 10754–10773. Available at https://doi.org/10.3390/molecules170910754SouthKorea, http://www.mdpi.com/1420-3049/17/9/10754/pdf.

Watts, D., Palombo, E.A., Jaimes Castillo, A., Zaferanloo, B., 2023. Endophytes in agriculture: potential to improve yields and tolerances of agricultural crops. Microorganisms 11 (5), 1276. Available at https://doi.org/10.3390/microorganisms11051276.

White, J.F., Kingsley, K.L., Zhang, Q., Verma, R., Obi, N., Dvinskikh, S., et al., 2019. Review: endophytic microbes and their potential applications in crop management. Pest management science 75 (10), 2558–2565. Available at https://doi.org/10.1002/ps.5527, http://onlinelibrary.wiley.com/journal/10.1002/(ISSN)1526-4998.

Yang, Z., Yuan, L., Zhu, H., Jiang, J., Yang, H., Li, L., 2022. Small RNA profiling reveals the involvement of microRNA-mediated gene regulation in response to symbiosis in raspberry. Frontiers in Microbiology 13. Available at https://doi.org/10.3389/fmicb.2022.1082494.

Yuan, Y., Zuo, J., Wan, X., Zhou, R., Xing, W., Liu, S., 2024. Multi-omics profiling reveal responses of three major *Dendrobium* species from different growth years to medicinal components. Frontiers in Plant Science 15. Available at https://doi.org/10.3389/fpls.2024.1333989, https://www.frontiersin.org/journals/plant-science.

Zhou, X., Zhang, J., Shi, J., Khashi u Rahman, M., Liu, H., Wei, Z., et al., 2024. Volatile-mediated interspecific plant interaction promotes root colonization by beneficial bacteria via induced shifts in root exudation. Microbiome 12 (1). Available at https://doi.org/10.1186/s40168-024-01914-w, https://microbiomejournal.biomedcentral.com/.

CHAPTER

7

Strigolactone-mediated signaling pathways in sorghum: molecular insights into *Striga* resistance

Piyush Panchabhai[1,*], *Sirisha Kaniganti*[2,*], *Hithaishy Aduru*[1], *and Sudhakar Reddy Palakolanu*[1]

[1]International Crops Research Institute for the Semi-Arid Tropics (ICRISAT), Hyderabad, Telangana, India [2]International Institute of Tropical Agriculture (IITA), Nairobi, Kenya

7.1 Introduction

Sorghum (*Sorghum bicolor* L.) is the fifth most important cereal crop worldwide, producing about 5.3 million metric tons annually (Khalid et al., 2022). Global consumption is expected to reach 26.5 million metric tons by 2026, with Nigeria, India, China, and Ethiopia as the leading consumers. Due to its low glycemic index, high antioxidant content, and gluten-free status, sorghum is increasingly popular in developed countries as a healthy alternative to major cereals (Thakur et al., 2024). Different types of sorghum—including grain, dual-purpose, fodder, forage, and sweet stalk—are cultivated globally (Reddy et al., 2012), providing food and income for millions in Africa (Andiku et al., 2021). It is rich in macro- and micronutrients and is used to make porridges, breads, snacks, malt products, and beverages (Wanga, Shimelis, and Mengistu, 2022). As a C4, stress-tolerant crop capable of thriving in saline, drought-prone, and nutrient-deficient environments, sorghum is also a promising source of forage and biofuel (Liu et al., 2019). Despite its resilience, low-input farming and biotic challenges continue to restrict its productivity (Reddy et al., 2004).

A major biotic obstacle is the parasitic weed *Striga*, which can cause yield reductions ranging from 10% to complete crop failure in severe cases. *Striga hermonthica* is the most damaging species on sorghum, maize, and millets, followed by *S. asiatica, S. aspera,* and *S. forbesii* (Kaniganti et al., 2022; Kaniganti et al., 2025). Low soil fertility, drought, and

[*] Contributed equally.

Harnessing Plant-Stimulating Molecules
DOI: https://doi.org/10.1016/B978-0-443-44994-9.00003-1

intensive farming on marginal lands have sped up the spread of *Striga* across Sub-Saharan Africa (Beyene, Gudeta, and Chalchisa, 2025). In eastern Africa, vulnerable sorghum varieties lose about 59% of their yield to *Striga*, while in West Africa and India, losses range from 10% to 35% (Obilana and Ramaiah, 1992). The weed infests millions of hectares and threatens the livelihoods of over 100 million farmers (Begna, 2021). *Striga* is an obligate hemiparasite that relies on its host for water, nutrients, and carbohydrates. Its tiny seeds can stay viable in the soil for 14–20 years and need specific signals from host roots to germinate (Sibhatu, 2016). Seed germination is closely linked to environmental factors such as moisture and temperature, but it critically depends on the presence of strigolactones (SLs)—carotenoid-derived compounds released by host roots (Bouwmeester et al., 2003; Kaniganti et al., 2022). Once germinated, *Striga* seedlings must attach to the host within 3–7 days through a specialized organ called the haustorium, which establishes a vascular connection and extracts resources during a 4- to 7-week underground phase (Agt, 2007). This hidden parasitism causes stunting, reduced photosynthesis, poor grain filling, and significant yield loss before the parasite even emerges (Gurney et al., 2006; Jamil, Kountche, and Al-Babili, 2021).

SLs are crucial for both normal sorghum growth and *Striga* parasitism. In addition to controlling shoot branching, root structure, and symbiosis with arbuscular mycorrhizal (AM) fungi (Gomez-Roldan et al., 2008; Umehara et al., 2008), SLs serve as important germination signals for *Striga* at very low concentrations. Nutrient stress—especially low nitrogen and phosphorus levels—significantly increases SL secretion, unintentionally encouraging *Striga* germination and attachment (Yoneyama et al., 2007). Canonical and noncanonical SLs differ in structure and function, and sorghum genotypes vary considerably in the types and amounts they produce (Chesterfield, Vickers, and Beveridge, 2020; Yoneyama et al., 2018). These differences explain the variation in *Striga* resistance among sorghum varieties. Therefore, understanding how SLs are formed, regulated, and detected is essential for developing sorghum varieties resistant to *Striga*. Current control methods—including nitrogen fertilizers, trap crops, herbicides, ethylene treatment, and manual weeding—offer only limited and short-term relief because of *Striga's* high seed output and long seed viability (Kanampiu et al., 2018). Host resistance and manipulation of SL biology present more sustainable, long-lasting options.

This chapter provides an overview of SL biosynthesis, regulation, and function in sorghum, with emphasis on their role in *Striga* germination and parasitism. We synthesize evidence on the genetic and biochemical pathways governing SL production, how nutrient stress and drought shape SL secretion, and how SL variation among sorghum genotypes influences *Striga* attachment and development. By integrating molecular, physiological, and ecological perspectives, we highlight opportunities for breeding and agronomic innovations that can strengthen sustainable *Striga* management.

7.2 Life cycle and management practices of the *Striga* parasite weed

"Striga" species are obligate hemiparasitic plants that attach to their host's roots to extract water, nutrients, and carbohydrates. "Striga" is the Latin word for "witch." It is called witchweed because plants infected by *Striga* exhibit stunted growth and a drought-like phenotype long before the *Striga* plant emerges. *Striga* infestations have been reported in at least 25 African countries and pose a significant obstacle to the production of sorghum, maize, pearl millet, and upland rice, leading to severe yield reductions and, in many cases, total crop failure (Teka, 2014). Morphologically, *Striga* plants are characterized by green, opposite leaves and irregular flowers that can be pink, red, white, or yellow. Its seeds are tiny and easily dispersed by wind, water, livestock, or human activity, and their viability in soil can last 14–20 years, making *Striga* one of the most persistent weeds in agriculture (Sibhatu, 2016).

The life cycle of *Striga* depends on its host. Seeds can remain dormant for years, germinating only when conditions are ideal, such as proper temperature and moisture. During this time, they respond to germination signals from host roots, with SLs as key cues. Without these signals, seeds will not sprout even in perfect conditions (Cardoso, Ruyter-Spira and Bouwmeester, 2011). Once germinated, seedlings have 3–7 days to find a host root. When within 2–3 mm, they detect cues that trigger haustorium formation, which attaches to the root, penetrates the tissue, and fuses with the xylem to extract water and nutrients. After this, *Striga* grows underground for 4–7 weeks, drawing resources while hidden. It then emerges, weakening the host. Infested plants display stunted growth, reduced photosynthesis, poor grain filling, and early aging, leading to up to 100% yield loss. Mature plants produce seeds for future infestations (Sibhatu, 2016). SLs, including synthetic strigol from cotton, trigger germination (Cardoso, Ruyter-Spira and Bouwmeester, 2011). *Striga* infestation also reduces soil fertility and alters the rhizosphere, making fields less productive over time (Ejeta and Gressel, 2007). The parasitic interaction disrupts the host's hormonal signaling, especially auxin, cytokinin, and abscisic acid (ABA), hampering root development and nutrient uptake (Yoshida and Shirasu, 2009).

Striga infection causes oxidative stress, reduces chlorophyll content, and lowers photosynthetic capacity, accelerating plant aging (Gurney et al., 2006). Its persistence is worsened by dust-like seeds—up to 50,000 per plant—that can stay viable for decades. Besides cereals, some species, such as *S. gesnerioides,* parasitize legumes, including cowpea, thereby broadening their impact. Its adaptability to poor soils and low-input farming makes it especially destructive in subsistence agriculture with limited control. SLs play a central role in *Striga* biology, influencing sorghum growth and mycorrhizal interactions and serving as germination signals at low levels (Bouwmeester et al., 2003; Wu et al., 2022). Synthetic strigol mimics these signals, triggering germination (Cardoso, Ruyter-Spira, and Bouwmeester, 2011). Dependence on host SLs offers a vulnerability, allowing disruption through genetic, biochemical, or agricultural methods in sorghum.

Controlling *Striga* is slow but possible (Ejeta and Gressel, 2007). *Striga* species are among the most difficult parasitic plants to control. Their ability to adapt to various hosts and environmental conditions makes them one of the most widespread and successful parasitic plants (Jamil, Kountche, and Al-Babili, 2021; Mohamed et al., 2006). The complexity of *Striga's* life cycle and its high seed production pose challenges for management. Preventive measures are the first line of defense. Using clean, certified sorghum seed free of *Striga* contamination helps prevent the parasite from invading new fields. Sanitizing farm tools and machinery is also essential to prevent the transfer of contaminated soil between fields. Livestock can also spread *Striga* seeds via fur and feces, making control of grazing practices another key preventive step (Jamil, Kountche, and Al-Babili, 2021; Begna, 2021). Effective management requires a comprehensive, integrated approach that targets multiple stages of its life cycle. Table 7.1 outlines management strategies including prevention, cultural

TABLE 7.1 Summary of cultural, biological, and chemical strategies for *Striga* control in crops.

Management category	Practices/methods	Key details/benefits	References
Prevention	Use clean, certified sorghum seeds free of *Striga* seeds	Prevents initial infestation and seed dispersal to new fields	Jamil, Kountche, and Al-Babili (2021)
	Sanitation of farm tools and machinery	Limits the spread of contaminated soil between fields	Jamil Kountche and Al-Babili (2021)
	Control of grazing by animals	Livestock grazing can transmit *Striga* seeds via fur and feces	Begna (2021)
Cultural practices	Crop rotation with nonhost or trap crops (e.g., legumes like cowpea, groundnut)	Reduces the *Striga* seed bank through suicidal germination or the absence of a host	Mrema et al. (2020); Sibhatu (2016); Begna (2021)
	Intercropping sorghum with legumes	Improves soil fertility and suppresses *Striga* infestation	Mrema et al. (2020); Jamil, Kountche, and Al-Babili (2021)
	Early planting	Allows sorghum to establish before peak *Striga* emergence	Sibhatu (2016)
	Hand weeding and uprooting	Removes *Striga* plants before seed production; labor-intensive but effective in low-input systems	Begna (2021)
Host plant resistance	Use of *Striga*-resistant or tolerant sorghum varieties	Resistance reduces haustorial attachment and parasite development	Mrema et al. (2020); Jamil, Kountche, and Al-Babili (2021)
	Postattachment resistance in sorghum	Mitigates damage even after *Striga* establishes itself	Begna (2021)
Chemical control	Application of selective herbicides (e.g., 2,4-D) on sorghum	Controls emerged for *Striga*; requires caution to avoid crop damage	Jamil, Kountche, and Al-Babili (2021)
	Use of synthetic germination stimulants (e.g., GR24 analogs)	Induces suicidal germination of *Striga* seeds in the absence of a sorghum host	Begna (2021)
	Fertilizer application (NPK and organic amendments)	Improves sorghum growth, reduces strigolactone exudation, and suppresses *Striga* emergence	Sibhatu (2016); Mrema et al. (2020)
Biological control	*Fusarium oxysporum* f.sp. *Striga*e (FOS) strain treatments	Causes *Striga* seedling wilt and death, reducing infestation	Begna (2021)
	Vesicular arbuscular mycorrhizal (VAM) fungi inoculation	Enhances nutrient uptake and host resistance in sorghum	Jamil, Kountche, and Al-Babili (2021)
Integrated *Striga* management	Combining resistant varieties, crop rotation, fertilization, biocontrol, and selective herbicides	Provides synergistic and sustainable suppression of *Striga* in sorghum cropping systems	Begna (2021); Mrema et al. (2020); Jamil, Kountche, and Al-Babili (2021)

practices, host plant resistance, chemical control, biological control, and integrated *Striga* management (ISM). Prevention focuses on limiting initial spread through clean planting materials and sanitation. Cultural practices such as crop rotation, intercropping, and timely hand weeding help reduce the seed bank and infestation levels. Host plant resistance offers genetic solutions to limit *Striga* attachment and damage. Chemical control methods, such as herbicides and synthetic germination stimulants, provide direct control but entail costs and ecological considerations. Biological agents like *Fusarium oxysporum* and mycorrhizal fungi enhance natural suppression and crop resilience. The most effective method to date is ISM. This strategy combines resistant varieties, crop rotation, fertilization, biocontrol, and targeted herbicide use into a coordinated system. Evidence shows that such integration provides synergistic suppression of *Striga*, boosts sorghum yields, and offers a sustainable solution for smallholder farmers (Begna, 2021; Mrema et al., 2020; Jamil, Kountche, and Al-Babili, 2021). While no single method can fully eliminate *Striga*, the strategic combination of preventive, cultural, genetic, chemical, and biological measures presents the best path toward long-term management of this persistent agricultural pest.

7.3 Strigalactone biosynthetic pathway in sorghum

The biosynthesis pathway of SL in sorghum is tightly regulated, supporting plant growth, fostering beneficial relationships with soil fungi, and defending against parasitic weeds. Understanding its molecular and genetic bases has enabled targeted sorghum breeding strategies to enhance resistance to *Striga*, a crucial step toward food security in affected regions. Research into these mechanisms has also led to innovative approaches to modifying SL profiles, thereby improving resistance to parasitic weeds such as *Striga*. Environmental factors, such as phosphate and nitrogen shortages, significantly enhance SL biosynthesis and exudation, thereby promoting symbiosis with AM fungi and improving nutrient uptake (Yoneyama et al., 2012). During phosphate starvation, genes involved in SL production—including *D27, CCD7, CCD8*, and *MAX1*—are strongly activated, resulting in higher SL levels in roots and exudates (Umehara et al., 2010).

At the biochemical level, SL biosynthesis initiates in plastids from carotenoid precursors and proceeds through sequential enzymatic conversions that generate diverse SL molecules, which are subsequently perceived by a dedicated signaling complex to regulate downstream gene expression (Fig. 7.1).

The biosynthesis of SLs in sorghum begins with the isomerization of *all-trans-β-carotene to 9-cis-β-carotene*, a process catalyzed by the enzyme *D27*. This is followed by sequential cleavage by the carotenoid cleavage dioxygenases *CCD7* and *CCD8*, producing the intermediate carlactone (CL), a key precursor shared in SL biosynthesis across plant species (Arens, 2013; Wu and Li, 2021). Sorghum uniquely uses the cytochrome P450 enzyme *SbMAX1a* to catalyze a four-step oxidation, converting carlactone to 18-hydroxy-carlactonoic acid (18-OH-CLA), an essential intermediate (Wu and Li, 2021). Notably, sorghum has the low-germination-stimulant *1 (LGS1)* gene, which encodes a sulfotransferase enzyme that acts on 18-OH-CLA to produce the main SLs, 5-deoxystrigol (5DS) and 4-deoxyorobanchol (4DO). This sulfonation step is uncommon among plants and is critical for producing SL

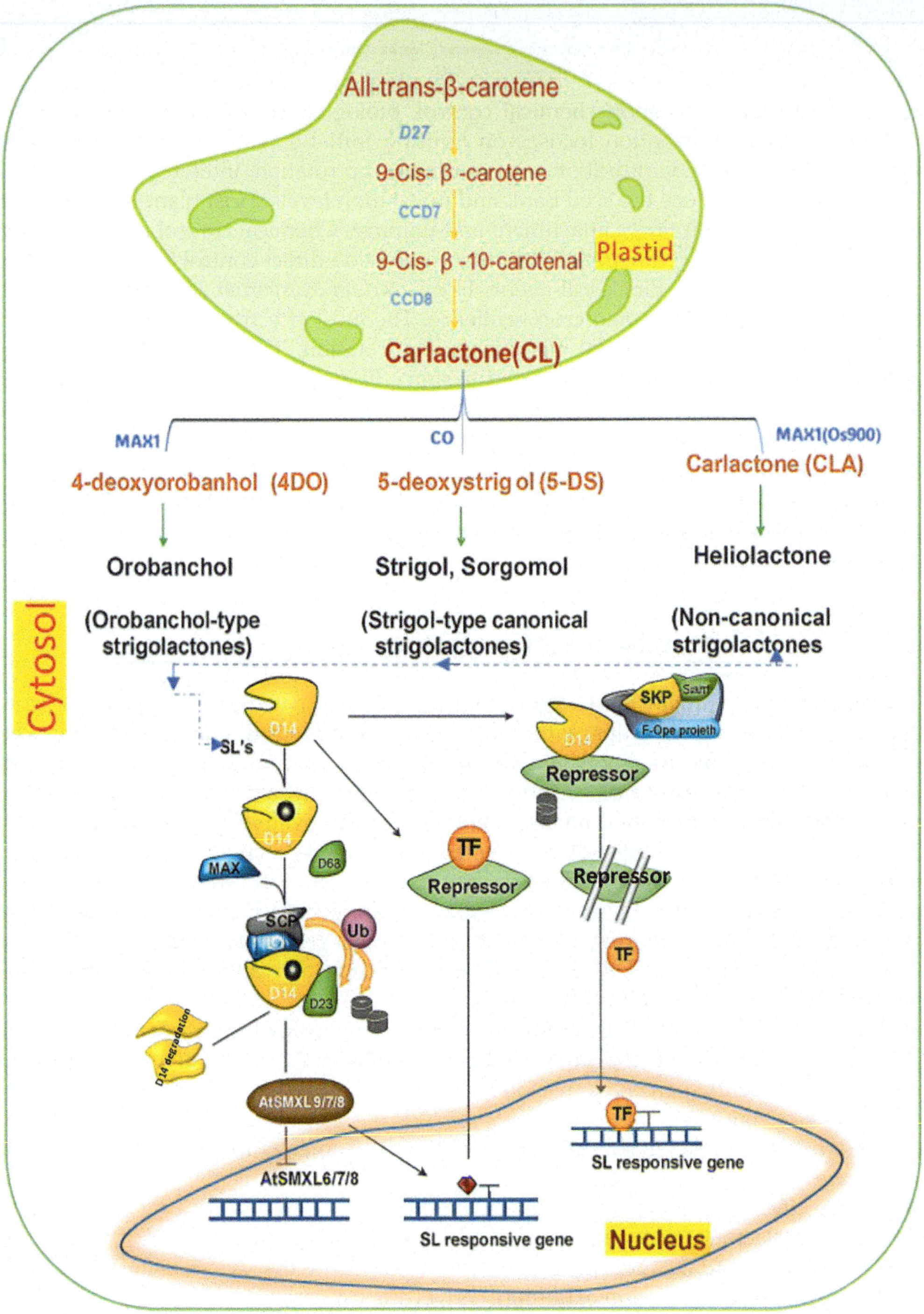

FIGURE 7.1 Strigolactone-mediated biosynthesis and signaling pathways in sorghum contributing to *Striga* resistance. This figure illustrates the key enzymatic steps involved in strigolactone (SL) biosynthesis, including the roles of D27, CCD7, CCD8, and MAX1 in converting carotenoid precursors into bioactive SLs. It also depicts the SL perception and signaling cascade, highlighting the receptor D14, the F-box protein MAX2, and downstream transcriptional regulators that mediate SL-responsive gene expression. The pathway integrates SL functions in regulating plant architecture and modulating root exudation, which influences Striga seed germination and host–parasite interactions. Together, these components provide a molecular framework for understanding SL-associated mechanisms underlying Striga resistance in sorghum.

molecules that stimulate *Striga* seed germination (Yoda et al., 2021; Wu and Li, 2021). Downstream, the cytochrome P450 enzyme CYP728B35 hydroxylates 5DS to produce sorgomol, diversifying the SL profile exuded by sorghum roots (Wu and Li, 2021). Additionally, a 2-oxoglutarate-dependent dioxygenase, Sb3500, has recently been identified as a stereoselective regulator that, together with *LGS1* and *SbMAX1a*, promotes the production of 5DS over 4DO, thereby refining the bioactive SL composition necessary for *Striga* germination (Yoda et al., 2023). Importantly, natural variation in the *LGS1* gene significantly influences sorghum's resistance to *Striga*. Loss-of-function alleles of *LGS1* alter the SL exudate profile, shifting from the potent *Striga* germination stimulant 5DS to orobanchol, a structurally different SL with much weaker activity in stimulating *Striga* seeds. This change offers increased resistance to *Striga* without severely affecting sorghum's growth or development, making it a valuable genetic resource for breeding resistant cultivars (Mutinda et al., 2025; Yoda et al., 2021).

Synthetic SL analogs are developed as alternatives to natural SLs due to their enhanced bioactivity and ease of synthesis (Zorrilla et al., 2022). Nonnatural SLs show greater structural flexibility (Boyer et al., 2012). These analogs mimic plant hormones called SLs, aiding biological research to understand their roles. Synthesis involves creating either full SL molecules or simpler analogs that retain bioactivity (Oancea et al., 2017; Zwanenburg et al., 2015). Producing SL analogs is promising, especially after identifying the bioactive D-ring. The A-ring's stereochemistry is also important, though its contribution to activity is minor. The D-ring mainly determines bioactivity, while side-chain groups influence function. These analogs help in studying SL roles and in developing solutions for weed control and plant growth enhancement (Aliche et al., 2020). Noncanonical SLs lack one or more fused bicyclic or tricyclic ABC rings but retain the active butenolide D-ring. Early noncanonical SLs, such as carlactone and derivatives such as carlactonoic acid and methyl carlactonoate, were first synthesized as racemic mixtures due to the complexity of the synthesis (Alder et al., 2012; Seto et al., 2014). Analogs are made via total or partial synthesis from available molecules, with partial synthesis reducing the number of steps by starting with related molecules. The most studied SL analog, GR24, was first synthesized by Johnson et al. (1981). It triggers parasitic weed seed germination, useful for suicidal germination. Used as a racemic mixture at the C-20 stereocenter, it is the standard control in germination assays, effective across genera like *Striga*, *Orobanche*, and *Phelipanche*. Recent halogenated derivatives, such as 7-fluoro-GR24 (7FGR24), exhibit higher potency and receptor affinity (Chen et al., 2021). Compounds MP1, MP3, MP13, and MP26 have shown promise in stimulating *Striga hermonthica* germination via suicidal germination (Jamil et al., 2019). The key genes and enzymes involved in SL biosynthesis in sorghum, along with their functions and products, are summarized in Table 7.2.

Taken together, the biosynthetic pathway of SLs in sorghum is a finely balanced process that integrates growth regulation, symbiotic signaling, and susceptibility to parasitic weeds. A detailed understanding of this pathway, from core enzymes such as *D27*, *CCD7*, *CCD8*, and *MAX1* to unique regulators such as *LGS1* and *Sb3500*, provides valuable targets for breeding and genetic engineering. By strategically modifying SL profiles, it is possible to reduce *Striga* germination while maintaining beneficial symbioses, making SL biosynthesis a key focus in developing durable *Striga*-resistant sorghum varieties. Developmentally, SL production is influenced by plant age and tissue differentiation, with higher expression of

TABLE 7.2 Strigolactone biosynthetic genes, their enzymatic functions, and resulting products in sorghum.

Gene/enzyme	Function	Product or role	Reference
D27	β-Carotene isomerase	Converts all-trans-β-carotene to 9-cis-β-carotene	Arens (2013); Wu and Li (2021)
CCD7 & CCD8	Carotenoid cleavage dioxygenases	Cleave 9-cis-β-carotene to produce carlactone (CL)	Arens (2013); Wu and Li (2021)
MAX1a	Cytochrome P450 enzyme (CYP711A family)	Performs four-step oxidation of carlactone to 18-hydroxy-carlactonoic acid (18-OH-CLA)	Wu and Li (2021)
LGS1	Sulfotransferase	Converts 18-OH-CLA to 5-deoxystrigol (5DS) and 4-deoxyorobanchol (4DO) via sulfonation	Yoda et al. (2021); Wu and Li (2021)
CYP728B35	Cytochrome P450 enzyme	Hydroxylates 5DS to form sorgomol	Wu and Li (2021)
Sb3500	2-Oxoglutarate-dependent dioxygenase	Stereoselective regulator favoring 5DS production over 4DO	Yoda et al. (2023)

SL biosynthetic genes observed in roots and senescing tissues, where they regulate shoot branching and resource allocation (Waters et al., 2017). Hormonal crosstalk further refines SL biosynthesis; auxin positively regulates the expression of *CCD7* and *CCD8*, while gibberellins can suppress SL production through feedback interactions (Yoneyama et al., 2018). Additionally, the external application of SL analogs, such as GR24, suppresses endogenous SL biosynthetic genes, indicating feedback inhibition that helps maintain hormonal balance (Mashiguchi, Seto, and Yamaguchi, 2021).

7.4 Hormonal crosstalk with strigolactones in sorghum

Plant hormones are vital in controlling plant growth and development. While metabolism provides the energy needed for survival, hormones regulate the speed of growth. Phytohormones are plant-produced growth regulators involved in many stages of the plant life cycle. Well-known examples include auxin, gibberellin, cytokinin, ethylene, and ABA. Recently, other compounds have been identified as important phytohormones, such as brassinosteroids (BRs), a class of steroidal hormones, as well as jasmonic acid, salicylic acid (SA), and, most recently, SLs (Faizan et al., 2020). SLs are key regulators of plant development, but their activity is part of a complex network of hormonal interactions that influence growth, stress responses, and resistance to parasitic weeds like *Striga*. This hormonal crosstalk enables plants to fine-tune their development and defense mechanisms, but it also makes manipulating SL pathways challenging without causing unintended effects. Therefore, understanding how SLs interact with other phytohormones in sorghum is essential for developing effective resistance strategies.

7.4.1 Auxins

Auxin plays a vital role in regulating bud outgrowth. It is primarily produced in the shoot apex and young leaves (Ljung, Bhalerao, and Sandberg, 2001). The export of auxin from buds is linked to bud outgrowth initiation, suggesting that buds must export auxin to become activated (Müller and Leyser, 2011). Decapitation and girdling experiments suggest that growing axillary branches or buds influences auxin sink strength and bud responsiveness to SLs. In the stem, auxin levels inversely regulate bud outgrowth by maintaining high local SL and low cytokinin levels. Once buds are activated, auxin is exported into the stem to support vascular development (Ferguson and Beveridge, 2009). In roots, high auxin concentrations tend to promote cell division, while lower levels favor cell expansion (Doerner, 2008). The initiation of root hairs is believed to be directly mediated by optimal auxin levels and signaling. Lateral root initiation depends on auxin, which recirculates within the root and is derived from both the shoot and the root. The development of lateral roots is sustained by shoot-derived auxin transported to the parent root and into the lateral root primordia via the PAT stream (Cheng, Ruyter-Spira, and Bouwmeester, 2013).

7.4.2 Cytokinins

Cytokinin signaling is vital for plant growth, influencing key processes (Faizan et al., 2020). SLs and CKs interact in buds to control outgrowth via the TCP transcription factor BRC1 (Dun et al., 2013). They promote parasitic seed germination without SLs (Logan and Stewart, 1991; Babiker et al., 1993), with cytokinin enhancing germination by boosting ethylene (Babiker et al., 1993). They act independently in adventitious rooting (Rasmussen et al., 2012), antagonistically during bud activation and shoot branching (Dun et al., 2013), but synergistically regulate lateral root development, requiring cytokinin signaling components for G(R)24-induced suppression (Jiang et al., 2015). *SHY2* links auxin, cytokinin, and SL signaling to manage lateral roots under high phosphate conditions (Omoarelojie et al., 2019).

7.4.3 Abscisic acid

SLs and ABA share structural similarities, indicating a possible link in their biosynthesis. Both SLs and ABA originate from the carotenoid pathway, where all-trans isomers are produced through a biosynthetic process starting with all-trans-β-carotene and leading to all-trans-violaxanthin (Faizan et al., 2020). Therefore, it is suggested that the homeostasis of both hormones depends on each other. The metabolism of SLs and their influence on ABA homeostasis differ between roots and shoots under stress. The interaction between SL and ABA is antagonistic when regulating specific processes, such as shoot architecture and fruit ripening. ABA suppresses the expression of genes involved in SL biosynthesis and signaling, which promotes tiller formation (Wang et al., 2018). ABA acts as a negative regulator of seed germination, while gibberellin and cytokinin serve as positive regulators (Faizan et al., 2020). During seed preconditioning, ABA levels decrease (Chae et al., 2004). It appears that crosstalk between SL and gibberellin is limited or may not involve direct interactions between signaling components, and the inhibition of SL synthesis by gibberellin mainly functions to modulate SL levels (Omoarelojie et al., 2019). *MAX2* interacts with gibberellin and ABA signaling to mediate light-induced seed germination and with constitutive photomorphogenesis *1 (COP1)* to regulate photomorphogenesis (Nelson and Steber, 2016). SL action is antagonized by suppressing SL-induced upregulation of specific ABC-type transporters (Lantzouni, Klermund, and Schwechheimer, 2017).

7.4.4 Ethylene

In light, SLs inhibit hypocotyl elongation by upregulating *HY5* expression in a *MAX2*-dependent process (Jia et al., 2014), whereas ethylene enhances *COP1*-mediated degradation of *HY5* to promote hypocotyl elongation (Yu et al., 2013). This suggests that an antagonistic interaction between SLs and ethylene signaling may regulate hypocotyl growth. In sorghum and other plants, SL control of root hair elongation is mediated through ethylene signaling. Studies in *Arabidopsis* have shown that ethylene signaling mutants (*At-ein* and *At-etr*) display reduced sensitivity to GR24 treatment, and inhibition of ethylene production completely abolishes SL-induced root hair elongation (Kapulnik et al., 2011). Furthermore, GR24 application upregulates *AtACS2*, a gene involved in ethylene biosynthesis, indicating that SLs may promote ethylene production to mediate this developmental response. SLs have also been shown to induce ethylene biosynthesis in *Striga* seeds before germination (Sugimoto et al., 2003). Additionally, ethylene may act downstream of SLs to control adventitious root (AR) initiation sites in the Arabidopsis hypocotyl, a process independent of SLs (Rasmussen et al., 2017).

7.4.5 Brassinosteroids

A variety of other compounds have recently been recognized as important phytohormones, including the class of steroidal plant hormones known as BRs, as well as other organic molecules such as jasmonic acid and SA (Smith and Li, 2014). Crosstalk between SL and BR signaling pathways has only recently been discovered, and a complete understanding of this inter-hormonal interaction is still developing. The E3 ubiquitin ligase *MAX2*, a key SL signaling component and inhibitor of shoot branching, directly interacts with *BZR1* and *BES1* transcription factors, which are central regulators of BR-dependent gene expression (Li et al., 2005). *BZR1* and *BES1* interact constantly with *MAX2*, and *MAX2* binding promotes their degradation. SL signaling promotes *MAX2*-mediated degradation of *BZR1/BES1*, thereby suppressing shoot branching. Consequently, it has been proposed that BRs and SLs regulate the same developmental process by modulating *the* stability of BZR1/BES1 (Wang et al., 2013). It acts synergistically with other growth regulators, such as auxins, particularly by promoting lateral root growth, which is essential for water and nutrient uptake (de Souza et al., 2023).

7.4.6 Salicylic acid

SA plays a crucial role in how plants perceive and defend against pathogens, as well as in their tolerance to abiotic stressors. Its regulatory function in these processes mainly depends on its influence on reactive oxygen species (ROS) levels (Prodhan et al., 2018). SA has been shown to regulate drought tolerance (Askari and Ehsanzadeh, 2015), senescence (Ji, Liu, and Xing, 2016), and stomatal closure and conductance (Prodhan et al., 2018). Treatment with the synthetic SL analog GR24 triggers SA accumulation, while the SL signaling mutant *max2* exhibits reduced SA levels (Rozpądek et al., 2018). Overall, the interaction between SLs and other hormones reveals a highly interconnected signaling network. Auxin and SLs coordinate the developmental suppression of branching but increase susceptibility when SL exudation is high; cytokinins counteract this suppression but must be carefully managed to prevent excessive tillering. ABA links abiotic stress tolerance to increased *Striga* risk, ethylene and SA boost defenses after infection, and BRs modulate the balance between growth and resistance. This hormonal complexity indicates that breeding sorghum for *Striga* resistance cannot focus solely on SLs but must consider their integration into broader hormonal networks. Careful manipulation of SL biosynthesis and signaling, while accounting for their hormonal interactions, offers the potential to develop sorghum varieties that are both resistant to parasitism and adaptable to various stress conditions.

7.5 Strigolactone signaling

Similar to other plant hormones, SLs are detected by a specific receptor complex. An F-box protein—*MAX2* in *Arabidopsis thaliana*, *D3* in *Oryza sativa*, and *RMS4* in pea—serves as a key part of this system (Aquino, Bradley, and Lumba, 2021; Wani et al., 2023). Like auxin, ABA, and gibberellin signaling pathways, SL signaling also relies on hormone-triggered protein degradation (Omoarelojie et al., 2019). The process likely begins with SL perception, leading to the formation of a complex that involves a Skp–Cullin–F-box (SCF)–E2 ubiquitination module. This SCF–E2 complex tags proteins for destruction by the 26S proteasome. In SL signaling, the F-box protein interacts with the SCF–E2 complex to ubiquitinate transcriptional repressors such as SMXLs in *Arabidopsis* and *D53* in rice (Waters et al., 2017). A key feature of SL signaling is that SCF-dependent ubiquitination is directed by a hormone-dependent interaction between the F-box protein and its targets (the only known SL receptor to date is *DWARF14* (D14), which contains a conserved serine–histidine–aspartate catalytic triad that is essential for its hydrolase activity (Faizan et al., 2020). When SLs bind to D14, a catalytic reaction produces a covalently bound D-ring–derived intermediate (Yoshida et al., 2019). This reaction causes a change in D14's shape, allowing it to interact with the *D3/MAX2*-containing SCF complex and recruit *D53* or SMXL proteins for ubiquitination. Functional *MAX2/D3* variants are essential for SL signaling, and mutations in these genes lead to phenotypes similar to SL biosynthesis mutants (Yang et al., 2019). However, applying the synthetic analog GR24 externally does not correct the defects in *max2/d3* mutants. In *d3* mutants, elevated SL levels are associated with negative feedback regulation of SL biosynthesis (Rani et al., 2023). *MAX2* also influences traits beyond SL signaling, including reduced seedling photomorphogenesis and increased seed dormancy, indicating its role in non-SL pathways. In the absence of SLs, *D53* and *D53*-like SMXL proteins act as repressors by interacting with TPL/TPR corepressors, thereby blocking downstream transcription factors (Sun et al., 2022).

7.6 Genetic determinants of resistance in sorghum

Sorghum has coevolved with *Striga* spp. throughout Sub-Saharan Africa, leading to the development of various resistance mechanisms. These mechanisms are supported by specific genetic factors that function at different stages of the host–parasite interaction. Generally, they can be grouped into genes controlling pre- and post-attachment resistance, quantitative trait loci (QTLs), findings from genome-wide association studies (GWAS), successful applications of marker-assisted selection (MAS), and, more recently, insights from CRISPR/Cas9 and functional genomics approaches.

7.7 Preattachment resistance genes

Pre-attachment resistance is the initial barrier to *Striga* infection, occurring before the parasite makes physical contact with host roots. This type of resistance mainly involves adjusting SL exudation, as SLs are germination stimulants released by host roots. Sorghum accessions with mutations in the low germination stimulant *1 (LGS1)* locus produce orobanchol-type SLs instead of strigol-type, leading to very low *Striga* germination. Similar findings in rice and maize emphasize the conserved role of the *cytochrome P450 gene* family, especially *MAX1* homologs, in determining SL structural diversity (Lucido et al., 2022; Cardoso et al., 2014). For example, rice has *Os900* and *Os1400*, which catalyze separate steps in SL biosynthesis, while natural deletions of other *MAX1*-like genes (*SLB1* and *SLB2*) greatly reduce exudation and parasite stimulation. These results mirror sorghum's *lgs1* polymorphisms and support the idea that preattachment resistance relies on conserved biosynthetic gene families across cereals. By altering SL exudation profiles, these genes play key roles in host signaling and are important targets for resistance breeding (Gobena et al., 2017; Fishman and Shirasu, 2021).

7.8 Postattachment resistance genes

When preattachment resistance is insufficient, postattachment resistance mechanisms determine whether the parasite successfully establishes itself within host roots. In sorghum, these mechanisms include hypersensitive responses, formation of mechanical barriers, and impaired vascular connections that limit nutrient flow to the parasite. Comparative studies between sorghum and rice show conservation of defense-related gene families such as cytochrome P450s and WRKY transcription factors (Yakubu and Price, 2024). For example, *WRKY45*, a key regulator of defense signaling in rice, has conserved orthologs in sorghum that are involved in resistance responses after *Striga* attachment. These genes likely work by activating host defense pathways or disrupting parasite haustorium development. The cross-species conservation of these families highlights their potential for use in translational breeding and genome-editing strategies aimed at enhancing postattachment resistance (Mutuku et al., 2015; Yakubu and Price, 2024).

7.9 Quantitative trait loci mapping and genome-wide association study findings

Quantitative genetics has played a key role in understanding the inheritance of *Striga* resistance in sorghum. Early research using recombinant inbred populations led to detailed mapping of the *lgs* locus on chromosome SBI-05, situated between SSR markers SB3344 and SB3352 and confined to a 5.8 cM interval (Satish et al., 2012). Broader QTL studies by Haussmann et al. (2004) identified resistance loci in populations derived from IS9830 × E36-1 and N13 × E36-1, encompassing both preattachment (low germination-stimulant activity) and postattachment (mechanical barriers) forms of resistance. Several of these loci remained consistent across different years and environments, making them promising candidates for MAS.

Recently, GWAS approaches have expanded these insights by utilizing high-density markers. Kavuluko et al. (2021) identified overlapping QTLs on chromosomes 1, 2, 9, and 10 linked to cell wall modification, defense compound secretion, and pathogen-resistance gene activation. Mohamed et al. (2014) also reported QTLs in the resistant landrace N13 and developed EST-SSR markers that improve the efficiency of MAS. Together, these studies provide a strong genetic framework for resistance breeding, combining major-effect loci like *lgs1* with polygenic contributions identified through QTL and GWAS analyses.

7.10 Marker-assisted selection: success stories

MAS has facilitated the practical deployment of resistance genes into farmer-preferred sorghum varieties. Using recombinant inbred populations and SSR markers, Yohannes et al. (2015) introgressed five resistance QTLs from donor parent N13 into the susceptible variety Hugurtay, successfully combining foreground and background selection to recover the recurrent genome while maintaining resistance. In Sudan, Ali and Hash (2016) used marker-assisted backcrossing (MABC) to transfer resistance from N13 into three elite cultivars—Tabat, Wad, and Ahmed. Their BC_3S_3 populations showed significant reductions in *Striga* emergence, from nearly 920 plants per plot to fewer than 10, along with yield increases of up to 189% under infestation. Similarly, Mohamed et al. (2014) developed backcross-derived populations carrying multiple resistance QTLs and identified four high-yielding, *Striga*-resistant lines across multiple locations. These were released as the first DNA marker-assisted sorghum varieties in sub-Saharan Africa, demonstrating that MAS can provide durable resistance in farmer-preferred backgrounds.

7.11 Gene editing and functional genomics

The advent of CRISPR/Cas9 genome editing has transformed the study of *Striga* resistance in sorghum, providing precise tools to investigate the SL biosynthetic pathway and develop resistant germplasm. Unlike traditional breeding, which is slow and limited by natural variation, CRISPR enables targeted editing of key resistance genes in a single generation. Several recent studies highlight the potential and trade-offs of this approach. Kaniganti et al. (2025) targeted core SL biosynthetic genes, including *CCD7, CCD8,* and *MAX1,* as well as an uncharacterized *DUF* gene within the fine-mapped *lgs1* region. Using *Agrobacterium*-mediated transformation in two sorghum cultivars, they achieved transformation efficiencies of about 70% and editing efficiencies of 17.5%. The edited lines carried stable loss-of-function mutations that reduced SL pathway expression, decreased SL exudation, and significantly delayed or prevented *Striga* emergence. Similarly, Hao et al. (2023) disrupted the *CCD8* genes in sorghum, which are essential for orobanchol production. The resulting mutants showed a sharp decline in *Striga* germination but also displayed increased tillering, reduced grain yield, and notable changes in rhizosphere microbial communities. These findings emphasize the double-edged nature of targeting SL biosynthesis: while resistance is enhanced, plant architecture, productivity, and ecological interactions may be affected. More recently, Shi et al. (2025) identified xylem-localized SL transporters (*SbSLT1* and *SbSLT2*) and demonstrated through CRISPR knockouts that reducing SL export into the rhizosphere significantly improved resistance to *S. hermonthica* under both greenhouse and field conditions.

Collectively, these studies demonstrate that CRISPR editing of SL-related genes can be a powerful approach to developing resistance, but they also emphasize the importance of balancing resistance gains with agronomic performance. Future approaches might include promoter engineering, allele fine-tuning, or combining pre- and post-attachment resistance genes to reduce trade-offs and accelerate the development of farmer-preferred, *Striga*-resistant sorghum varieties.

7.12 Natural variation in strigolactone biosynthesis and signaling genes

Natural variation in SL biosynthesis and signaling genes greatly contributes to the diversity of SL composition, abundance, and biological functions across different plant species and genotypes. Variations in genes such as *D27, CCD7, CCD8, and MAX1,* which encode essential enzymes in the SL biosynthetic pathway, cause differences in SL profiles and influence plant architecture, symbiotic relationships, and resistance to parasitic weeds (Wakabayashi et al., 2019; Yoneyama, 2020). For instance, allelic variation in the low germination stimulant *1 (LGS1)* gene in sorghum changes the stereochemical configuration of SLs, shifting production from potent germination stimulants like 5DS to less-active orobanchol-type SLs—providing *Striga* resistance without harming beneficial AM associations (Gobena et al., 2017; Yoda et al., 2023). Likewise, natural polymorphisms in SL signaling components, such as *D14, MAX2,* and SMXL/D53 proteins, affect how plants perceive SLs and regulate downstream transcription (Jiang et al., 2015). Such variations can alter shoot branching, root growth, and stress responses, demonstrating how the SL pathway has evolved to suit different ecological environments. Understanding this genetic diversity offers valuable insights for breeding and biotechnological efforts to improve SL-related traits, including nutrient use efficiency and parasite resistance.

7.13 Conclusion

Striga threatens sorghum yields in tropical areas, causing losses for smallholders. The complex interaction between SLs and rhizosphere signaling makes management difficult. Recent genetic discoveries, including *LGS1* allelic variation and QTLs, enable breeding sorghum with altered SL profiles, reducing *Striga* stimulation with little impact on growth, yields, or beneficial microbes. Advances in molecular breeding, mapping, and farming practices improve *Striga* resistance, but climate change may increase its spread, emphasizing the need for sustainable solutions. Techniques such as CRISPR/Cas9 and genomic selection target SL biosynthesis and defense pathways to achieve durable resistance. Long-term control will require combining genetic, cultural, chemical, and biological strategies for safe, effective management.

7.14 Future prospects

Advances in understanding SL biology are key to developing sorghum that promotes growth and reduces *Striga* germination. Using CRISPR/Cas9 to modify SL genes (e.g., *LGS1, CCD7, CCD8*) can reduce *Striga* susceptibility and support beneficial fungi. However, lowering SL might limit fungal colonization or alter root morphology, thereby affecting nutrient availability and plant health. Future efforts should combine molecular biology, genomics, breeding, and agronomy to optimize SL without harming plants. Combining genetic resistance with better soil, water, and seed management will enhance success. Beneficial fungi and bacteria support resistance by improving soil health and reducing *Striga* germination. Ultimately, integrating breeding, biotech, and farmer practices offers a path for sustainable *Striga* control, requiring collaboration among researchers, breeders, extension agents, and farmers to build resilient sorghum systems.

References

Agt, B., 2007. *Striga*: the spreading scourge in Africa. Regulation of Plant Growth & Development 42 (1), 74–87.

Alder, A., Jamil, M., Marzorati, M., Bruno, M., Vermathen, M., Bigler, P., et al., 2012. The path from β-carotene to carlactone, a strigolactone-like plant hormone. American Association for the Advancement of Science, Germany Science 335 (6074), 1348–1351. Available at https://doi.org/10.1126/science.1218094, http://www.sciencemag.org/content/335/6074/1348.full.pdf.

Aliche, E.B., Screpanti, C., De Mesmaeker, A., Munnik, T., Bouwmeester, H.J., 2020. Science and application of strigolactones. New Phytologist 227 (4), 1001–1011. Available at https://doi.org/10.1111/nph.16489, http://onlinelibrary.wiley.com/journal/10.1111/(ISSN)1469-8137.

Ali, R., Hash, C.T., Damaris, O., Elhussein, A., Mohamed, A.H., 2016. Introgression of Striga resistance into popular sudanese sorghum varieties using marker assisted selection. World Journal of Biology and Biotechnology 1 (1), 49. Available at https://doi.org/10.33865/wjb.001.01.0006.

Andiku, C., Shimelis, H., Laing, M., Shayanowako, A.I.T., Adrogu Ugen, M., Manyasa, E., et al., 2021. Assessment of sorghum production constraints and farmer preferences in Uganda: implications for nutritional quality breeding. Acta Agriculturae Scandinavica, Section B – Soil & Plant Science 71 (7), 620–632.

Aquino, B., Bradley, J.M., Lumba, S., 2021. On the outside looking in: roles of endogenous and exogenous strigolactones. Plant Journal 105 (2), 322–334. Available at https://doi.org/10.1111/tpj.15087, http://onlinelibrary.wiley.com/journal/10.1111/(ISSN)1365-313X.

Arens, K., 2013. Strigolactone profiles in *Sorghum bicolor*: In relation to arbuscular mycorrhizal fungal symbioses. Wageningen University & Research.

Askari, E., Ehsanzadeh, P., 2015. Drought stress mitigation by foliar application of salicylic acid and their interactive effects on physiological characteristics of fennel (*Foeniculum vulgare* Mill.) genotypes. Acta Physiologiae Plantarum 37 (2). Available at https://doi.org/10.1007/s11738-014-1762-y.

Babiker, A.G.T., Butler, L.G., Ejeta, G., Woodson, W.R., 1993. Enhancement of ethylene biosynthesis by 1-aminocyclopropane-1-carboxylic acid and cytokinins in *Striga asiatica* seeds. Physiologia Plantarum 89, 21–26. Available at https://doi.org/10.1111/j.1399-3054.1993.tb01781.x.

Begna, T., 2021. Effect of *Striga* species on sorghum (*Sorghum bicolor* L. Moench) production and its integrated management approaches. International Journal of Research Studies in Agricultural Sciences 7 (7), 10–22.

Beyene, D., Gudeta, M., Chalchisa, D., 2025. Breeding and genetics of sorghum for *Striga* resistance: future perspectives. American Journal of Biomedical and Life Sciences 13 (3), 59–66. Available at https://doi.org/10.11648/j.ajbls.20251303.11.

Bouwmeester, H.J., Matusova, R., Zhongkui, S., Beale, M.H., 2003. Secondary metabolite signalling in host-parasitic plant interactions. Current Opinion in Plant Biology 6 (4), 358–364. Available at https://doi.org/10.1016/S1369-5266(03)00065-7, www.elsevier.com/inca/publications/store/6/0/1/3/1/4/index.htt.

Boyer, F.D., de Saint Germain, A., Pillot, J.P., Pouvreau, J.B., Chen, V.X., Ramos, S., et al., 2012. Structure–activity relationship studies of strigolactone-related molecules for branching inhibition in garden pea. Plant Physiology 159, 1524–1544.

Cardoso, C., Ruyter-Spira, C., Bouwmeester, H.J., 2011. Strigolactones and root infestation by plant-parasitic *Striga*, *Orobanche* and *Phelipanche* spp. Plant Science 180 (3), 414–420. Available at https://doi.org/10.1016/j.plantsci.2010.11.007.

Cardoso, C., Zhang, Y., Jamil, M., Hepworth, J., Charnikhova, T., Dimkpa, S.O.N., et al., 2014. Natural variation of rice strigolactone biosynthesis is associated with the deletion of two MAX1 orthologs. Proceedings of the National Academy of Sciences 111 (6), 2379–2384. Available at https://doi.org/10.1073/pnas.1317360111.

Chae, S.H., Yoneyama, K., Takeuchi, Y., Joel, D.M., 2004. Fluridone and norflurazon, carotenoid-biosynthesis inhibitors, promote seed conditioning and germination of the holoparasite *Orobanche minor*. Physiologia Plantarum 120 (2), 328–337. Available at https://doi.org/10.1111/j.0031-9317.2004.0243.x.

Cheng, X., Ruyter-Spira, C., Bouwmeester, H., 2013. The interaction between strigolactones and other plant hormones in the regulation of plant development. Frontiers Research Foundation, Netherlands Frontiers in Plant Science 4. Available at https://doi.org/10.3389/fpls.2013.00199, http://journal.frontiersin.org/Journal/10.3389/fpls.2013.00199/pdf.

Chen, Y., Kuang, Y., Shi, L., Wang, X., Fu, H., Yang, S., et al., 2021. Synthesis and evaluation of new halogenated GR24 analogs as germination promotors for *Orobanche cumana*. Frontiers in Plant Science 12. Available at https://doi.org/10.3389/fpls.2021.725949, https://www.frontiersin.org/journals/plant-science.

Chesterfield, R.J., Vickers, C.E., Beveridge, C.A., 2020. Translation of strigolactones from plant hormone to agriculture: achievements, future perspectives, and challenges. Trends in Plant Science 25 (11), 1087–1106. Available at https://doi.org/10.1016/j.tplants.2020.06.005, www.elsevier.com/inca/publications/store/3/0/9/6/0/index.htt.

Doerner, P., 2008. Plant roots: recycled auxin energizes patterning and growth. Current Biology (2), R72. Available at https://doi.org/10.1016/j.cub.2007.11.048.

Dun, E.A., De Saint Germain, A., Rameau, C., Beveridge, C.A., 2013. Dynamics of strigolactone function and shoot branching responses in *Pisum sativum*. Molecular Plant 6 (1), 128–140. Available at https://doi.org/10.1093/mp/sss131.

Ejeta, G., Gressel, J., 2007. Integrating New Technologies for *Striga* Control: Towards Ending the Witch-Hunt Integrating New Technologies for Striga Control: Towards Ending the Witch-Hunt. World Scientific Publishing Co, United States, pp. 1–345. Available at http://www.worldscientific.com/worldscibooks/10.1142/6470#t=toc, https://doi.org/10.1142/6470.

Faizan, M., Faraz, A., Sami, F., Siddiqui, H., Yusuf, M., Gruszka, D., et al., 2020. Role of strigolactones: signalling and crosstalk with other phytohormones. Open Life Sciences 15 (1), 217–228. Available at https://doi.org/10.1515/biol-2020-0022, http://www.degruyter.com/view/j/biol.

Ferguson, B.J., Beveridge, C.A., 2009. Roles for auxin, cytokinin, and strigolactone in regulating shoot branching1[C][W][OA]. American Society of Plant Biologists, Australia. Plant Physiology 149 (4), 1929–1944. Available at https://doi.org/10.1104/pp.109.135475, http://www.plantphysiol.org/cgi/reprint/149/4/1929.

Fishman, M.R., Shirasu, K., 2021. How to resist parasitic plants: pre- and post-attachment strategies. Current Opinion in Plant Biology 62, 102004. Available at https://doi.org/10.1016/j.pbi.2021.102004.

Gobena, D., Shimels, M., Rich, P.J., Ruyter-Spira, C., Bouwmeester, H., Kanuganti. S., et al., (2017). Mutation in sorghum LOW GERMINATION STIMULANT 1 alters strigolactones and causes Striga resistance. Proceedings of the National Academy of Sciences, 114(17), 4471–4476.

Gomez-Roldan, V., Fermas, S., Brewer, P.B., Puech-Pagès, V., Dun, E.A., Pillot, J.P., et al., 2008. Strigolactone inhibition of shoot branching. Nature 455 (7210), 189–194. Available at https://doi.org/10.1038/nature07271.

Gurney, A.L., Slate, J., Press, M.C., Scholes, J.D., 2006. A novel form of resistance in rice to the angiosperm parasite *Striga hermonthica*. New Phytologist 169 (1), 199–208. Available at https://doi.org/10.1111/j.1469-8137.2005.01560.x.

Hao, J., Yang, Y., Futrell, S., Kelly, E.A., Lorts, C.M., Nebie, B., et al., 2023. CRISPR/Cas9-mediated mutagenesis of carotenoid cleavage dioxygenase (CCD) genes in sorghum alters strigolactone biosynthesis and plant biotic interactions. Phytobiomes Journal 10 (10). Available at https://doi.org/10.1094/PBIOMES-08-22-0053-R, https://apsjournals.apsnet.org/doi/pdf/10.1094/PBIOMES-08-22-0053-R.

Haussmann, B.I.G., Hess, D.E., Omanya, G.O., Folkertsma, R.T., Reddy, B.V.S., Kayentao, M., et al., 2004. Genomic regions influencing resistance to the parasitic weed *Striga hermonthica* in two recombinant inbred populations of sorghum. Theoretical and Applied Genetics 109 (5), 1005–1016. Available at https://doi.org/10.1007/s00122-004-1706-9.

Jamil, M., Kountche, B.A., Al-Babili, S., 2021. Current progress in *Striga* management. Plant Physiology 185 (4), 1339–1352. Available at https://doi.org/10.1093/plphys/kiab040, https://academic.oup.com/plphys/article/185/4/1339/6129111?login=true.

Jamil, M., Kountche, B.A., Haider, I., Wang, J.Y., Aldossary, F., Zarban, R.A., et al., 2019. Methylation at the C-30 in the D-ring of strigolactone analogs reduces biological activity in root parasitic plants and rice. Frontiers in Plant Science.

Jiang, L., Matthys, C., Marquez-Garcia, B., De Cuyper, C., Smet, L., De Keyser, A., et al., 2015. Strigolactones spatially influence lateral root development through the cytokinin signaling network. Journal of Experimental Botany 67 (1), 379–389. Available at https://doi.org/10.1093/jxb/erv478.

Jia, K.P., Luo, Q., He, S.B., Lu, X.D., Yang, H.Q., 2014. Strigolactone-regulated hypocotyl elongation is dependent on cryptochrome and phytochrome signaling pathways in *Arabidopsis*. Molecular Plant 7 (3), 528–540. Available at https://doi.org/10.1093/mp/sst093.

Ji, Y., Liu, J., Xing, D., 2016. Low concentrations of salicylic acid delay methyl jasmonate-induced leaf senescence by up-regulating nitric oxide synthase activity. Journal of Experimental Botany 67 (17), 5233–5245. Available at https://doi.org/10.1093/jxb/erw280, http://jxb.oxfordjournals.org/.

Johnson, A.W., Gowda, G., Hassanali, A., Knox, J., Monaco, S., Razavi, Z., et al., 1981. The preparation of synthetic analogues of strigol. Transactions 1 1734–1743. Available at https://doi.org/10.1039/P19810001734.

Kanampiu, F., Makumbi, D., Mageto, E., Omanya, G., Waruingi, S., Musyoka, P., et al., 2018. Assessment of management options on *Striga* infestation and maize grain yield in Kenya. Weed Science 66 (4), 516–524. Available at https://doi.org/10.1017/wsc.2018.4.

Kaniganti, S., Bhattacharya, J., Petla, B.P., Reddy, P.S., 2022. Strigolactone, a neglected plant hormone, with a great potential for crop improvement: Crosstalk with other plant hormones. Environmental and Experimental Botany 204, 105072.

Kaniganti, S., Palakolanu, S.R., Thiombiano, B., Damarasingh, J., Bommineni, P.R., Che, P., et al., 2025. Developing *Striga* resistance in sorghum by modulating host cues through CRISPR/Cas9 gene editing. Plant cell Reports 44 (4). Available at https://doi.org/10.1007/s00299-025-03474-1, https://www.springer.com/journal/299.

Kapulnik, Y., Resnick, N., Mayzlish-Gati, E., Kaplan, Y., Wininger, S., Hershenhorn, J., et al., 2011. Strigolactones interact with ethylene and auxin in regulating root-hair elongation in *Arabidopsis*. Journal of Experimental Botany 62 (8), 2915–2924. Available at https://doi.org/10.1093/jxb/erq464.

Kavuluko, J., Kibe, M., Sugut, I., Kibet, W., Masanga, J., Mutinda, S., et al., 2021. GWAS provides biological insights into mechanisms of the parasitic plant (*Striga*) resistance in sorghum. BMC Plant Biology 21 (1). Available at https://doi.org/10.1186/s12870-021-03155-7, http://www.biomedcentral.com/bmcplantbiol/.

Khalid, W., Ali, A., Arshad, M.S., Afzal, F., Akram, R., Siddeeg, A., et al., 2022. Nutrients and bioactive compounds of *Sorghum bicolor* L. used to prepare functional foods: a review on the efficacy against different chronic disorders. International Journal of Food Properties 25 (1), 1045–1062. Available at https://doi.org/10.1080/10942912.2022.2071293, https://www.tandfonline.com/loi/ljfp20.

Lantzouni, O., Klermund, C., Schwechheimer, C., 2017. Largely additive effects of gibberellin and strigolactone on gene expression in *Arabidopsis thaliana* seedlings. Plant Journal 92 (5), 924–938. Available at https://doi.org/10.1111/tpj.13729, http://www.blackwellpublishing.com/journals/TPJ.

Liu, H., Liu, H., Zhou, L., Lin, Z., 2019. Genetic architecture of domestication- and improvement-related traits using a population derived from *Sorghum virgatum* and *Sorghum bicolor*. Plant Science 283, 135–146. Available at https://doi.org/10.1016/j.plantsci.2019.02.013, http://www.journals.elsevier.com/plant-science/.

Li, L., Xu, J., Xu, Z.H., Xue, H.W., 2005. Brassinosteroids stimulate plant tropisms through modulation of polar auxin transport in Brassica and *Arabidopsis*. The Plant Cell 17 (10), 2738–2753. Available at https://doi.org/10.1105/tpc.105.034397, http://www.plantcell.org/.

Ljung, K., Bhalerao, R.P., Sandberg, G., 2001. Sites and homeostatic control of auxin biosynthesis in *Arabidopsis* during vegetative growth. Plant Journal 28 (4), 465–474. Available at https://doi.org/10.1046/j.1365-313X.2001.01173.x.

Logan, D.C., Stewart, G.R., 1991. Role of ethylene in the germination of the hemiparasite *Striga hermonthica*. Plant Physiology 97 (4), 1435–1438. Available at https://doi.org/10.1104/pp.97.4.1435, http://www.plantphysiol.org/.

Lucido, A., Basallo, O., Sorribas, A., Marin-Sanguino, A., Vilaprinyo, E., Alves, R., 2022. A mathematical model for strigolactone biosynthesis in plants. Frontiers in Plant Science 13. Available at https://doi.org/10.3389/fpls.2022.979162, https://www.frontiersin.org/journals/plant-science.

Mashiguchi, K., Seto, Y., Yamaguchi, S., 2021. Strigolactone biosynthesis, transport and perception. Plant Journal 105 (2), 335–350. Available at https://doi.org/10.1111/tpj.15059, http://onlinelibrary.wiley.com/journal/10.1111/(ISSN)1365-313X.

Mohamed, A., Ali, R., Elhassan, O., Suliman, E., Mugoya, C., Masiga, C.W., et al., 2014. First products of DNA marker-assisted selection in sorghum released for cultivation by farmers in sub-saharan Africa. Journal of Plant Science and Molecular Breeding 3 (1), 3. Available at https://doi.org/10.7243/2050-2389-3-3.

Mohamed, K.I., Papes, M., Williams, R., Benz, B.W., Peterson, A.T., 2006. Global invasive potential of 10 parasitic witchweeds and related orobanchaceae. AMBIO: A Journal of the Human Environment 35 (6), 281–288. Available at https://doi.org/10.1579/05-r-051r.1.

Mrema, E., Shimelis, H., Laing, M., Mwadzingeni, L., 2020. Integrated management of *Striga hermonthica* and *S. asiatica* in sorghum: a review. Australian Journal of Crop Science 14 (1), 36–45. Available at https://doi.org/10.21475/ajcs.20.14.01.p1749, https://www.cropj.com/mrema_14_1_2020_36_45.pdf.

Mutinda, S., Jamil, M., Wang, J.Y., Berqdar, L., Ateka, E., Bellis, E.S., et al., 2025. Strigolactone biosynthesis *lgs1* mutant alleles mined from the sorghum accession panel are a promising resource of resistance to witchweed (*Striga*) parasitism. JPlants People Planet 7 (2), 382–395. Available at https://doi.org/10.1002/ppp3.10442, nph.onlinelibrary.wiley.com/journal/25722611.

Mutuku, J.M., Yoshida, S., Shimizu, T., Ichihashi, Y., Wakatake, T., Takahashi, A., et al., 2015. The WRKY45-dependent signaling pathway is required for resistance against *Striga hermonthica* parasitism. Plant Physiology 168 (3), 1152–1163. Available at https://doi.org/10.1104/pp.114.256404, http://www.plantphysiol.org/content/168/3/1152.full.pdf.

Müller, D., Leyser, O., 2011. Auxin, cytokinin and the control of shoot branching. Annals of Botany 107 (7), 1203–1212. Available at https://doi.org/10.1093/aob/mcr069.

Nelson, S.K., Steber, C.M., 2016. Gibberellin hormone signal perception: down-regulating DELLA repressors of plant growth and development. Annual Plant Reviews: The Gibberellins 49, 153–188. Available at https://doi.org/10.1002/9781119210436.ch6, http://onlinelibrary.wiley.com/book/10.1002/9781119210436.

Oancea, F., Georgescu, E., Matusova, R., Georgescu, F., Nicolescu, A., Raut, I., et al., 2017. New strigolactone mimics as exogenous signals for rhizosphere organisms. Molecules 22 (6). Available at https://doi.org/10.3390/molecules22060961, http://www.mdpi.com/1420-3049/22/6/961/pdf.

Obilana, A.T., Ramaiah, K.V., 1992. *Striga* (witchweeds) in sorghum and millet: knowledge and future research needs.

Omoarelojie, L.O., Kulkarni, M.G., Finnie, J.F., Van Staden, J., 2019. Strigolactones and their crosstalk with other phytohormones. Annals of Botany 124 (5), 749–767. Available at https://doi.org/10.1093/aob/mcz100, http://aob.oxfordjournals.org/.

Prodhan, M.Y., Munemasa, S., Nahar, M.N.E.N., Nakamura, Y., Murata, Y., 2018. Guard cell salicylic acid signaling is integrated into abscisic acid signaling via the Ca2+/CPK-dependent pathway. Plant Physiology 178 (1), 441–450. Available at https://doi.org/10.1104/pp.18.00321, http://www.plantphysiol.org/content/plantphysiol/178/1/441.full.pdf.

Rani, Sengar, Garg, Mishra, Shukla, 2023. RETRACTED ARTICLE: physiological and molecular role of strigolactones as plant growth regulators: a review. Molecular Biotechnology.

Rasmussen, A., Hu, Y., Depaepe, T., Vandenbussche, F., Boyer, F.D., Van Der Straeten, D., et al., 2017. Ethylene controls adventitious root initiation sites in *Arabidopsis* hypocotyls independently of strigolactones. Journal of Plant Growth Regulation 36 (4), 897–911. Available at https://doi.org/10.1007/s00344-017-9692-8, http://www.springer.com/life+sci/plant+sciences/journal/344.

Rasmussen, A., Mason, M.G., de Cuyper, C., Brewer, P.B., Herold, S., Agusti, J., et al., 2012. Strigolactones suppress adventitious rooting in *Arabidopsis* and pea. Plant Physiology 158 (4), 1976–1987. Available at https://doi.org/10.1104/pp.111.187104, http://www.plantphysiol.org/content/158/4/1976.full.pdf+html.

Reddy, B.V.S., Ramesh, S., Reddy, P.S., 2004. Sorghum breeding research at ICRISAT—goals, strategies, methods and accomplishments. International Sorghum and Millets Newsletter 45, 5–12.

Reddy, B.V., Reddy, P.S., Sadananda, A.R., Dinakaran, E., Ashok Kumar, A., Deshpande, S.P., et al., 2012. Postrainy season sorghum: constraints and breeding approaches. Journal of SAT Agricultural Research 10 (1), 1–12.

Rozpądek, P., Domka, A.M., Nosek, M., Ważny, R., Jędrzejczyk, R.J., Wiciarz, M., et al., 2018. The role of strigolactone in the cross-talk between *Arabidopsis thaliana* and the endophytic fungus mucor sp. Frontiers in Microbiology 9. Available at https://doi.org/10.3389/fmicb.2018.00441.

Satish, K., Gutema, Z., Grenier, C., Rich, P.J., Ejeta, G., 2012. Molecular tagging and validation of microsatellite markers linked to the low germination stimulant gene (lgs) for *Striga* resistance in sorghum [*Sorghum bicolor* (L.) Moench]. Theoretical and Applied Genetics 124 (6), 989–1003. Available at https://doi.org/10.1007/s00122-011-1763-9, link.springer.de/link/service/journals/00122/index.htm.

Seto, Y., Sado, A., Asami, K., Hanada, A., Umehara, M., Akiyama, K., et al., 2014. Carlactone is an endogenous biosynthetic precursor for strigolactones. Proceedings of the National Academy of Sciences 111 (4), 1640–1645. Available at https://doi.org/10.1073/pnas.1314805111.

Shi, J., Mei, C., Ge, F., Hu, Q., Ban, X., Xia, R., et al., 2025. Resistance to *Striga parasitism* through reduction of strigolactone exudation. Cell 188 (7), 1955. Available at https://doi.org/10.1016/j.cell.2025.01.022.

Sibhatu, B., 2016. Review on *Striga* weed management. International Journal of Life Sciences Scientific Research 2 (2), 110–120.

Smith, S.M., Li, J., 2014. Signalling and responses to strigolactones and karrikins. Current Opinion in Plant Biology 21, 23–29. Available at https://doi.org/10.1016/j.pbi.2014.06.003, www.elsevier.com/inca/publications/store/6/0/1/3/1/4/index.htt.

Souza, A.C.D.de, Reis, B.S.S., Bacin, D.F., Oliveira, L.R.de, Santos, W.S.dos, Borges, L.P., et al., 2023. Importance of brassinosteroids for mitigating water stress in sorghum. Australian Journal of Crop Science 17 (9), 721–726. Available at https://doi.org/10.21475/ajcs.23.17.09.p3950, http://www.cropj.com/.

Sugimoto, Y., Ali, A.M., Yabuta, S., Kinoshita, H., Inanaga, S., Itai, A., 2003. Germination strategy of *Striga hermonthica* involves regulation of ethylene biosynthesis. Physiologia Plantarum 119 (1), 137–145. Available at https://doi.org/10.1034/j.1399-3054.2003.00162.x.

Sun, H., Li, W., Burritt, D.J., Tian, H., Zhang, H., Liang, X., et al., 2022. Strigolactones interact with other phytohormones to modulate plant root growth and development. Crop Journal 10 (6), 1517–1527. Available at https://doi.org/10.1016/j.cj.2022.07.014, http://www.elsevier.com/journals/the-crop-journal/2214-5141.

Teka, H.B., 2014. Advance research on *Striga* control: a review. African Journal of Plant Science 8 (11), 492–506.

Thakur, N.R., Gorthy, S., Vemula, A., Odeny, D.A., Ruperao, P., Sargar, P.R., et al., 2024. Genome-wide association study and expression of candidate genes for Fe and Zn concentration in sorghum grains. Scientific Reports 14 (1), 12729.

Umehara, M., Hanada, A., Magome, H., Takeda-Kamiya, N., Yamaguchi, S., 2010. Contribution of strigolactones to the inhibition of tiller bud outgrowth under phosphate deficiency in rice. Plant and Cell Physiology 51 (7), 1118–1126. Available at https://doi.org/10.1093/pcp/pcq084.

Umehara, M., Hanada, A., Yoshida, S., Akiyama, K., Arite, T., Takeda-Kamiya, N., et al., 2008. Inhibition of shoot branching by new terpenoid plant hormones. Nature 455 (7210), 195–200. Available at https://doi.org/10.1038/nature07272.

Wakabayashi, T., Hamana, M., Mori, A., Akiyama, R., Ueno, K., Osakabe, K., et al., 2019. Direct conversion of carlactonoic acid to orobanchol by cytochrome P450 CYP722C in strigolactone biosynthesis. Science Advances 5 (12). Available at https://doi.org/10.1126/sciadv.aax9067, https://advances.sciencemag.org/content/5/12/eaax9067/tab-pdf.

Wanga, M.A., Shimelis, H., Mengistu, G., 2022. Sorghum production in northern namibia: farmers' perceived constraints and trait preferences. Sustainability 14 (16), 10266. Available at https://doi.org/10.3390/su141610266.

Wang, H., Chen, W., Eggert, K., Charnikhova, T., Bouwmeester, H., Schweizer, P., et al., 2018. Abscisic acid influences tillering by modulation of strigolactones in barley. Journal of Experimental Botany 69 (16), 3883–3898. Available at https://doi.org/10.1093/jxb/ery200.

Wang, Q., Zhu, Z., Ozkardesh, K., Lin, C., 2013. Phytochromes and phytohormones: the shrinking degree of separation. Molecular Plant 6 (1), 5–7. Available at https://doi.org/10.1093/mp/sss102.

Wani, K.I., Zehra, A., Choudhary, S., Naeem, M., Khan, M.M.A., Khan, R., et al., 2023. Exogenous strigolactone (GR24) positively regulates growth, photosynthesis, and improves glandular trichome attributes for enhanced artemisinin production in *Artemisia annua*. Journal of Plant Growth Regulation 42 (8), 4606–4615. Available at https://doi.org/10.1007/s00344-022-10654-w, https://www.springer.com/journal/344.

Waters, M.T., Gutjahr, C., Bennett, T., Nelson, D.C., 2017. Strigolactone signaling and evolution. Annual Review of Plant Biology 68, 291–322. Available at https://doi.org/10.1146/annurev-arplant-042916-040925, http://arjournals.annualreviews.org/loi/arplant.

Wu, F., Gao, Y., Yang, W., Sui, N., Zhu, J., 2022. Biological functions of strigolactones and their crosstalk with other phytohormones. Frontiers in Plant Science 13. Available at https://doi.org/10.3389/fpls.2022.821563, https://www.frontiersin.org/journals/plant-science.

Wu, S., Li, Y., 2021. A unique sulfotransferase-involving strigolactone biosynthetic route in sorghum. Frontiers in Plant Science 12, 793459. Available at https://doi.org/10.3389/fpls.2021.793459.

Yakubu, M.N., Price, A.H., 2024. Variability in *Striga* resistance among a collection of diverse rice cultivars. Rice Science. Available at https://doi.org/10.1016/j.rsci.2024.03.003.

Yang, T., Lian, Y., & Wang, C. (2019). Comparing and contrasting the multiple roles of butenolide plant growth regulators: strigolactones and karrikins in plant development and adaptation to abiotic stresses. International Journal of Molecular Sciences, 20(24), 6270.

Yoda, A., Mori, N., Akiyama, K., Kikuchi, M., Xie, X., Miura, K., et al., 2021. Strigolactone biosynthesis catalyzed by cytochrome P450 and sulfotransferase in sorghum. New Phytologist 232 (5), 1999–2010. Available at https://doi.org/10.1111/nph.17737, http://onlinelibrary.wiley.com/journal/10.1111/(ISSN)1469-8137.

Yoda, M., et al., 2023. Role of Sb3500 dioxygenase as a stereoselective regulator in sorghum strigolactone biosynthesis. Plant Cell Physiology.

Yohannes, T., Abraha, T., Kiambi, D., Folkertsma, R., Tom Hash, C., Ngugi, K., et al., 2015. Marker-assisted introgression improves *Striga* resistance in an eritrean farmer-preferred sorghum variety. Field Crops Research 173, 22–29. Available at https://doi.org/10.1016/j.fcr.2014.12.008.

Yoneyama, K., Xie, X., Yoneyama, K., Kisugi, T., Nomura, T., Nakatani, Y., et al., 2018. Which are the major players, canonical or non-canonical strigolactones? Journal of Experimental Botany 69 (9), 2231–2239. Available at https://doi.org/10.1093/jxb/ery090, http://jxb.oxfordjournals.org/.

Yoneyama, K., Xie, X., Kusumoto, D., Sekimoto, H., Sugimoto, Y., Takeuchi, Y., et al., 2007. Nitrogen deficiency as well as phosphorus deficiency in sorghum promotes the production and exudation of 5-deoxystrigol, the host recognition signal for arbuscular mycorrhizal fungi and root parasites. Planta 227 (1), 125–132. Available at https://doi.org/10.1007/s00425-007-0600-5.

Yoneyama, K., Xie, X., Kim, H.I., Kisugi, T., Nomura, T., Sekimoto, H., Yokota, T., Yoneyama, K. (2012). How do nitrogen and phosphorus deficiencies affect strigolactone production and exudation? Planta. 235(6):1197–1207. Available at https://doi.org/10.1007/s00425-011-1568-8.

Yoshida, S., Kim, S., Wafula, E.K., Tanskanen, J., Kim, Y.-M., Honaas, L., et al., 2019. Genome sequence of *Striga asiatica* provides insight into the evolution of plant parasitism. Current Biology 29 (18), 3041. Available at https://doi.org/10.1016/j.cub.2019.07.086.

Yoshida, S., Shirasu, K., 2009. Multiple layers of incompatibility to the parasitic witchweed, *Striga hermonthica*. New Phytologist 183 (1), 180–189. Available at https://doi.org/10.1111/j.1469-8137.2009.02840.x.

Yu, Y., Wang, J., Zhang, Z., Quan, R., Zhang, H., Deng, X.W., et al., 2013. Ethylene promotes hypocotyl growth and HY5 degradation by enhancing the movement of COP1 to the nucleus in the light. PLoS genetics 9 (12), e1004025. Available at https://doi.org/10.1371/journal.pgen.1004025.

Yoneyama, K. (2020) Recent progress in the chemistry and biochemistry of strigolactones. Journal of Pesticide Science, 45(2), 45–53.

Zorrilla, J.G., Rial, C., Varela, R.M., Molinillo, J.M.G., Macías, F.A., 2022. Strategies for the synthesis of canonical, non-canonical and analogues of strigolactones, and evaluation of their parasitic weed germination activity. Phytochemistry Reviews 21 (5), 1627–1659. Available at https://doi.org/10.1007/s11101-022-09801-8, https://www.springer.com/journal/11101.

Zwanenburg, B., Zeljkovic, C.S., Pospisil, T., 2015. Synthesis of strigolactones, a strategic account. Pest Management Science 72, 15–19.

CHAPTER

8

Role of reactive carbonyl compounds in plant stress signaling

Prajjal Dey[1], *Venkata H. Limmada*[2], *Preethi Vijayaraghavareddy*[3], *Sapalika Dogra*[1], *Rajneesh Kumar*[4], *Ratnakumar Pasala*[5], *Chita Sudhakar*[6], *and Amaranatha R. Vennapusa*[7,*]

[1]Division of Basic Sciences and Humanities, Faculty of Agriculture, Sher-e-Kashmir University of Agricultural Sciences and Technology, Wadura, Jammu and Kashmir, India [2]Department of Agronomy, Purdue University, IN, United States [3]School of Biosciences and Centre for Sustainability, Chanakya University, Devanahalli, Karnataka, India [4]Division of Genetics and Plant Breeding, Faculty of Agriculture, Sher-e-Kashmir University of Agricultural Sciences and Technology, Wadura, Jammu and Kashmir, India [5]ICAR-Indian Institute of Oilseeds Research, Hyderabad, Telangana, India [6]Mangalayatan University, Jabalpur, Madhya Pradesh, India [7]Department of Agriculture and Natural Resources, Delaware State University, Dover, DE, United States

8.1 Introduction

In recent decades, molecular mediators of stress responses and adaptive mechanisms have attracted greater attention in plant science. Among these, a class of metabolites known as reactive carbonyl compounds (RCCs) has emerged as a link between toxicity and signaling. RCCs were previously considered cytotoxic end products of lipid peroxidation, sugar fragmentation, and Maillard reactions but are now recognized to play an essential role in cellular redox homeostasis and stress signaling (Jahan et al., 2025; Sultana et al., 2024). This change indicates that understanding of plant stress biology is becoming increasingly complex, with carbonyl and oxidative stress influencing defense, development, and death under various stresses (Patel et al., 2025; Sultana et al., 2024; Sun et al., 2023).

[*]Corresponding author. email address: avennapusa@desu.edu

Harnessing Plant-Stimulating Molecules
DOI: https://doi.org/10.1016/B978-0-443-44994-9.00012-2

As sessile organisms, plants face numerous abiotic stresses, including drought, UV radiation, salinity, and temperature extremes. All these forms of environmental stress are known to induce excessive reactive oxygen species (ROS) production, thereby causing oxidative stress. The least well-studied effects of ROS accumulation are the production of RCCs, such as malondialdehyde (MDA), 4-hydroxynonenal (4-HNE), acrolein, and methylglyoxal (MG). These small, highly reactive molecules are electrophilic and cause structural and functional damage to proteins, DNA, and lipids by reacting with nucleophilic sites on these macromolecules. However, emerging data indicate that RCC molecules likely regulate transcription factors, signaling kinase cascades, and plant hormones through their involvement in plant stress-response signaling pathways (Yamauchi et al., 2015; Farmer & Mueller, 2013).

One key example of the active role of RCCs is "reactive short-chain leaf volatiles" (RSLVs), a subgroup of RCCs that induce stress-response gene expression in *Arabidopsis*. For example, (E)-2-hexenal activates the transcriptional regulators HSFA2 and ZAT12, thereby increasing tolerance to heat and oxidative stress (Yamauchi et al., 2015). This finding challenges the conventional view of RCCs as passive by-products of oxidative stress and demonstrates their importance in plant adaptation and survival.

Furthermore, protein carbonylation, one of the most common posttranslational modifications under oxidative stress, has become increasingly linked to both stress signaling and programmed cell death (PCD) (Chaki et al., 2015). RCC-mediated carbonylation may function as a redox switch, influencing protein activity, stability, or localization. Notably, a particular group of transcription factors and chaperones is activated by carbonylation, thereby connecting pathways into an integrated response. Although ROS and antioxidant systems have been studied extensively, RCCs have been largely neglected. This chapter repositions RCCs to the front line of redox biology and stress physiology. We discuss their biogenesis, molecular targets, interactions with phytohormones, and potential as biomarkers for stress tolerance. Critical gaps in current knowledge are also highlighted, along with proposals for integrating RCC signaling into crop improvement strategies for climate-resilient crops.

8.2 Biogenesis and sources of reactive carbonyl compounds

Stress-induced oxidative conditions increase several metabolic pathways in plants, which yield RCCs. RCCs in plants are formed through various metabolic pathways that involve lipids, carbohydrates, amino acids, and secondary metabolites. Abiotic stress stimulates these pathways, thereby enhancing RCC formation. RCCs primarily induce oxidative degradation reactions: those that arise directly from ROS-dependent cellular damage or indirectly through enzymatic and nonenzymatic activities that target cellular macromolecules.

Polyunsaturated fatty acids (PUFAs) in cell membranes are major contributors to RCCs through lipid peroxidation (Allakhverdiev et al., 2008; Sharma et al., 2023). Under oxidative stress, ROS attack the double bonds of PUFAs, leading to the formation of α,β-unsaturated aldehydes such as MDA, 4-HNE, and acrolein (Fig. 8.1). These aldehydes are highly electrophilic and react with biomolecules containing nucleophilic sites, causing their toxic

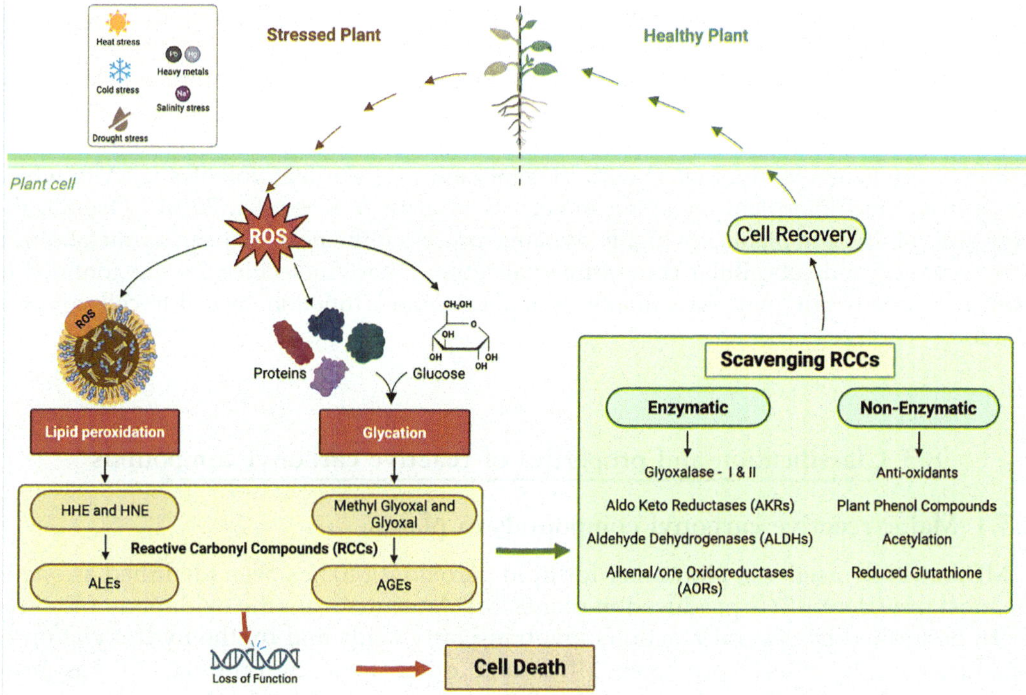

FIGURE 8.1 Overview of the reactive carbonyl compounds (*RCCs*) biogenesis and detoxification pathways in plants. The figure depicts the external and internal factors that induce RCC production and the formation of AGEs and ALEs, as well as the scavenging mechanisms. *AGEs*, Advanced glycation end products; *ALEs*, advanced lipid peroxidation end products.

properties and signaling effects (Allakhverdiev et al., 2008; Farmer & Mueller, 2013). Other notable RCCs include reactive dicarbonyl compounds like MG and glyoxal; these are produced from glycolytic intermediates, namely, DHAP and G3P, and tend to increase during water stress, reflecting disturbances in cellular metabolism under drought or heat stress (Sharma et al., 2023).

MG is the most widely studied RCC due to its high cellular abundance, its broad spectrum of effects on protein structure and function, and its central role in glycation (Sharma et al., 2023). While oxidative deamination of amino acids to produce propanol and butanal, the further breakdown of aromatic amino acids (e.g., phenylalanine) can yield highly reactive substances under stress conditions, including aromatic aldehydes such as vanillin and benzaldehyde (Hassan and Said, 2020). In addition to these oxidative processes, reducing sugars that react with amino acids can undergo Maillard reactions, generating advanced glycation end products (AGEs) containing carbonyl functional groups, thereby increasing the total cellular carbonyl load. In addition, environmental factors including drought, temperature, heavy metals, increased light intensities (such as that associated with increased use of sunlight), and ultraviolet B (UV-B) radiation increase the rate of lipid peroxidation that causes an increased formation of volatile aldehydes (responsible for the "foul smell" in plant processing) in chloroplast membranes and increase the quantity of reactive carbonyls (e.g., MG) (Kaur et al., 2015).

It is also important to note that a portion of RCCs originates from secondary metabolite pathways. For example, phenylpropanoid metabolism significantly contributes to both volatile and nonvolatile carbonyl pools under biotic and abiotic stress conditions. These carbonyls are generally associated with the emission of green leaf volatiles (GLVs), which act as intra- and interplant signaling molecules (Matsui & Koeduka, 2016) . Overall, the biogenesis of RCCs in plants is a highly dynamic process influenced by primary metabolism, stress severity, and subcellular compartmentalization. These molecules are not confined to damaged tissues but may accumulate systemically and function in cell-to-cell communication.

8.3 Classification and properties of reactive carbonyl compounds

8.3.1 Major reactive carbonyl compounds in plants

MDA, widely used as a biomarker for lipid peroxidation, has been identified as one of the most prevalent RCCs produced in plants. MDA results from the oxidation of PUFAs. It can be formed by several reactions involving fatty acids and by the hydroxylation of unsaturated fats, which produces MDA and other toxic by-products. It has been demonstrated that MDA can form cross-links between proteins and deoxyribonucleic acid (DNA), thereby causing cell death and adversely affecting plant growth and development through genotoxicity and cytotoxicity (Hasanuzzaman et al., 2022; Del Río, 2015). In addition to MDA, acrolein is an important carbonyl compound formed during the oxidative degradation of PUFAs as well as during the oxidative breakdown of polyamines. Acrolein, being an α,β-unsaturated aldehyde, readily forms stable covalent linkages to proteins and DNA. During the plant's response to pathogen attack, activation of the hypersensitive response occurs due to membrane damage caused by acrolein (Biswas & Mano, 2021).

4-Hydroxy-2-nonenal (HNE), formed during the lipid peroxidation of ω-6 fatty acids, represented by linoleic and linolenic acids, has a dual role. In low concentrations, HNE acts as a potent signaling molecule, initiating antioxidant defense mechanisms and modulating gene expression. On the other hand, at higher concentrations, it provokes PCD and serious oxidative damage. Recently, the formation and accumulation of HNE in chloroplast membranes under stress conditions, such as high light and ionizing radiation, have been reported; these findings point to its important role as a retrograde signal in redox regulation (Sonmez et al., 2024; Ksas et al., 2024). Among lipid-derived aldehydes, glyoxal and MG belong to the class of dicarbonyl compounds, which are mainly produced from sugar and amino acid metabolism. Glyoxal is produced from sugar autoxidation, ascorbate breakdown, and lipid peroxidation, whereas MG is formed as a by-product of triose phosphate metabolism in glycolysis. Stress conditions increase the levels of dicarbonyls, which induce carbonyl stress, leading to glycation of proteins and nucleic acids and the formation of advanced glycosylation end products. At the same time, MG acts as a signaling molecule, and recent studies have evidenced that it induces transcriptional

TABLE 8.1 Classification, sources, and roles of reactive carbonyl compounds (*RCCs*) in plants.

RCC type	Common compounds	Major sources	Physiological role
α,β-Unsaturated aldehydes	Acrolein, 4-hydroxynonenal (4-HNE)	Lipid peroxidation of polyunsaturated fatty acids	Highly toxic; disrupts membranes, modifies proteins; also acts as a signal in HR (hypersensitive response)
Dicarbonyls	Methylglyoxal (MG), glyoxal	Glycolysis overflow, triose phosphate degradation	Induces glyoxalase system; regulates redox and signaling; cytotoxic via protein glycation
Aromatic aldehydes	Vanillin, benzaldehyde	Phenylpropanoid pathway metabolism	Possible allelopathic and defense roles; involved in secondary metabolism
Saturated aldehydes	Propanal, butanal	Oxidative cleavage of fatty acids	Less reactive; markers of lipid peroxidation may play roles in carbonyl signaling

reprogramming in response to drought and salinity stresses. The development of fluorescent probes has enabled real-time visualization of MG accumulation under plant stress conditions (Hasanuzzaman et al., 2022; Yadav and Singla-Pareek, 2005a; Yadav and Singla-Pareek, 2005b) (Table 8.1).

8.3.2 Physicochemical properties and reactivity

The central feature of RCCs is the presence of an electrophilic carbonyl group that interacts with nucleophilic sites on biomolecules. The electrophilic nature of carbonyls means they can react with nucleophilic sites on proteins (e.g., lysines, cysteines, or histidines), nucleic acids, and phospholipids. Two main types of reactions define the biochemical activity of RCCs: Schiff base formation (when a carbonyl combines with an amino group to produce an imine) and Michael addition (when a carbonyl reacts with a thiol or amino group). The sugar-amine reaction forms a Schiff base, which undergoes the Amadori rearrangement to form stable Amadori products (e.g., fructose-lysine). Plants undergo a natural protein-sugar reaction called glycation, which forms glycoconjugates that enhance protein stability, functionality, and bioactivity. It also changes protein structure and overall function in plants (Nisarga et al., 2017). The biochemical effects of both types of reactions are similar to those seen when proteins experience irreversible covalent modifications (immobilization and cross-linking), enzyme inactivation, and disruption of normal metabolic processes (Dalle-Donne et al., 2006; Biswas & Mano, 2021). RCC produced through lipoxidation and glycoxidation reactions leads to the formation of AGEs or advanced lipoxidation end products (ALEs). AGEs/ALEs influence cell signaling pathways, serve as biomarkers of aging and mediators of stress, and impact the function of proteins and other biomolecules, which can result in cell death (Nareshkumar et al., 2020).

8.3.3 Stability, diffusibility, and half-life in plant tissues

RCCs are much more stable and mobile compared to ROS, ranging from seconds to minutes. While hydroxyl radicals, an example of ROS, exist for nanoseconds to microseconds, RCCs are more stable and can persist in plant tissues for minutes to hours. This chemical stability varies with their structures, particularly the presence of α,β-unsaturated carbonyl groups. RCCs such as MDA, acrolein, and HNE are relatively unstable due to their high electrophilicity and propensity to react with nucleophilic groups on macromolecules (Ramu et al., 2020). Due to their greater stability and diffusibility, RCCs can migrate out of their sites across cellular and subcellular membranes and serve as second messengers in the stress signaling pathway, thereby enabling long-distance transmission of oxidative stress signals. This characteristic feature accounts for their role as mediators of systemic oxidative damage and regulators of stress-responsive transcriptional networks in plants (Mano, 2012; Hasanuzzaman et al., 2022; Nisarga et al., 2017). However, this diffusivity also depends on molecular size, polarity, lipophilicity, and intercellular scavenging capacity.

8.4 Toxicological effects of reactive carbonyl compounds at the cellular level

8.4.1 Protein carbonylation and enzyme inactivation

One significant toxicological effect of RCCs in plants is their ability to modify proteins via carbonylation. Carbonyl groups in both aldehydes and dicarbonyl compounds are electrophiles that react with amino acid side chains in proteins (e.g., lysine, arginine, cysteine, and histidine), forming stable adducts. Formation of these stable carbonyl-based adducts leads to structural changes and alterations in catalytic site structure/functionality, potentially resulting in impaired or complete loss of function of numerous proteins crucial for metabolic, photosynthetic, and defense responses (in response to biotic and abiotic factors). Examples of this include studies demonstrating that MDA, HNE, and MG all induced irreversible changes to Rubisco and/or the antioxidant enzymes, as well as glyoxalase (GLY)-type enzymes under abiotic stress conditions, which ultimately resulted in decreased stress tolerance (Mano, 2012; Hasanuzzaman et al., 2022; Dalle-Donne et al., 2006; Mulozi et al., 2023; Limmada et al., 2026). Further, the process of accumulating carbonylated proteins becomes a feedback loop of sorts, because as more and more of the proteins become carbonylated, the plant is left with increasingly fewer options to reverse the carbonylation process, frequently resulting in the proteasomal degradation of the proteins (as opposed to their degradation through other protein degradation pathways) (Fig. 8.1).

8.4.2 Reactive carbonyl compounds' effect on nucleic acids

RCCs are just as effective as genotoxic agents because they can form adducts with DNA bases. The aldehydic products of lipid peroxidation, such as MDA and acrolein, react with deoxyguanosine and deoxyadenosine, creating exocyclic DNA adducts that interfere with DNA replication and transcription. These lesions increase the risk of mutations and genomic instability, which in plants can lead to stunted growth, impaired reproductive development, and decreased stress resilience. MGs have been shown to glycate DNA, causing structural distortions and potentially blocking polymerase progression. This genotoxicity has been demonstrated in plants exposed to high salinity, drought, and radiation stress, where an imbalance between RCC production and detoxification results in the accumulation of DNA damage (Hasanuzzaman et al., 2022; Singla-Pareek et al., 2003; Del Río, 2015).

8.4.3 Lipid and membrane disruption

RCCs produced from lipid peroxides have numerous damaging effects, including promoting further lipid damage and disrupting membrane structure. In addition to their role in lipid peroxidation, MDA, acrolein, and HNE are also considered second- to third-generation propagators of peroxidation because they can generate additional free radicals on membrane lipids. This results in decreased membrane fluidity, loss of lipid bilayer structural integrity, increased lipid bilayer permeability, and, if prolonged, irreversible cell death. In plants, it is critical in chloroplasts and mitochondria, where carbonyls have damaged membranes, as carbonyl-induced membrane destruction adversely affects photosynthetic and respiratory electron transport systems and compounds the oxidative stress cycle by generating additional ROS and forming new carbonyls (Mano, 2012; Vennapusa et al., 2022; Ksas et al., 2024; Limmada et al., 2026).

8.4.4 Cellular redox imbalance

The buildup of RCCs disrupts cellular redox balance by covalently modifying thiol-containing proteins. Carbonyls can interfere with antioxidant defense systems, including glutathione (GSH), peroxiredoxins, thioredoxins, and ascorbate peroxidases, thereby decreasing the plant cell's ability to detoxify ROS and creating a vicious cycle of oxidative stress. Additionally, reactive carbonyls may act as secondary messengers; when uncontrolled, they can lead to PCD and hypersensitive responses. Recently, the connection between ROS and reactive carbonyls has been recognized as a key factor in determining whether stress causes cellular damage or triggers adaptation. Recent studies on plant acclimation to ionizing radiation and high light stress highlight its importance (Sonmez et al., 2024; Sheela et al., 2023; Yadav et al., 2024; Biswas & Mano, 2021; Ksas et al., 2024).

8.4.5 Reactive carbonyl compounds' effect on plant physiological processes

Under normal conditions and in moderation, RCC levels act as signaling molecules for plant physiological processes and regulate plant growth and development (Mano et al., 2019; Sengupta et al., 2015). However, higher production of RCCs under plant stress treatments or RCC-induced stress significantly impaired plant physiological growth and morphology, as well as the development of plant parts (Mano et al., 2019). MG treatments in tobacco and cucumber seedlings reduced seed germination and shoot and root growth (Nareshkumar et al., 2020; Nisarga et al., 2017). Higher production of RCC under glucose, NaCl, and methyl viologen (MV)–induced oxidative stress treatments reduced plant seedling growth and fresh weight and was accompanied by enhanced production of ROS and RCCs, protein carbonyls (PCs), and AGEs like N-carboxymethyl-lysine (AGE-CML)

(Nareshkumar et al., 2020). RCC production under accelerated-aging-induced stress in rice and tobacco seedlings reduced seed germination and vigor and increased detoxification enzyme production (Tian et al., 2025; Narayana et al., 2017). Induction of RCCs under abiotic stress has been reported to reduce plant biomass, chlorophyll content, and root biomass, and to enhance the production of ROS, antioxidant enzymes, and osmolytes (Sonmez et al., 2024; Vemanna et al., 2017a). Collectively, these studies highlight that RCCs play a significant role in linking environmental factors to physiological decline and emphasize the importance of RCC detoxification and redox balance in maintaining typical plant physiological functions.

8.5 Signaling functions of reactive carbonyl compounds

8.5.1 Reactive carbonyl compounds as secondary messengers

RCCs function as secondary messengers, integrating oxidative signals and modulating plant responses to stress. They remain more chemically stable and less reactive than ROS. As a result, they can persist for a few seconds to a minute, allowing transmission to more distant regions of the cell than the site of generation. RCCs such as HNE and MDA modulate cellular signaling mainly through covalent modification of nucleophilic amino acid residues like Cys, His, and Lys in target proteins. Studies conducted in *Arabidopsis* and other model plants have indicated that increased production of carbonyl compounds from lipids, including HNE and acrolein, occurs at membrane sites when plants are exposed to environmental stressors, primarily through light intensity, drought, or radiation. These carbonyls migrate from their sites of accumulation to the cytoplasm, thereby modifying subcellular molecules, including sensor proteins and transcription factors. As these two forms of information sharing occur, they enable plants to convert an immediate response to transitory bursts of ROS into longer-defined signals, thereby fine-tuning plant defensive actions and responses (Sonmez et al., 2024; Mano et al., 2012; Biswas & Mano, 2021).

8.5.2 Transcriptional regulation of stress genes

RCCs regulate gene expression by altering the activity of transcriptional regulators or by initiating retrograde signaling pathways. Electrophilic carbonyls can react with redox-sensitive proteins and transcription factors, thereby inducing conformational changes that upregulate or downregulate the expression of target genes. MGs have been shown to increase expression of GLY enzymes, aldehyde dehydrogenases (ALDHs), aldo-keto reductases (AKRs), and other detoxification systems in *Arabidopsis* and rice (Hasanuzzaman et al., 2022; Yadav et al., 2008; Narayana et al., 2017). In addition, the accumulation of HNE in response to stress stimulates increases in antioxidant enzyme levels, including ascorbate peroxidase and superoxide dismutase (Hasanuzzaman et al., 2022). Studies of ionizing radiation stress have also demonstrated the activation of TGAII transcription factors in response to carbonyl stress, which reprogram the expression of nuclear-encoded genes to restore cellular redox homeostasis (Ksas et al., 2024). Thus it can be concluded that RCCs mediate coupling between oxidative events at membranes and nuclear responses through their ability to induce transcriptional reprogramming.

8.5.3 Reactive carbonyl compound interplay with phytohormone signaling

Carbonyl compounds also interact with phytohormone signaling pathways, contributing to the integration of multiple stress signals. There is growing evidence that RCCs influence abscisic acid (ABA) synthesis and sensitivity, particularly under drought and salinity stress, thereby enhancing stomatal closure and water conservation. Similarly, lipid-derived aldehydes, such as HNE, have been shown to interact with jasmonic acid (JA) and salicylic acid (SA) signaling, thereby amplifying the expression of pathogenesis-related genes and activating systemic acquired resistance (SAR). It has also been reported that reactive carbonyl accumulation can accelerate ethylene production and promote stress-induced senescence. This interplay positions RCCs as intermediaries that amplify ROS signaling and modulate hormone-dependent stress responses (Hasanuzzaman et al., 2022; Biswas & Mano, 2021).

8.5.4 Role in programmed cell death and hypersensitive response

A key function of the RCC signaling pathway is regulating PCD in response to both biotic and abiotic stressors. High levels of RCCs, particularly HNE, acrolein, and MG, can trigger PCD initiation. PCD is a tightly controlled process that induces cell death to limit stress spread. Elevated reactive carbonyls are also linked to inducing oxidative bursts and boosting the hypersensitive response during fungal and other pathogenic infections (Sakthivel et al., 2025; Limmada et al., 2026). When reactive carbonyls accumulate in infected cells, they cause membrane collapse, chromatin condensation, and activate proteolytic cascades, ultimately destroying the infected cells so that neighboring uninfected cells can mount a defense. In this way, RCC-mediated signaling serves as a checkpoint determining whether impaired cells adapt and recover or undergo PCD under severe stress.

8.6 Detoxification mechanisms

The cellular toxicity of RCCs can be controlled by various enzymatic and nonenzymatic processes. These are collectively known as the plant's carbonyl detoxification network. These systems prevent the buildup of reactive carbonyl species by scavenging and converting them to less-reactive metabolites, thereby helping maintain redox balance under stress conditions (Table 8.2).

8.6.1 Enzymatic detoxification pathways

8.6.1.1 Aldehyde dehydrogenases

ALDHs are NAD(P)+-dependent enzymes that catalyze the oxidation of aldehydes. These aldehydes, including MDA, HNE, and other lipid-derived carbonyls, are converted into

TABLE 8.2 Detoxification enzymes and pathways for RCC homeostasis in plants.

Detoxification enzyme/ system	Primary function	Stress relevance
Aldo-keto reductases (AKRs)	Reduce aldehydes and ketones to alcohols	Broad substrate specificity; induced under cold and oxidative stress
Aldehyde dehydrogenases (ALDHs)	Oxidize aldehydes to corresponding acids	Plays a central role in heat, drought, and salt stress adaptation
Glutathione S-transferases (GSTs)	Conjugate RCCs with glutathione (GSH)	Also detoxifies xenobiotics and herbicides; key in redox balance maintenance
Carbonyl reductases (CRs)	Convert RCCs to less-reactive alcohols	Important in maintaining carbonyl balance under metal stress
Glyoxalase I & II system	Detoxify methylglyoxal to D-lactate via GSH	Rapidly upregulated under osmotic stress; key in salinity and drought tolerance
Nonenzymatic antioxidants	Scavenge and neutralize RCCs	Includes ascorbate, tocopherols, and flavonoids; supports enzymatic detoxification during ROS surge

ROS, Reactive oxygen species; *RCC*, reactive carbonyl compounds.

their corresponding carboxylic acids. This reaction reduces the reactivity of these toxic intermediates and facilitates their incorporation into primary metabolism. ALDHs have been reported to enzymatically scavenge aldehydes (RCCs) derived from stress-related lipid peroxidation in plants and to play important roles in signaling, stress tolerance, and plant development and metabolism (Tola et al., 2021). ALDH gene families are stress-inducible in plants, and these have an essential role in mitigating oxidative and carbonyl stress. For example, in *Arabidopsis*, strong expression of ALDH3I1 under drought and high-salinity stress has been shown to improve tolerance by reducing aldehyde toxicity (Singh et al., 2013; Kotchoni et al., 2010). In *Arabidopsis*, knockout of ALDH genes results in increased sensitivity to drought and salinity due to impaired detoxification of lipid-derived aldehydes. In contrast, overexpression of the ALDH3I1 reduced lipid-derived aldehyde levels and conferred resistance to drought and oxidative stress (Kotchoni et al., 2010).

8.6.1.2 Aldo-keto reductases

AKRs, as multitasking soldier enzymes, have a significant role in detoxifying reactive carbonyls by reducing them to less-reactive alcohols, using NADPH as a cofactor, and also have multiple roles in development and stress tolerance. AKRs target a broad range of substrates, including MG, glyoxal, and lipid peroxidation-derived aldehydes and pesticides (Turóczy et al., 2011; Sengupta et al., 2015; Vemanna et al., 2017b). AKR expression in *Arabidopsis* is enhanced under drought, osmotic, and oxidative stresses, suggesting its importance in detoxifying electrophilic carbonyl compounds and maintaining cellular redox balance (Simpson et al., 2009). Overexpression of AKRs in plants showed improved tolerance to salt, heat, herbicides, drought, glucose, and oxidative stress by detoxifying the RCCs compounds and showed enhanced production of antioxidant enzymes and reduced phytotoxicity in plants (Vennapusa et al., 2022; Sheela et al., 2023; Nareshkumar et al., 2020; Vemanna et al., 2017a; Narayana et al., 2017).

8.6.1.3 Glutathione S-transferases

Glutathione S-transferases (GSTs) detoxify reactive carbonyls by conjugating them with GSH, forming water-soluble adducts. These adducts can be transported into vacuoles for sequestration. GST-mediated conjugation is particularly effective for α,β-unsaturated aldehydes like acrolein and HNE. GSTs are transcriptionally regulated by stress and play a dual role in detoxification and signaling by removing electrophilic species and influencing redox-sensitive pathways (Cummins et al., 2011). Besides detoxification, GSTs also have a role in RCC-mediated signaling by regulating the intensity, duration, and spatial spread of RCC signals. Due to their quick scavenging ability, GSTs establish a signaling threshold that allows RCCs to function as secondary messengers at low concentrations while preventing cytotoxic levels of accumulation.

8.6.1.4 Glyoxalases

GLYs play a significant role in detoxifying MG in plants through the GLY pathway, which comprises two enzymes: glyoxalase I (GLYI) and glyoxalase II (GLYII). MG rapidly reacts with GSH to form adducts, which GLYI metabolizes, and GLYII subsequently converts these adducts into nontoxic forms. Glyoxalase III (GLYIII) directly converts MG into nontoxic forms in a single step, without requiring the GSH pathway, and also has chaperone and protease activities. GLYs are reported in plants to scavenge RCCs, reducing cytotoxicity and improving stress tolerance to various abiotic and biotic stresses (Singla-Pareek et al., 2020). Numerous transgenic experiments have shown that by manipulating the enzymes involved in RCC detoxification, plants can become more tolerant of various abiotic stresses. For instance, overproduction of GLYI and GLYII in tobacco plants demonstrated a 50%–60% decrease in the amount of MG present during the salinity stress when compared to nontransgenic plants, as well as superior growth and photosynthetic efficiency when compared to the wild-type tobacco (Singla-Pareek et al., 2003). Rice has also shown similar findings: a decrease in MG levels and an increase in grain yield when GLYI is overexpressed under moderate saline conditions (Mustafiz et al., 2011). Mutant and transgenic lines with altered expression of detoxification enzymes provide direct evidence of RCC involvement in stress physiology. For example, overexpression of GLYI and GLYII in rice decreased MG levels and improved salinity tolerance, confirming the importance of MG detoxification in stress adaptation (Singla-Pareek et al., 2003). Quantitative trait locus (QTL) mapping and Genome-wide association studies have both indicated that carbonyl detoxification and stress tolerance are correlated. For example, in barley (*Hordeum vulgare*), QTLs for low MG accumulation during drought stress colocalized with the chromosomal locations of two genes (*HvGLYI-1* and *HvGLYII-2*) that encode the enzymes GLYI and GLYII, respectively (Kaur et al., 2023). GLYI and GLYII catalyze the detoxification of MG to D-lactate via GSH-dependent pathways. Additionally, marker-assisted selection of these QTL enables the identification of barley lines with superior tolerance to osmotic stress and reduced carbonyl cytotoxicity, thereby increasing grain yield under water-deficient conditions.

8.6.2 Nonenzymatic scavenging

8.6.2.1 Glutathione and ascorbate

In addition to enzymatic systems, nonenzymatic antioxidants, such as GSH and ascorbate (vitamin C), act as direct scavengers of reactive carbonyls. GSH is a ubiquitous tripeptide present in plants at approximately millimolar concentrations. The HSH/GSSG ratio is a critical indicator of cellular redox status and directly influences redox-sensitive signaling pathways. Ascorbate is the most abundant water-soluble antioxidant in plants. The thiol group of GSH can react with electrophilic carbonyls, while ascorbate reduces oxidative intermediates. GSH also conjugates RCCs either enzymatically via GSTs. These processes prevent the propagation of lipid peroxidation. This antioxidant buffering system is closely linked to the ascorbate–GSH cycle, which also detoxifies ROS (Hasanuzzaman et al., 2022).

8.6.2.2 Phenolic compounds and flavonoids

Phenolics and flavonoids, which are plentiful in plant tissues, also help detoxify carbonyl compounds. The nucleophilic hydroxyl groups of these secondary metabolites can form stable adducts with reactive aldehydes and neutralize them. Additionally, their antioxidant properties mainly lower ROS levels and inhibit lipid peroxidation, the primary sources of RCCs, thereby indirectly reducing carbonyl formation from lipid peroxidation (Gill and Tuteja, 2010a, 2010b). Their strong radical-scavenging abilities limit the spread of lipid-derived peroxyl radicals, decreasing the formation of unsaturated carbonyls. Moreover, flavonoids offer localized protection in membranes and apoplastic regions, which are susceptible to oxidative stress.

8.6.2.3 Acetylation

Acetylation is a type of posttranslational modification that modifies how RCCs signal and detoxify cells within a plant system, affecting protein behavior (such as retaining, remaining intact, or being damaged through carbonylation). RCCs strongly prefer nucleophilic molecules as their targets (i.e., nucleophiles), especially lysine, an amino acid. Acetylating lysine lowers its nucleophilic potential and prevents most RCC–protein adduct formation, thereby protecting proteins from excessive carbonyl damage. When a specific lysine residue undergoes acetylation (e.g., site-specific), it can change the activity of enzymes involved in RCC detoxification (like GSTs, ALDHs, and AKRs), leading to variations in the concentrations, half-lives, and signaling strength of RCC-mediated pathways (Ramu et al., 2020). Additionally, histone acetylation promotes chromatin accessibility and the expression of genes related to RCC detoxification and stress responses. Moreover, protein acetylation connects the cell's overall metabolic state (reflected by levels of acetyl-CoA and acetyl-carnitine) to its ability to interpret metabolic signals within redox and RCC-driven stress signaling systems. This creates an integrated system that helps the cell maintain adaptive responses and manage cytotoxic effects.

Besides direct detoxification, plants use compartmentalization to prevent the spread of carbonyl stress. Vacuolar transporters sequester GSH-conjugated adducts, while peroxisomes and mitochondria keep local detox systems to stop carbonyls from building up in these organelles. This spatial control is crucial because carbonyls' stability and ability to diffuse allow them to move away from where they originate. This effective compartmentalization prevents damage to sensitive organelles like chloroplasts and nuclei over long distances (Mano, 2012; Hasanuzzaman et al., 2022). Therefore these processes are essential for managing RCC toxicity while maintaining their signaling functions.

8.7 Reactive carbonyl compounds in specific stress conditions

Due to various environmental stresses, RCCs accumulate in plant tissues, with distinct biochemical signatures and physiological outcomes that depend on the nature, intensity, and duration of the stress. RCC production often parallels the rise of ROS, particularly under stresses (both biotic and abiotic), due to the interlinked nature of redox and carbonyl stress. This section examines how various environmental cues modulate RCC levels and function.

8.7.1 Reactive carbonyl compounds under drought and salinity stress

Drought and salinity are major abiotic stressors that induce elevated oxidative and carbonyl stress in plants. Stomatal closure due to moisture deprivation and/or the presence of salt limits the CO_2 available to plants, thereby reducing their capacity to perform photosynthesis, and results in increased production of reactive ROS in the chloroplasts and mitochondria of plants. This oxidative explosion degrades the lipid component of the cell membrane, producing cellular-damaging aldehydes, including MDA, HNE, and acrolein. These RCCs disrupt the fluidity of cell membranes, inhibit enzymes, and disrupt metabolism. The presence of high levels of MG, formed by the spontaneous conversion of dihydroxyacetone phosphate (DHAP) and glyceraldehyde-3-phosphate (G3P) due to excessive accumulation of triose phosphates, is characteristic of osmotic stress (Yadav and Singla-Pareek, 2005a; Yadav and Singla-Pareek, 2005b). MG has a dual role in plant cells: it is a well-known cytotoxic compound that promotes the formation of AGEs in plant proteins, and it is an important signaling molecule that induces the expression of GLY enzymes responsible for the detoxification of MG. The action of GLYI and GLYII is to convert MG into the nontoxic metabolite D-lactate via a GSH-dependent enzymatic reaction. Under drought and salt stress, GLY gene expression is markedly increased in *Arabidopsis*, rice, and mustard, indicating that effective RCC management is critical for successful adaptation to stress (Hasanuzzaman et al., 2022). Salt and drought stress induce different types of RCCs and ROS in plants, and overexpression of AKRs protects plants from stress-induced RCCs by enhancing the production of antioxidant enzymes and transcription factors (Sheela et al., 2023; Vemanna et al., 2017b).

8.7.2 Reactive carbonyl compounds in heat and cold stress

Excess temperatures induce oxidative and carbonyl stress via distinct mechanisms. When exposed to heat, the fluidity of thylakoid membranes increases and ROS production is elevated due to impaired activity in both the chloroplast and mitochondria (Sonmez et al.,

2024). Lipid peroxidation of membranes produces high levels of the RCCs, MDA, HNE, and other aldehydes, which can inactivate key proteins involved in photosynthesis, damage ATP synthase, and cause greater cellular dehydration through loss of membrane integrity (Gill and Tuteja, 2010a, 2010b). Heat also increases MG production through metabolic imbalances and the stimulation of altered carbohydrate metabolic pathways. Cold stresses, however, decrease the activity of metabolic enzymes and carbon fixation in photosynthesis, resulting in increased rates of photoinhibition even under low-light conditions. The degree of photoinhibition leads to increased ROS formation, primarily within chloroplasts, and to the formation of carbonyl by-products from lipid peroxidation. Under cold-stress conditions, the accumulation of RCCs induces upregulation of cold-response detoxification enzymes and antioxidant genes, including AKR, ALDH, and GSH-dependent enzymes (Mano, 2012). Thus it is likely that RCCs function as secondary messengers that initiate signaling pathways for cold acclimation, including those associated with CBF/DREB transcriptional networks. Similarly, under heat-stress conditions, RCCs and reactive oxygen species (ROS) are induced in plants. The enzymatic detoxification and scavenging of these harmful compounds help reduce stress-induced damage in rice plants and enhance overall stress tolerance (Sheela et al., 2023; Turóczy et al., 2011).

8.7.3 Reactive carbonyl compounds in heavy metal and herbicide toxicity

Heavy metals such as cadmium (Cd), arsenic (As), and lead (Pb) can induce RCCs via multiple pathways. One pathway is via Fenton reactions. Hydroxyl radicals are generated during these reactions, which then initiate lipid peroxidation, resulting in the production of large quantities of reactive carbonyls, such as MDA, HNE, and acrolein, as well as through the oxidation of lipid membranes (Somagattu et al., 2024; Nareshkumar et al., 2015). Reactive carbonyls can work together with heavy metals to inactivate thiol-proteins, denature enzymes, and inflict DNA damage. In addition, heavy metals reduce mitochondrial respiration and glycolytic energy production, thereby promoting the accumulation of branched-chain fatty acids, which lead to the formation of reactive carbonyls via Michael addition reactions (Singh et al., 2013).

To reduce elevated levels of RCCs, plants have developed several mechanisms for carbonyl detoxification. These mechanisms include GLY pathways, ALDHs, and the conjugation of reactive aldehydes by GSH. The induction of GSTs and decomposition of GSH promoted by heavy metal exposure have aided in detoxifying reactive carbonyls and reducing oxidative stress. Recent studies using transcriptome profiling have identified that the expression of several key genes involved in the detoxification of RCCs in plants is coordinated with the activation of metal-responsive regulatory mechanisms (Hasanuzzaman et al., 2022).

Herbicides disrupt plant metabolic processes, thereby inducing the production of ROS and RCC compounds, which lead to cell death. Scavenging of RCCs molecules by enzymatic detoxification via AKR overexpression significantly enhanced herbicide resistance and reduced the phytotoxicity in rice and tobacco plants (Vennapusa et al., 2022; Vemanna et al., 2017b).

8.7.4 Reactive carbonyl compounds in pathogen interactions and biotic stress

RCCs also play a vital role in plant–pathogen interactions. During effector-triggered immunity (ETI) and pattern-triggered immunity (PTI), plants respond to pathogens by triggering an oxidative burst at infection sites, which increases ROS, including superoxide and hydrogen peroxide. The rise in ROS causes peroxidation of membrane lipids, generating large amounts of acrolein, highly reactive aldehydes (HNE), and related compounds (Limmada et al., 2026). RCCs contribute to an additional aspect of the hypersensitive response by inducing localized cell death, thus limiting pathogen growth (Biswas et al., 2019).

In addition to their toxic effects at high concentrations, RCCs can also act as signal transduction molecules in plants during biotic stress. Sublethal levels of RCCs trigger the expression of genes involved in defense responses, boost SA and JA signaling pathways, and promote SAR. Temporary increases in MG production during bacterial and fungal infections can also influence MAPK cascade activation and activate WRKY transcription factors in plants. The exact timing and location of RCCs during biotic stress define the line between signaling and toxic effects. Pathogens may also exploit host-produced RCCs to suppress immunity or facilitate colonization of the plant. Therefore RCCs represent a constantly changing, biochemically dynamic interface of conflict between invading microbes and plants (Sonmez et al., 2024).

8.8 Conclusion

There is a range of RCCs that act as intermediaries in the chain of events linking plant physiological stress to growth and development, stress signaling, and tolerance. RCCs are cytotoxic at high concentrations; they also play essential roles in signaling processes that mediate the transition from stress-induced oxidative damage to the overall plant response through multiple pathways, including acclimation, detoxification, modulation of gene expression, and hormone signaling depending on age. RCCs impair rates of photosynthesis, cause protein denaturation, lead to irreversible alterations of the primary structure of DNA, and even disrupt lipid biogenesis. However, at moderate levels, RCCs and their related metabolites may play a significant role in activating cellular antioxidant systems and triggering processes like angiogenesis and apoptosis. Ultimately, the balance between RCC production and their detoxification largely determines a plant's capacity to endure and adapt to its environment.

Detoxification of RCCs is mediated by a variety of enzymatic and nonenzymatic systems, including GLYs, ALDHs, and AKRs, as well as antioxidants like GSH and ascorbate. These systems are upregulated under stress conditions, emphasizing their importance in reducing carbonyl toxicity. Notably, transgenic overexpression of GLY, ALDHs, and AKRs genes in both model and crop species has enhanced tolerance to salinity, drought, and metal stress, reinforcing the significance of RCC detoxification mechanisms. RCC biology is expected to play a key role in climate-smart agriculture moving forward. Accelerating the identification of robust genotypes can be achieved by incorporating RCC profiling into high-throughput phenotyping systems. Advances in metabolomic and imaging technologies will enable real-time tracking of RCC dynamics in plants, revealing spatial patterns and stress-responsive fingerprints. Better understanding of how RCC interacts with hormonal pathways and plant–microbe interactions could lead to novel strategies for establishing lasting stress resilience.

In conclusion, RCCs act as a biochemical link between how plants perceive environmental stress and their adaptive responses. Their accurate regulation decides whether the cellular outcomes are protective or harmful. Utilizing the diagnostic, signaling, and toxicological properties of RCCs provides promising opportunities to develop resilient crops for future agricultural systems facing climate change.

References

Allakhverdiev, S.I., Kreslavski, V.D., Klimov, V.V., Los, D.A., Carpentier, R., Mohanty, P., 2008. Heat stress: an overview of molecular responses in photosynthesis. Photosynthesis Research 98 (1-3), 541–550. Available at https://doi.org/10.1007/s11120-008-9331-0.

Biswas, M.S., Mano, J., Iwata, T., 2019. Reactive carbonyl species in plants: their biosynthesis, metabolism, and role in stress responses. International Journal of Molecular Sciences 20 (22), 5562.

Biswas, M.S., Mano, J.I., 2021. Lipid peroxide-derived reactive carbonyl species as mediators of oxidative stress and signaling. Frontiers in Plant Science 12, 720867.

Chaki, M., Alvarez de Morales, P., Ruiz, C., Begara-Morales, J.C., Barroso, J.B., Corpas, F.J., et al., 2015. Ripening of pepper (Capsicum annuum) fruit is characterized by an enhancement of protein tyrosine nitration. Annals of Botany 116 (4), 637–647.

Cummins, I., Dixon, D.P., Freitag-Pohl, S., Skipsey, M., Edwards, R., 2011. Multiple roles for plant glutathione transferases in xenobiotic detoxification. Drug Metabolism Reviews 43 (2), 266–280.

Dalle-Donne, I., Aldini, G., Carini, M., Colombo, R., Rossi, R., Milzani, A., 2006. Protein carbonylation, cellular dysfunction, and disease progression. Journal of Cellular and Molecular Medicine 10 (2), 389–406.

Farmer, E.E., Mueller, M.J., 2013. ROS-mediated lipid peroxidation and RES-activated signaling. Annual review of plant biology 64, 429–450.

Gill, S.S., Tuteja, N., 2010a. Polyamines and abiotic stress tolerance in plants. Plant Signaling & Behavior 5 (1), 26–33.

Gill, S.S., Tuteja, N., 2010b. Reactive oxygen species and antioxidant machinery in abiotic stress tolerance in crop plants. Plant Physiology and Biochemistry 48 (12), 909–930. Available at https://doi.org/10.1016/j.plaphy.2010.08.016.

Hassan, A., Said, A., 2020. Importance of the applicability of O-vanillin Schiff base complexes. Advanced Journal of Chemistry, Section A 4, 87–103.

Hasanuzzaman, M., Bhuyan, M.H.M.B., Anee, T.I., Parvin, K., Nahar, K., Fujita, M., 2022. Reactive carbonyl species and their involvement in plant abiotic stress responses. Antioxidants 11 (3), 556.

Jahan, I., Islam, M.M., Nakamura, T., Nakamura, Y., Munemasa, S., Mano, J.'ichi, et al., 2025. Reactive carbonyl species function downstream of reactive oxygen species in chitosan-induced stomatal closure. Physiologia Plantarum 177 (1). Available at https://doi.org/10.1111/ppl.70094.

Kaur, R., Kaur, G., Bains, N.S., 2023. Molecular breeding for methylglyoxal detoxification: a promising strategy for drought resilience. Plant Science Today 10 (1), 50–62.

Kaur, C., Kushwaha, H.R., Mustafiz, A., Pareek, A., Sopory, S.K., Singla-Pareek, S.L., 2015. Analysis of global gene expression profile of rice in response to methylglyoxal indicates its possible role as a stress signal molecule. Frontiers in Plant Science 6, 682.

Kotchoni, S.O., Jimenez-Lopez, J.C., Gao, D., Edwards, V., Gachomo, E.W., Margam, V.M., et al., 2010. Modeling-dependent protein characterization of the rice aldehyde dehydrogenase (ALDH) superfamily reveals distinct functional and structural features. PloS one 5 (7), e11516.

Ksas, B., Chiarenza, S., Dubourg, N., Ménard, V., Gilbin, R., Havaux, M., 2024. Plant acclimation to ionising radiation requires activation of a detoxification pathway against carbonyl-containing lipid oxidation products. Plant, Cell & Environment 47 (10), 3882–3898.

Limmada, V.H., Vennapusa, A.R., Moore, M., Abrams, M.E., Elavarthi, S., Melmaiee, K., 2026. Trichomes as a barrier: understanding the strawberry resistance to anthracnose caused by *Colletotrichum acutatum*. Scientia Horticulturae 355, 114528.

Mano, J., 2012. Reactive carbonyl species: their production from lipid peroxides, action in environmental stress, and the detoxification mechanism. Plant Physiology and Biochemistry 59, 90–97. Available at https://doi.org/10.1016/j.plaphy.2012.03.010.

Mano, J., Biswas, M.S., Sugimoto, K., 2019. Reactive carbonyl species: a missing link in ROS signaling. Plants 8 (10). Available at https://doi.org/10.3390/plants8100391, https://www.mdpi.com/2223-7747/8/10/391/pdf.

Mano, J.I., 2012. Reactive carbonyl species: their production from lipid peroxides, action in environmental stress, and the detoxification mechanism. Plant Physiology and Biochemistry 59, 90–97.

Matsui, K., Koeduka, T., 2016. Green leaf volatiles in plant signaling and response. In Lipids in plant and algae development (pp. 427-443). Cham: Springer International Publishing.

Mustafiz, A., Singh, A.K., Pareek, A., Sopory, S.K., Singla-Pareek, S.L., 2011. Genome-wide analysis of rice and Arabidopsis identifies two glyoxalase genes that are highly expressed in abiotic stresses. Functional & Integrative Genomics 11 (2), 293–305.

Mulozi, L., Vennapusa, A.R., Elavarthi, S., Jacobs, O.E., Kulkarni, K.P., Natarajan, P., et al., 2023. Transcriptome profiling, physiological, and biochemical analyses provide new insights towards drought stress response in sugar maple (*Acer saccharum* Marshall) saplings. Frontiers in Plant Science 14, 1150204.

Narayana, N.K., Vemanna, R.S., Kodekallu Chandrashekar, B., Rao, H., Vennapusa, A.R., Narasimaha, A., et al., 2017. Aldo-ketoreductase 1 (AKR1) improves seed longevity in tobacco and rice by detoxifying reactive cytotoxic compounds generated during ageing. Rice 10 (1), 11. Available at https://doi.org/10.1186/s12284-017-0148-3, http://www.springerlink.com/content/1939-8425.

Nareshkumar, A., Veeranagamallaiah, G., Pandurangaiah, M., Kiranmai, K., Amaranathareddy, V., Lokesh, U., et al., 2015. Pb-stress induced oxidative stress caused alterations in antioxidant efficacy in two groundnut (*Arachis hypogaea* L.) cultivars. Agricultural Sciences. 6 (10), 1283–1297. Available at https://doi.org/10.4236/as.2015.610123.

Nareshkumar, A., Subbarao, S., Vennapusa, A.R., Ashwin, V., Banarjee, R., Kulkarni, M.J., et al., 2020. Enzymatic and non-enzymatic detoxification of reactive carbonyl compounds improves the oxidative stress tolerance in cucumber, tobacco and rice seedlings. Journal of Plant Growth Regulation 39 (3), 1359–1372. Available at https://doi.org/10.1007/s00344-020-10072-w, http://www.springer.com/life+sci/plant+sciences/journal/344.

Nisarga, K.N., Vemanna, R.S., Kodekallu Chandrashekar, B., Rao, H., Vennapusa, A.R., Narasimaha, A., et al., 2017. Aldo-ketoreductase 1 (AKR1) improves seed longevity in tobacco and rice by detoxifying reactive cytotoxic compounds generated during ageing. Rice 10 (1), 11.

Patel, J., Sharma, R., Khan, M.S., Ahmad, A., 2025. UV-C-induced reactive carbonyl species are better detoxified in halophyte plants (*Salicornia brachiata* and *Arthrocnemum macrostachyum*) than in Sarcocornia fruticosa. The Plant Journal 122 (4), e70239. Available at https://doi.org/10.1111/tpj.70239.

Ramu, V.S., Preethi, V., Nisarga, K.N., Srivastava, K.R., Sheshshayee, M.S., Mysore, K.S., et al., 2020. Carbonyl cytotoxicity affects plant cellular processes and detoxifying enzymes scavenge these compounds to improve stress tolerance. Journal of Agricultural and Food Chemistry 68 (23), 6237–6247. Available at https://doi.org/10.1021/acs.jafc.0c02005, http://pubs.acs.org/journal/jafcau.

Ramu, V.S., Preethi, V., Nisarga, K.N., Srivastava, K.R., Sheshshayee, M.S., Mysore, K.S., et al., 2020. Carbonyl cytotoxicity affects plant cellular processes and detoxifying enzymes scavenge these compounds to improve stress tolerance. Journal of Agricultural and Food Chemistry 68 (23), 6237–6247.

Del Río, L.A., 2015. ROS and RNS in plant physiology: an overview. Journal of Experimental Botany 66 (10), 2827–2837.

Sharma, P., Lakra, N., Goyal, A., Ahlawat, Y.K., Zaid, A., Siddique, K.H., 2023. Drought and heat stress mediated activation of lipid signaling in plants: a critical review. Frontiers in Plant Science 14, 1216835.

Simpson, P.J., Tantitadapitak, C., Reed, A.M., Mather, O.C., Bunce, C.M., White, S.A., et al., 2009. Characterization of two novel aldo–keto reductases from Arabidopsis: expression patterns, broad substrate specificity, and an open active-site structure suggest a role in toxicant metabolism following stress. Journal of Molecular Biology 392 (2), 465–480.

Sonmez, M.C., Yirmibesoglu, S.S.S., Ozgur, R., Uzilday, B., Turkan, I., 2024. Roles of reactive carbonyl species (RCS) in plant response to abiotic stress. ROS signaling in plants: Methods and protocols. Springer US, New York, NY, pp. 101–130.

Sakthivel, S.K., Vennapusa, A.R., Melmaiee, K., 2025. Enhancing quality and climate resilient traits in vegetatively propagated polyploids: transgenic and genome editing advancements, challenges and future directions. Frontiers in Genetics 16. Available at https://doi.org/10.3389/fgene.2025.1599242, https://www.frontiersin.org/journals/genetics#.

Sengupta, D., Naik, D., Reddy, A.R., 2015. Plant aldo-keto reductases (AKRs) as multi-tasking soldiers involved in diverse plant metabolic processes and stress defense: a structure-function update. Journal of Plant Physiology 179, 40–55. Available at https://doi.org/10.1016/j.jplph.2015.03.004, www.urbanfischer.de/journals/jpp/p_physio.htm.

Sheela, H.S., Vennapusa, A.R., Melmaiee, K., Prasad, T.G., Reddy, C.P., 2023. Pyramiding of transcription factor, PgHSF4, and stress-responsive genes of p68, Pg47, and PsAKR1 impart multiple abiotic stress tolerance in rice (*Oryza sativa* L). Frontiers in Plant Science 14. Available at https://doi.org/10.3389/fpls.2023.1233248, https://www.frontiersin.org/journals/plant-science.

Singh, H.P., Batish, D.R., Kaur, G., Arora, K., Kohli, R.K., 2013. Reactive carbonyl species in plants: from stress to signaling. Environmental and Experimental Botany 91, 1–14.

Singla-Pareek, S.L., Reddy, M.K., Sopory, S.K., 2003. Genetic engineering of the glyoxalase pathway in tobacco leads to enhanced salinity tolerance. Proceedings of the National Academy of Sciences of the United States of America 100 (25), 14672–14677. Available at https://doi.org/10.1073/pnas.2034667100.

Singla-Pareek, S.L., Kaur, C., Kumar, B., Pareek, A., Sopory, S.K., 2020. Reassessing plant glyoxalases: large family and expanding functions. New Phytologist 227 (3), 714–721. Available at https://doi.org/10.1111/nph.16576, http://onlinelibrary.wiley.com/journal/10.1111/(ISSN)1469-8137.

Somagattu, P., Chinnannan, K., Yammanuru, H., Reddy, U.K., Nimmakayala, P., 2024. Selenium dynamics in plants: uptake, transport, toxicity, and sustainable management strategies. Science of the Total Environment 949. Available at https://doi.org/10.1016/j.scitotenv.2024.175033, https://www.sciencedirect.com/science/journal/00489697.

Sonmez, H.B., Tokdemir, E., Avci, U., 2024. Aldehyde accumulation and detoxification during plant-pathogen interactions: emerging roles of reactive carbonyl species in defense. Frontiers in Plant Science (in press).

Sultana, M.S., Sakurai, C., Biswas, M.S., Szabados, L., Mano, J., 2024. Accumulation of reactive carbonyl species in roots as the primary cause of salt stress-induced growth retardation of *Arabidopsis thaliana*. Physiologia Plantarum 176 (1). Available at https://doi.org/10.1111/ppl.14198, http://onlinelibrary.wiley.com/journal/10.1111/(ISSN)1399-3054.

Sun, M., Wang, X., Gao, C., Chen, W., Liu, Y., 2023. Genome-wide analysis and expression profiling of reactive dicarbonyl compounds, including methylglyoxal (MG), glyoxal (GO), and 3-deoxyglucosone (3-DG), as main toxic metabolites under stress. Frontiers in Plant Science 14, 1215084. Available at https://doi.org/10.3389/fpls.2023.1215084.

Tian, J., Yin, M., Huang, K., Zhou, S., Chen, T., Tang, N., et al., 2025. The expression of Aldo-keto reductase gene OsKOB1 coordinates ABA homeostasis and antioxidant defense to maintain seed vigor under osmotic stress. Scientific Reports 15 (1). Available at https://doi.org/10.1038/s41598-025-13633-9.

Tola, A.J., Jaballi, A., Germain, H., Missihoun, T.D., 2021. Recent development on plant aldehyde dehydrogenase enzymes and their functions in plant development and stress signaling. Genes 12 (1), 1–18. Available at https://doi.org/10.3390/genes12010051, https://www.mdpi.com/2073-4425/12/1/51.

Turóczy, Z., Kis, P., Török, K., Cserháti, M., Lendvai, Á., Dudits, D., et al., 2011. Overproduction of a rice aldo–keto reductase increases oxidative and heat stress tolerance by malondialdehyde and methylglyoxal detoxification. Plant Molecular Biology 75 (4-5), 399–412. Available at https://doi.org/10.1007/s11103-011-9735-7.

Vemanna, R.S., Babitha, K.C., Solanki, J.K., Amarnatha Reddy, V., Sarangi, S.K., Udayakumar, M., 2017a. Aldo-keto reductase-1 (AKR1) protect cellular enzymes from salt stress by detoxifying reactive cytotoxic compounds. Plant Physiology and Biochemistry 113, 177–186. Available at https://doi.org/10.1016/j.plaphy.2017.02.012, http://www.journals.elsevier.com/plant-physiology-and-biochemistry/.

Vemanna, R.S., Vennapusa, A.R., Easwaran, M., Chandrashekar, B.K., Rao, H., Ghanti, K., et al., 2017b. Aldo-keto reductase enzymes detoxify glyphosate and improve herbicide resistance in plants. Plant Biotechnology Journal 15 (7), 794–804. Available at https://doi.org/10.1111/pbi.12632, http://www.wiley.com/bw/journal.asp?ref=1467-7644.

Vennapusa, A.R., Agarwal, S., Rao Hm, H., Aarthy, T., Babitha, K.C., Thulasiram, H.V., et al., 2022. Stacking herbicide detoxification and resistant genes improves glyphosate tolerance and reduces phytotoxicity in tobacco (*Nicotiana tabacum* L.) and rice (*Oryza sativa* L. Plant Physiology and Biochemistry 189, 126–138. Available at https://doi.org/10.1016/j.plaphy.2022.08.025, http://www.journals.elsevier.com/plant-physiology-and-biochemistry/.

Yadav, S.K., Singla-Pareek, S.L., Reddy, M.K., Sopory, S.K., 2005b. Methylglyoxal detoxification by glyoxalase system: a survival strategy during environmental stresses. Physiology and Molecular Biology of Plants 11 (1), 1–11.

Yadav, S.K., Singla-Pareek, S.L., Ray, M., Reddy, M.K., Sopory, S.K., 2005a. Methylglyoxal levels in plants under salinity stress are regulated by the glyoxalase pathway. Plant Physiology 134 (1), 170–180.

Yadav, S.K., Singla-Pareek, S.L., Sopory, S.K., 2008. An overview on the role of methylglyoxal and glyoxalases in plants. Drug Metabolism and Drug Interactions 23 (1-2), 51–68. Available at https://doi.org/10.1515/DMDI.2008.23.1-2.51, http://www.reference-global.com/loi/dmdi.

Yadav, S., Preethi, V., Dadi, S., Seth, C.S., G, K., Chandrashekar, B.K., et al., 2024. Small chemical molecules regulating the phytohormone signalling alter the plant's physiological processes to improve stress adaptation, growth and productivity. Physiology and Molecular Biology of Plants 30 (10), 1593–1610. Available at https://doi.org/10.1007/s12298-024-01514-w, https://www.springer.com/journal/12298.

Yamauchi, S., Obara, K., Uchibori, K., Kamimura, A., Azumi, K., Kihara, A., 2015. Opt2 mediates the exposure of phospholipids during cellular adaptation to altered lipid asymmetry. Journal of Cell Science 128 (1), 61–69.

CHAPTER

9

Integrating gene editing and plant stimulants to develop virus-resilient crops

Ashwini Talakayala[1,2,4], Are Harini[4], Mohana Durga Prabhakaran[3], Sandhya Jagtap[4], Lali Lingfa[4], Mallikarjuna Garladinne[2], and Srinivas Ankanagari[4,*]

[1]International Crops Research Institute for the Semi-Arid Tropics (ICRISAT), Patancheru, Telangana, India [2]Plant Molecular Biology Laboratory, Agri Biotech Foundation, Hyderabad, Telangana, India [3]School of Life Sciences, Bharatidasan University, Tiruchirapalli, Tamilnadu, India [4]Department of Genetics, Osmania University, Hyderabad, Telangana, India

9.1 Introduction

9.1.1 Global significance of plant viral diseases

Plant viral diseases remain among the most pervasive biological constraints on global crop productivity, and their increasing incidence underscores the need to develop molecular-based control strategies that extend beyond currently used management approaches. Some of the viral diseases that continue to affect the productivity of food and horticultural crops include the tomato yellow leaf curl virus (TYLCV), the potato virus Y (PVY), the cucumber mosaic virus (CMV), and the rice tungro virus complex, and their continued incidence contributes significantly to yeild losses in vegetables and other crops, worldwide (Jones, 2021). The problem of viral diseases is also compounded by the growing effects of climate-related vector population and activity dynamics, which continue to accelerate viral disease incidence beyond control, as observed in regional impacts (Jones, 2021). While emerging strategies for high-throughput diagnostics, RNA interference (RNAi), and the use of the CRISPR/Cas system have continued to expand the control strategies for viral diseases (Varanda et al., 2021; Tatineni and Hein, 2023), the biological and

*Corresponding Author E-mail: asrinivas@osmania.ac.in.

Harnessing Plant-Stimulating Molecules
DOI: https://doi.org/10.1016/B978-0-443-44994-9.00010-9

physiological complexities underlying host–viral pathogenic interactions continue to hinder the development of efficient resistance for these viral diseases. This is because studies on the incidence of TYLCV among tomatoes, for example, continue to reveal physiological bases of stress that, despite their biological origins, affect plant productivity (Shteinberg et al., 2021).

9.1.2 Limitations of traditional breeding and opportunities for genome editing

In this respect, the use of CRISPR/Cas systems for genome editing offers a progressively stronger and more targeted approach to inducing antiviral traits, particularly compared with the time and genetic constraints of traditional breeding. Resistance breeding, by tradition, is hampered not only by the long generation cycles of the plants but also by the comparatively fewer resistance alleles that segregate, not to mention the rapid evolution of viral mechanisms that target single-gene resistance. Using these strategies, targeted manipulation of viral genomes, disruption of viral susceptibility (S) genes , or modification of immunity-regulatory pathways represent a versatile platform for addressing key vulnerabilities at various stages of the viral life cycle (Zaidi et al., 2016). In particular, the modification of translation initiation factors eIF4E and eIF4G—key components of recessive resistance mechanisms—has been shown to confer durable resistance across diverse plant species (Hashimoto et al., 2016). When integrated with RNAi and advanced diagnostic tools, CRISPR-based interventions enable the development of broader, more robust strategies to address antiviral resistance. However, effective deployment will depend on continuous refinement, in parallel with evolving biosafety requirements and regulatory frameworks (Zhan and Ding, 2024).

9.1.3 Integration of genome editing with immune-modulating plant molecules

An emerging strategy toward improved antiviral resistance in plants is the integration of CRISPR/Cas-based genome editing with plant-derived immune-modulating molecules. The main mechanism comprises a set of differentially regulated defense signaling pathways, mediated by hormonal responses such as salicylic acid (SA), jasmonic acid (JA), and ethylene (ET), together with regulation of reactive oxygen species (ROS) metabolism. These components have been suggested to contribute to antiviral signaling, transcriptional control, and growth–immunity trade-offs. These pathways have also been reported to functionally interact with chromatin and redox regulation, thereby shaping cellular contexts that reflect relationships among immune status, genome-editing efficiency, and cellular regenerability (Avramova, 2019; Kanyuka et al., 2022; Aerts, Pereira Mendes and Van Wees, 2021). While current findings increasingly support connections among hormonal signaling, chromatin priming, and CRISPR system efficiency, only a limited number of studies have examined the integrated priming method for optimizing CRISPR-mediated editing efficiency.

9.1.4 Toward a unified framework for durable antiviral resistance

Taken together, the promising potential supports the conclusion that combining precise genome editing with the CRISPR tool kit and targeted immune priming is well supported by emerging data. Realizing the full potential will require progress across many areas, from molecular engineering and plant defense biology and breeding to biosafety and environmental assessment. This chapter advances these efforts by showing how virus stress physiology, antiviral approaches using CRISPR technology, and the functions of immune-modulatory substances and chromatin regulation must be integrated to enable practical deployment.

9.2 Viral stress physiology in plants

Viral infections pose quite serious challenges to the physiological systems of infected plants. Therefore, the success of all gene-editing-based antiviral strategies will depend on a deep understanding of the physiological stresses caused by viral infections in plants. These stresses are a consequence of interactions among oxidative stress responses, metabolic reprogramming, host susceptibility factors, and immune signaling pathways—all factors that together influence disease progression and modulate the robustness of plant defense responses.

9.2.1 Mechanisms of virus-induced stress: oxidative burst and metabolic disruption

The rapid onset of the oxidative burst and the concomitant production of ROS are hallmarks of early plant–virus interactions. Although excessive ROS accumulation can cause oxidative damage to cellular membranes, proteins, and nucleic acids—impairing metabolic function and genomic integrity, respectively (Mittler, 2017). ROS also serve as necessary secondary messengers in the activation and orchestration of plant immune signaling pathways. In close association with this oxidative gradient, viral infection often involves extensive reprogramming of primary metabolism. Such metabolic changes include inhibition of photosynthesis, disruption of nutrient allocation, and a shift of cellular energy toward viral replication and virion assembly; these factors collectively lead to a decline in the host plant's vigor and productivity (Cao et al., 2020; Robertson et al., 2022). Additionally, autophagy—an important mechanism for cellular homeostasis—further contributes to the complexity of virus-induced physiological responses. While it facilitates degradation of viral components and limits viral replication, various viral pathogens hijack the autophagy pathway for viral replication or facilitate viral movement within cells (Yang et al., 2020).

9.2.2 Host responses and vulnerability factors

The success of viral infection depends heavily on host factors, which the virus exploits to advance its life cycle. Translation initiation factors eIF4E and eIF4G, are among the most well-studied susceptibility (*S*) genes, since many types of RNA viruses require these proteins to facilitate viral protein synthesis. Other components of the ubiquitin-proteasome system and the RNA degradation apparatus also regulate viral protein turnover and modulate the host antiviral state (García-Ruiz, 2019). Plants defend themselves against these attacks using complex immunological systems, including the hypersensitive reaction, pathogenesis-related protein induction, systemic acquired resistance, and small-RNA-mediated antiviral silencing. However, viral suppressors of RNA silencing (VSRs) often target these mechanisms, thereby facilitating both systemic spread and viral infection establishment. However, the fundamental role of the S-gene family in viral compatibility strategies makes it a highly desirable target for genome engineering. CRISPR-based techniques for disrupting and modifying these loci have repeatedly demonstrated success not only in reducing viral replication but also in inducing transgenerational resistance to these pathogens (Zaidi et al., 2016; Robertson et al., 2022).

9.2.3 Role of reactive oxygen species, phytohormones, and secondary metabolites in antiviral defense

Plant antiviral immunity is regulated by a highly integrated network that also includes ROS signaling, phytohormone-mediated control, and the induction of secondary metabolites. The controlled burst of ROS activates the transcription of defense genes, enhances callose deposition, and triggers localized programmed cell death, thereby acting as both a signal and a control mechanism that suppresses viral spread (Mittler, 2017). Better control of these redox-mediated processes is provided by various phytohormones: SA remains a critical regulator of antiviral defense responses, and JA, ethylene, and abscisic acid reinforce antiviral responses typically in complex, conditionally defined signaling interactions. Secondary metabolites, including phenolics, flavonoids, alkaloids, and terpenoids, contribute an additional layer of antiviral defense by suppressing viral replication, strengthening physical barriers, regulating cellular redox balance, and serving as signaling molecules within plant defense networks (Piasecka et al., 2015; Giordano et al., 2023). Superoxide dismutase, catalase, and ascorbate peroxidase, as antioxidant enzymes, reduce ROS accumulation during infections and are strongly associated with greater stress resilience. These responses are mediated through the regulation of mitogen-activated protein kinase (MAPK) cascades, calcium-mediated signaling pathways, and the integration of local and antiviral immunity systems (Meng & Zhang, 2013). Even with the advent of precision gene-modification efforts using the antiviral CRISPR platform, modifying *S* genes and regulators of these complex processes remains hindered by viral defense and their inherent complexity (Cao et al., 2020; Robertson et al., 2022).

9.3 CRISPR–Cas as a molecular defense mechanism

By leveraging these identified physiological vulnerabilities, current CRISPR/Cas systems provide a flexible platform for genetic engineering of antiviral immunity in plants. Following the design logic that originated in the prokaryotic adaptive immunity system, current CRISPR-Cas systems enable targeted cleavage of viral genomes, modification of host susceptibility factors, and transcriptional or epigenetic reprogramming of the immune response. Each of these approaches can, of course, also be used cumulatively: direct attacks on viral nucleic acids reduce viral replication capability, host modification interferes with host–viral interaction factors, and transcriptional or epigenetic approaches increase immunological readiness.

9.3.1 CRISPR/Cas-based genome editing

Cas9, the prototype of the first widely used programmable nuclease, continues to facilitate double-strand breaks by single-guide RNA (sgRNA)-directed DNA targeting and remains the preeminent tool for genome editing in plants. In prevailing plant systems, the predominant mode of repair is nonhomologous end joining (NHEJ), whereas homology-directed repair (HDR) is less frequent. This difference spurred the development of approaches to enhance donor integration, including cell cycle synchronization, modulation of DNA repair pathways, and optimal donor-template design (Jinek et al., 2012; Chen et al., 2022). Beyond the widely used Cas9 system, Type V nucleases—exemplified by Cas12a—-have opened up more avenues for genome editing by recognizing thymine-rich protospacer adjacent motifs (PAMs), generating staggered DNA double-strand breaks that facilitate s precise repair, and autonomously processing CRISPR arrays to enable efficient multiplex genome editing (Bandyopadhyay et al., 2020; Zhang et al., 2021). Building on these advances in DNA targeting, Type VI nucleases, exemplified by Cas13, have recently expanded the CRISPR toolkit to RNA targeting. This new ability of Cas13 provided a conceptual and practical framework for controlling RNA viruses by programmable transcript degradation (Aman et al., 2018; Kavuri et al., 2022).

In parallel development with nuclease-based genome editing, base editing, and prime editing technologies introduce targeted changes at the nucleotide level without the need to induce double-strand breaks (DSBs). The methodologies enable fine-scale, allele-specific modulation of genes involved in immunity or resistance with minimal perturbation of biological function (Azameti and Dauda, 2021; Vu et al., 2024). However, as the technology becomes more versatile, a major structural limitation is the delivery of CRISPR components into target plant species. This is an area of ongoing research in which newly developed DNA-free nanoparticles and viral vectors facilitate the spread of mobile guide RNAs (gRNAs), thereby lowering barriers to tissue- and species-specific targeting (Demirer et al., 2019; Ellison et al., 2020). In this context, Table 9.1 summarizes various CRISPR platforms, including their targeting mechanisms and the advantages and challenges for antiviral design.

9.3.2 Targeting viral genomes: cleavage, mutation, and suppression

To date, CRISPR-mediated antiviral approaches have relied almost exclusively on nucleases to target viral genomes. For DNA viruses in the genus *Geminivirus*, Cas9-guided cleavage at conserved regions, including replication-associated proteins and viral coat protein-encoding genes, has been shown to substantially reduce viral titers and alleviate disease symptoms across a wide range of hosts (Ali et al., 2015; Tashkandi et al., 2018). Building on this approach, simultaneous targeting of multiple sites within viral genomes further enhances antiviral efficacy while suppressing the emergence of escape mutants—an important concern given the generally high evolutionary potential of plant viruses (Zaidi et al., 2016). By contrast, in RNA viruses, CRISPR-mediated interference using the Cas13 system constitutes a distinct, orthogonal strategy in which viral RNAs are selectively degraded, thereby reducing viral titers under controlled experimental conditions (Aman et al., 2018; Kavuri et al., 2022).

Additionally, transcriptional interference using catalytically inactive Cas proteins with a repressor domain (CRISPR interference [CRISPRi]) provides a secondary method to modify host or viral gene expression without causing genomic breaks. This approach enables reversible knockdown of viral or host genes essential for infection, thereby preventing secondary off-target modifications to the genome while the targeted pathway remains affected (Qi et al., 2013; Karlson et al., 2021). It is now clear that various antiviral strategies,

TABLE 9.1 CRISPR methods are used to improve virus resistance in plants.

CRISPR modality	Primary target	Mechanism of action	Advantages	Limitations
Cas13	RNA viruses	RNA cleavage without DNA editing	Ideal for RNA viruses; transient or stable delivery	Risk of viral escape; specificity constraints
CRISPRa	Defense gene promoters	Transcriptional activation	Nondestructive; tunable; reversible	Requires constant expression of activator complexes
CRISPRi	Host/viral promoters	Transcriptional repression	Useful when gene knockdown is preferable	Lower repression than gene knockout
Base/prime editing	Susceptibility genes and regulatory motifs	Precise nucleotide changes without DSBs	High precision; minimal genomic disruption	Lower efficiency; PAM constraints
dCas–Epigenetic tools	Chromatin regions	Demethylation/acetylation to modify expression	Mimics natural priming, reversible	Off-target marks; incomplete editing; unstable effects; delivery challenges

CRISPRa, CRISPR activatior; *CRISPRi*, CRISPR interference.

including genomic cleavage, gene suppression, and potentially combined approaches, can effectively suppress viral infection.

9.3.3 Editing plant susceptibility genes to reinforce viral immunity

Targeted modulation of host *S*-genes via CRISPR technology represents one of the most durable and broadly applicable antiviral approaches. Translation initiation factors eIF4E and eIF(iso)4E are among the best-validated targets for S-gene-mediated antiviral resistance. Precise gene knockouts or allelic modifications of these genes in cucumber, tomato, and other plants abrogate the critical VPg–host interaction essential for postviral replication, conferring long-lasting resistance without transgene integrations (Chandrasekaran et al., 2016; Noureen et al., 2022; Pyott et al., 2016). More recent studies have identified host factors, including UbEF1B and CCR4/NOT3, that participate in viral replication or movement as new, highly effective targets for broad-spectrum geminivirus resistance *via* CRISPR technology, indicating a rapidly growing number of efficacious targets among the *S*-genes (Li et al., 2024).

Besides the protein-coding genes, regulators of RNA modifications and chromatin structure, such as RNA methyltransferases, have been identified as candidates that could contribute to susceptibility, as they affect viral RNA stability and host transcriptional structure (Wnuk et al., 2020). Base editing and prime editing are well-suited to modifying target genes because they enable precise alterations within specific domains without necessarily compromising organismal functions. However, the use of the strategy to modify *S*-genes, for example, requires consideration of its effects on organismal fitness, as the genes often regulate growth, development, and responses to environmental stresses (Pyott et al., 2016). Based on the physiological context elucidated above, host-gene modification represents an important bridge between direct antiviral targeting and immune modulation.

9.4 Role of plant-stimulating molecules in enhancing CRISPR-induced resistance

The integration of the CRISPR/Cas genome-editing system with plant-based immunomodulators appears highly promising, given that the outcomes of the former depend on cellular factors such as chromatin accessibility, redox state, DNA repair, and regenerability. During viral infections, these factors become imbalanced due to disturbances in ROS metabolism, hormone signaling pathways, and transcriptional programs; therefore, efficient editing requires restoring these factors. The use of plant-based stimulators, therefore provides a conceptual platform for boosting the efficiency of transformation, regenerability, and the establishment of edited phenotypes. Although the experimental integration of genomic editing with hormonal or metabolic stimulators is still in its preliminary stages, the growing literature nonetheless highlights their use to enhance the immune system's readiness and the cellular environment, thereby facilitating precise editing (Cardi et al., 2023; Chen et al., 2024). A conceptual overview of the interactions between CRISPR editing processes and plant-derived signaling molecules is presented in Fig. 9.1.

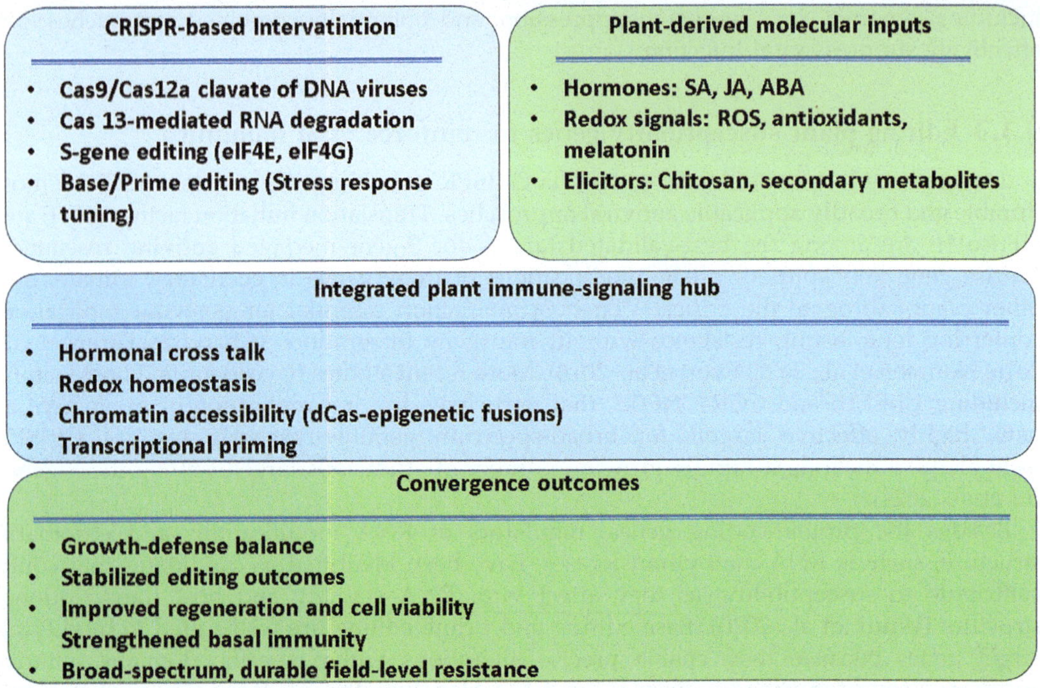

FIGURE 9.1 CRISPR–molecule synergy in antiviral defense. Conceptual framework illustrating the convergence of the CRISPR/Cas system and the plant-derived signal compounds salicylic acid, jasmonic acid, abscisic acid, melatonin, antioxidants, and secondary metabolites and their combined effects on chromatin opening, DNA repair mechanisms, redox state, and immunopriming. This convergence highlights the complementary mechanisms of action, whereby the CRISPR/Cas system modulates viral infectivity/cellular susceptibility factors, as well as the stabilizing effects of the cellular environment.

9.4.1 Hormonal modulation: effects on immunity and editing outcomes

Phytohormones are central regulators of antiviral immunity in plants, integrating cell physiology that directly affects CRISPR editing. SA is the central regulator of the systemic acquired immunity of plants, which alters the landscape of immune transcriptional networks, requiring growth–immunity trade-offs, especially during tissue culture and regeneration, where the metabolic state of cells and the balance of hormone homeostasis are decisive for the success of transformation (Aerts, Pereira Mendes and Van Wees, 2021; van Butselaar & Van den Ackerveken, 2020). JA, the central regulator of transcriptional immunity upon various stresses, provides a major contribution to antiviral immunity because of the enhancement of RNA silencing and the induction of AGO18 in rice, emerging as a clear link between immunity regulated by JA and the antiviral objective of the use of the CRISPR-based resistance approach (Yang et al., 2020). Abscisic acid, a regulator of immunity to abiotic stresses, alters immune outputs in a context-dependent manner, influencing cellular resilience during regeneration and potentially increasing cellular tolerance requirements, which are essential for successful genome editing (Aerts, Pereira Mendes and Van Wees, 2021).

Melatonin, acting both as a signaling agent and a potent antioxidant, enables cells to attain higher tolerance to various stresses, thereby protecting tissues during cellular reprogramming events (Erland et al., 2017). Even though presently, there are no cases where any of the existing hormonal pathways have been incorporated within the existing frameworks of the CRISPR approach, current studies clearly indicate that the regulation of both the hormonal and redox aspects of cells during the critical phases of the regeneration process could ensure the generation of more stable edited genotypes.

9.4.2 Secondary metabolites: defense enhancement and cellular context

Secondary metabolites, phenolics, flavonoids, alkaloids, and terpenoids are also critical components of plant secondary antiviral immunity and exert diverse effects on the cellular environment relevant to genome editing. Secondary metabolites enhance barrier strength, regulate membrane fluidity, and reduce oxidative damage, thereby facilitating survival of living cells during transformation or viral challenges (Piasecka et al., 2015). The role of these compounds is extended by the use of plant-derived elicitors of secondary metabolism, which trigger SA, JA, and ethylene pathways, stimulate the synthesis of lipid mediators, and activate secondary metabolism, reducing pathogen growth (Suarez-Fernandez et al., 2020). While no indication exists for the use of these compounds for improvements of the cutting efficiency or the accuracy of repair during genome editing using CRISPR, the enhancement of tolerance, resistance, or restriction of pathogen pressure, as well as other favorable physiological adjustments, make these compounds useful ancillaries of breeding approaches using antiviral-related traits provided by the use of the CRISPR technology. In a field environment, where plants are subject to variable abiotic and biotic factors, the induction of secondary metabolites can help preserve their vigor and maintain the fidelity of their gene expression. Table 9.2 summarizes the biological relevance of botanical compounds to the CRISPR mechanism.

9.4.3 Molecular priming, chromatin state, and editing consistency

Molecular priming with SA or other elicitors promotes immune responsiveness and often induces chromatin modifications at defense-related loci. Priming transgenic plants reveals changes in histone marks and transcriptional memory that promote more rapid activation of immune responses upon challenge (Aerts, Pereira Mendes and Van Wees, 2021). On the other hand, recent work on genome editing suggests that chromatin openness is a universally important factor for CRISPR/Cas system targeting efficiency, as open chromatin promotes efficient cleavage and repair (Chen et al., 2024). Related studies on plant epigenomics reveal that modification of chromatin-modifying enzymes or complexes can promote transformation or gene editing, suggesting that epigenome engineering offers strategies to overcome existing barriers (Cardi et al., 2023).

Although the integration of priming-mediated chromatin relaxation with the CRISPR system has not been investigated, the intersection of these results offers a compelling conceptual framework. Epigenetic regulation, which is quantifiable by chromatin profiling *via* ATAC-seq, thus represents a promising approach to increasing both the precision and potency of CRISPR-based genome editing systems.

TABLE 9.2 Mechanistic roles of plant signaling molecules that influence CRISPR editing and antiviral immunity.

Molecular class	Representative compounds	Antiviral functions	Influence on CRISPR editing/regeneration
Phytohormones	SA, JA, ABA	Immune activation, RNA-silencing enhancement, transcriptional reprogramming	Modulate chromatin accessibility; influence stress tolerance during transformation
Antioxidants	Melatonin, flavonoids, ascorbate	ROS buffering, membrane protection, redox signaling	Improve DNA-repair balance, enhance cell viability in culture
Elicitors	Chitosan, β-glucans	Trigger defense pathways (SA/JA/ET), induce secondary metabolites	Assist acclimation of regenerated lines; may reduce transformation stress
Secondary metabolites	Phenolics, terpenoids, alkaloids	Viral replication inhibition, structural reinforcement	Support tissue health; enhance survival of edited cells
Redox regulators	ROS-modulating enzymes (superoxide dismutase [SOD], catalase [CAT], and ascorbate peroxidase [APX])"	Prevent oxidative damage, strengthen stress tolerance	Maintain nuclear integrity; support precise editing outcomes

SA, Salicylic acid; *JA*, jasmonic acid; *ROS*, reactive oxygen species; *ABA*, abscisic acid.

9.4.4 Hormonal crosstalk and physiological trade-offs after editing

CRISPR-based approaches that target either *S* genes or immune regulators may restore physiological homeostasis but may also entail trade-offs between defense activation and plant growth. For instance, enhancing SA-mediated immune responses is often associated with dwarfism in plants, illustrating the importance of detailed assessment of how specific allelic variants affect immune activation and development (van Butselaar & Van den Ackerveken, 2020). Application-oriented studies provide further empirical evidence for these complex interactions. In watermelon, CRISPR/Cas9-mediated editing of the eIF4E gene conferred strong resistance to potyviruses; however, edited genotypes differed in growth performance in an allele-dependent manner, indicating tight linkage between antiviral defense mechanisms and growth regulation (Li et al., 2024). These observations collectively emphasize the need for precision at the level of alleles, which can be achieved through base-editing technologies that induce targeted amino acid substitutions.

Overall, the role of hormone interactions is twofold: it both facilitates and potentially limits the functionality of the antiviral resistance conferred by the CRISPR system in plants. When appropriately harnessed, hormone- and metabolite-based modulation could enhance the effects of CRISPR gene editing and stabilize immune phenotypes.

9.5 Transcriptional and epigenetic reprogramming *via* CRISPR and plant molecules

Recent advances in CRISPR-based gene regulation and epigenome editing provide novel, precise approaches for designing antiviral responses. Unlike the established use of the CRISPR system, which introduces permanent gene modifications, CRISPR activation (CRISPRa), CRISPRi, and dCas-based epigenetic fusions provide mechanisms for reversible, environment-responsive control of gene transcription. This approach conceptually complements the functions of plant-derived signaling compounds, specifically phytohormones and immune–priming–related compounds, which alter chromatin structure, transcriptional regulatory circuits, and RNA silencing pathways that are central to antiviral immunity. As previously explained, immune priming occurs through transcriptional and epigenetic modifications that promote sensitivity to subsequent exposure to environmental stresses (Hickman et al., 2019; Aerts, Pereira Mendes and Van Wees, 2021). Therefore, integrating CRISPR-based regulator systems into these processes is a promising yet nascent approach that can support environmentally responsive antiviral strategies.

9.5.1 CRISPRa and CRISPRi for defense gene modulation

CRISPRa and CRISPRi systems, which leverage nuclease-deactivated Cas enzymes fused with transcriptional activators or repressors, provide a strategy for controlling endogenous gene expression without inducing DNA damage. Such approaches, which include the use of activator motifs VP64, VPR, or SunTag, and repressor motifs SRDX or KRAB, have demonstrated efficient single-gene and multigene regulation in *Arabidopsis thaliana*, *Oryza sativa*, and *Solanum lycopersicum* (Lowder et al., 2015; Karlson et al., 2021). These systems can activate the expression of multiple genes simultaneously, as illustrated by the use of Act2.0 and Act3.0 systems, which coordinately regulate transcriptional cascades involved in stress responses (Lowder et al., 2015; Pan and Qi, 2023; Pan et al., 2021).

These approaches appear capable of supporting immunological readiness. For example, activation of the tomato *PR1* gene *via* CRISPRa led to improved resistance to bacteria infection, confirming that transcriptional priming can be specifically targeted using dCas approaches (García-Murillo et al., 2023). Although the range of antiviral targets for CRISPRa is limited, the fact that because these approaches do not alter the underlying DNA sequence they represent promising alternatives to permanent genome editing strategies.

9.5.2 Epigenetic modulation using CRISPR–dCas fusions

Epigenome editing extends the functionality of the CRISPR system by enabling precise control over chromatin, thereby regulating gene expression. One of the major applications of this technique used a SunTag-TET1 fusion protein to target DNA demethylation in the Arabidopsis model, thereby reactivating genes silenced by endogenous regulatory mechanisms(Gallego-Bartolomé et al., 2018). Subsequent studies have demonstrated the broad utility of dCas fusions, ranging from DNA demethylases to HAT fusions, for modulating stress responses and antiviral defense pathways(Gallego-Bartolomé, 2020; Gallego-Bartolomé et al., 2018).

Although various HAT-dependent systems, such as dCas9-p300, were initially demonstrated in mammalian cells, recent findings in plant systems suggest that similar methods for chromatin modification could enhance the flexibility of defense loci and promote transcriptional activation (Xie and Duan, 2023). These programmable systems enable the development of "synthetic priming" states that mimic the transcriptional memory associated with natural priming. However, there is an increasing need for empirical analysis of the long-term stability, reversibility, and off-target effects of chromatin modifications, particularly under environmental change.

9.5.3 Hormonal crosstalk and transcriptomic shifts in molecule–CRISPR synergy

Major plant hormones, including SA, JA, and ABA, regulate not only the induction of immunity but also the transcriptional and chromatin environment in which the CRISPR-based regulatory tool operates. These hormones affect the activity of transcription factors, RNA silencing, and chromatin enzymes, thus influencing the transcriptional environment that directs the activation of antiviral genes (Hickman et al., 2019; Aerts, Pereira Mendes and Van Wees, 2021). Transcriptional profiling reveals that treatment of plants with these hormones induces large-scale gene networks that shape the chromatin environment and the transcriptional memory of the secondary immune response (Aerts, Pereira Mendes and Van Wees, 2021).

While there are curently limited studies on integrating the CRISPRa/CRISPRi system with hormone-based treatment approaches, the possibility that hormonal signaling pathways could stabilize CRISPR-mediated regulatory states is intriguing. Nevertheless, the use of hormone-mediated defense often incurs growth costs, suggesting that integrating signaling compounds and CRISPR regulation must be carefully balanced.

9.5.4 Advances in CRISPR–hormone and epigenetic synergy

Advances in technology have strengthened the potential of integrated strategies using the CRISPR-hormone-epigenetics approach. Successful targeted demethylation using TET1 provides a clear proof of concept for the activation of stress genes *via* programmable epigenetics (Gallego-Bartolomé et al., 2018). CRISPR-Combo, a system that combines genome editing with a CRISPRa-based transcriptional activation system, provides insight into the integration of engineered and native regulatory mechanisms, enabling trait stacking and multifaceted activation of defense genes (Pan, Qi 2023).

Evaluations of plant epigenetic immunity have shown that hormone-responsive chromatin profiles are highly amenable to transcriptional modulation *via* the CRISPR approach, suggesting that the development of conditionally active, optogenetically controlled, or environment-responsive CRISPRa systems could allow precise control of antiviral immunity (Xie and Duan, 2023). Even at this early stage, the paradigm that appears to emerge is that of a complex regulatory system, wherein hormone-mediated regulation of transcriptional and epigenetic states can be coordinated with CRISPR-based regulatory systems.

9.6 Advanced CRISPR strategies to strengthen crop immunity

The faster pace of viral evolution, the growing prevalence of mixed infections, and the range expansion of various host-associated phytopathogens require the development of antiviral approaches that are precise, flexible, and resilient. Next-generation CRISPR toolkits , ranging from multiplex gene editing, base editing, and prime editing to integrated approaches combining CRISPR-RNAi, offer precise biological tools that will enable the design of multilayer antiviral systems suited to address the transcriptional and epigenetic complexities of plant–viral interactions. Although species-optimized tools, delivery challenges, and risk analysis for their use within the environment will remain important considerations for antiviral approaches beyond the laboratory settings, the existing literature suggests that current and next-generation CRISPR toolkits will provide an indispensable antiviral platforms for the development of biologically informed, physiologically robust and durable antiviral strategies.

9.6.1 Multiplex genome editing for resistance to multiple viruses

Multiplex genome editing enables simultaneous editing of multiple host or viral targets, thereby expanding the resistance spectrum and increasing the mutational barrier to viral escape. Architectures that exploit indigenous processing machinery, such as polycistronic tRNA-gRNA (PTG) vectors, transcribe multiple gRNAs and have already been validated for major crops such as rice and tomato, indicating the feasibility of multiplexing (Xie et al., 2015). More recent "drive-and-process" array designs incorporate multiplexing approaches for base and prime editing, indicating a viable pathway for integrating various nucleotide modifications in eukaryotic organisms (Chen et al., 2024). In practical terms, targeting genes in redundant pathways, such as different eIF4E paralogs or viral genomic regions, reduces the adaptive advantage conferred by single-site mutants, thereby improving the durability of the engineered resistance. However, the advantages of multiplex target editing must be balanced against potential pleiotropic effects and fitness costs, therefore requiring careful allele selection., phenotypic characterization, and hierarchical validation under diverse developmental and environmental conditions (Abdelrahman et al., 2021; Yoon et al., 2020).

9.6.2 Base editing and prime editing for precise stress-response tuning

Base editors and prime editors have expanded the utility of the CRISPR tool kit, enabling precise nucleotide changes or short insertions at specific loci without generating canonical DSBs. This reduces the likelihood of large indels and complex genomic rearrangements. These approaches can readily be applied to fine-tune regulatory domains, untranslated regions, and pivotal residues of susceptibility factors, thereby creating allelic variants that preserve the primary gene activity while blocking viral exploitation. Plant-based applications of these tools are presently advancing very rapidly. For example, prime editing has been used to modify promoter motifs of disease-responsive genes in rice (Gupta et al., 2023), and orchestrated base-editing protocols have been developed to create diverse eIF4E1 alleles in tomato, conferring resistance to various potyviruses (Kuroiwa et al., 2023). While off-target editing and the efficiency of these processes remain actively explored, current base- and prime-editing systems possess sufficient precision for allele-level modifications of antiviral genes and for mitigating the fitness loss imposed by gene knockouts (Vu et al., 2024).

9.6.3 Integration with RNAi mechanisms for layered defense

By integrating CRISPR-based genomic and transcriptomic methods with RNAi pathways, a synergistic or supplementary resistance mechanism can be established, thereby protecting against the emergence of escape mutants. For instance, in a resistance model, CRISPR interacts with plant RNAi processes to reduce viral transcript levels, thereby limiting viral replication. The idea of combining CRISPR and RNAi technologies is supported by empirical evidence, such as multiplexing genomic edits on *S* genes to enhance viral tolerance in cassava (Gomez et al., 2019) and using CRISPRi to target specific viruses, which has led to the selection of resistant viral genotypes. This underscores the importance of designing redundant targeting strategies and assessing potential evolutionary threats (Mehta et al., 2019). These comprehensive strategies—including editing, degradation, and multiplex targeting approaches—provide a strong foundation for antiviral immunity, provided that their effects on viral dynamics are thoroughly tested (Robertson et al., 2022).

9.6.4 CRISPR strategies in practice

Recent, translational studies have demonstrated the feasibility of sophisticated CRISPR-based strategies. Editing genes in the eIF4E family conferred substantial resistance to potyviruses in tomato and other plants, with allele- and background-dependent developmental phenotypes that can often be harnessed for allele optimization and breeding (Yoon et al., 2020; Kuroiwa et al., 2023). Prime and base editing have expanded the spectrum of potential modifications in major cereal crops, enabling modifications to promoter structure and the precision adjustment of susceptibility loci without compromising gene integrity (Gupta et al., 2023; Vu et al., 2024). Concurrently, improvements in delivery systems, notably the use of high-aspect-ratio nanomaterials for the delivery of DNA-free ribonucleoprotein complexes and RNA into mature plants, have lessened reliance on transgenic approaches and improved adaptability for use on recalcitrant accessions (Demirer et al., 2019). Integration of multiplexed targeting, precision editing, and layered silencing approaches offers a realistic and adaptable roadmap for achieving durable crop resistance, requiring optimization of allele selection, delivery, and regeneration; comprehensive phenotypic analysis; and integration with regulatory and ecological safeguards.

9.7 Delivery systems and bioinformatics-guided design

Successful implementation of CRISPR systems in plants depends on more than the choice of nuclease and the nature of the target. It also depends on the delivery mechanism employed to deliver the editing components and requires computational tools to facilitate gRNA design. Advances in nanocarrier-mediated transport, viral- and *Agrobacterium*-mediated delivery, and continuously improved gRNA design platforms have enabled the precise and efficient application of genome editing in plants. However, recent studies suggest that the optimal delivery systems, combined with computer-aided design, significantly improve the deployment of CRISPR systems in plants (Demirer et al., 2019; Azizi-Dargahlou et al., 2024; Demirer et al.,2021b). The cellular and chromatin factors, including hormone signaling and regeneration competency that determine editing efficiency in plant systems, are summarized in Fig. 9.2.

9.7.1 Nanocarriers, *Agrobacterium*, and virus-based vectors for CRISPR delivery

Agrobacterium tumefaciens remains widely used for plant transformation owing to its efficiency, stable transgene integration, and rich toolkit. However, species-, gene-, and genotype-related challenges, along with the transgene-related regulatory hurdles, have

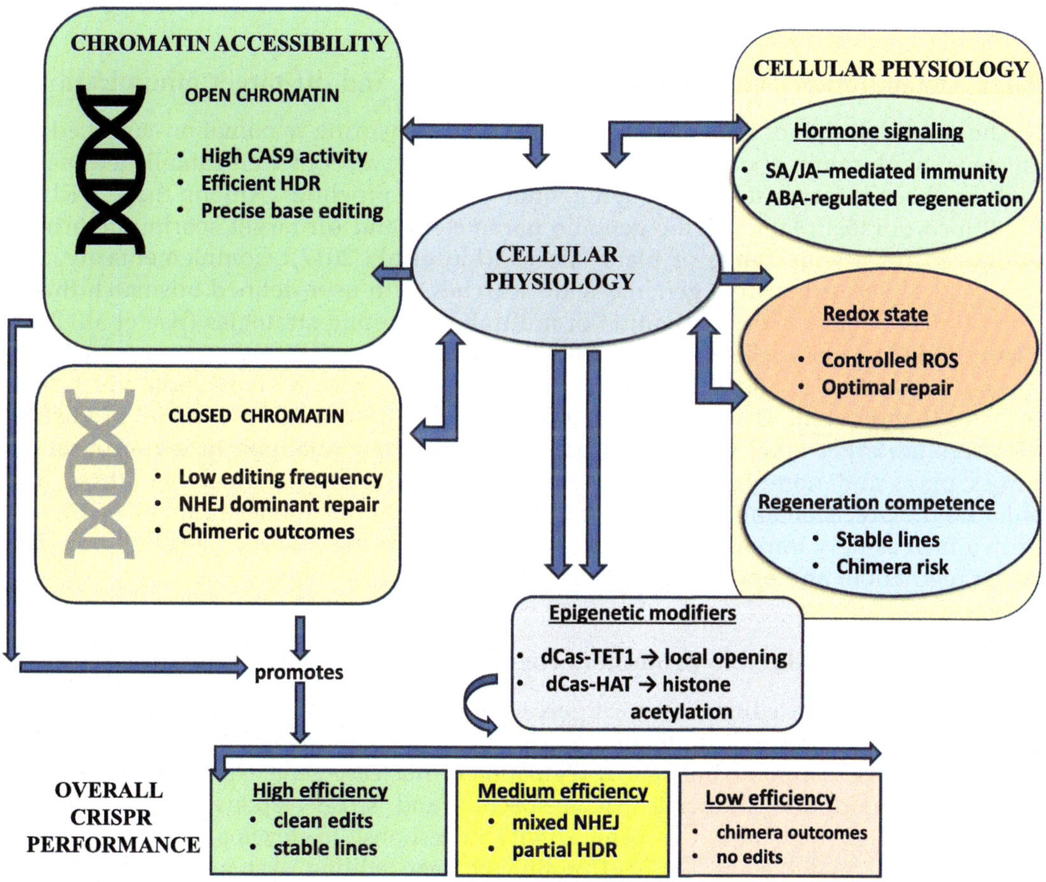

FIGURE 9.2 Cellular and chromatin contexts governing CRISPR editing efficiency. Conceptual framework showing the relationship between chromatin status, redox homeostasis, hormone signaling, and regeneration competency and how these parameters affect the outcome of the CRISPR/Cas editing process for antiviral plant modifications.

contributed substantially to the rapid advancement of alternative approaches (Azizi-Dargahlou et al., 2024). Among alternative approaches, DNA-free delivery systems have become increasingly popular. Carbon nanotube nanocarriers, for example, enable the direct delivery of plasmid DNA, single-stranded DNA, and even RNA to mature tissues, thereby facilitating transient CRISPR activity without the need for genomic integration (Demirer et al., 2019). Larger-scale nanotechnology studies have focused on identifying lipid nanoparticles, nanogels, and biodegradable polymers for ribonucleoprotein delivery, which offer promising avenues for editing difficult-to-transform monocots and dicots (Demirer et al., 2021b).

Virus-based vectors offer a significant approach for systemic delivery. Tobacco rattle virus (TRV)- and mobile sgRNA format-based vectors have demonstrated highly efficient somatic and germline editing in plants that harbor Cas nucleases (Ellison et al., 2020). Along these lines, geminiviral replicons, specifically BeYDV, can increase the number of template copies, thereby improving homology-directed repair and prime editing (Baltes et al., 2015). *Agrobacterium*, nanocarriers, and viral vectors provide complementary delivery systems.

9.7.2 Computational tools for guide RNA design and off-target minimization

The careful design of gRNAs is a critical factor in ensuring maximal on-target editing efficiency with minimal off-targeting. To facilitate this, a set of well-established *in silico* tools has been incorporated into plant genome-editing pipelines. Among these, CRISPR-P 2.0 incorporates plant-specific genome parameters and off-target scoring algorithms optimized for a wide range of plant species (Liu et al., 2017). Complementarily, Cas-OFFinder allows exhaustive genome-wide searches with user-defined mismatch thresholds, thereby enabling the evaluation of multiplex targeting strategies (Bae et al., 2014). CHOPCHOP v3 allows flexible gRNA design for multiplex CRISPR systems—including Cas9, Cas12a, Cas13, and prime editors—through advanced visualization and scoring features (Labun et al., 2019). Similarly, CRISPOR integrates multiple model predictions to assess on-target efficiency and off-target risk, thereby enabling broad applicability across plant and non-plant systems (Haeussler et al., 2016). Altogether, these tools enhance the precision and reliability of CRISPR-based applications, a capability that has grown increasingly important with the expansion of multiplex-based editing, transcriptional regulation, and epigenetic engineering.

9.7.3 Hormone-inducible promoters for temporal and spatial control

CRISPR/Cas-based editing often targets essential or defense-related genes, leading to growth or developmental trade-offs. Greater emphasis is therefore being placed on achieving precise temporal and spatial regulation of nuclease gene expression to minimize these effects. Hormone-inducible, tissue-specific, and stress-responsive promoters are practical tools for regulating Cas gene activity in response to developmental or environmental cues (Lowder et al., 2015). Conditional gene editing systems demonstrate the feasibility of alleviating pleiotropic phenotypes, facilitating tissue regeneration, and optimizing editing efficiency in sensitive tissues by activating the Cas9 gene during specific developmental phases (Chen et al., 2019).

Promoters that respond to SA, abscisic acid, or JA are receiving increasing attention because of the need to precisely control the timing of genome editing to prevent off-target effects that could affect growth or reproduction. The structure of these promoters aligns well with the conceptual frameworks for hormonal and chromatin-based regulation described in Sections 9.4 and 9.5 . As next-generation CRISPR technology advances toward environmentally responsive, inducible, and tissue-specific approaches, hormone-responsive promoters will play a central role in the precise engineering of the plant genome (Lowder et al., 2015; Chen et al., 2019).

9.8 CRISPR–plant molecule synergy in enhancing viral resistance

The increasing complexity of plant viral diseases necessitates resistance mechanisms that require the integration of precise genome engineering approaches with broader physiological resistance. This will enable the combined use of the CRISPR system's antiviral mechanism and plant-based immunomodulatory compounds, with the former targeting viral genomes and *S* genes, while the latter modulates the defense-related transcription, redox homeostasis, and stress priming. To date, there is very little empirical evidence demonstrating the combined use of the proposed antiviral strategies. However, evidence from virology, genome-editing systems, and plant physiology strongly suggests the feasibility of the integrated systems that combine antiviral genome engineering with immune modulation (Herrera-Vásquez et al., 2015; Hickman et al., 2019).

9.8.1 Complementary roles of CRISPR editing and plant immune modulators

CRISPR-based tools offer powerful approaches for viral interference and host-factor modification. Biologically, studies in rice have shown that targeted disruption of eIF4G efficiently suppress viral populations of rice tungro spherical virus, demonstrating the antiviral potential of allelic variation in translation initiation factors for the resistant trait (Macovei et al., 2018). Studies in tomato, cucumber, and other plant species have identifiedr eIF4E and eIF4E isoforms as the common susceptibility loci, modifications of which restrict replication of various potyviruses (Zahir Ali et al., 2015; Zaidi et al., 2016). Direct targeting of viral genomes also results in considerable antiviral effects, where Cas9 cleavage of replication-associated genes of TYLCV decreases the severity of the disease, and Cas13a-induced degradation of PVY transcripts inhibits the spread of the virus throughout the plant (Aman et al., 2018; Tashkandi et al., 2018). Taken together, these studies provide insight into the host or viral strategies that target fundamental processes of the viral life cycle.

9.8.2 CRISPR targeting of viral genomes and host susceptibility factors

Previously discussed plant signaling molecules, SA, JA, and antioxidant mechanisms, regulate transcriptional and redox networks that form the basis of antiviral immunity. SA mediates the regulation of ROS, pathogenesis-related genes, and redox homeostasis, which enhances basal defense mechanisms and restricts virus spread within the plant tissue (Herrera-Vásquez et al., 2015). JA-mediated signaling pathways impinge on key transcription factors controlling induced defense responses, thereby enhancing antiviral immunity (Hickman et al., 2019). Other complementary ways to enhance antiviral immunity include application of antioxidants, while the elicitor chitosan improves resistance by enhancing transcriptional priming. Together, these may favorably modulate the intracellular environment necessary for long-term persistence of CRISPR-based immunity under physiological conditions of infection.

9.8.3 Plant-derived signaling molecules as reinforcers of antiviral immunity

The integration of gene-editing methodologies with immune-modulating compounds provides a conceptual framework for developing multilayered antiviral resistance in plants. In this concept, CRISPR-mediated edits either disrupt viral replication directly or modulate host resistance pathways, while immune priming enhances the responsiveness and robustness of plant defense systems. This integrated approach is especially relevant for viral pathogens with high evolutionary rates, frequent coinfections, and environmental conditions that modulate the selective pressure of viruses and the physiological states of the host.

9.8.4 Toward an integrated multilayered framework for durable viral resistance

The combined effects of these mechanisms can reinforce the intracellular environment supporting strong CRISPR-mediated antiviral resistance under physiological conditions typical of viral infection. Future studies are thus warranted to move beyond the testing of single strategies in isolation and focus on the deliberate integration of CRISPR genome editing with treatments that modulate immune functions, including applications of phytohormones, ROS management, or the use of defense elicitors. Achieving this integration involves consideration not only of the trade-offs between improved immunity and the potential loss of plant growth, but also of trait stability across various developmental stages and environmental conditions. With regard to this area, continued advances in inducible CRISPR systems, hormone-responsive promoters, and chromatin-based priming approaches position physiologically activated genome editing as a promising strategy for durable viral resistance in crops.

9.9 Regulatory, environmental, and ethical perspectives

As the use of CRISPR for genome editing becomes more widespread and enters mainstream agriculture, regulatory, environmental, and ethical frameworks will increasingly affect the deployment of these edited, virus-resistant plants. While the precision and adaptability of the technology offer a potentially superior strategy for antiviral crop protection compared with conventional approaches, their ultimate application will, of course, be regulated by specific jurisdictional frameworks, environmental risk factors, and public interests. Because the edited plants will require no foreign DNA, they may fall into a distinct and potentially favorable regulatory category.

9.9.1 Regulatory advantages relative to chemical pesticides and transgenic GMOs

The advantages of CRISPR-based resistance approaches are also evident when compared with chemical pesticides and transgenic genetically modified organisms (GMOs). In fact, the need to control virus vectors with insecticides imposes a heavy burden on the environment, including biodiversity loss and soil and water pollution. Genome editing enables modification of resistance loci that may obviate the use of these environmental toxins, provided that the genetic modifications confer long-lasting resistance at the field level (Chen et al., 2019b).

One of the major regulatory advantages of the CRISPR-based editing is the promise of developing plants that do not contain exogenous DNA. Some countries assess these products on the basis of their genetic makeup, not on the technology used for their development. This means that the majority of the plants developed using the CRISPR technique, especially for small insertions, deletions, or nucleotide substitutions that can also be achieved through traditional breeding, often fall outside strict GMO regulatory frameworks. This has contributed to a relatively positive perception compared with transgenic plants that introduce foreign genes across species (Zhang et al., 2020). On the other hand, rigorous field testing continues to play a critical role in assessing their ultimate agronomic and sustainable impacts (Kanchiswamy, 2016).

9.9.2 Regulatory diversity, biosafety standards, and public acceptance

Regulatory frameworks for genome-edited crops also differ among countries worldwide. Countries actively developing genome-edited crops, the United States, Japan, Argentina, and Brazil, take a product-focused or case-by-case approach, where some genome-edited organisms are exempted from GMO regulations if foreign DNA is not introduced and the gene modification can also be obtained through traditional breeding. However, the USDA SECURE framework, Argentina's case-by-case approach, and the 2019 Japanese approach indicate a shift in exemption strategies. For example, the European Union takes a process-based approach, under which gene-edited plants are GMOs regardless of the presence or absence of foreign DNA, following the 2018 judgment of the Court of Justice of the European Union (Sprink et al., 2019).

Public acceptance is equally important for the use of gene-edited plants. Evidence from the Eurobarometer study suggests that gene editing is cautiously accepted, particularly when the modifications entail nontransgenic modifications and when specific social benefits, such as sustainability or reduced chemical use, are specified (*European Commission, 2022*). This indicates that building public trust will require open communication, risk assessment, and clarity regarding off-target mutagenesis, environmental interactions, and long-term management.

9.9.3 Ethical considerations: diversity, ecosystem interactions, and equity

Ethical considerations associated with CRISPR-based approaches to improve stress resistance extend beyond the analysis of technological risks. The magnitude and extent of genome editing within the available pools of resistance genes—which eventually contribute to susceptibility—may carry unforeseen consequences for genetic diversity and, by extension, for ecologically important plant traits. These challenges are most pronounced in viral diseases due to their high evolutionary dynamics and sensitivity to ecological feedback mechanisms (Ishii, Araki, 2017).

At the same time, equity considerations significantly influence the ethical dimensions of genome editing. High licensing fees for key CRISPR components, coupled with unequal access to advanced biotechnological infrastructure and expertise, tend to widen the gap between large-scale breeding programs and smallholder or resource-poor farmers. Despite this, CRISPR-based approaches hold significant potential for public-sector breeding, open innovation platforms, and conservation-oriented breeding, with potential benefits for global food security (Jaganathan et al., 2018). These ethical considerations collectively underpin the necessity for a governance framework that balances scientific rigor with social equity and environmental sustainability. iIn this context, effective monitoring, transparency, and meaningful public engagement must accompany technological innovation to ensure responsible development and deployment of genome editing technologies.

9.10 Challenges and limitations

Despite these recent advances, significant biological, technological, and regulatory challenges remain that hinder the reproducibility of integrated CRISPR-molecule approaches. Furthermore, editing efficiencies vary considerably across plant species and genomic loci and are influenced by factors such as chromatin accessibility, engagement of DNA repair pathways, redox signaling, and transformation efficiency. Together, these biological variables create heterogeneity in the resulting mutation profiles. If such mutations occur in *S* genes or their regulatory regions, this heterogeneity can significantly impact the consistency and overall efficacy of antiviral resistance strategies (Chen et al., 2019b). Moreover, the requirement for multiplexing to prevent viral escape and suppress resistant viral variants is likely to increase the risk of off-target effects, especially when genomic sequences are repetitive or similar to other genes, and therefore necessitates a comprehensive genome-wide assessment (Modrzejewski et al., 2020). This challenge is compounded by limitations associated with transgene delivery. *Agrobacterium*-mediated delivery exhibits substantial genotype-dependent variation (Altpeter et al., 2016). Carbon nanotube delivery systems require species-specific biocompatibility studies, particularly because of potential toxicity-related concerns (Demirer et al., 2021a,b). Viral replicons and mobile gRNA vectors exhibit varying efficiencies among plant species. Lastly, the interaction between the hormone signaling and redox pathways may influence CRISPR system performance, because immune priming pathways may impose growth or metabolic penalties (Hickman et al., 2019).

Epigenetic editing introduces a new level of uncertainty regarding the stability of engineered DNA methylation and histone modifications, which can exhibit variable stability or heritability under different environmental and developmental conditions (Gallego-Bartolomé, 2020). Such variability in regulation also affects the translational applicability of the technology: whereas some jurisdictions regulate DNA-free CRISPR edits similarly to conventionally bred organisms. The European Union applies broader regulatory oversight to treat such edits as GMOs, with implications for commercialization and investment (Chen et al., 2019b; Modrzejewski et al., 2020). Taken together, these biological, technological, physiological, and regulatory constraints underscore that integrated CRISPR-based antiviral strategies remain an emerging and transformative technology.

9.11 Future directions

Future developments in CRISPR-based antiviral engineering are likely to focus on improving precision by incorporating newer tools, such as prime editing, various RNA-guided CRISPR platforms, and epigenetic regulators. Integrative studies linking CRISPR-mediated modifications to plant signaling—in particular those mediated by phytohormones and ROS—may further enhance the ability of plants to temporally regulate immune responses. In the future, it will be important to determine the long-term sustainability of these approaches in light of possible viral evolution driven by editing-mediated selective pressure, and to develop robust regulatory framework that facilitate responsible and effective deployment worldwide.

Acknowledgments

GM sincerely thanks the Council of Scientific Industrial Research for funding (CSIR-EMRII-38 (1450)/18/EMRII) awarded to GM. LL acknowledges the National Fellowship for Higher Education of ST Students (NFST) of the Ministry of Tribal Affairs, Government of India, for the fellowship. AS acknowledges the Rashtriya Uchchatar Shiksha Abhiyan (RUSA) 2.0 Program, under the Ministry of Human Resources Development, Government of India, and CAS, DST-PURSE-II, UPE-FAR, and DST-FIST, for funding and providing the necessary facilities to carry out the present work. We thank the Executive Director, Agri Biotech Foundation, Hyderabad, for the facilities and support.

Conflicts of interest

The authors declare no conflict of interest.

References

Abdelrahman, M., Wei, Z., Rohila, J.S., Zhao, K., 2021. Multiplex genome-editing technologies for revolutionizing plant biology and crop improvement. Frontiers in Plant Science 12. Available at https://doi.org/10.3389/fpls.2021.721203.

Aerts, N., Pereira Mendes, M., Van Wees, S.C.M., 2021. Multiple levels of crosstalk in hormone networks regulating plant defense. Plant Journal 105 (2), 489–504. Available at https://doi.org/10.1111/tpj.15124.

Ali, Z., Abul-faraj, A., Li, L., Ghosh, N., Piatek, M., Mahjoub, A., et al., 2015. Efficient virus-mediated genome editing in plants using the CRISPR/Cas9 system. Molecular Plant 8 (8), 1288–1291. Available at https://doi.org/10.1016/j.molp.2015.02.011.

Ali, Z., Abulfaraj, A., Idris, A., Ali, S., Tashkandi, M., Mahfouz, M.M., 2015. CRISPR/Cas9-mediated viral interference in plants. Genome Biology 16 (1). Available at https://doi.org/10.1186/s13059-015-0799-6.

Altpeter, F., Springer, N.M., Bartley, L.E., Blechl, A.E., Brutnell, T.P., Citovsky, V., et al., 2016. Advancing crop transformation in the era of genome editing. The Plant Cell 28 (7), 1510–1520. Available at https://doi.org/10.1105/tpc.16.00196.

Aman, R., Ali, Z., Butt, H., et al., 2018. RNA virus interference via CRISPR/Cas13a in plants. Nature Biotechnology 36 (9), 932–936. Available at https://doi.org/10.1038/nbt.4020.

Avramova, Z., 2019. Defense-related priming and chromatin memory in plants: mechanisms and applications. Plant, Cell & Environment 42 (3), 761–774. Available at https://doi.org/10.1111/pce.13515.

Azameti, M.K., Dauda, W.P., 2021. Base editing in plants: applications, challenges, and future prospects. Frontiers in Plant Science 12. Available at https://doi.org/10.3389/fpls.2021.664997.

Azizi-Dargahlou, M., Raji, R., Karimi, M., 2024. Recent advances in plant transformation technologies for genome editing. Plant Cell Reports 43 (1), 1–17. Available at https://doi.org/10.1007/s00299-023-03050-7.

Bae, S., Park, J., Kim, J.-S., 2014. Cas-OFFinder: a fast and versatile algorithm for off-target search. Bioinformatics 30 (10), 1473–1475. Available at https://doi.org/10.1093/bioinformatics/btu048.

Baltes, N.J., Hummel, M., Scharenberg, A.M., Voytas, D.F., 2015. Geminivirus replicons for functional genomics in monocots and dicots. Plant Physiology 169 (1), 313–323. Available at https://doi.org/10.1104/pp.15.00646.

Bandyopadhyay, A., Kancharla, N., Javalkote, V.S., Dasgupta, S., Brutnell, T.P., 2020. CRISPR-Cas12a (Cpf1): a versatile tool in the plant genome-editing toolbox. Plant Cell Reports 39 (10), 1219–1230. Available at https://doi.org/10.1007/s00299-020-02559-2.

Cao, Y., Zhou, H., Zhou, X., Li, F., 2020. Control of plant viruses by CRISPR/cas system-mediated adaptive immunity. Frontiers in Microbiology 11. Available at https://doi.org/10.3389/fmicb.2020.593700.

Cardi, T., Murovec, J., Bakhsh, A., Boniecka, J., Bruegmann, T., Bull, S.E., et al., 2023. CRISPR/Cas-mediated plant genome editing: outstanding challenges a decade after implementation. Trends in Plant Science 28 (10), 1144–1165. Available at https://doi.org/10.1016/j.tplants.2023.05.012.

Chen, J., Li, S., He, Y., Li, J., Xia, L., 2022. An update on precision genome editing by homology-directed repair in plants. Plant Physiology 188, 1780–1794. Available at https://doi.org/10.1093/plphys/kiac037.

Chandrasekaran, J., Brumin, M., Wolf, D., et al., 2016. Development of broad virus resistance in non-transgenic cucumber using CRISPR/Cas9 to target eIF4E. Plant Biotechnology Journal 14 (12), 2190–2201. Available at https://doi.org/10.1111/pbi.12609.

Chen, F., Chen, L., Yan, Z., Xu, J., Feng, L., He, N., et al., 2024. Recent advances of CRISPR-based genome editing for enhancing staple crops. Frontiers in Plant Science 15. Available at https://doi.org/10.3389/fpls.2024.1478398.

Chen, K., Wang, Y., Zhang, R., Zhang, H., Gao, C., 2019. CRISPR/Cas genome editing and precision plant breeding in agriculture. Annual Review of Plant Biology 70 (1), 667–697. Available at https://doi.org/10.1146/annurev-arplant-050718-100049.

Chen, K., Wang, Y., Zhang, R., Zhang, H., Gao, C., 2019b. CRISPR genome editing and its applications in crop improvement. Nature Reviews Genetics 20 (10), 674–690. Available at https://doi.org/10.1038/s41576-019-0150-9.

Demirer, G.S., González-Grandío, E., Jackson, C.T., Landry, M.P., 2021a. Carbon nanotube biocompatibility in plants is determined by their surface chemistry and route of entry. Journal of Nanobiotechnology 19, 431. Available at https://doi.org/10.1186/s12951-021-01147-3.

Demirer, G.S., Silva, T.N., Jackson, C.T., Thomas, J.B., W. Ehrhardt, D., Rhee, S.Y., et al., 2021b. Nanotechnology to advance CRISPR–Cas genetic engineering of plants. Nature Nanotechnology 16 (3), 243–250. Available at https://doi.org/10.1038/s41565-021-00854-y.

Demirer, G.S., Zhang, H., Goh, N.S., Cunningham, F.J., et al., 2019. Carbon-nanotube–mediated DNA and RNA delivery into plants. Nature Nanotechnology 14 (5), 456–464. Available at https://doi.org/10.1038/s41565-019-0382-5.

Ellison, E.E., Nagalakshmi, U., Gamo, M.E., et al., 2020. Multiplexed heritable gene editing using RNA viruses and mobile guide RNAs. Plant Physiology 183 (2), 557–570. Available at https://doi.org/10.1104/pp.20.00238.

Erland, L.A.E., Saxena, P.K., Murch, S.J., 2017. Melatonin in plant signalling and behaviour. Functional Plant Biology 45, 58–69. Available at https://doi.org/10.1071/fp16384.

European Commission, 2022. Attitudes of Europeans towards the use of biotechnology in agriculture (Special Eurobarometer 526).

Gallego-Bartolomé, J., Gardiner, J., Liu, W., Papikian, A., Ghoshal, B., Kuo, H.Y., et al., 2018. Targeted DNA demethylation of the arabidopsis genome using the human TET1 catalytic domain. Proceedings of the National Academy of Sciences of the United States of America 115 (9) E2125,. Available at https://doi.org/10.1073/pnas.1716945115.

Gallego-Bartolomé, J., 2020. DNA methylation in plants: mechanisms and tools for targeted manipulation. New Phytologist 227 (1), 38–44. Available at https://doi.org/10.1111/nph.16529.

García-Murillo, L., Valencia-Lozano, E., Priego-Ranero, N.A., Cabrera-Ponce, J.L., Duarte-Aké, F.P., Vizuet-de-Rueda, J.C., Rivera-Toro, D.M., Errera-Ubaldo, H., de Folter, S., Alvarez-Venegas, R., 2023. CRISPRa-mediated transcriptional activation of the SlPR-1 gene in edited tomato plants. Plant Sci 329, 111617. Available at 10.1016/j.plantsci.2023.111617.

Giordano, D., Facchiano, A., Carbone, V., 2023. Food plant secondary metabolites antiviral activity and their possible roles in SARS-CoV-2 treatment: an overview. Molecules 28, 2470. Available at 10.3390/molecules28062470.

García-Ruiz, H., 2019. Host factors against plant viruses. Molecular Plant Pathology 20 (5), 1023–1030. Available at https://doi.org/10.1111/mpp.12836.

Gomez, M.A., Lin, Z., Caro, M., 2019. Simultaneous CRISPR/Cas9-mediated editing of cassava genes improves tolerance to viral disease. Molecular Plant Pathology 20 (5), 829–843. Available at https://doi.org/10.1111/mpp.12791.
Gupta, A., Liu, B., Chen, Q.J., Yang, B., 2023. High-efficiency prime editing enables new strategies for broad-spectrum resistance to bacterial blight of rice. Plant Biotechnology Journal 21 (7), 1454–1464. Available at https://doi.org/10.1111/pbi.14049.
Haeussler, M., Schönig, K., Eckert, H., et al., 2016. Evaluation of off-target and on-target scoring algorithms and integration into CRISPOR. Genome Biology 17, 148. Available at https://doi.org/10.1186/s13059-016-1012-2.
Herrera-Vásquez, A., Salinas, P., Holuigue, L., 2015. Salicylic acid and reactive oxygen species interplay in the transcriptional control of defense genes expression. Frontiers in Plant Science 6. Available at 10.3389/fpls.2015.00171.
Hashimoto, M., Neriya, Y., Yamaji, Y., Namba, S., 2016. Recessive resistance to plant viruses: Potential resistance genes beyond eIF4E and eIF4G. Frontiers in Plant Science 7, 1695. Available at https://doi.org/10.3389/fpls.2016.01695.
Hickman, R., Hill, C., Penfold, C.A., et al., 2019. Chromatin-mediated transcriptional priming in plant immunity. Cell Reports 27 (13), 3803–3815. Available at https://doi.org/10.1016/j.celrep.2019.05.077 e7.
Ishii, T., Araki, M., 2017. A future scenario of genome-edited crops. GM Crops & Food 8 (1), 44–56. Available at https://doi.org/10.1080/21645698.2017.1289303.
Jaganathan, D., Ramasamy, K., Sellamuthu, G., et al., 2018. CRISPR for crop improvement: an ethical and societal perspective. Plant Cell Reports 37 (10), 1465–1477. Available at https://doi.org/10.1007/s00299-018-2293-z.
Jinek, M., Chylinski, K., Fonfara, I., et al., 2012. A programmable dual-RNA-guided DNA endonuclease. Science 337 (6096), 816–821. Available at https://doi.org/10.1126/science.1225829.
Jones, R.A.C., 2021. Global plant virus disease pandemics and epidemics. Plants 10 (2), 1–41. Available at https://doi.org/10.3390/plants10020233.
Kanchiswamy, C.N., 2016. DNA-free genome editing methods for targeted crop improvement. Plant Cell Reports 35 (7), 1469–1474. Available at https://doi.org/10.1007/s00299-016-1982-2.
Kanyuka, K., Hammond-Kosack, K., Cooper, A., Ton, J., 2022. Immune priming in plants: from the onset to transgenerational maintenance. Essays in Biochemistry 66 (5), 635–646. Available at https://doi.org/10.1042/ebc20210082.
Karlson, D., Liu, Y., Qi, Y., 2021. CRISPR-mediated gene regulation in plants. Trends in Plant Science 26 (10), 1002–1021. Available at https://doi.org/10.1016/j.tplants.2021.05.012.
Kavuri, N.R., Ramasamy, M., Qi, Y., Mandadi, K., 2022. Applications of CRISPR/Cas13-based RNA editing in plants. Cells 11 (17), 2665. Available at https://doi.org/10.3390/cells11172665.
Kuroiwa, K., Toki, S., Endo, M., 2023. Iterative gene-editing broadens eIF4E1 diversity. Plant Cell Reports 42 (2), 119–132. Available at https://doi.org/10.1007/s00299-022-02961-x.
Labun, K., Montague, T.G., Krause, M., Torres Cleuren, Y.N., Tjeldnes, H., Valen, E., 2019. CHOPCHOP v3: expanding the CRISPR web toolbox beyond genome editing. Nucleic Acids Research 47 (1), W171. Available at https://doi.org/10.1093/nar/gkz365.
Li, H., Gong, P., Xu, X., et al., 2024. Knockout of UbEF1B and CCR4/NOT3 confers resistance to geminiviruses. Plant Biotechnology Journal 22 (4), 793–795. Available at https://doi.org/10.1111/pbi.14265.
Li, M., Qiu, Y., Zhu, D., et al., 2024. Editing eIF4E in watermelon confers resistance to ZYMV. International Journal of Molecular Sciences 25 (21), 11468. Available at https://doi.org/10.3390/ijms252111468.
Liu, H., Ding, Y., Zhou, Y., Jin, W., Xie, K., Chen, L.-L., 2017. CRISPR-P 2.0: an Improved CRISPR-Cas9 tool for genome editing in plants. Molecular Plant 10, 530–532. Available at 10.1016/j.molp.2017.01.003.
Lowder, L.G., Zhang, D., Baltes, N.J., et al., 2015. A CRISPR/Cas9 toolbox for multiplexed plant genome editing. Nature Plants 1 (12), 1–8. Available at https://doi.org/10.1038/nplants.2015.202.
Macovei, A., Sevilla, N.R., Cantos, C., et al., 2018. Novel alleles of rice eIF4G confer virus resistance. Plant Biotechnology Journal 16 (11), 1918–1927. Available at https://doi.org/10.1111/pbi.12927.
Meng, X., Zhang, S., 2013. MAPK cascades in plant disease resistance signaling. Annual Review of Phytopathology 51, 245–266. Available at 10.1146/annurev-phyto-082712-102314.
Mehta, D., Stürchler, A., Anjanappa, R.B., et al., 2019. Evolution of editing-resistant geminiviruses. Genome Biology 20, 80. Available at https://doi.org/10.1186/s13059-019-1678-3.
Mittler, R., 2017. ROS are good. Trends in Plant Science 22 (1), 11–19. Available at https://doi.org/10.1016/j.tplants.2016.08.002.

Modrzejewski, D., Hartung, F., Sprink, T., et al., 2020. Evidence for genome-editing applications. Environmental Evidence 9, 23. Available at https://doi.org/10.1186/s13750-020-00201-8.

Noureen, A., Khan, M.Z., Amin, I., et al., 2022. CRISPR/Cas9 targeting of eIF4E enhances PVY resistance. Frontiers in Genetics 13, 922019. Available at https://doi.org/10.3389/fgene.2022.922019.

Pan, C., Qi, Y., 2023. CRISPR-Combo-mediated orthogonal genome editing. Nature Protocols 18 (6), 1760–1794. Available at https://doi.org/10.1038/s41596-023-00823-w.

Pan, C., Wu, X., Markel, K., et al., 2021. CRISPR-Act3.0. Nature Plants 7, 944–953. Available at https://doi.org/10.1038/s41477-021-00960-w.

Piasecka, A., Jedrzejczak-Rey, N., Bednarek, P., 2015. Secondary metabolites in plant immunity. New Phytologist 206 (3), 948–964. Available at https://doi.org/10.1111/nph.13325.

Pyott, D.E., Sheehan, E., Molnar, A., 2016. Engineering of CRISPR/Cas9-mediated potyvirus resistance in transgene-free *Arabidopsis* plants. Molecular Plant Pathology 17, 1276–1288. Available at 10.1111/mpp.12417.

Qi, L.S., Larson, M.H., Gilbert, L.A., et al., 2013. Repurposing CRISPR for gene regulation. Nature Biotechnology 31 (12), 1198–1204. Available at https://doi.org/10.1038/nbt.2679.

Robertson, N.R., Canto, T., Palukaitis, P., 2022. CRISPR/Cas-based control of plant viruses. Molecular Plant Pathology 23 (6), 804–821. Available at https://doi.org/10.1111/mpp.13252.

Sprink, T., Eriksson, D., Schiemann, J., Hartung, F., 2019. Regulatory hurdles for genome editing. Plant Biotechnology Journal 17 (9), 1770–1774. Available at https://doi.org/10.1111/pbi.13122.

Shteinberg, M., Mishra, R., Anfoka, G., Altaleb, M., Brotman, Y., Moshelion, M., Gorovits, R., Czosnek, H., 2021. Tomato Yellow Leaf Curl Virus (TYLCV) Promotes Plant Tolerance to Drought. Cells 10, 2875. Available at https://doi.org/10.3390/cells10112875.

Suarez-Fernandez, M., Marhuenda-Egea, F.C., Lopez-Moya, F., Arnao, M.B., Cabrera-Escribano, F., Nueda, M.J., Gunsé, B., Lopez-Llorca, L.V., 2020. Chitosan induces plant hormones and defenses in tomato root exudates. Frontiers in Plant Science 11. Available at 10.3389/fpls.2020.572087.

Tashkandi, M., Ali, Z., Aljedaani, F., Shami, A., Mahfouz, M.M., 2018. Engineering resistance against Tomato yellow leaf curl virus via the CRISPR/Cas9 system in tomato. Plant Signaling & Behavior 13 (10), e1525996. Available at https://doi.org/10.1080/15592324.2018.1525996.

Tatineni, S., Hein, G.L., 2023. Plant viruses of agricultural importance: current and future perspectives of virus disease management strategies. Phytopathology® 113, 117–141. Available at https://doi.org/10.3390/cells10112875.

Varanda, C.M.R., Félix, M.R.F., Soares, C.M., et al., 2021. Plant RNA viruses. Critical Reviews in Plant Sciences 40 (3), 199–243. Available at https://doi.org/10.1080/07352689.2021.1898560.

van Butselaar, T., Van den Ackerveken, G., 2020. Salicylic acid steers the growth–immunity tradeoff. Trends in Plant Science 25, 566–576. Available at 10.1016/j.tplants.2020.02.002.

Vu, T.V., Nguyen, N.T., Kim, J., Hong, J.C., Kim, J.Y., 2024. Prime editing: mechanism insight and recent applications in plants. Plant Biotechnology Journal 22 (1), 19–36. Available at https://doi.org/10.1111/pbi.14188.

Wnuk, M., Slipek, P., Dziedzic, M., Lewinska, A., 2020. RNA methyltransferases in viral infection. International Journal of Molecular Sciences 21 (21), 8176. Available at https://doi.org/10.3390/ijms21218176.

Xie, S.S., Duan, C.G., 2023. Epigenetic regulation of plant immunity. Abiotech 4 (2), 124–139. Available at https://doi.org/10.1007/s42994-023-00101-z.

Xie, K., Minkenberg, B., Yang, Y., 2015. Boosting multiplex editing with tRNA processing. PNAS 112 (11), 3570–3575. Available at https://doi.org/10.1073/pnas.1420294112.

Yang, Z., Huang, H., Zhang, Y., et al., 2020. Jasmonate signaling enhances antiviral defense. Nature Communications 11, 2811. Available at https://doi.org/10.1038/s41467-020-16627-3.

Yoon, Y.J., Kim, H., Kim, S.J., 2020. Genome editing of eIF4E1 confers resistance. Frontiers in Plant Science 11, 1098. Available at https://doi.org/10.3389/fpls.2020.01098.

Zaidi, S.S.-e-A., Tashkandi, M., Mansoor, S., Mahfouz, M.M., 2016. Engineering plant immunity by CRISPR/Cas-mediated gene modification. Frontiers in Plant Science 7, 1673. Available at https://doi.org/10.3389/fpls.2016.01673.

Zhang, D., Hussain, A., Manghwar, H., et al., 2020. Genome editing in plants. Plant Cell Reports 39 (10), 1219–1235. Available at https://doi.org/10.1007/s00299-020-02556-5.

Zhang, Y., Ren, Q., Tang, X., et al., 2021. Improving Cas12a editing efficiency. Nature Plants 7 (1), 25–33. Available at https://doi.org/10.1038/s41477-020-00819-3.

Zhan, X., Ding, X.S., 2024. CRISPR/Cas-mediated antiviral strategies in plants. Trends in Plant Science 29 (2), 150–164. Available at https://doi.org/10.1016/j.tplants.2023.09.003.

CHAPTER

10

Biotechnological innovations for sustainable agriculture

Dubbala Prshanthi Divya Vani[1], *Narkhede Gopal Wasudeo*[1], *and Shabir H. Wani*[2]

[1]Department of Genetics and Plant Breeding, School of Agriculture, SR University, Warangal, Telangana, India [2]National Seed Project Laboratory, Sher-e-Kashmir University of Agricultural Sciences and Technology of Kashmir, Srinagar, Jammu and Kashmir, India

10.1 Introduction

Sustainable agriculture is an integrated approach to farming that emphasizes the long-term viability of food production while safeguarding the health of the environment, ensuring economic profitability, and promoting social and ethical responsibility. The term was first popularized in the 1980s in response to growing concerns about the environmental degradation caused by conventional farming practices, including soil erosion, biodiversity loss, water pollution, and the excessive use of chemical fertilizers and pesticides (Canton, 2021; Kamakaula, 2024). The Food and Agriculture Organization (FAO) defines sustainable agriculture as the management and conservation of the natural resource base and the orientation of technological and institutional change in such a manner as to ensure the attainment and continued satisfaction of human needs for present and future generations (Ewert, Baatz, and Finger, 2023). This includes maintaining or enhancing environmental quality, conserving biodiversity, optimizing input use efficiency, and ensuring equitable access to resources and innovations. Sustainable agriculture spans diverse systems including organic farming, conservation agriculture, agro-ecology, regenerative agriculture, and precision farming. These systems are increasingly being supported and enhanced by biotechnological interventions (Arora, 2018; Abobatta and Fouad, 2024).

10.1.1 Role of biotechnology in enhancing agricultural sustainability

Biotechnology, in the context of agriculture, refers to the use of scientific techniques and tools including genetic engineering, molecular markers, tissue culture, and microbial

Harnessing Plant-Stimulating Molecules
DOI: https://doi.org/10.1016/B978-0-443-44994-9.00008-0

technologies to improve the productivity, efficiency, and resilience of agricultural systems (Rather et al., 2025). It plays a transformative role in addressing the major challenges that threaten sustainable agriculture, such as climate change, land degradation, water scarcity, and a growing global population (Abdul Aziz and Masmoudi, 2025; Limenie and Alehegn, 2025; Sibanda, Selvarajan, and Reinhold-Hurek, 2025). Genetic modification and molecular breeding allow the introduction of traits such as high yield, early maturity, and enhanced nutrient use efficiency (Nazir et al., 2023; Mahajan *et al.*, 2025). Biotechnological tools can develop crops that are resistant to drought, salinity, and extreme temperatures, thereby enabling cultivation in marginal environments (Nazir et al., 2024). The use of pest- and disease-resistant genetically modified (GM) crops and microbial biopesticides reduces dependence on synthetic agrochemicals, lowering environmental and health risks. Biofertilizers and plant-growth-promoting rhizobacteria (PGPR) help reduce the need for synthetic fertilizers while improving soil health. Marker-assisted selection (MAS) and genomic selection (GS) facilitate faster and more precise development of superior varieties (Henkrar, Udupa, 2020; Angon et al., 2023; Cai and Dimopoulos, 2025; Nabati et al., 2025). Biotechnology complements and enhances traditional knowledge systems and agroecological approaches, enabling farmers to adopt more sustainable practices without compromising productivity or profitability. Moreover, biotechnology offers solutions at multiple levels—genes, organisms, and ecosystems—that make it a versatile tool in designing climate-resilient and resource-efficient farming systems (Adefila et al., 2024; Alemu et al., 2024; Vikas and Ranjan, 2024).

10.1.2 Historical perspectives and milestones in agri-biotechnology

The evolution of agricultural biotechnology is rooted in humanity's long-standing efforts to improve crop and livestock productivity through selection and breeding. In its earliest form, biotechnology involved the use of fermentation and artificial selection, practiced by ancient civilizations to produce better food and beverages. The 20th century marked the beginning of scientific breeding, with Gregor Mendel's work on inheritance laying the foundation for modern genetics. The post–World War II era, particularly the Green Revolution of the 1960s and 1970s, brought significant agricultural gains through high-yielding varieties, synthetic fertilizers, and chemical pesticides. However, these gains came at the cost of environmental degradation and resource depletion, prompting the need for more sustainable solutions. The modern era of agricultural biotechnology began in the 1970s with the advent of recombinant DNA technology, enabling scientists to insert specific genes into organisms (Chen et al., 2024; Hayat et al., 2025; Riaz et al., 2025; Zhu, Li and Gao, 2020). The 1980s saw the development of GM crops, and by 1994, the first GM crop, the *Flavr Savr* tomato, was commercially released. This was followed by the widespread adoption of Bt cotton, herbicide-tolerant soybean, and virus-resistant papaya. Entering the 21st century, rapid advancements in genomics, molecular markers, and genome editing tools like CRISPR/Cas9 have revolutionized the field, allowing precise and efficient manipulation of plant and microbial genomes. These innovations have not only enhanced crop productivity and resilience but also paved the way for environmentally responsible farming practices, positioning biotechnology as a cornerstone of sustainable agriculture in the modern era (Ansori et al., 2023; Prado et al., 2024; Noack et al., 2024).

10.2 Modern biotechnological tools in agriculture

Modern biotechnological tools in agriculture include genetic engineering, genome editing (e.g., CRISPR/Cas9), MAS, tissue culture, and microbial biotechnology. These tools enhance crop yield, stress tolerance, and resource efficiency, contributing to sustainable farming by reducing chemical inputs, improving genetic gains, and promoting eco-friendly agricultural practices (Singh et al., 2025).

10.2.1 Genetic engineering and transgenic crops

Genetic engineering involves the direct manipulation of an organism's DNA to introduce desirable traits. In agriculture, it enables the development of transgenic crops with enhanced resistance to pests, diseases, herbicides, and environmental stresses. Notable examples include Bt cotton and herbicide-tolerant soybean. These crops reduce reliance on chemical inputs, increase productivity, and support sustainable farming by improving resource use efficiency and minimizing environmental impact (Table 10.1).

10.2.2 Genome editing technologies

Genome editing technologies, such as CRISPR/Cas9, TALENs, and ZFNs, enable precise modification of DNA at specific genomic locations. These tools allow the insertion, deletion, or alteration of genes without introducing foreign DNA. Widely used in crop improvement, they help develop disease-resistant, climate-resilient, and nutrient-rich plants, supporting sustainable agriculture by accelerating breeding and reducing reliance on agrochemicals (Table 10.2).

10.2.3 Marker-assisted selection and genomic selection

MAS uses DNA markers linked to desirable traits to accelerate the selection process in crop breeding, improving efficiency and accuracy for traits like disease resistance and drought tolerance. GS, a more advanced approach, utilizes genome-wide markers to predict the breeding value of individuals without phenotyping each trait. GS is particularly useful for complex, quantitative traits such as yield and stress tolerance. Both MAS and GS reduce breeding cycles, enhance genetic gain, and support the development of climate-resilient and high-yielding crop varieties, making them powerful tools for sustainable and precision agriculture (Hasan et al., 2021) (Table 10.3).

10.2.4 RNA interference and gene silencing

RNA interference (RNAi) is a natural gene-silencing mechanism that regulates gene expression by degrading specific messenger RNA (mRNA) molecules. In agriculture, RNAi is used to silence undesirable genes in pests, pathogens, or plants, offering precise control over traits. It has been successfully applied to develop virus-resistant papaya, insect-resistant maize, and reduced acrylamide potatoes. RNAi offers a nontransgenic alternative to traditional GM approaches and minimizes environmental risks. By targeting specific

TABLE 10.1 Mechanisms of genetic transformation and case studies of transgenic crops.

Aspect	*Agrobacterium*-mediated transformation	References	Gene Gun (Biolistic Transformation)	References
Mechanism	Natural DNA transfer from *Agrobacterium tumefaciens* T-DNA to plant genome	Laforest and Nadakuduti (2022); Rahman *et al.* (2024); Zhong et al. (2024)	Physical delivery of DNA-coated microprojectiles into plant cells	Lacroix and Citovsky (2020); Ozyigit and Yucebilgili Kurtoglu (2020); Hu and Liu (2025)
Plant types	Primarily dicots (e.g., cotton, tomato, and soybean)		Mainly monocots (e.g., maize, rice), also some dicots	
DNA integration	Stable, precise integration with fewer copies		Random integration; may have multiple copies	
Advantages	Efficient, less tissue damage, precise integration		Applicable to wide plant species including cereals	
Limitations	Lower efficiency in monocots, requires wounding		Can cause tissue damage; costly equipment	
Key examples/ case studies	Bt cotton: Insect-resistant cotton expressing *cry* genes		Glyphosate-resistant maize and soybean	
	Golden rice: Beta-carotene-enriched rice for vitamin A deficiency		Insect-resistant rice varieties	

TABLE 10.2 Genome editing tools and their applications in crop improvement.

Genome editing tool	Mechanism	Precision	Ease of use	Applications in crops	References
CRISPR/Cas9	Uses guide RNA to direct Cas9 nuclease to a specific DNA sequence	High	Simple, widely adopted	Drought-tolerant rice (knockout of stress-related genes)	Santosh Kumar et al. (2020)
				High-yield tomato (modification of growth regulators)	Yu et al. (2017)
				Disease-resistant wheat and banana	Xu et al. (2020)
TALENs	Uses customizable transcription activator-like effector domains to bind DNA	Very high	Moderate complexity	Improved shelf life in tomato	Ahmar et al. (2020); Namata et al. (2025)
				Rice resistant to bacterial blight	
				Enhanced oil content in oilseed crops	
ZFNs	Uses engineered zinc-finger proteins fused with nucleases to target DNA	High but less specific	Technically complex	Herbicide-tolerant maize	Gosavi et al. (2022); Yadav et al. (2023)
				Fungal disease resistance in soybean	
				Early flowering in *Arabidopsis*	

TABLE 10.3 Molecular markers and their integration in plant breeding programs.

Feature	SSR (Simple sequence repeats)	SNP (Single nucleotide polymorphisms)	References
Concept	Tandem repeats of 1–6 base pairs in DNA	Single base pair variation at a specific genomic position	Hasan et al. (2021); Testolin et al. (2023); Xu et al. (2025)
Polymorphism level	High	Moderate to high	
Abundance in genome	Moderate	Extremely abundant	
Genomic distribution	Scattered but less uniform	Evenly distributed across the genome	
Detection method	PCR-based, gel or capillary electrophoresis	High-throughput platforms (e.g., SNP arrays, sequencing)	
Cost per assay	Relatively low	Initially high, now reducing with NGS technologies	
Reproducibility	High	Very high	
Automation and throughput	Moderate	High	
Integration in breeding	Marker-assisted selection (MAS)	Genomic selection (GS)	
	QTL mapping	Genome-wide association studies (GWAS)	
	Genetic diversity studies	High-resolution mapping	
Examples in crops	MAS in rice for blast resistance (Pi genes)	GS for drought tolerance in maize	
	Wheat rust resistance	Oil content prediction in soybean	

genes without altering the entire genome, RNAi enhances crop protection, improves quality, and supports sustainable agriculture through reduced chemical usage and enhanced crop resilience (Table 10.4).

10.3 Microbial biotechnology for sustainable farming

Microbial biotechnology harnesses beneficial microorganisms such as biofertilizers, biopesticides, and PGPR to enhance soil fertility, crop productivity, and stress tolerance. These microbes improve nutrient availability, suppress pathogens, and reduce the need for chemical inputs. Integrating microbial solutions into farming systems supports sustainable agriculture by promoting eco-friendly practices, improving soil health, and ensuring long-term agricultural productivity (de Andrade et al., 2023).

TABLE 10.4 RNA interference (RNAi): mechanism and agricultural applications.

Aspect	Details	References
Mechanism	Double-stranded RNA (dsRNA) is processed into small interfering RNAs (siRNAs) by the Dicer enzyme. These siRNAs guide the RNA-induced silencing complex (RISC) to degrade complementary mRNA, silencing target gene expression.	Hernández-Soto and Chacón-Cerdas (2021); Singh et al. (2021); He, Huang, and Tang (2022); Yin et al. (2023); Luo et al. (2024); Basso et al. (2025); Tardin-Coelho et al. (2025)
Target type	mRNA of endogenous plant genes, insect pests, or plant pathogens	
Delivery methods	Transgenic expression in plants, topical application of dsRNA, or host-induced gene silencing (HIGS)	
Application: virus resistance	**Example:** Virus-resistant papaya (against *Papaya ringspot virus* via RNAi of viral coat protein gene)	
Application: pest control	**Example:** RNAi-based maize-targeting *Diabrotica virgifera* (Western corn rootworm), silencing essential pest genes	
Application: trait improvement	**Example:** Low-acrylamide potato (silencing asparagine synthetase gene to reduce carcinogen formation during frying)	
Advantages	High specificity, environmentally friendly, nontoxic, does not involve permanent genome modification	
Limitations	Stability of dsRNA in the environment, delivery challenges, possible off-target effects	

10.3.1 Biofertilizers

Biofertilizers are formulations containing beneficial microorganisms that enhance nutrient availability to plants by fixing nitrogen, solubilizing phosphorus, and mobilizing potassium. Common examples include *Rhizobium*, *Azotobacter*, and *Phosphobacteria*. They improve soil fertility, reduce dependence on chemical fertilizers, and support sustainable, eco-friendly agricultural practices (Table 10.5).

10.3.2 Biopesticides

Biopesticides are natural pest control agents derived from microorganisms, plants, or natural substances. They include bacterial (*Bacillus thuringiensis*), fungal (*Trichoderma*), and viral formulations that target specific pests and pathogens. Biopesticides are

TABLE 10.5 Microbial biofertilizers and their functions.

Type of microorganism	Function	Examples	Application in crops	References
Nitrogen-fixing bacteria (symbiotic)	Fix atmospheric nitrogen in association with leguminous plants	*Rhizobium* spp.	Pulses (e.g., chickpea, soybean, pea)	Wani et al. (2015); Adedeji, Häggblom, and Babalola (2020); Gautam et al. (2022); Kumar (2022); de Andrade et al. (2023)
Nitrogen-fixing bacteria (free-living)	Fix atmospheric nitrogen independently in the rhizosphere	*Azotobacter*, *Azospirillum*	Cereals, vegetables, oilseeds	
Phosphate-solubilizing microorganisms	Convert insoluble phosphorus to plant-available forms	*Pseudomonas*, *Bacillus*, *Aspergillus*	Wheat, maize, rice, horticultural crops	
Potassium-mobilizing microorganisms	Mobilize nonexchangeable potassium into soluble forms	*Frateuria aurantia*, *Bacillus mucilaginosus*	Sugarcane, banana, vegetables	

TABLE 10.6 Biopesticides—*Bacillus thuringiensis* and *Trichoderma spp.*

Biopesticide	Mode of action	Benefits over chemical pesticides	References
Bacillus thuringiensis (Bt)	Produces insecticidal crystal proteins (Cry toxins) that disrupt the gut lining of target insect larvae, causing death.	Highly specific to target pests, safe for humans and nontarget organisms, biodegradable, reduces chemical residues in environment.	Kumar et al. (2021)
Trichoderma spp.	Acts as a mycoparasite by colonizing plant roots and antagonizing soilborne fungal pathogens through competition, antibiosis, and induced systemic resistance.	Controls fungal diseases sustainably, promotes plant growth, reduces chemical fungicide use, and enhances soil health, environmentally friendly.	Alqahtani and Almanzalawi (2025)

environmentally safe, biodegradable, and reduce the need for synthetic chemicals, supporting sustainable and eco-friendly crop protection (Table 10.6).

10.3.3 Microbiome eengineering

Microbiome engineering involves manipulating the plant-associated microbial communities to enhance crop health, growth, and stress resilience. By selecting or introducing beneficial microbes, this approach improves nutrient uptake, disease resistance, and tolerance to abiotic stresses. Advances in metagenomics and synthetic biology enable precise design of microbiomes, promoting sustainable agriculture through natural, eco-friendly methods that reduce chemical inputs and improve soil and plant ecosystem functions (Table 10.7).

10.4 Plant tissue culture and in vitro technologies

Plant tissue culture and in vitro technologies involve the aseptic cultivation of plant cells, tissues, or organs under controlled laboratory conditions. These techniques enable rapid multiplication of disease-free, genetically uniform plants through methods like micropropagation, somatic embryogenesis, and organogenesis. They support conservation of endangered species, production of novel hybrids, and genetic transformation. In vitro approaches facilitate the development of synthetic seeds and secondary metabolite production. By ensuring year-round plant production independent of environmental constraints, tissue culture enhances crop improvement, accelerates breeding programs, and contributes significantly to sustainable agriculture and biotechnology (Chadipiralla et al., 2020; Kumar et al., 2022) (Table 10.8).

TABLE 10.7 Rhizosphere microbiome and synthetic microbial consortia.

Aspect	Details	References
Role of rhizosphere microbiome in plant health	Enhances nutrient availability (N, P, and K). Suppresses soilborne pathogens via competition and antibiosis. Produces plant-growth-promoting hormones (e.g., auxins). Improves plant tolerance to abiotic stresses (drought, salinity, etc.). Supports soil structure and fertility through organic matter decomposition.	Marín, González, and Poupin (2021); Gholap et al. (2024)
Synthetic microbial consortia	Designed combinations of beneficial microbes tailored for specific crops or soils. Synergistic interactions enhance nutrient cycling, pathogen suppression, and stress resilience. Enables stable and predictable performance compared to single strains. Used to bioaugment soils for improved productivity and sustainability.	

10.5 Bioinformatics and data-driven agriculture

Bioinformatics and data-driven agriculture leverage computational tools to analyze biological data and enhance agricultural decision-making. By integrating genomics, transcriptomics, and phenomics data, bioinformatics supports precision breeding, trait mapping, and genome editing. Big data analytics and machine learning enable predictive modeling for yield forecasting, disease outbreaks, and resource optimization. These technologies accelerate crop improvement, reduce trial and error in breeding, and support climate-resilient agriculture. Data-driven approaches promote sustainability by optimizing input use, improving productivity, and ensuring informed, real-time decisions in modern farming systems (Table 10.9).

10.6 Innovations in precision agriculture

Precision agriculture involves the use of advanced technologies to optimize field-level management regarding crop farming. Innovations in precision agriculture include the application of GPS, GIS, remote sensing, drones, Internet of Things (IoT), and automated machinery to monitor and manage variability in crops, soils, and environmental conditions. These tools enable site-specific decisions for irrigation, fertilization, and pest control, thereby increasing efficiency,

TABLE 10.8 Plant tissue culture techniques and applications.

Aspect	Description	Key features/benefits	Applications/examples	References
Micro-propagation	Rapid multiplication of plants from small explants (shoot tips, nodal segments) under sterile conditions.	Produces large numbers of uniform plants Fast propagation Year-round production	Commercial propagation of banana, orchids, sugarcane, and potato	Hussain et al. (2012); Kumar (2022); Gautam et al. (2024)
Somatic embryogenesis	Development of embryos from somatic cells, mimicking zygotic embryogenesis in vitro.	Enables synthetic seed production Regenerates plants from single cells Useful for genetic transformation	Synthetic seed production, propagation of conifers, and coffee	
Germplasm conservation	In vitro preservation of genetic resources via slow growth, cryopreservation, or tissue culture banking.	Long-term storage Space saving Protects against environmental threats	Conservation of rare/endangered species, crop genetic resources	
Clonal propagation	Production of genetically identical plants from a single elite parent using tissue culture techniques.	Maintains genetic fidelity Rapid multiplication Uniform quality	Fruit trees (mango, apple), plantation crops, ornamentals	
Production of pathogen-free planting material	Use of meristem culture and micropropagation to eliminate viruses and pathogens from planting material.	Generates disease-free "clean" plants Improves crop health and yield Reduces chemical usage	Virus-free banana, potato, sugarcane, and citrus plantlets	

TABLE 10.9 Bioinformatics and data-driven tools in sustainable agriculture.

Technology	Role in agriculture	Applications / Benefits	References
Genomics	Study of whole genomes to identify genes associated with key traits.	Marker-assisted selection, genome editing, crop improvement, resistance gene discovery	Wolfert et al. (2017); Caponetti et al. (2018); Kosová et al. (2018); Liakos et al. (2018); Kamilaris and Prenafeta-Boldú (2018); Bhat and Huang (2021); Razzaq et al. (2022); Manickam et al. (2023); Zhang et al. (2024)
Proteomics	Analysis of plant proteins and their functions to understand responses to biotic and abiotic stresses.	Stress tolerance analysis, development of resilient crops, identification of functional proteins	
Metabolomics	Study of plant metabolites for understanding biochemical responses and pathways.	Enhancing nutritional quality, secondary metabolite profiling, stress and defense mechanisms	
Big data analytics	Integration and processing of large datasets from diverse sources (field, lab, sensors).	Yield prediction, decision support systems, precision agriculture, resource optimization	
Artificial intelligence (AI)	Simulation of human intelligence for pattern recognition and decision-making in complex datasets.	Pest and disease detection, soil health monitoring, automated phenotyping	
Machine learning (ML)	Algorithms that learn from data to predict outcomes and uncover hidden patterns.	Genomic prediction, trait mapping, disease forecasting, climate-smart agriculture planning	

reducing input costs, and minimizing environmental impacts. Variable rate technology (VRT) and real-time sensors allow precise application of inputs based on crop needs. Unmanned aerial vehicles (UAVs) provide high-resolution imagery for crop health assessment, while IoT devices continuously monitor soil moisture and weather conditions. Data collected from these technologies feed into AI-driven models to support timely and informed decision-making. As a result, precision agriculture enhances productivity, conserves natural resources, and supports sustainable farming practices by enabling smarter, data-driven interventions (Sishodia, Ray, and Singh, 2020; Iaksch, Fernandes, and Borsato, 2021) (Table 10.10).

10.7 Environmental and socioeconomic impacts

Biotechnological innovations in agriculture offer significant environmental and socioeconomic benefits. Environmentally, they reduce reliance on chemical fertilizers and pesticides, lower greenhouse gas emissions, and promote biodiversity through sustainable practices.

TABLE 10.10 Innovations in precision agriculture and role of biotechnology.

Technology/approach	Description	Applications/benefits	References
Soil and environmental sensors	Devices that measure soil moisture, temperature, pH, and nutrient levels in real time.	Site-specific irrigation and fertilization, improved input efficiency, and reduced waste.	Vullaganti, Ram, and Sun (2025)
Drones/UAVs (unmanned aerial vehicles)	Aerial imaging tools equipped with multispectral and thermal cameras for crop monitoring.	Crop health assessment, early disease/pest detection, plant stand mapping, yield estimation.	Guebsi, Mami, and Chokmani (2024)
IoT (Internet of Things) devices	Networked smart devices that collect and transmit real-time data across farms.	Integrated farm management, automated alerts, and data-driven decision-making.	Rajak et al. (2023)
Variable rate technology (VRT)	Applies inputs (fertilizers, pesticides, seeds) variably based on spatial data.	Reduces costs, minimizes environmental impacts, and increases efficiency.	He (2023)
Biotech: drought-tolerant crops	Genetically engineered crops with enhanced water-use efficiency and root traits.	Adapted to sensor-based irrigation systems, thrive under water-limited conditions.	Xing and Wang (2024)
Biotech: nutrient-efficient crops	Crops modified for better uptake and use of nutrients (e.g., N and P).	Better performance with precision fertilizer application guided by sensors and VRT.	Ali et al. (2025)

Socioeconomically, biotechnology enhances crop yields, food security, and farmer income, especially in resource-limited regions. It enables smallholders to adopt climate-resilient varieties, reducing crop failure risks. However, concerns persist regarding biosafety, intellectual property rights (IPRs), and equitable access to technologies. Addressing these challenges through transparent regulation, public awareness, and inclusive policies ensures that biotechnology contributes positively to both ecological sustainability and rural development (Das et al., 2023; Tyczewska, Twardowski, and Woźniak-Gientka, 2023).

10.7.1 Environmental sustainability

Biotechnology contributes to environmental sustainability by promoting eco-friendly agricultural practices. GM crops with pest and disease resistance reduce the need for chemical pesticides, minimizing soil and water contamination. Biofertilizers and biopesticides enhance soil health and biodiversity, while stress-tolerant crops reduce excessive water and fertilizer use. Tissue culture and precision farming further support resource-efficient cultivation. By lowering greenhouse gas emissions and conserving natural resources, biotechnology aligns with climate-smart agriculture. Additionally, in vitro conservation and microbiome engineering aid in preserving genetic diversity and soil ecosystems, making agriculture more resilient to environmental challenges and climate change (Table 10.11).

TABLE 10.11 Environmental benefits of agricultural biotechnology.

Aspect	Biotechnological intervention	Impact/benefits	Examples	References
Reduction in agrochemical use	Bt crops expressing insecticidal proteins (e.g., Bt cotton)	Decreased use of synthetic pesticides and herbicides	Bt cotton reduces insecticide sprays by ~50%.	Brookes (2022); Chaudhary et al. (2022); Das et al. (2023); Tyczewska, Twardowski, and Woźniak-Gientka (2023); Arora and Fatima (2024)
	Herbicide-tolerant (HT) crops	Reduced chemical runoff and water contamination	Biopesticides like *Bacillus thuringiensis*, *Trichoderma* spp.	
	Use of biopesticides and biofertilizers	Lower input costs for farmers		
Soil health conservation	Use of nitrogen-fixing biofertilizers (*Rhizobium* and *Azotobacter*)	Enhances nutrient cycling and availability	Legume inoculation with *Rhizobium* improves N content.	
	Phosphate solubilizing microorganisms (PSMs)	Maintains natural soil microbial balance	PSMs increase P uptake in cereals and vegetables.	
		Minimizes soil acidification and salinization		
Biodiversity conservation	Microbiome engineering	Preserves genetic resources	Cryopreservation of endangered species.	
	In vitro germplasm conservation	Maintains ecosystem services and beneficial microbial communities	Climate-resilient rice or wheat conserving marginal ecosystems.	
	Stress-resilient crop varieties	Promotes agroecological balance		

TABLE 10.12 Economic benefits, adoption trends, and policy needs in agricultural biotechnology.

Aspect	Details	Examples/case studies	References
Case study: developing countries	Adoption of Bt cotton by smallholder farmers improved yield, reduced pesticide use, and increased income.	India: Bt cotton increased yield by 30%–60%, reduced pesticide costs by 50%, benefiting over 7 million farmers.	Brookes and Barfoot (2020); Daniel et al. (2022); Omari et al. (2024)
	Tissue culture–based banana production improved quality and disease resistance.	Kenya: Smallholder farmers using tissue-cultured bananas achieved 40% higher market value.	
Case study: developed countries	GM crops like herbicide-tolerant soybean and maize-boosted farm efficiency and profits.	USA: Over 90% adoption of GM soybeans; savings of over $10/ acre in weed control costs.	
	Precision agriculture tools coupled with biotech crops optimized input use.	Canada: Combined use of HT canola and GPS-guided VRT improved profitability and sustainability.	
Barriers to adoption	Limited access to biotech seeds and inputs Lack of awareness and technical knowledge Regulatory uncertainty Market and trade restrictions	Smallholder farmers in sub-Saharan Africa face access and affordability issues; public perception remains a hurdle in Europe.	
Policy needs	Strengthening biosafety and regulatory frameworks Farmer training and extension services Public–private partnerships Inclusive access strategies	Policies enabling smallholder access to biotech in countries like Bangladesh (Bt brinjal) show effective results.	

GM, Genetically modified.

10.7.2 Economic benefits and farmer adoption

Biotechnological innovations offer substantial economic benefits by enhancing crop productivity, reducing input costs, and improving resource use efficiency. GM and improved varieties enable higher yields, better resistance to pests and stress, and reduced need for agrochemicals, translating into greater profitability for farmers. Technologies like tissue culture and biofertilizers promote cost-effective and sustainable farming practices. Smallholder farmers benefit from increased income stability and reduced crop losses. Widespread farmer adoption is driven by demonstrable economic returns, training programs, and access to biotech products. However, equitable access, regulatory support, and market acceptance remain critical for broader and inclusive technology adoption (Table 10.12).

TABLE 10.13 Regulatory framework and public perception of GM crops.

Aspect	Description	Examples/notes	References
GM crop regulations	Strict biosafety assessments before approval	USA: Product-based regulation, faster approvals	Rippe and Willemsen (2018); Hrbáčková et al. (2020); Dessie and Zegeye (2024)
	Field trials and environmental impact studies	EU: Process-based, precautionary principle, slower approvals	
	Labeling and traceability requirements	India: Regulatory body GEAC (Genetic Engineering Appraisal Committee)	
	Varying regulatory stringency across countries		
Public perception and acceptance	Influenced by knowledge, cultural beliefs, and trust in authorities	Higher acceptance in Americas and parts of Asia	
	Concerns over food safety, environmental risks, and corporate control	Resistance is stronger in Europe and some African countries	
	Education and transparent communication improve acceptance	NGOs and media shape public opinion significantly	

10.7.3 Ethical, legal, and social implications

Ethical, legal, and social implications (ELSI) of agricultural biotechnology address concerns about biosafety, IPRs, and equitable access. Ethical debates focus on GM crop safety, environmental impacts, and the rights of farmers versus corporations. Legal issues involve patenting of biotech traits, regulatory approvals, and trade policies. Social implications include public acceptance, cultural values, and the impact on smallholder farmers. Transparent regulation, stakeholder engagement, and public awareness are essential to balance innovation benefits with societal concerns. Addressing ELSI fosters responsible technology deployment, builds trust, and ensures that biotechnological advances contribute fairly to sustainable agriculture (Singh, 2016) (Table 10.13).

10.8 Future perspectives and challenges

As the global population grows and climate change intensifies, the role of biotechnology in sustainable agriculture will become even more critical. Emerging technologies such as synthetic biology and next-generation plant factories offer revolutionary tools for transforming agricultural systems. Synthetic biology enables the design and construction of novel biological pathways and organisms with enhanced traits, such as crops capable of fixing atmospheric nitrogen or synthesizing essential nutrients. Plant factories, including vertical and controlled-environment agriculture, can be engineered to produce high-value compounds, pharmaceuticals, or food with minimal environmental impact, independent of soil and weather conditions (Avgoustaki and Xydis, 2020; Ndudzo et al., 2024; Goold, Wright, and Hailstones, 2018). Climate-resilient agriculture will rely heavily on biotechnology to develop crops that withstand extreme weather, pests, and declining soil fertility. Drought-tolerant, salt-tolerant, and heat-resistant crop varieties developed through genetic engineering, genome editing (like CRISPR), and MAS will help stabilize yields and secure food supplies in vulnerable regions (Shamseldin, 2022). However, realizing the full potential of these innovations requires overcoming several challenges. Regulatory frameworks often lag behind scientific advancements, resulting in inconsistent or overly cautious policies that delay the deployment of safe and beneficial technologies. IPRs can restrict access to biotech innovations for smallholder farmers and public-sector researchers, especially in developing countries. Balancing innovation incentives with equitable access remains a pressing concern (Kumar et al., 2024). Furthermore, global disparities in infrastructure, investment, and technical capacity limit the adoption of biotechnology in low-income regions. Public skepticism, fueled by misinformation and lack of transparency, continues to influence biotech policy and acceptance. To move forward, international cooperation is essential to harmonize regulations, build public trust through transparent risk assessments, and ensure fair benefit-sharing. Encouraging open-source biotechnology, strengthening biosafety capacity, and fostering inclusive innovation will be crucial to ensuring that future biotechnological advances contribute equitably to global food and environmental security (Piergentili et al., 2021; Ibrahim et al., 2023; Al-Tawaha et al., 2022; Vatistas, Avgoustaki, and Bartzanas, 2022).

10.9 Conclusion

Biotechnological innovations are transforming agriculture into a more sustainable, productive, and resilient system. From genetic engineering and genome editing to microbial biotechnology, tissue culture, and data-driven farming, these tools offer immense potential to address pressing global challenges such as food insecurity, climate change, and environmental degradation. By reducing dependency on agrochemicals, enhancing stress tolerance, and improving crop yields and quality, biotechnology contributes significantly to sustainable agricultural practices. However, realizing the full benefits requires supportive regulatory frameworks, equitable access, and public engagement. Ethical, legal, and social considerations must be addressed transparently to build trust and ensure responsible innovation. The integration of synthetic biology, climate-smart crops, and precision agriculture will define the next frontier of agricultural biotechnology. A collaborative global effort involving scientists, policymakers, farmers, and the public is essential to harness biotechnology's potential for inclusive and sustainable development. With continued innovation and inclusive policy support, biotechnology can play a pivotal role in securing the future of agriculture.

References

Abdul Aziz, M., Masmoudi, K., 2025. Molecular breakthroughs in modern plant breeding techniques. Horticultural Plant Journal 11 (1), 15–41. Available at https://doi.org/10.1016/j.hpj.2024.01.004.

Abobatta, W.F., Fouad, F.W., 2024. Sustainable Agricultural Development: Introduction and Overview Achieving Food Security Through Sustainable Agriculture. IGI Global, Egypt, pp. 1–27. Available at https://www.igi-global.com/book/achieving-food-security-through-sustainable/338377, https://doi.org/10.4018/979-8-3693-4240-4.ch001.

Adedeji, A.A., Häggblom, M.M., Babalola, O.O., 2020. Sustainable agriculture in Africa: plant growth-promoting rhizobacteria (PGPR) to the rescue. Scientific African 9. Available at https://doi.org/10.1016/j.sciaf.2020.e00492, https://www.journals.elsevier.com/scientific-african.

Adefila, A.O., Ajayi, O.O., Toromade, A.S., Sam-Bulya, N.J., 2024. Integrating traditional knowledge with modern agricultural practices: a socio-cultural framework for sustainable development. World Journal of Biology, Pharmacy and Health Sciences 20 (2), 125–135.

Ahmar, S., Saeed, S., Khan, M.H.U., Khan, S.U., Mora-Poblete, F., Kamran, M., et al., 2020. A revolution toward gene-editing technology and its application to crop improvement. International Journal of Molecular Sciences 21 (16), 1–28. Available at https://doi.org/10.3390/ijms21165665, https://www.mdpi.com/1422-0067/21/16/5665/pdf.

Al-Tawaha, A.R.M.S., Mrabet, R., Bayanati, M., Santhosh, B., Benkeblia, N., Imran, et al., 2022. Islam, Role of biotechnology in climate-resilient agriculture Climate Change and Agriculture: Perspectives, Sustainability and Resilience. wiley, Jordan, pp. 78–94. Available at https://www.wiley.com/en-us/Climate+Change+and+Agriculture:+Perspectives,+Sustainability+and+Resilience-p-9781119789772, https://doi.org/10.1002/9781119789789.ch4.

Alemu, A., Åstrand, J., Montesinos-López, O.A., Isidro y Sánchez, J., Fernández-Gónzalez, J., Tadesse, W., et al., 2024. Genomic selection in plant breeding: key factors shaping two decades of progress. Molecular Plant 17 (4), 552–578. Available at https://doi.org/10.1016/j.molp.2024.03.007, https://www.sciencedirect.com/science/journal/16742052.

Ali, A., Jabeen, N., Farruhbek, R., Chachar, Z., Laghari, A.A., Chachar, S., et al., 2025. Enhancing nitrogen use efficiency in agriculture by integrating agronomic practices and genetic advances. Frontiers in Plant Science 16. Available at https://doi.org/10.3389/fpls.2025.1543714, https://www.frontiersin.org/journals/plant-science.

Alqahtani, T.M., Almanzalawi, E.A., 2025. Trichoderma spp. as bio-pesticides: exploring diverse modes of action. Romanian Agricultural Research 2025 (42), 433–548. Available at https://doi.org/10.59665/rar4246, https://www.incda-fundulea.ro/rar/nr42/rar42.46.pdf.

de Andrade, L.A., Santos, C.H.B., Frezarin, E.T., Sales, L.R., Rigobelo, E.C., 2023. Plant growth-promoting rhizobacteria for sustainable agricultural production. Microorganisms 11 (4). Available at https://doi.org/10.3390/microorganisms11041088, www.mdpi.com/journal/microorganisms.

Angon, P.B., Mondal, S., Akter, S., Sakil, M.A., Jalil, M.A., 2023. Roles of CRISPR to mitigate drought and salinity stresses on plants. Plant Stress 8. Available at https://doi.org/10.1016/j.stress.2023.100169, www.journals.elsevier.com/plant-stress.

Ansori, A.N.M., Antonius, Y., Susilo, R.J.K., Hayaza, S., Kharisma, V.D., Parikesit, A.A., et al., 2023. Application of CRISPR-Cas9 genome editing technology in various fields: a review. Indonesia Narra J 3 (2). Available at https://doi.org/10.52225/narra.v3i2.184, https://narraj.org/main/article/view/184/143.

Arora, N.K., 2018. Biodiversity conservation for sustainable future. Environmental Sustainability 1 (2), 109–111. Available at https://doi.org/10.1007/s42398-018-0023-1.

Arora, N.K., Fatima, T., 2024. Environmental sustainability and biotechnology: opportunities and challenges. Environmental Sustainability 7 (2), 115–119. Available at https://doi.org/10.1007/s42398-024-00317-9.

Avgoustaki, D.D., Xydis, G., 2020. Plant factories in the water-food-energy Nexus era: a systematic bibliographical review. Food Security 12 (2), 253–268. Available at https://doi.org/10.1007/s12571-019-01003-z, http://www.springer.com/life+sci/agriculture/journal/12571.

Basso, M.F., Vásquez, D.D.N., Campos-Pinto, E.R., Pinheiro, D.H., Cruz, B., Maktura, G.C., et al., 2025. Progress and opportunities of in planta and topical RNAi for the biotechnological control of agricultural pests. Agronomy 15 (4). Available at https://doi.org/10.3390/agronomy15040859, http://www.mdpi.com/journal/agronomy/.

Bhat, S.A., Huang, N.F., 2021. Big data and AI revolution in precision agriculture: survey and challenges. IEEE Access 9, 110209–110222. Available at https://doi.org/10.1109/ACCESS.2021.3102227, http://ieeexplore.ieee.org/xpl/RecentIssue.jsp?punumber=6287639.

Brookes, G., 2022. Genetically modified (GM) crop use 1996–2020: environmental impacts associated with pesticide use change. GM Crops & Food 13 (1), 262–289. Available at https://doi.org/10.1080/21645698.2022.2118497.

Brookes, G., Barfoot, P., 2020. Environmental impacts of genetically modified (GM) crop use 1996–2018: impacts on pesticide use and carbon emissions. GM Crops & Food 11 (4), 215–241. Available at https://doi.org/10.1080/21645698.2020.1773198.

Cai, P., Dimopoulos, G., 2025. Microbial biopesticides: a one health perspective on benefits and risks. One Health 20, 100962. Available at https://doi.org/10.1016/j.onehlt.2024.100962.

Canton, H., 2021. Food and Agriculture Organization of the United Nations—FAO. The Europa directory of international organizations. Routledge, pp. 297–305.

Caponetti, J.D., Gray, D.J., Trigiano, R.N., 2018. History of plant tissue and cell culture. Plant tissue culture concepts and laboratory exercises. CRC Press, pp. 11–17.

Chadipiralla, K., Gayathri, P., Rajani, V., Reddy, P.V.B., 2020. Plant tissue culture and crop improvement. In: Roychowdhury, In. R., Choudhury, S., Hasanuzzaman, M., Srivastava, S. (Eds.), Sustainable agriculture in the era of climate change. Springer, Cham, pp. 391–412.

Chaudhary, P., Singh, S., Chaudhary, A., Sharma, A., Kumar, G., 2022. Overview of biofertilizers in crop production and stress management for sustainable agriculture. Frontiers in Plant Science 13. Available at https://doi.org/10.3389/fpls.2022.930340.

Chen, F., Chen, L., Yan, Z., Xu, J., Feng, L., He, N., et al., 2024. Recent advances of CRISPR-based genome editing for enhancing staple crops. Frontiers in Plant Science 15. Available at https://doi.org/10.3389/fpls.2024.1478398.

Daniel, A.I., Fadaka, A.O., Gokul, A., Bakare, O.O., Aina, O., Fisher, S., et al., 2022. Biofertilizer: the future of food security and food safety. Microorganisms 10 (6). Available at https://doi.org/10.3390/microorganisms10061220, https://www.mdpi.com/2076-2607/10/6/1220/pdf?version=1655217795.

Das, S., Ray, M.K., Panday, D., Mishra, P.K., 2023. Role of biotechnology in creating sustainable agriculture. PLOS Sustainability and Transformation 2 (7). Available at https://doi.org/10.1371/journal.pstr.0000069, https://journals.plos.org/sustainabilitytransformation/.

Dessie, A., Zegeye, Z., 2024. Review on: public perception of biotechnology on genetically modified crops, bio policy and intellectual property rights. American Journal of Polymer Science and Technology 10 (2), 26–35. Available at https://doi.org/10.11648/j.ajpst.20241002.11.

Ewert, F., Baatz, R., Finger, R., 2023. Agroecology for a sustainable agriculture and food system: from local solutions to large-scale adoption. Annual Review of Resource Economics 15 (1), 351–381. Available at https://doi.org/10.1146/annurev-resource-102422-090105.

Gautam, S.S., Gondwal, M., Gautam, B.P.S., Soni, R., 2022. Potash biofertilizers: Current development, formulation, and applications Trends of Applied Microbiology for Sustainable Economy. Elsevier, India, pp. 481–500. Available at https://www.sciencedirect.com/book/9780323915953, https://doi.org/10.1016/B978-0-323-91595-3.00014-8.

Gautam, Y., Mishra, R., Bishnoi, L., Bhardwaj, S., Mohbe, G., Sharma, A., 2024. Application of plant tissue culture in crop improvement. In Advances in plant biotechnology (Vol. 1, pp. 93–114). Cornous Publications LLP.

Gholap, A.D., Pardeshi, S.R., Khuspe, P.R., Sayyad, S.F., Chavan, M., Hatvate, N.T., et al., 2024. Manipulating the rhizosphere microbiome for plant health. Microbiome Engineering: The New Dimension of Biotechnology. CRC Press, India, pp. 175–193. Available at https://www.routledge.com/Microbiome-Engineering-The-New-Dimension-of-Biotechnology/Srivastava-Ibrahim-Nasr/p/book/9781032479606, https://doi.org/10.1201/9781003394662-13.

Goold, H.D., Wright, P., Hailstones, D., 2018. Emerging opportunities for synthetic biology in agriculture. Genes 9 (7), 341. Available at https://doi.org/10.3390/genes9070341.

Gosavi, G., Ren, B., Li, X., Zhou, X., Spetz, C., Zhou, H., 2022. A new Era in herbicide-tolerant crops development by targeted genome editing. ACS Agricultural Science & Technology 2 (2), 184–191. Available at https://doi.org/10.1021/acsagscitech.1c00254.

Guebsi, R., Mami, S., Chokmani, K., 2024. Drones in precision agriculture: a comprehensive review of applications, technologies, and challenges. Drones 8 (11), 686. Available at https://doi.org/10.3390/drones8110686.

Hasan, N., Choudhary, S., Naaz, N., Sharma, N., Laskar, R.A., 2021. Recent advancements in molecular marker-assisted selection and applications in plant breeding programmes. Journal of Genetic Engineering and Biotechnology 19 (1). Available at https://doi.org/10.1186/s43141-021-00231-1, https://jgeb.springeropen.com/.

Hayat, M.K., Khan, T.Y., Sial, T.N., Khan, H.A., Ashraf, Z.U., Sheraz, M., et al., 2025. The biotech and precision agriculture revolution: feeding the world sustainably. Policy Research Journal 3 (3), 293–316.

He, L., 2023. Variable Rate Technologies for Precision Agriculture Encyclopedia of Digital Agricultural Technologies. Springer International Publishing, United States, pp. 1533–1542. Available at https://link.springer.com/referencework/10.1007/978-3-031-24861-0, https://doi.org/10.1007/978-3-031-24861-0_34.

Henkrar, F., Udupa, S., 2020. Marker assisted selection in plant breeding. Moroccan Journal of Agricultural Sciences 1 (5), 237–247.

Hernández-Soto, A., Chacón-Cerdas, R., 2021. RNAi crop protection advances. International Journal of Molecular Sciences 22 (22), 12148. Available at https://doi.org/10.3390/ijms222212148.

He, L., Huang, Y., Tang, X., 2022. RNAi-based pest control: production, application and the fate of dsRNA. Frontiers in Bioengineering and Biotechnology 10. Available at https://doi.org/10.3389/fbioe.2022.1080576.

Hrbáčková, M., Dvořák, P., Takáč, T., Tichá, M., Luptovčiak, I., Šamajová, O., et al., 2020. Biotechnological Perspectives of omics and genetic engineering methods in alfalfa. Frontiers in Plant Science 11. Available at https://doi.org/10.3389/fpls.2020.00592, https://www.frontiersin.org/journals/plant-science.

Hussain, A., Qarshi, I.A., Nazir, H., Ullah, I., 2012. Plant tissue culture: Current status and opportunities. Recent advances in plant in vitro culture. IntechOpen.

Hu, M., Liu, D., 2025. Unlocking the potential of genome editing in agriculture with tissue culture-free techniques. New Phytologist 246 (4), 1478–1484. Available at https://doi.org/10.1111/nph.70046.

Iaksch, J., Fernandes, E., Borsato, M., 2021. Digitalization and Big data in smart farming – a review. Journal of Management Analytics 8 (2), 333–349. Available at https://doi.org/10.1080/23270012.2021.1897957.

Ibrahim, M., Nabeel, M., Raza, M.Q., Hameed, N., Rafiq, R., Zaheer, M., et al., 2023. The role of technology and innovation in enhancing food security. Journal of Physical and Biomedical Biological Sciences 2023 (1), 14 –14.

Kamakaula, Y., 2024. Sustainable Agriculture practices: economic, ecological, and social approaches to enhance farmer welfare and environmental sustainability. West Science Nature and Technology 2 (02), 47–54. Available at https://doi.org/10.58812/wsnt.v2i02.964.

Kamilaris, A., Prenafeta-Boldú, F.X., 2018. Deep learning in agriculture: a survey. Computers and Electronics in Agriculture 147, 70–90. Available at https://doi.org/10.1016/j.compag.2018.02.016, http://www.elsevier.com/inca/publications/store/5/0/3/3/0/4.

Kosová, K., Vítámvás, P., Urban, M.O., Prášil, I.T., Renaut, J., 2018. Plant abiotic stress proteomics: the major factors determining alterations in cellular proteome. Frontiers in Plant Science 9. Available at https://doi.org/10.3389/fpls.2018.00122, http://journal.frontiersin.org/article/10.3389/fpls.2018.00122/full.

Kumar, N., 2022. Biotechnology and Crop Improvement: Tissue Culture and Transgenic Approaches. CRC Press. Available at https://doi.org/10.1201/9781003239932.

Kumar, R., Das, S.P., Choudhury, B.U., Kumar, A., Prakash, N.R., Verma, R., et al., 2024. Advances in genomic tools for plant breeding: harnessing DNA molecular markers, genomic selection, and genome editing. Biological Research 57 (1). Available at https://doi.org/10.1186/s40659-024-00562-6, https://biolres.biomedcentral.com/.

Kumar, P., Kamle, M., Borah, R., Mahato, D.K., Sharma, B., 2021. Bacillus thuringiensis as microbial biopesticide: uses and application for sustainable agriculture. Egyptian Journal of Biological Pest Control 31 (1). Available at https://doi.org/10.1186/s41938-021-00440-3, https://link.springer.com/journal/41938.

Kumar, S., Sindhu, S.S., Kumar, R., 2022. Biofertilizers: an eco friendly technology for nutrient recycling and environmental sustainability. Current Research in Microbial Sciences 3, 100094.

Lacroix, B., Citovsky, V., 2020. Biolistic approach for transient gene expression studies in plants. Methods in Molecular Biology 2124, 125–139. Available at https://doi.org/10.1007/978-1-0716-0356-7_6, http://www.springer.com/series/7651.

Laforest, L.C., Nadakuduti, S.S., 2022. Advances in delivery mechanisms of CRISPR gene-editing reagents in plants. Frontiers in Genome Editing 4. Available at https://doi.org/10.3389/fgeed.2022.830178, https://www.frontiersin.org/journals/genome-editing/about.

Liakos, K., Busato, P., Moshou, D., Pearson, S., Bochtis, D., 2018. Machine learning in agriculture: a review. Sensors (Basel, Switzerland) 18 (8), 2674. Available at https://doi.org/10.3390/s18082674.

Limenie, A.D., Alehegn, M., 2025. Genetic engineering for cereal crop yield improvement and disease resistant breeding. Scientific World Journal 2025 (1). Available at https://doi.org/10.1155/tswj/6743917, https://onlinelibrary.wiley.com/journal/8086.

Luo, X., Nanda, S., Zhang, Y., Zhou, X., Yang, C., Pan, H., 2024. Risk assessment of RNAi-based biopesticides. New Crops 1, 100019. Available at https://doi.org/10.1016/j.ncrops.2024.100019.

Mahajan, M., Poor, P., Kaur, H., Aher, R.R., Palakolanu, S.R., Khan, M.I.R., 2025. Salt stress tolerance and abscisic acid in plants: associating role of plant growth regulators and transcription factors. Plant Physiology and Biochemistry 228. Available at https://doi.org/10.1016/j.plaphy.2025.110303, https://www.sciencedirect.com/science/journal/09819428.

Manickam, S., Rajagopalan, V.R., Manoharan, B., Natesan, S., & Muthurajan, R. (2023). Genomics-assisted breeding for climate-resilient crops. In Molecular marker techniques: A potential approach of crop improvement (pp. 357–368). Springer Nature Singapore.Kamakaula, Y., 2024. Sustainable agriculture practices: economic, ecological, and social approaches to enhance farmer welfare and environmental sustainability. West Sci. Nat. Technol. 2(02), 47-54.

Marín, O., González, B., Poupin, M.J., 2021. From Microbial dynamics to functionality in the rhizosphere: A systematic review of the opportunities with synthetic microbial communities. Frontiers in Plant Science 12. Available at https://doi.org/10.3389/fpls.2021.650609, https://www.frontiersin.org/journals/plant-science.

Nabati, J., Nezami, A., Yousefi, A., Oskoueian, E., Oskoueian, A., Ahmadi-Lahijani, M.J., 2025. Biofertilizers containing plant growth promoting rhizobacteria enhance nutrient uptake and improve the growth and yield of chickpea plants in an arid environment. Scientific Reports 15 (1). Available at https://doi.org/10.1038/s41598-025-93070-w, https://www.nature.com/srep/.

Namata, M.J., Xu, J., Habyarimana, E., Palakolanu, S.R., Wang, L., Li, J., 2025. Genome editing in maize and sorghum: a comprehensive review of CRISPR/Cas9 and emerging technologies. Plant Genome 18 (2). Available at https://doi.org/10.1002/tpg2.70038, https://acsess.onlinelibrary.wiley.com/journal/19403372.

Nazir, F., Khan, L., Umar, S., Khan, M.I.R., 2024. Omics-based strategies for improving salt tolerance in rice. Current Omics Advancement in Plant Abiotic Stress Biology. Academic Press, pp. 63–70.

Nazir, F., Mahajan, M., Khatoon, S., Albaqami, M., Ashfaque, F., Chhillar, H., et al., 2023. Sustaining nitrogen dynamics: a critical aspect for improving salt tolerance in plants. Frontiers in Plant Science 14. Available at https://doi.org/10.3389/fpls.2023.1087946, https://www.frontiersin.org/journals/plant-science.

Ndudzo, A., Sibanda Makuvise, A., Moyo, S., Bobo, E.D., 2024. CRISPR-Cas9 genome editing in crop breeding for climate change resilience: Implications for smallholder farmers in Africa. Journal of Agriculture and Food Research 16. Available at https://doi.org/10.1016/j.jafr.2024.101132. Available from: https://www.sciencedirect.com/science/journal/26661543.

Noack, F., Engist, D., Gantois, J., Gaur, V., Hyjazie, B.F., Larsen, A., et al., 2024. Environmental impacts of genetically modified crops. Science 385 (6712). Available at https://doi.org/10.1126/science.ado9340, https://www.science.org/toc/science/385/6712.

Omari, E.N., Mucheru-Muna, M., Mburu, B.K., Odawa, A.A., 2024. Assessing the advantages of tissue culture bananas technology production of banana farmers in kisii county, Kenya. Environmental Challenges 14, 100843. Available at https://doi.org/10.1016/j.envc.2024.100843.

Ozyigit, I.I., Yucebilgili Kurtoglu, K., 2020. Particle bombardment technology and its applications in plants. Molecular Biology Reports 47 (12), 9831–9847. Available at https://doi.org/10.1007/s11033-020-06001-5, http://www.springer.com/journal/11033.

Piergentili, R., Del Rio, A., Signore, F., Umani Ronchi, F., Marinelli, E., Zaami, S., 2021. CRISPR-cas and its wide-ranging applications: from human genome editing to environmental implications, technical limitations, hazards and bioethical issues. Cells 10 (5), 969. Available at https://doi.org/10.3390/cells10050969.

Prado, G.S., Rocha, D.C., Santos, L.N.dos, Contiliani, D.F., Nobile, P.M., Camargo Martinati-Schenk, J., et al., 2024. CRISPR technology towards genome editing of the perennial and semi-perennial crops citrus, coffee and sugarcane. Frontiers in Plant Science 14. Available at https://doi.org/10.3389/fpls.2023.1331258.

Rahman, S.U., Khan, M.O., Ullah, R., Ahmad, F., Raza, G., 2024. Agrobacterium-mediated transformation for the development of transgenic crops; present and future prospects. Molecular Biotechnology 66 (8), 1836–1852. Available at https://doi.org/10.1007/s12033-023-00826-8, https://www.springer.com/journal/12033.

Rajak, P., Ganguly, A., Adhikary, S., Bhattacharya, S., 2023. Internet of things and smart sensors in agriculture: scopes and challenges. Journal of Agriculture and Food Research 14, 100776. Available at https://doi.org/10.1016/j.jafr.2023.100776.

Rather, B.A., Xu, W., Tantray, A.Y., Mahajan, M., Sun, H., Cong, H., et al., 2025. Biotechnological advances in sanguinarine and chelerythrine production from plume poppy (*Macleaya cordata*): a gene editing perspective. Plants 14 (17). Available at https://doi.org/10.3390/plants14172667, http://www.mdpi.com/journal/plants.

Razzaq, A., Wishart, D.S., Wani, S.H., Hameed, M.K., Mubin, M., Saleem, F., 2022. Advances in metabolomics-driven diagnostic breeding and crop improvement. Metabolites 12 (6). Available at https://doi.org/10.3390/metabo12060511, https://www.mdpi.com/2218-1989/12/6/511/pdf?version=1654164954.

Riaz, M., Yasmeen, E., Saleem, B., Hameed, M.K., Saeed Almheiri, M.T., Saeed Al Mir, R.O., et al., 2025. Evolution of agricultural biotechnology is the paradigm shift in crop resilience and development: a review. Frontiers in Plant Science 16. Available at https://doi.org/10.3389/fpls.2025.1585826, https://www.frontiersin.org/journals/plant-science.

Rippe, K.P., Willemsen, A., 2018. The idea of precaution: ethical requirements for the regulation of new biotechnologies in the environmental field. Frontiers in Plant Science 871. Available at https://doi.org/10.3389/fpls.2018.01868, https://www.frontiersin.org/journals/plant-science.

Santosh Kumar, V.V., Verma, R.K., Yadav, S.K., Yadav, P., Watts, A., Rao, M.V., et al., 2020. CRISPR-Cas9 mediated genome editing of drought and salt tolerance (OsDST) gene in indica mega rice cultivar MTU1010. Physiology and Molecular Biology of Plants 26 (6), 1099–1110. Available at https://doi.org/10.1007/s12298-020-00819-w, http://www.springerlink.com/content/120962/.

Shamseldin, A., 2022. Future outlook of transferring biological nitrogen fixation (BNF) to cereals and challenges to retard achieving this dream. Current Microbiology 79 (6). Available at https://doi.org/10.1007/s00284-022-02852-2.

Sibanda, T., Selvarajan, R., Reinhold-Hurek, B., 2025. Enhancing crop resilience to drought and salinity: the potential role of extremophiles in mitigating food insecurity in sub-Saharan Africa. Journal of Crop Health 77 (1). Available at https://doi.org/10.1007/s10343-024-01090-9.

Singh, J., Cobb-Smith, D., Higgins, E., Khan, A., 2021. Comparative evaluation of lateral flow immunoassays, LAMP, and quantitative PCR for diagnosis of fire blight in apple orchards. Journal of Plant Pathology 103 (S1), 131–142. Available at https://doi.org/10.1007/s42161-020-00644-w.

Singh, K.K., 2016. Intellectual property rights in agricultural biotechnology and access to technology: a critical appraisal. Reasearch and Infornmation Syatem for Developing Countries RIS. India Asian Biotechnology and Development Review 18 (3), 3–23. Available from: http://www.ris.org.in.

Singh, M., Singh, R.P., Patra, A., Singh, H., Ram, R.M., 2025. Application and recent trends of biotechnology in sustainable agriculture: a brief review. Journal of Advances in Biology & Biotechnology 28 (1), 846–858. Available at https://doi.org/10.9734/jabb/2025/v28i11939.

Sishodia, R.P., Ray, R.L., Singh, S.K., 2020. Applications of remote sensing in precision agriculture: a review. Remote Sensing 12 (19), 1–31. Available at https://doi.org/10.3390/rs12193136, https://www.mdpi.com/2072-4292/12/19/3136/pdf.

Tardin-Coelho, R., Fletcher, S., Manzie, N., Gunasekara, S.N., Fidelman, P., Mitter, N., et al., 2025. A systematic review on public perceptions of RNAi-based biopesticides: developing social licence to operate. npj Sustainable Agriculture 3 (1). Available at https://doi.org/10.1038/s44264-025-00057-1, https://www.nature.com/npjsustainagric/.

Testolin, R., Messina, R., Cipriani, G., De Mori, G., 2023. SSR-based DNA fingerprinting of fruit crops. Crop Science 63 (2), 390–459. Available at https://doi.org/10.1002/csc2.20896.

Tyczewska, A., Twardowski, T., Woźniak-Gientka, E., 2023. Agricultural biotechnology for sustainable food security. Trends in Biotechnology 41 (3), 331–341. Available at https://doi.org/10.1016/j.tibtech.2022.12.013.

Vatistas, C., Avgoustaki, D.D., Bartzanas, T., 2022. A systematic literature review on controlled-environment agriculture: how vertical farms and greenhouses can influence the sustainability and footprint of urban microclimate with local food production. Atmosphere 13 (8). Available at https://doi.org/10.3390/atmos13081258, http://www.mdpi.com/journal/atmosphere.

Vikas, Ranjan, R., 2024. Agroecological approaches to sustainable development. Frontiers in Sustainable Food Systems 8. Available at https://doi.org/10.3389/fsufs.2024.1405409.

Vullaganti, N., Ram, B.G., Sun, X., 2025. Precision agriculture technologies for soil site-specific nutrient management: a comprehensive review. Artificial Intelligence in Agriculture 15 (2), 147–161. Available at https://doi.org/10.1016/j.aiia.2025.02.001, https://www.sciencedirect.com/science/journal/25897217.

Wani, F.S., Ahmad, L., Ali, T., Mushtaq, A., 2015. Role of microorganisms in nutrient mobilization and soil health - a review. Journal of Pure and Applied Microbiology 9 (2), 1401–1410. Available at http://www.microbiologyjournal.org/.

Wolfert, S., Ge, L., Verdouw, C., Bogaardt, M.J., 2017. Big data in smart farming – a review. Agricultural Systems 153, 69–80. Available at https://doi.org/10.1016/j.agsy.2017.01.023, http://www.elsevier.com/inca/publications/store/4/0/5/8/5/1.

Xing, Y., Wang, X., 2024. Precision agriculture and water conservation strategies for sustainable crop production in arid regions. Plants 13 (22), 3184. Available at https://doi.org/10.3390/plants13223184.

Xu, Y., Chen, S., Chen, S., Wei, X., Shang, H., Zhang, Q., et al., 2025. Genome-wide development of SSR molecular markers for modern sugarcane cultivars. Frontiers in Plant Science 16. Available at https://doi.org/10.3389/fpls.2025.1573967.

Xu, X., Yuan, Y., Feng, B., Deng, W., 2020. CRISPR/Cas9-mediated gene-editing technology in fruit quality improvement. Food Quality and Safety 4 (4), 159–166. Available at https://doi.org/10.1093/fqsafe/fyaa028.

Yadav, R.K., Tripathi, M.K., Tiwari, S., Tripathi, N., Asati, R., Chauhan, S., et al., 2023. Genome editing and improvement of abiotic stress tolerance in crop plants. Life 13 (7). Available at https://doi.org/10.3390/life13071456, www.mdpi.com/journal/life/.

Yin, J., Zhao, H., Wu, X., Ma, Y., Zhang, J., Li, Y., et al., 2023. SSR marker based analysis for identification and of genetic diversity of non-heading Chinese cabbage varieties. Frontiers in Plant Science 14. Available at https://doi.org/10.3389/fpls.2023.1112748.

Yu, Q.H., Wang, B., Li, N., Tang, Y., Yang, S., Yang, T., et al., 2017. CRISPR/Cas9-induced targeted mutagenesis and gene replacement to generate long-shelf life tomato lines. Scientific Reports 7 (1). Available at https://doi.org/10.1038/s41598-017-12262-1, www.nature.com/srep/index.html.

Zhang, X., Ibrahim, Z., Khaskheli, M.B., Raza, H., Zhou, F., Shamsi, I.H., 2024. Integrative approaches to abiotic stress management in crops: combining bioinformatics educational tools and artificial intelligence applications. Sustainability 16 (17), 7651. Available at https://doi.org/10.3390/su16177651.

Zhong, H., Li, C., Yu, W., Zhou, Hp, Lieber, T., Su, X., et al., 2024. A fast and genotype-independent in planta Agrobacterium-mediated transformation method for soybean. Plant Communications 5 (12). Available at https://doi.org/10.1016/j.xplc.2024.101063, https://www.sciencedirect.com/science/journal/25903462.

Zhu, H., Li, C., Gao, C., 2020. Applications of CRISPR–cas in agriculture and plant biotechnology. Nature Reviews Molecular Cell Biology 21 (11), 661–677. Available at https://doi.org/10.1038/s41580-020-00288-9.

CHAPTER

11

Mobile signals in systemic stress response: RNA, peptides, and hormonal interactions

Vaishnavi Tripathi[1], *Gargi Bhardwaj*[1], *Sunil Kumar Jaiswal*[1], *and Sunil Kumar Yadav*[1,2]

[1]School of Bioscience and Technology, Galgotias University, Greater Noida, Uttar Pradesh, India
[2]Department of Bioscience and Biotechnology, Banasthali Vidyapith, Niwai, Rajasthan, India

11.1 Introduction

Plants are the living but nonmotile organisms, which, during their lifespan, undergo many stress factors (Mahajan et al., 2025a). Plant stress factors are mainly categorized into two main groups: abiotic factors and biotic factors. The abiotic factors include different environmental factors such as water scarcity (drought), flood, extreme temperatures (hot or cold), salinity (extreme salt), light, and heavy metal toxicity (Khatoon et al., 2025b), while the biotic factors are the other organisms that share the common environment and interact with the plants such as pests and pathogens. The extreme environmental conditions under abiotic factors cause abiotic stresses on plants and can have a devastating impact on growth and productivity, potentially resulting in the collapse of the entire ecosystems (Zandalinas et al., 2020). These abiotic stresses cause various physiological and molecular changes in plants through specific sensing mechanisms, depending on the type of stress. The resulting signal transduction cascade often relies on similar signaling modules (Mahajan et al., 2023; Zhang et al., 2020). Generally, plants are unable to overcome or tolerate stress conditions by just changing their environmental conditions like body temperature and other stress stimuli. Response to such stresses usually involves complex molecular signaling mechanisms, including alteration in gene expression and regulatory networks.

Common part of these responses involves stress-induced systemic signaling and systemic acquired acclimation, which are essential for plant adaptation to environmental stress (Zandalinas et al., 2020). It involves a wide range of mobile signals such as RNAs, peptides,

Harnessing Plant-Stimulating Molecules
DOI: https://doi.org/10.1016/B978-0-443-44994-9.00009-2

and hormones. Interestingly, these mobile signals not only help plants to overcome stresses but also play critical roles in regulating other essential life processes like growth and development.

11.1.1 Stress responses in plants

Plants encounter a variety of stresses, known as biotic and abiotic types, which include pathogenic infections, herbivore damage, extreme temperatures, drought, salinity, and heavy metals, and negatively impact overall fitness (Mahajan et al., 2025b). In response to such challenges, plants have evolved highly regulated systems to sense, signal, and adapt to various biotic and abiotic stresses (Kashtoh et al., 2025). One of the fundamental characteristics of plant stress responses is their ability to perceive and distinguish between various stress signals. Plants acquire specialized receptors and sensors that can detect changes in environmental conditions in the form of stresses such as extreme temperature, salinity, drought, and the presence of pathogens or herbivores (Nazir et al., 2023; Kumari et al., 2025). Upon perceiving these stress signals, plants initiate a cascade of molecular, cellular, and physiological responses aimed at minimizing damage and enhancing their chances of survival or death (Li et al., 2024a).

11.1.2 Local versus systemic response

Plants have both local and systemic stress responses against biotic and abiotic factors. Local stress response occurs at or near the site of stress, like wounding and pathogen infection. Local stress response is triggered by immediate detection of stress signals such as mechanical damage, pathogen invasion, or abiotic stress (like salinity or heat) (Du et al., 2024). It produces signaling molecules like reactive oxygen species (ROS), calcium ions (Ca^{2+}), salicylic acid (SA), or jasmonic acid (JA), depending on the type of stress. Local systemic stress leads to the strengthening of cell walls, expression of defense-related genes, and accumulation of secondary metabolites. However, systemic stress refers to the plant response that affects the whole plant instead of certain parts. So systemic stress response spreads throughout the plant, even to parts that are not directly affected by the initial stress. It is involved in long-distanced signaling and prepares unaffected tissues for future stress. Systemic stress responses involve the production and movement of various mobile signaling molecules, including electrical signals, peptides/proteins, hormonal signals, small RNAs (sRNAs), and hydraulic or ROS waves (Zandalinas et al., 2020). It leads to systemic acquired resistance (SAR) against pathogens and systemic induced resistance, which is triggered by beneficial microbes and activation of defense genes in remote tissues.

11.1.3 Plant immune system

Plants have an ability to respond against various types of infections and this ability is known as plant immunity. During biotic stress, especially against pathogenic infections, plants respond with its natural defense system known as innate immune response, which is essential for its survival. Plant innate immune system relies on two interconnected layers: PTI (pattern-triggered immunity) and ETI (effector-triggered immunity) (Jones and Dangl,

2006). PTI is the first line of defense against pathogens in plants. It is activated when a plant cell detects microbe-associated molecular patterns (MAMPs), pathogen-associated molecular patterns (PAMPs), or damage-associated molecular patterns (DAMPs). This triggers the cascade signaling in plants, which usually involves MAP kinase pathway. This signaling pathway leads to the activation of various defense mechanisms in plants, like antimicrobial compounds production, strengthening of cell walls, and apoptosis in the infected area to prevent areas near it. ETI is a specific defense mechanism of plants used when pathogens secrete virulence factor or effector (Yuan et al., 2021).

ETI is generally activated when pathogen effectors successfully suppress or evade PTI, thereby requiring a stronger, more targeted immune response. It acts as backup immunity in plants when PTI is ineffective against pathogens. Activation of ETI typically results in a rapid, robust defense that includes hypersensitive response (HR), which is a localized programmed cell death at infection sites to restrict pathogen spread, a surge in SA and ROS to signal and mediate further defense processes, and transcriptional reprogramming to activate a broad set of defense genes, including *pathogenesis-related* (PR) genes. ETI can trigger SAR, conferring long-lasting, broad-spectrum immunity to distal tissues in the plant (Yuan et al., 2021).

11.2 Role of immunity in stress signaling

Plants' innate immune system helps them to sense and respond against both harmful organisms (like bacteria and insects) as well as tough environmental conditions (like drought or heat). Unlike animals, plants do not have immune cells that move around, so each plant cell must protect itself and also communicate with other parts of the plant. When stress occurs, plants use special receptors on their cell surfaces or inside cells to detect danger signals such as molecules from pathogens (PAMPs) or damaged tissues (DAMPs). This recognition activates internal signaling pathways involving molecules like Ca^{2+}, ROS, and plant hormones (Yuan et al., 2021). These signals help the plant build defenses at the local site of stress and also alert distant parts of the plant through mobile signal elements to prepare for possible trouble, a process known as systemic signaling. Key hormones like SA, JA, and abscisic acid (ABA) through mobile signaling help the plant to balance its defense and growth, depending on the type of stress (Aerts et al., 2021). This whole system allows plants not just to defend themselves immediately, but also to "prime" or prepare other parts of the plant to respond more quickly if the same problem occurs again.

11.2.1 Importance of mobile signals

Generally, plants utilize a wide array of signal molecules, e.g., ions, phytohormones, proteins/peptides, metabolites, and various forms of RNAs, to mediate local and systemic responses to various types of stimuli. In this regard, sRNAs are important signaling molecules that function in the regulation of both plant developmental processes and abiotic/biotic stress responses. These sRNAs are 21–24 nucleotides in size and can be categorized into three major groups: small interfering RNAs (siRNAs), microRNAs (miRNAs), and transfer RNA–derived fragments (Yan and Ham, 2022). Signaling peptides (SPs), also known as "peptide hormones," are emerging growth hormones. These are small biological molecules of the proteome present in plants (usually less than 20 amino acids when matured and rarely more than 120 amino acids as a full-length precursor). SPs occur in very low concentrations at the physiological level. SPs play an essential role in plant reproduction, growth, and development processes, including homeostasis maintenance, stomatal aperture control, and the development of seeds, shoots, roots, pollen, and yield, through specific mobile cell-to-cell communication. SPs and receptor kinase–mediated cell-to-cell communication play important roles in the initial stages of plant growth (Pandita et al., 2023). Plant hormones are also important mobile signal molecules, which are biosynthesized within a plant. These hormones include auxins (IAA), gibberellins (GA), cytokinin (CK), ABA, JA, and SA, which are synthesized at different locations and are transported to the site of action when they are required or have some stressors encountered.

The responses to biotic stressors, such as pathogens and herbivores, often involve the activation of defense mechanisms. These mechanisms can include the production of defensive chemicals, such as phytochemicals and secondary metabolites, the reinforcement of cell walls, and the induction of programmed cell death (apoptosis) at the site of infection. Additionally, plants can emit volatile organic compounds to attract predators or parasitoids that can help mitigate herbivore attacks. In contrast, abiotic stress responses involve adaptations to adverse environmental conditions that are nonliving in nature. For example, during drought stress, plants may close stomata to reduce water loss, accumulate osmolytes to maintain cell turgor, and modify root architecture to explore deeper soil layers in search of moisture. When exposed to extreme temperatures, plants may undergo changes in membrane lipid composition, produce heat shock proteins (HSPs), and engage in antioxidant defense mechanisms to protect cellular structures (Li et al., 2024b).

11.3 Small RNAs as mobile signals

sRNAs, as mobile signaling elements, play major roles in the regulation of plant growth, development, reproduction, and biotic and abiotic stress responses. sRNAs are typically 21–24 nucleotides in length and classified into miRNA and siRNA (Yan and Ham, 2022; Zhao and Zheng, 2025). sRNAs in plants are generated through several pathways from either endogenous or exogenous sources and processed further from longer RNA precursors through the activity of ribonuclease III–like nucleases called Dicer proteins. The main difference between these two classes of sRNA is their point of origin: miRNAs are encoded by their own genes, which are distinct from recognized protein-coding genes, whereas siRNAs are produced from long double-stranded RNA (dsRNA) precursors arising from mRNAs, transposons, viruses, or heterochromatic DNA. These sRNAs particularly down-regulate expression of genes by targeting mRNA and this mechanism is known as gene/RNA silencing.

11.3.1 Mechanisms of small RNA movement

One of the most fascinating aspects of gene silencing found in plants is its mobile nature. Due to small size, sRNAs possess the remarkable ability to travel both short and long distances. It has an ability to spread from the cell where it has been produced to neighboring cells. This movement is facilitated by plasmodesmata (the pores interconnecting plant cells) for cell-to-cell communication and phloem for systemic transport (Yan and Ham, 2022). However, upon local induction, RNA silencing spreads from the site of initiation to the surrounding 10–15 neighboring cells through plasmodesmata. Occasionally, RNA silencing can be amplified through conversion of target RNAs into new dsRNA by the RNA-dependent RNA polymerase RDR6 and the production of secondary siRNAs (Maida and Masutomi, 2011). This amplification process, known as "transitivity," results in a more extensive cell-to-cell spread of silencing through reiterated short-distance signaling events. Dicer-like 4 (DCL4) protein was identified in plants for this short-range silencing movement, suggesting a key role of DCL4 in the generation of 21-nucleotide-long siRNAs. Subsequently, these siRNAs were directly identified as the cell-to-cell silencing signal by using a combination of transgenic reporter systems (Dunoyer et al., 2013; Yan and Ham, 2022). In addition to DCL4, several mutations in the RNA-directed DNA methylation pathway compromised cell-to-cell silencing movement.

Mobile sRNAs in plants are generated through several pathways from either endogenous or exogenous sources. During plant development, sRNA movement establishes gradients of target mRNA concentrations used in specifying developmental patterns. For example, movement of endogenous siRNAs from the top of the leaves, where they are produced, to the bottom cell layers, generates a gradient that functions in the establishment of top-to-bottom leaf pattern formation (Yan and Ham, 2022). Certain miRNAs also act as mobile RNA and coordinate plant development process. For instance, miR165/166 is produced in specific root cells and moves to adjacent cells, where it targets mRNAs encoding multiple transcription factors. The resulting gradient of target mRNA determines vascular cell types. Interestingly, some miRNAs move long distances in plants. During phosphate starvation, miRNA mir399 is systemically transported from the top of the plant to the root, where it degrades its target mRNA that encodes a suppressor of phosphate uptake (Dunoyer et al., 2013).

11.3.2 Role of small RNAs in abiotic stress tolerance in plants

Plants face numerous environmental adversities every day, which affect them physiologically, leading to delayed emergence of seeds, stunted plant growth, reduced rates of photosynthesis, oxidative stress, etc. Abiotic stress factors lead to alterations in expression profiles of the various genes. In plants, the miRNAs are significantly involved in the posttranscriptional regulation of genes related to various stress tolerance.

11.3.2.1 Drought stress tolerance

Drought stress is one of the most common environmental challenges faced by plants, leading to reduced photosynthetic efficiency and significant yield losses. sRNAs, particularly miRNAs, act as key mobile signaling molecules that coordinate drought stress responses at the cellular and systemic levels. Seven conserved and six novel miRNAs were identified in mustard (*Brassica juncea)*, which are upregulated under abiotic stress conditions, among which two (Bju-N21 and Bju-N29) specifically respond to drought stress. Similarly, 11 miRNAs were reported in cowpea that exhibit altered expression during drought conditions (Li et al., 2024b; Zhang et al., 2022b). These stress-induced miRNAs function by repressing negative regulators of drought tolerance or by tuning metabolic and developmental pathways. For instance, miRNAs such as miR394, miR408, and miR2118, when overexpressed in *Arabidopsis* (*Arabidopsis thaliana)*, rice (*Oryza sativa*), creeping bentgrass (*Agrostis stolonifera*), chickpea (*Cicer arietinum*), and tobacco (*Nicotiana tabacum*), enhance drought tolerance (Liang et al., 2024; Zhang et al., 2022b). These miRNAs commonly target transcription factors (e.g., MYB, NAC, and HD-ZIP), enzymes regulating ROS detoxification, and genes linked to osmolyte biosynthesis, thereby strengthening protective mechanisms against dehydration. Moreover, tissue-specific studies in rice have shown that both the upregulation and downregulation of different miRNAs contribute to drought tolerance, indicating that the movement of miRNAs between different tissues helps the plant adjust its stress response.

11.3.2.2 Salinity stress tolerance

Similar to drought stress, salinity stress significantly limits crop productivity by disturbing physiology and metabolism (Khatoon et al., 2025a,c). Various miRNAs were found regulating salt stress tolerance by targeting TFs, ion transporters, and stress-related genes at posttranscriptional levels. For example, in Arabidopsis and maize, salt stress induces the expression of miR396, which regulates growth-related TFs, while in rice, abnormal expression of miR169 alters expression of NFYA, a transcription factor involved in stress adaptation. In mustard, various miRNAs such as miR395_2, miR390_1, miR172_2, Bju-N29, and Bju-N21 display differential expression patterns under saline conditions, indicating their role in tissue-specific signaling and systemic stress regulation (Zhang et al., 2022b; Samynathan et al., 2023). Similarly, in *Populus trichocarpa*, reduced expression of miR530a, miR1445, miR1446a-e, and miR1447 has been observed in response to salt stress (Zhang et al., 2022b).

In cotton, salt-sensitive LM-6 genotypes exhibit a distinct regulatory pattern, characterized by upregulation of miR156, miR169, miR535, and miR827, as well as downregulation of miR167, miR397, and miR399. It highlights how sRNAs modulate both tolerance- and sensitivity-related mechanisms (Samynathan et al., 2023). Transgenic approaches have further confirmed the importance of miRNAs in conferring salinity tolerance. Overexpression of osaMIR393 in Arabidopsis enhances salt tolerance by modulating auxin signaling through TIR1 repression, while the expression of mTIR1 (a miR393-resistant *TIR1* gene) improves tolerance by balancing auxin perception (Wang et al., 2013). Similarly, gma-miR172 in soybean and osa-miR528 in creeping bentgrass act as systemic regulators and control expression of TFs and redox-related genes to improve salt tolerance and germination rates (Samynathan et al., 2023). By targeting TFs, hormone pathways, and genes linked to ion homeostasis and oxidative stress detoxification, sRNA coordinates adaptive mechanisms that enhance salt stress tolerance.

11.3.2.3 Temperature stress tolerance

To cope with temperature stress including both cold and heat stress, plants utilize various miRNAs as mobile signaling molecules that move between cells and tissues to regulate stress responses at the systemic level. Temperature stress severely disturbs plant growth by impairing photosynthesis, enzyme activity, and water balance.

Under cold stress, several miRNAs have been reported to act as systemic regulators. In rice, 18 miRNAs were found to be downregulated, while miR812q was upregulated in response to cold stress (Li et al., 2024b; Zhang et al., 2022b). In turnip rape (*Brassica rapa*), Bra-novel-miR3936-5p, miR319e, and miR166m-2 were identified as overexpressed under cold conditions. Similarly, in trifoliate orange (*Poncirus trifoliata*), miR396b was reported regulating growth and defense signaling to enhance cold tolerance (Zhang et al., 2022b). Transgenic studies further confirmed that OsmiR156 and ScmiR393 significantly improved cold stress tolerance by regulating development pathways and hormonal signaling in Arabidopsis (Wang et al., 2024). Large-scale profiling in tea plants (*Camellia sinensis*) has revealed upregulation of 31 miRNAs and downregulation of 43 miRNAs in response to cold stress (Qian et al., 2024). Similarly, in tomato (*Solanum lycopersicum*), 192 miRNAs were upregulated and 205 miRNAs were downregulated under cold stress (Qian et al., 2024).

Similarly, high temperature triggers a distinct miRNA-mediated systemic response. In wheat (*Triticum aestivum*), 12 miRNAs have been identified as heat-responsive, while in sunflower (*Helianthus annuus*), miR396 expression is downregulated under heat stress, influencing growth-regulating TFs. Furthermore, in switch grass (*Panicum virgatum*), seven miRNA families were reported to be downregulated under high-temperature conditions. These studies highlighted that miRNAs function as systemic messengers and mobilized to integrate stress perception with systemic adaptive responses (Chaudhary et al., 2021). By targeting TFs such as NYFA, MYB, HD-ZIP, and hormone signaling pathways, specially auxin and ABA, miRNAs ensure that plants can cope with stress conditions (Zhang et al., 2022b).

11.3.2.4 Hypoxia and oxidative stress

Hypoxia is a condition when there is a prevalence of low oxygen concentration in the environment. It might happen due to waterlogging and scarcity of oxygen in the roots. This can alter the mode of respiration from aerobic to anaerobic and cause severe damage to the plants. Similarly, oxidative stress, commonly triggered by the overproduction of ROS, disrupts cellular homeostasis and impairs growth. By regulating TFs, ROS-detoxifying enzymes, and metabolic pathways at the posttranscriptional level, miRNAs help plants maintain redox balance and improve survival under stress (Zhang et al., 2022b).

In Arabidopsis, hypoxia induces the upregulation of several miRNAs, including miR156g, miR157d, miR158a, miR159a, miR172a/b, miR391, and miR775. These miRNAs function as mobile messengers, targeting transcription factors such as SPL, AP2, and MYB, which control developmental reprogramming, stress signaling, and carbohydrate metabolism (Li et al., 2022; Chaudhary et al., 2021). This systemic regulation ensures that energy resources are diverted toward anaerobic survival strategies. Under oxidative stress, miR398 plays a particularly critical role. In Arabidopsis, miR398 is downregulated during ROS accumulation, which relieves repression of superoxide dismutase (CSD1 and CSD2) enzymes. This upregulation of antioxidant enzymes enhances the plant's capacity to scavenge ROS and limit cellular damage (Li et al., 2022).

Similarly, in rice seedlings, hydrogen peroxide treatment alters the expression of several mobile miRNAs. miR169, miR397, miR827, and miR1425 are upregulated, whereas miR528 is downregulated (Samynathan et al., 2023). These miRNAs regulate transcription factors such as NFYA, enzymes involved in lignin biosynthesis, and components of nutrient homeostasis, ultimately helping the plant adjust metabolism, strengthen cell walls, and maintain ROS detoxification capacity.

11.3.3 Role of small RNAs in biotic stress tolerance in plants

sRNAs are critical regulators of plant immune responses under biotic stress as well, somewhat similar to abiotic stress, where they control defense-related genes, hormone signaling, and responses during pathogen infections. Various findings confirm that both miRNAs and siRNAs as mobile signals are essential for tuning plant immune signaling and conferring resistance to diverse pathogens. For example, miR393 is one of the first plant miRNAs shown to be induced by the bacterial PAMP flg22 in Arabidopsis, where it represses the auxin receptor TIR1, thereby strengthening PTI (Samynathan et al., 2023). Similarly, miR160, miR167, and miR393 are highly induced by *Pseudomonas syringae*, which regulate auxin signaling by targeting auxin receptor genes, thereby restricting pathogen proliferation (Jiang et al., 2023). Also, miR398 was observed downregulated during infection, releasing Cu/Zn superoxide dismutases (CSD1 and CSD2) from repression, which enhances ROS detoxification and strengthens defense (Samynathan et al., 2023; Jiang et al., 2023). Certain mutants deficient in miRNAs like dcl1 and hen1 exhibit increased susceptibility to bacterial pathogens and allow enhanced bacterial growth (Khraiwesh et al., 2012). This indicates that miRNAs are crucial regulators of plant defense, as they modulate immune-related gene expression and strengthen resistance against infection. Beyond bacteria, miRNAs are also involved in other pathogen infections such as viruses, fungi, and nematodes. In case of viral infections, miR156 and miR164 are induced by turnip yellow mosaic virus and tobacco mosaic virus, reshaping host gene regulation to mount effective antiviral defenses in *Brassica* sp. (Khraiwesh et al., 2012). In fungal infections also, a large number of sRNAs are involved. Three specific miRNAs, miR482, miR1444, and miR1448, were found to regulate plant biotic and abiotic stress tolerance by targeting polyphenol oxidase genes and disease resistance–related genes (*R* genes) during fungal infections (Khraiwesh et al., 2012). Powdery mildew infections in wheat modulate a variety of miRNAs like miR156, miR159, miR164, miR171, miR396, miR393, miR444, and miR827. These miRNAs, particularly miR393, have been shown to be involved in basal defense mechanisms related to auxin signaling pathways by downregulating the transport inhibitor response 1 (TIR1), which is crucial for disease resistance during fungal infections (Samynathan et al., 2023).

Similarly, in nematode infection, various miRNAs such as miR161, miR164, miR167a, miR172c, miR396a/b, and miR398a are modulated to enhance defense responses (Khraiwesh et al., 2012). Alongside miRNAs, siRNAs also play vital roles in immune response during biotic stress. The first identified biotic stress–related siRNA (nat-siRNAATGB2) is induced by *P. syringae* and strengthens ETI (Jiang et al., 2023). Moreover, these miRNAs and siRNAs influence SAR by acting as mobile signals that spread defense information throughout the plant. They regulate pathways involving SA, JA, and ethylene (ET), which are central to plant immune signaling (Chaudhary et al., 2021).

11.4 Peptides as mobile signals

Peptide molecules involved in mobile signaling are composed of short chains of amino acids ranging from 3 to 200 residues. They function as crucial regulatory molecules that coordinate many physiological processes in plants. Peptide signaling is well known for its role in regulating plant growth and development, and it is particularly important under stress circumstances, where diverse classes of peptides are produced to activate specific signaling pathways linked to environmental adaptation and defense (Skripnikov, 2023). Peptide signaling is a vital mechanism of cellular communication in plants, regulating developmental control and allowing responses to environmental stimuli and pathogenic threats. These peptides modulate intercellular signaling networks that coordinate various physiological outcomes, including organogenesis, immunological responses, and tolerance to abiotic stress.

11.4.1 Classification of plant peptides

11.4.1.1 Secreted peptides

Secreted peptides function as intercellular messengers and are crucial for cell-to-cell communication in plant development and stress responses. Examples include CLE9, CLE25, IDA, ProPSK, PSK, CEP5, CAP-TAC1, RALF, IDL6, IDL7, T20K, and thionin. These peptides regulate essential developmental activities, including root and shoot morphogenesis, nodulation, pollen tube guidance, pollen development, and stomatal movement. Moreover, many released peptides significantly contribute to the mediation of responses to environmental stress and plant immunity (Chang and Xiao, 2025). An exemplary instance is systemin, an SP in tomato, which is produced as a precursor (pro-systemin) and later converted into its mature form. Mature systemin triggers jasmonate-mediated signaling pathways to evoke defense responses against herbivores and pathogens, therefore enhancing resilience to both biotic and abiotic stresses (Cirillo et al., 2022).

11.4.1.2 Nonsecreted peptides

Nonsecreted peptides operate intracellularly and in general encoded by short open reading frames (sORFs). These peptides have distinct regulatory roles, such as plant immunity, development, especially root architecture, and cellular responses to stress. The CYSTM family of nonsecreted peptides is linked with responses to abiotic stress (Xu et al., 2018). ROTUNDIFOLIA4 (ROT4) belongs to a family of Arabidopsis nonsecreted small peptides that acts as a regulator of plant immunity by modulating the signaling pathways (Li et al., 2024b). These peptides are involved in the exact modulation of gene expression at both the transcriptional and translational stages, underscoring their importance in post-transcriptional control under fluctuating environmental conditions.

11.4.2 Peptides role in growth and development

Plant peptides are vital in regulating various physiological and developmental processes, including cell proliferation, stem cell maintenance, and organogenesis. Generally, plant peptides perform their functions by interacting specifically with membrane-bound receptors, by that activating intracellular signaling cascades. The CLE peptides represent a well-defined family that regulate stem cell homeostasis in the shoot apical meristem, hence preserving the equilibrium between stem cell proliferation and differentiation. CLE peptides are involved in several facets of shoot development, including vascular patterning, shoot proliferation, leaf development, and stomatal differentiation. CLV3, a prototypical CLE peptide, acts as a crucial regulatory element in the shoot apical meristem. CLV3 promotes stem cell differentiation, whereas CLV1 acts downstream to regulate this response (Ren et al., 2024). CLE peptides act as negative regulators of root growth, while they positively govern the directionality and patterning of root development. However, the molecular processes by which CLE peptides engage with their corresponding receptors and inhibit root development remain partially elucidated (Kim et al., 2021; Kashtoh et al., 2025).

Phytosulfokine (PSK) and plant peptide–containing sulfated tyrosine (PSY) are sulfated peptides that promote cellular proliferation and growth. These peptides are crucial for tissue growth and regeneration by triggering certain receptor kinases and consequent signaling pathways (Yamada and Hsiao, 2021).

Root meristem growth factor (RGF) peptides, or golven (GLV) peptides, are important for root development. They sustain the stem cell niche in the root meristem by composing positional signaling gradients that regulate meristem growth and cellular differentiation. RGF peptides positively modulate root meristem development by accelerating the stability of plethora (PLT1/2) transcription factors through ROS. The RGF–RGFR–RITF1 signaling module is involved in an ROS-dependent cascade that regulates PLT1/2 proteins at the translational level. Still, the specific downstream regulatory components in this pathway are yet to be discovered. ROS have emerged as crucial regulators in maintaining the equilibrium between cell division and differentiation in the root meristem, subsequently influencing root architecture (Yamada and Hsiao, 2021).

The rapid alkalinization factor (RALF) peptide family participates in calcium-mediated signaling and regulates root development. RALF peptides cooperate with receptor-like kinases, such as FERONIA (FER), and alter cell expansion and general plant development by several mechanisms, such as Ca^{2+} signaling, MAP kinase cascades, and pH control. The exact role of pH in RALF-mediated signaling is ambiguous; however, it is postulated that variations in pH could alter the binding affinity of RALF peptides to their receptors, thereby altering the effectiveness of signal transduction (Yamada and Hsiao, 2021; Ren et al., 2024).

Members of the epidermal patterning factor (EPF) peptide family are important regulators of stomatal development and patterning. EPF1 and EPF2 act as negative regulators of stomatal development, preventing excessive stomatal production. Conversely, stomagen (also known as EPFL9) acts as a positive regulator and operates as an antagonist to EPF1/2. stomagen is expressed in the leaf mesophyll and its increased expression results in profound stomatal density (Xiong et al., 2022; Li et al., 2025).

Notably, the concurrent overexpression of both CLE and EPF family peptides has been shown to curb the development of many organs, such as roots, shoots, stomata, and other plant structures, underscoring the precisely maintained equilibrium necessary for normal organogenesis. Altogether, these peptides demonstrate the complex peptide-mediated signaling networks that govern plant development by synthesizing positional information, hormone signals, and environmental stimuli.

11.4.3 Peptide signaling in abiotic stress

Peptide signaling is also important in plant responses to stress. In response to various biotic and abiotic challenges, plants initiate various signaling pathways that promote protection and survival. Within the CLE peptide family, various members are either elevated or downregulated in response to many situations (Kim et al., 2021). The CLE45–SKM1/SKM2 signaling pathway promotes pollen functionality and seed development in conditions of high-temperature stress. In apple (*Malus domestica*), elevated temperatures facilitate the expression of MdCLE6/21 and MdCLE18, but in grapevine (*Vitis vinifera*), VvCLE6 is likewise increased (Wang et al., 2025; Zhang et al., 2022a). On the contrary, cold stress induces the expression of MdCLE6/21, MdCLE10, and MdCLE16, while reducing MdCLE17 levels (Wang et al., 2025). CLE25 participates in drought response by promoting stomatal closure. Moreover, NtCLE3, which was discovered in tobacco, has showed the capacity to emend drought tolerance in three *Solanaceae* species: tobacco, pepper (*Capsicum annuum*), and eggplant (*Solanum melongena*) (Wang et al., 2025). The increased expression of OSS1Fa1, a peptide derived from rice, markedly increases drought tolerance in Arabidopsis. Similarly, the small paraquat resistance protein (SPQ), particularly LcSPQ, which possesses an N-terminal signal peptide, has showed an augmentation of drought resistance when overexpressed in plants (Li et al., 2025). The peptide PIP3, accelerated by PAMPs, specifically binds receptor-like kinase 7 (RLK7), a receptor characterized by leucine-rich repeat motifs, especially in conditions of salt stress. The activation of RLK7 by PIP3 boosts salt stress signaling by initiating the MPK3/6 pathway, therefore facilitating ion homeostasis and augmenting salt tolerance. Moreover, AtPEP3, a hormone-mimicking peptide, augments salt tolerance by its interaction with the PEPR1 receptor (Li et al., 2025). In response to salt stress, enhanced cellular damage triggers the production of AtPEP3 that activates consequent defensive signaling pathways, therefore enhancing plant resilience to salinity. The specific mechanisms governing AtPEP3-mediated signal transduction in salt stress tolerance are predominantly unidentified. Pathogen-induced AtPEP3 triggers both PEPR1 and PEPR2 receptors, but salt-induced AtPEP3 primarily interacts with PEPR1, attributing that salt stress may generate conformational alterations in PEPR2 that involve additional investigation (Cao et al., 2025). CYSTM, a newly discovered cysteine-rich transmembrane protein module, promotes stress tolerance in certain eukaryotic systems. CYSTM proteins in plants govern tolerance to heat and UV light. CYSTM3 has been found as a negative regulator of salt stress tolerance in Arabidopsis. It reduces salt tolerance by impeding Na^+ efflux, decreasing ROS-scavenging enzyme activity, and downregulating the expression of salt-responsive genes, so undermining the plant's overall capacity to tolerate salinity stress (Ren et al., 2024). Nitrogen and phosphorus are crucial elements that affect root formation. In Arabidopsis, under nitrogen-deficient conditions, nitrogen-responsive CLE peptides, including AtCLE1, AtCLE3, AtCLE4, and AtCLE7, accumulate in the root vasculature, regulating lateral root development in an N-dependent way

(Yamada and Hsiao, 2021; Ren et al., 2024; Cao et al., 2025). The overexpression of these genes stops lateral root emergence during nitrogen deprivation, whereas clv1 mutants display an augmented quantity and length of lateral roots, along with heightened transcript levels of AtCLE2/3/4/7 (Yamada and Hsiao, 2021; Kim et al., 2021). This demonstrates a feedback regulatory mechanism that involves the CLV1 receptor. In instances of phosphate deficit, AtCLE14 expression is increased in the root apical meristem and promotes root meristem termination via the CLV2 and PEPR2 signaling pathways under phosphorus-limiting circumstances (Cao et al., 2025). On the contrary, in several plant species like Miyakogusa *(Lotus japonicas)*, purple false brome *(Brachypodium distachyon)*, and barrel medic *(Medicago truncatula)*, specific CLE genes are activated in response to increased phosphate (Pi) concentrations (Nakagami et al., 2024; Wang et al., 2016). In barrel medic, MtCLE33 is upregulated under elevated Pi circumstances in roots, which results in a MtSUNN-mediated suppression of strigolactone biosynthesis genes, resulting in reduced strigolactone levels relative to untreated controls (Nakagami et al., 2024). Strigolactones are essential phytohormones that regulate plant responses to phosphorus deprivation by modulating root and shoot architecture and promoting plant–microbe interactions. Consequently, phosphorus-regulated CLE signals such as MtCLE33 may alter developmental and symbiotic responses, connecting internal phosphorus levels with adaptive signaling.

Moreover, IMA1 and IMA3 peptides enhance tolerance to cadmium (Cd) toxicity by promoting iron (Fe) accumulation. Cadmium exposure additionally increases the expression of BrCEP8 and BrCEP19 peptides, showing that BrCEPs promote cadmium uptake and provide amplified cadmium tolerance in Brassicaceae species (Wang et al., 2023).

11.4.4 Peptide signaling in response to biotic stress

Biotic stress negatively regulates plant growth and development by introducing pathogens including fungi, bacteria, viruses, and nematodes. Fungal diseases may be necrotrophic or biotrophic, which results in vascular wilts, leaf spots, and cankers. Nematodes, which consume plant tissues, mostly cause soilborne illnesses, resulting in nutrient shortages, stunted development, and wilting. Viruses cause local and systemic infections, showing as chlorosis and stunted growth. Plants have a passive initial defense mechanism comprising structural barriers, including the cuticle, wax layers, and trichomes, which prevent the access of pathogens and insects. They also synthesize antimicrobial chemicals or peptides to encounter microbial invasion. Apart from this, plant-derived small peptides also play multifaceted roles in plant defense responses against infections as well as in symbiotic relationships. Cysteine-rich peptides (NCRs, RALFs, DEFs, and nsLTPs) and posttranslationally modified peptides (CLEs, CEPs, GLV/RGF, and PSKs) are well known for their role in plant defense mechanisms (Song et al., 2025). Systemin, the first peptide identified in plants, plays an important role in regulating plant wound response signaling mediated by pathogens or insects. Microbe-derived peptides such as lipopeptides (surfactins, fengycins, and iturins) from *Bacillus* species are also reported to contribute in pathogen inhibition and plant growth promotion activities (Song et al., 2025).

Ca^{2+} signaling is vital in plant immunological responses, as transient fluctuations in cytosolic Ca^{2+} concentrations can trigger gene reprogramming for effective defense. Recent reports highlighted the role of CLE peptides in root nodulation and symbiosis through Ca^{2+}

signaling (Nakagami et al., 2024). In barrel medic, the expression of MtCLE53 is significantly increased during arbuscular mycorrhizal (AM) colonization and its absence enhances the expression of AM symbiosis marker genes, indicating its role as a negative regulator (Nakagami et al., 2024). Fungal CLEs have been recognized in phylogenetically distinct AM fungi, with foreign RiCLE1 from *Rhizophagus irregularis* promoting mycorrhization in barrel medic. The CLV3 peptide is detected by the flagellin receptor FLS2, which triggers MAPK and increases the expression of PTI marker genes (Nakagami et al., 2024). Arabidopsis mutants, such as clv1, clv2, and crn, have enhanced resistance to *Ralstonia solanacearum* and *Hyaloperonospora arabidopsidis*, but not against necrotrophic fungi (Nakagami et al., 2024; Wang et al., 2025). Recent studies demonstrate that CLE-like peptides are also found in other organisms that interact with plants. Some nematodes release CLE precursors that use host secretion routes and stimulate plant CLEs to promote infection. The Arabidopsis receptors CLV2 and CRN identify these nematode CLE effectors, which are crucial for efficient parasitism (Nakagami et al., 2024).

Certain other peptides such as PEPs, PIPs, and SCOOPs have been found to activate immune signaling pathways and enhance resistance against *Pseudomonas syringae* in Arabidopsis and *Phytophthora infestans* in tomato (Xiao et al., 2025). Several RALF peptides also play crucial roles in plant immunity. RALF23 and RALF33 negatively regulate responses to hemi-biotrophic bacterial and fungal pathogens. They achieve this through their interactions with receptor-like kinase FERONIA (FER). This interaction facilitates the assembly of receptor complexes including other receptor kinases such as EFR, FLS2, and BAK1, thereby dampening defense responses against certain pathogens (Xiao et al., 2025). In contrast, RALF22 enhances immune signaling by amplifying the responses induced by PEPs peptide and enhances resistance to the necrotrophic fungus *Sclerotinia sclerotiorum* in *Brassica* sp.

Furthermore, PSK has been shown to regulate immune responses in tomato by elevating cytosolic Ca^{2+} levels, therefore activates auxin-dependent defense pathways and enhancing resistance to *Botrytis cinerea* (Cao et al., 2025). The IDA peptides, which are known for their role in organ separation, also enhance immune responses by inducing cytosolic Ca^{2+} release, promoting ROS production, and upregulating defense-related genes to provide additional protection against pathogens.

Additionally, some plant-derived peptides known as antimicrobial peptides (AMPs) have gained their attention for their role in mediating innate defense mechanisms and systemic stress responses. These peptides exhibit broad-spectrum antimicrobial activities against various pathogens, including bacteria, viruses, and fungi. In particular, defensins and hevein-like peptides inhibit pathogen growth and work synergistically with other secreted peptides to enhance immune responses. For example, the alfalfa antifungal peptide (alfAFP), a defensin isolated from *Medicago sativa*, demonstrates potent antifungal activity against *Verticillium dahlia*, *Phytophthora infestans*, and *Fusarium solani* (Lacerda et al., 2014). Hevein-like peptides that target fungal cell walls play a crucial role in disrupting pathogen structure and providing resistance to fungal pathogens such as *Bipolaris sorokiniana* and *Thielaviopsis basicola* (Slavokhotova et al., 2017).

Furthermore, AMPs such as DmAMP1W derived from wheat impede fungal infections deliberates resistance to sharp eyespot and root rot diseases caused by, especially, *Rhizoctonia cerealis* and *Bipolaris sorokiniana* (Ren et al., 2024). Some more peptides, including those originating from papain-like cysteine proteases, defensin-related antifungal peptides, and small cysteine-rich (SCR) proteins (e.g., RsMf8HN), are pivotal for plant defense against various bacterial and fungal infections (Ren et al., 2024; Wang et al., 2025). The virus-induced small peptide 1 (VISP1) and stPIP1 have also emerged as important regulators of plant immunity. VISP1 acts at the crossroads of autophagy, RNA silencing, and viral pathogenesis to control antiviral defenses and disease outcomes.

11.5 Hormones

Plants, similar to animals and humans, produce and secrete particular chemical compounds that regulate many biological and cellular functions within specific cells. These chemicals are important for various tasks, such as peripheral development, metabolism, organogenesis, reproduction, defensive mechanisms, and adaptability to environmental stress. Among these chemicals, plant hormones, or phytohormones, assume a vital function. These signaling molecules are mostly produced in one tissue or organ and sent to other locations within the plant to perform their regulatory functions. Phytohormones are naturally occurring, low-molecular-weight organic molecules that act as essential chemical messengers, governing various physiological processes throughout the plant life cycle. The processes of phytohormone production, transport, perception, signal transduction, and subsequent downstream reactions jointly govern cell division, expansion, differentiation, and defense responses (Lee et al., 2019).

These hormones generally operate within a network of crosstalk, engaging either synergistically or antagonistically to coordinate developmental processes and environmental responses. There are a total of 12 unique categories of phytohormones that have been identified across diverse plant species including IAA, CK, GA, ABA, ET, SA, JA, brassinosteroids, polyamine (PA), melatonin, strigolactones, and peptide hormones. The structural variety of these hormones is crucial for their functional specialization and specific biosynthesis routes (Kurepa and Smalle, 2023).

11.5.1 Role of hormones in stress responses

Advancements in molecular biology have greatly improved our understanding of the molecular mechanism and regulation of hormone-mediated activities in plants. Various researches have demonstrated the complex interactions among different hormones in regulating plant growth, development, and responses to biotic and abiotic stresses (Jaillais and Chory, 2010). In this section, we will discuss the important phytohormones implicated in biotic as well as abiotic stress signaling.

11.5.1.1 Abscisic acid

ABA is a 15-carbon sesquiterpenoid molecule mainly derived from the isoprenoid pathway, especially from xanthophylls. ABA was initially identified for its buildup in senescent and abscising leaves, and subsequently it was found to elevate under environmental stresses, such as extreme temperature, drought, salinity, and alterations in photoperiod. It is also known as a stress hormone for plants. ABA plays a crucial role in postponing leaf senescence and promoting seed dormancy, preventing premature germination, and promoting the production of storage proteins and lipids during embryogenesis. In viticulture, the exogenous use of ABA has showed an increase in freezing resistance among grape cultivars cultivated in colder areas (Ren et al., 2023).

Depending on the type and location of stress perception, ABA production is frequently seen in both roots and leaves under stress. For example, one of the main locations for ABA production during a soil water deficiency is the roots. The xylem can then carry the freshly produced ABA to the plant's aerial parts, where it can function as a long-distance signal to close stomata, lower transpiration, and trigger genes in leaves that respond to stress. On the other hand, in reaction to high temperatures or evaporative demands, ABA can also be produced by the leaves themselves. Furthermore, ABA can be dispersed via the phloem, especially in signaling from the leaves to the roots, which helps coordinate growth repression or stress acclimation responses across the plant and contributes to systemic signaling. ABA biosynthesis and signaling pathway are extremely dynamic and strictly regulated and control the growth and development in adverse environmental circumstances. 9-Cis-epoxycarotenoid dioxygenase 3 (NCED3) is a key enzyme in ABA biosynthesis, catalyzing the rate-limiting step in the conversion of 9-cis-epoxycarotenoids into ABA precursor xanthoxin. Stresses like drought and excessive salinity cause NCED3 to be transcriptionally activated, which increases the synthesis of ABA. ABA levels are precisely regulated in line with the physiological state of the plant through the modulation of NCED3 expression by both internal feedback signals and external inputs.

The functions of ABA in plant physiology have garnered significant interest over the last few years. It is currently recognized as a principal stress hormone owing to its important function in governing plant reactions to environmental stresses. ABA plays a central role as a signaling phytohormone in the systemic stress response in plants, particularly under abiotic stress conditions such as drought, salinity, and extreme temperatures. ABA demonstrates chemical stability, high solubility in hot water, and resistance to degradation, rendering it good for exogenous application and experimental studies (Ozturk et al., 2023). In conditions of drought, salt, or severe temperatures, ABA levels markedly enhance in plant tissues, activating a series of physiological responses that promote stress tolerance. This encompasses stomatal closure to decrease water loss via transpiration and the accumulation of suitable solutes that promote cellular water retention. Moreover, ABA also promotes cold tolerance by stimulating dehydration resistance–associated genes and manufactures a variety of protective proteins (Ng et al., 2014).

The perception of ABA is expedited by the PYR/PYL/RCAR family of intracellular receptors. Upon ABA binding, these receptors block clade A type 2 C protein phosphatases (PP2Cs), which serve as negative regulators of ABA signaling (Fidler et al., 2022). In the absence of ABA, PP2Cs suppress SNF1-related protein kinases 2 (SnRK2s; sucrose nonfermenting-1-related protein kinase 2), which are important for initiating downstream ABA reactions. The binding of ABA results in the assembly of a PYR/PYL/RCAR–A-BA–PP2C complex, consequently alleviating the suppression on SnRK2s, which undergo phosphorylation and activate downstream substrates, such as transcription factors and ion channels. Among these, bZIP transcription factors (e.g., AREB/ABFs and ABI5) regulate stress-responsive genes, including those involved in osmo-protection, detoxification, and protective protein synthesis, whereas S-type anion channels such as SLAC1 promote stomatal closure (Fidler et al., 2022).

HOS15, a negative regulator, interacts with the E3 ubiquitin ligase complex CUL4-DDB1 and involves in the proteasome-mediated degradation of ABA signaling components. HOS15 forms complexes with ABI1, ABI2, and OST1; however, ABA diminishes the HOS15–OST1 interaction, therefore augmenting OST1 activity and drought resistance. ABA signaling is compartmentalized within cellular sites, such as the nucleus, cytosol, and plasma membrane, promoting a coordinated response via transcriptional regulation and ion transport. This intricate signaling network indicates the essential function of ABA in providing abiotic stress resilience in plants (Ozturk et al., 2023). Crucially, intricate feedforward and feedback regulation loops incorporate the ABA production and signaling pathways. Excessive ABA can both promote its catabolism by inducing ABA 8'-hydroxylase (CYP707A) enzymes and suppress its own synthesis by downregulating the expression of the NCED gene. Transcriptional and posttranslational regulation of ABA signaling molecules guarantees prompt activation and deactivation of responses. Plants can save energy by avoiding long stress conditions and quickly restart growth processes when conditions improve through these feedback mechanisms.

Overall, ABA serves as a key player of environmental stress signals, coordinating systemic reactions that entail root-to-shoot communication. ABA makes sure that the plant is responding appropriately and effectively to environmental stresses through finely coordinated receptor signaling pathways, focused transport, and controlled biosynthesis and increases their resilience and chances of survival.

11.5.1.2 Salicylic acid

SA is a phenolic compound highly important for plant immunity and systemic stress responses. It contributes to responses to abiotic as well as biotic stresses, especially for defense mechanisms against pathogen infections. SA allows plants to sense local stresses and coordinates a systemic response across the plant body. SA was initially identified for its function in SAR. SA facilitates crosstalk that improves resilience and promotes survival under stress by interacting with other hormones and stress pathways.

SA is primarily synthesized via two different pathways: isochorismate (IC) and phenylalanine ammonia-lyase (PAL) pathways (Seyfferth and Tsuda, 2014). IC pathway is prevalent in Arabidopsis, where chorismate (a precursor metabolite for aromatic amino acids as well as vitamins) is converted to IC, followed by conversion to SA. Whereas the PAL pathway converts phenylalanine amino acid to trans-cinnamic acid and subsequently to SA. This pathway is more common in other plant species and mostly specific to abiotic stress conditions (Seyfferth and Tsuda, 2014).

After synthesis, SA can travel symplastically or through the apoplast and vascular tissues, especially through the phloem, enabling long-distance communication during SAR. Cellular transporters like EDS5 (enhanced disease susceptibility 5), which is in charge of SA export from the chloroplast, regulate its movement (Zeier, 2021). SA influences various physiological and molecular pathways that enable plants to better withstand adverse environmental conditions, specifically drought, salinity, temperature extremes, and heavy metal toxicity.

The perception and modulation occur through a variety of receptors and related proteins such as salicylic acid methyltransferase 1 (SAMT1), SA-binding protein 2 (SABP2), methyl salicylate esterase (MSE), carbonic anhydrase (CA), and NPR1 (nonexpressor of pathogenesis-related genes 1) (Khan and Khan, 2013; Khan et al., 2015). These components maintain a dynamic equilibrium between intracellular SA and its systemic signaling form methyl salicylate (MeSA). MeSA is volatile in nature and used in long-distance signaling, allowing signal transmission from the local to the distant parts of the plant. The external application of SA has also exhibited the ability to mitigate the impacts of abiotic stress by regulating endogenous SA levels and improving stress resilience mechanisms. Enhanced SA concentrations affect critical metabolic and signaling pathways, resulting in increased tolerance to abiotic stress (Khan and Khan, 2013; Khan et al., 2015).

Under drought and salinity stress conditions, SA enhances plant's adaptive responses by improving stomatal behavior and water-use efficiency (Khan et al., 2015). SA promotes the production of osmo-protectant molecules such as proline, glycine, betaine, and various soluble sugars, which help in maintaining cell turgor pressure and osmotic balance. Additionally, SA activates expression of antioxidant enzymes like superoxide dismutase (SOD), catalase (CAT), and ascorbate peroxidase (APX), which scavenge ROS produced under oxidative stress (Khan et al., 2015). This mitigates oxidative damage to cellular components, preserving membrane integrity and photosynthetic efficiency. SA response also accumulates various secondary metabolites, including phenolics, flavonoids, and terpenoids, contributing to both antioxidant defense and membrane integrity (Khan et al., 2015).

SA has a significant impact on the plant's response to cold stress, similar to heat stress. It enhances the expression of various HSPs, which act as molecular chaperones and prevent protein denaturation and aggregation under cold conditions. Under such conditions, SA also helps in stabilizing the photosynthetic apparatus, especially the photosystem II complex, thereby maintaining photosynthetic performance (Khan and Khan, 2013; Khan et al., 2015).

SA also contributes in plant's tolerance to heavy metal exposure like Cd, arsenic (As), and lead (Pb) (Yang et al., 2023). It modulates metal chelation mechanisms and ion transport pathways to detoxify heavy metal toxicity. Also induces the expression of metal-binding proteins such as phytochelatins and metallothioneins, which sequester toxic ions and reduce their bioavailability. Overall, SA enhances antioxidant defenses, counteracting oxidative stress induced by metal toxicity. By regulating both detoxification and redox balance, SA enables plants to mitigate the deleterious effects of heavy metal exposure (Yang et al., 2023).

In case of biotic stress, SA is a key hormone that induces resistance against biotrophic and hemi-biotrophic pathogens (Yun et al., 2024). SA level increases upon pathogen attack and modulates the immune response. Such modulations are mainly mediated by gene expression changes through NPR1, a central transcriptional coactivator. Upon receiving signals like increased SA, NPR1 is converted from an oligomeric to a monomeric form and transported into the nucleus, where it does not bind to DNA directly but instead interacts with TGA transcription factors. These TGAs are basic leucine zipper (bZIP) proteins that bind specific promoter elements of defense genes, recruiting NPR1 to facilitate transcriptional activation and activate the expression of downstream defense genes, most notably pathogenesis-related (PR) genes such as PR1 (Khan and Khan, 2013; Khan et al., 2015).

Various genome-wide studies indicated that NPR1 can directly activate a set of transcription factor genes such as WRKY family, which then regulates broader defense gene expression programs. Direct NPR1 targets are mainly transcription factor genes, while most defense genes are regulated through TGA interactions (Tomaž et al., 2022).

SA not only works in isolation but also engages in crosstalk with other stress-related phytohormones such as ABA (especially during drought and salt stress), JA, ET, and melatonin (Khan and Khan, 2013).

11.5.1.3 *Jasmonate*

JA is biosynthesized from oxylipins, formed from linolenic acid, and functions as a crucial regulator of plant defense and development. They are swiftly generated in reaction to mechanical injury, herbivory, and pathogen assault, triggering the manufacture of secondary metabolites such as phenolics and alkaloids that repel pests and pathogens (Wang et al., 2021; Kurepa and Smalle, 2023). Besides defense, jasmonates alter activities like root development, pollen maturation, and flower opening.

JA functions as a critical hormonal regulator coordinating systemic responses to both biotic and abiotic stresses. Its role involves complex signaling pathways that modulate gene expression and defense mechanisms across tissues (Wang et al., 2020). Upon stress perception (e.g., wounding, pathogen attack, drought, and salt), JA is synthesized de novo through the octadecanoid pathway from α-linolenic acid (plastid lipid) released from chloroplast membranes and converted into an active form called jasmonoyl-isoleucine (JA-Ile) specially in peroxisomes (Wang et al., 2021). Under systemic stress, JA and JA-Ile levels rise locally and can be transported via the phloem or xylem to remote tissues, mediating systemic signaling.

The JA signaling cascade involves the recognition and interaction of JA-Ile by the F-box protein COI1 (Zhao et al., 2024). This interaction leads to ubiquitination and degradation of JAZ repressors by the 26 S proteasome, releasing transcription factors (e.g., MYC2, MYC3, and MYC4) previously inhibited by the JAZ repressor protein in the absence of JA. Such TFs activate numerous target genes involved in defense, stress tolerance, secondary metabolite production (e.g., alkaloids and glucosinolates), ROS-detoxifying enzymes (e.g., APX, glutathione peroxidase [GPX], CAT, POD, and SOD), pathogen-related (e.g., chitinases, β-1,3-glucanases, and defensins), and osmolyte biosynthesis (e.g., proline, trehalose, and glycine betaine) (Wang et al., 2021; Zhao et al., 2024). MYC2 is a master regulator in this process, modulating responses to diverse stresses including drought, salinity, heavy metals, and pathogens. JA signaling also activates ERFs (ethylene response factors), NPR1, MAPKs, and WRKYs TFs, which mediate crosstalk with other hormonal pathways such as ET, SA, ABA, and GA (Wang et al., 2020; Khan and Khan, 2013).

JA signaling contributes to SAR by activating defense gene expression not only locally but also in distant tissues, allowing whole-plant adaptation to stress. Signal molecules and transcription factor cascades coordinate this systemic transcriptional reprogramming.

11.5.1.4 *Ethylene*

ET is a basic gaseous phytohormone that has a varied role in plant physiology. It regulates a wide array of developmental activities, such as inhibition of stem elongation, floral senescence, seed germination, fruit ripening, root nodulation, root hair formation, and organ abscission (Dubois et al., 2018). ET also serves as a pivotal signaling molecule, specially in the modulation of ROS in response to various environmental stresses such as salt, drought, low temperature, and heavy metal toxicity (Husain et al., 2020; Khatoon et al., 2025b).

ET is synthesized from methionine via a well-characterized pathway known as the Yang cycle with the help of SAM synthetase, ACC synthase (ACS), and ACC oxidase (ACO) enzymes (Pattyn et al., 2021). ACS and ACO gene families are differentially expressed in response to various developmental cues and stresses. Stress conditions such as flooding, wounding, and pathogen attack rapidly upregulate ACS genes. Although its production and signaling are meticulously governed via feedback processes that cover nitric oxide (NO), hydrogen sulfide (H_2S), and auxins (Husain et al., 2020; Dubois et al., 2018). NO is known to act as a negative regulator of ET synthesis, while auxins promote ET production via influencing the expression of the ACO gene. The ET signal transduction pathway follows a relatively linear cascade from receptor perception to gene activation. The key components involved in the perception pathway are ethylene receptor 1 (ETR1; located on the ER membrane), constitutive triple response 1 (CTR1), a Raf-like MAPKKK acting as a negative regulator, ethylene insensitive 2 (EIN2), ethylene insensitive 3/EIN3-Like 1 (EIN3/EIL1), and ethylene response factors (ERFs) (Husain et al., 2020; Dubois et al., 2018).

In the absence of ET, ETR1 remains active and activates CTR1, which subsequently phosphorylates EIN2, thereby inactivating it and blocking downstream signaling related to ET-responsive gene expression. However, upon ET binding, ETR1 becomes inactivated, leading to the inactivation of CTR1. As a result, EIN2 is no longer phosphorylated and undergoes proteolytic cleavage. The C-terminal fragment of EIN2 translocated to the nucleus and activates EIN3/EIL1. These transcription factors then induce the expression of AP2/ERFs, a large family of ET-responsive transcription factors that regulate a broad range of genes associated with stress responses and development (Husain et al., 2020; Dubois et al., 2018).

ET acts both locally and systemically to coordinate plant responses under various stress conditions. During drought and salt stress, ET regulates stomatal conductance through crosstalk with ABA and contributes to water conservation. It also promotes root elongation and the formation of aerenchymal tissue, thus enhancing water uptake and oxygen diffusion in roots. Additionally, ET enhances the activity of ROS-scavenging enzymes like SOD and CAT to mitigate oxidative stress. ROS functions as key intermediates in ET-mediated communication throughout all phases of stress from detection to response in plants. In Arabidopsis, ROS production gets intensified by ethylene overproducer 1 (ETO1) protein, which acts as a negative regulator to ET biosynthesis. ETO1 interacts with and inhibits the activity of the ACS enzyme, which is a key enzyme in ET biosynthesis. ETO1 also promotes ion homeostasis (Na^+/K^+) during salt stress conditions (Husain et al., 2020).

Under flooding or hypoxic conditions, ET signaling triggers the expression of submergence tolerance genes such as ERF-VII and SUB1A. These genes facilitate the development of adaptive structures like aerenchyma and adventitious roots to improve oxygen transport and survival under low-oxygen environments (Husain et al., 2020).

In extreme temperature stress, ET plays a regulatory role by promoting the expression of HSPs encoding genes, often through interaction with C-repeat binding factor/dehydration-responsive element-binding (CBF/DREB) transcription factors. ET also helps plants against heavy metal toxicity by activating antioxidant pathways and upregulating genes encoding metal transporter proteins, thereby limiting oxidative damage (Husain et al., 2020).

In biotic stress conditions, ET signaling interacts synergistically or antagonistically with JA and SA, depending on the type of pathogen. ET is particularly crucial for defense against necrotrophic pathogens, where it works in concert with JA to activate pathogenesis-related (PR) genes and other defense mechanisms. Through this integrated signaling network, ET ensures rapid and appropriate responses to both environmental and biological stresses, maintaining plant homeostasis and survival (Singh et al., 2022).

11.5.1.5 Melatonin

Melatonin acts as a key regulator in the systemic stress response of plants that enhances tolerances to both abiotic and biotic stresses (Lee et al., 2019). Melatonin's effects often extend beyond local treatment sites, as it can be transported within the plant to exert protective actions systemically.

Melatonin (N-acetyl-5-methoxytryptamine), an indoleamine synthesized from serotonin, was initially discovered in vascular plants. Melatonin production in plants begins with tryptophan, which is decarboxylated by tryptophan decarboxylase (TDC) to produce tryptamine. Tryptamine is hydroxylated by tryptamine 5-hydroxylase (T5H) to form serotonin (Liu et al., 2022).

It exhibited auxin-like characteristics and is known for its various physiological roles in plant growth and development. Originally invented as a powerful antioxidant, melatonin has subsequently been linked with increasing seed germination, root elongation, photosynthesis, and postponing leaf senescence. It is widely found in plant tissues, such as seeds, roots, leaves, and fruits, where it operates under both normal and stress situations (Arnao and Hernández-Ruiz, 2018; Ahmad et al., 2023). Melatonin functions as a multifunctional signaling molecule and protective agent. Its potent antioxidant activity is one of its most important functions. ROS and reactive nitrogen species (RNS), which are overproduced in environmental stresses, are directly scavenged by it (Li et al., 2022). Melatonin strengthens the plant's natural antioxidant defense system by promoting the activity of antioxidant enzymes such as POD, CAT, and SOD to neutralize these dangerous species. Melatonin not only has antioxidant properties but also stimulates the expression of genes that respond to stress to enhance plant resilience. Numerous transcription factors that are known to regulate different defense-related genes, including WRKY, NAC, MYB, bHLH, and HD-ZIP, are modulated by melatonin (Arnao and Hernández-Ruiz, 2018; Ahmad et al., 2023). These TFs stimulate the expression of various genes related to osmo-protection, detoxification, protein stabilization, and cell wall strengthening. These transcriptional control processes are all critical for plant stress tolerance. For instance, melatonin exposure frequently results in the upregulation of genes related to late embryogenesis abundant (LEA) proteins and proline biosynthesis, well-known osmo-protectants (Li et al., 2022). By enabling systemic and prolonged defense mechanisms, this ability of rewiring the plant transcriptome prepares both distant and local tissues to withstand stress more successfully.

Melatonin's effects are further expanded through hormonal crosstalk by interacting with different plant hormones like ABA, ET, SA, and JA (Arnao and Hernández-Ruiz, 2018).

Other hormone production and signaling mechanisms are both impacted by this relationship. For example, during drought and heat stress, melatonin promotes ABA synthesis and increases water-use efficiency of plants by inducing stomatal closure through its interactions with ABA and ROS. In addition, this control aids in preserving cellular hydration and turgor pressure, enabling plants to continue their physiological functions even under dry situations. Under stressful circumstances, melatonin is also known to sustain photosynthesis and encourage plant development. Under oxidative and osmotic stress, it stabilizes the photosynthetic machinery and helps maintain the amount of chlorophyll (Li et al., 2022). Additionally, it optimizes auxin signaling to support root development in salinity to improve uptake of nutrients and water (Ahmad et al., 2023). To prevent premature senescence, melatonin inhibits overproduction of ET during stressful conditions. Melatonin balances development and defensive responses by coordinating numerous levels of stress tolerance pathways through such intricate hormonal interactions (Arnao and Hernández-Ruiz, 2018).

Under biotic stress also, melatonin is essential for plant defense. It is known to activate important immune signaling pathways such as SAR, ETI, and PTI (Tiwari et al., 2022). These pathways involve the identification of particular effectors or PAMPs, which sets off a series of defensive reactions like callose deposition, hypersensitivity reaction, and the buildup of antimicrobial chemicals. Melatonin promotes the production of proteins linked to pathogenesis as well as secondary metabolites with antibacterial properties, such as flavonoids and phenolics. Melatonin is a key component of plant immunity because of its dual function in systemic immune priming and direct pathogen resistance (Tiwari et al., 2022).

In conclusion, melatonin is a multifunctional hormone that coordinates a variety of defense processes in plants. Melatonin is an essential mediator of the systemic stress response, supporting defense, development, and water conservation, regulating hormone responses, and neutralizing oxidative damage. It is a prospective target for improving crop resilience in the face of growing environmental problems.

11.5.1.6 Gibberellin

GA is a diterpenoid carboxylic acid that is majorly synthesized in developing tissues such as seeds, young leaves, and root tips. It plays an important role in overcoming seed dormancy by triggering hydrolytic enzymes that mobilize stored food reserves for initial growth phases. GA also promotes cell proliferation and elongation in the stem, serving as an important regulator of plant height and reproductive transition.

GA is synthesized via the terpenoid pathway, starting from geranylgeranyl diphosphate (GGDP) in the plastids and further proceeding through a series of oxidation steps in different cellular compartments (e.g., endoplasmic reticulum and cytoplasm), ultimately leading to the formation of bioactive GAs like GA1 and GA4 (Muhammad et al., 2024). GA signaling is mainly promoted by DELLA proteins, which act as a growth inhibitor. GA signals by binding to its receptor (GID1), which then interacts with DELLA proteins to promote its degradation via the ubiquitin-proteasome pathway, hence alleviating their suppression on transcriptional regulators and promoting the expression of growth and development-related genes (EL Sabagh et al., 2022; Muhammad et al., 2024).

Systemic stress like drought, salinity, and cold, often downregulate GA biosynthesis or signaling, leading to accumulation of DELLA proteins, which in turn activate or repress a wide range of stress-responsive transcription factors (e.g., WRKY, NAC, MYB, and PIFs) (Muhammad et al., 2024). Activation of defense genes and/or suppression of cell elongation and division as well as growth-related genes contributes to increased stress tolerance and adaptation. Elevated DELLA levels favor survival over plant growth to conserve energy and prioritize defense under limited resources conditions. In abiotic stress, GA biosynthesis is often suppressed either by downregulating its key enzymes involved in biosynthesis (e.g., GA20-oxidase and GA3-oxidase) or upregulating production of GA2-oxidase, which inactivates bioactive GAs. However, in biotic stress pathogens can directly manipulate GA levels via JA/SA pathways and suppress plant immunity (Aerts et al., 2021).

DELLA proteins serve as integration hubs connecting GA signaling with other hormone pathways, such as ABA and JA. For example, DELLAs can enhance ABA responses, which are central to drought and salt stress adaptation and modulate crosstalk with JA, thus broadly influencing plant defense, growth arrest, and metabolic reprogramming (Aerts et al., 2021). Plants can also boost GA signaling in certain contexts such as flooding or shade avoidance to promote rapid elongation and escape from stress. GA-mediated responses are systemically coordinated. Changes in GA levels or signaling in roots or shoots can trigger adjustments throughout the plant, modulating not just local but systemic stress responses and growth-restraint mechanisms.

11.5.1.7 Cytokinin

CK is an adenine nucleotide–derived phytohormone that mainly regulates essential processes of cell proliferation, apical dominance, and senescence but also plays a crucial role in systemic stress responses at the molecular level in plants (Lee et al., 2019). CK, mostly generated in root tissues, is transported to the aerial portions of the plant by xylem. In the shoot apical meristem, they facilitate cell division and promote organogenesis. Moreover, they postpone senescence by maintaining chlorophyll and protein levels in leaves. The ratio of CK to auxin is a vital factor in organ development, where enhanced auxin levels promote root growth, whereas elevated CK levels support shoot development.

CK biosynthesis is mainly regulated by isopentenyltransferases (IPTs), and its homeostasis is maintained by cytokinin oxidase/dehydrogenase (CKX) enzymes (Lee et al., 2019). During stress conditions such as drought or salinity, CK levels and distribution are altered. Under drought, CK biosynthesis is suppressed in roots, and transport from root to shoot is minimized, leading to adaptive growth suppression. Decline in CK levels during drought allows ABA-mediated responses to dominate (e.g., stomatal closure). Although, during salt or osmotic stress, CK biosynthesis genes (IPTs) are often upregulated. Transgenic plants overexpressing IPT genes exhibit delayed senescence and better stress tolerance by maintaining chlorophyll levels and photosynthesis (EL Sabagh et al., 2022). Under nutrient limitation, CKs act as root-to-shoot signaling molecules to regulate nutrient acquisition and allocation. Long-distance transport of CK via xylem and phloem enables systemic communication, allowing plants to coordinate responses at the whole-plant level.

CK sensing and signaling entail two-component systems comprising histidine kinase receptors (AHKs), histidine phosphotransferase, and response regulators. Upon CK binding, these receptors initiate a phosphorylation cascade that ultimately regulates transcription of CK-responsive genes, including numerous genes implicated in stress responses.

In response to various abiotic stresses (drought, salt, cold, nutrient deficiency), changes in CK levels modulate the expression of specific transporter genes, antioxidant defense genes, and those involved in osmotic adjustment (Cortleven et al., 2019). Under salt and osmotic stresses, CK enhances the expression of stress-responsive genes, such as HSPs, late embryogenesis abundant (LEA) proteins, and antioxidant enzymes. CK signaling often leads to the adjustment of root system architecture (RSA) and activation/repression of nutrient uptake genes (e.g., phosphate, sulfate, and iron transporters). By regulating root architecture and cell cycle, CK reduces Na^+ accumulation and improves K^+/Na^+ balance by modulating ion transporters (like HKT1 and SOS1) to maintain ion homeostasis, especially under salt stress. During extreme temperature stress, CK induces ROS-scavenging enzyme expression (e.g., SOD, CAT, APX), limiting oxidative stress. They interact with heat shock transcription factors (HSFs) and other regulators to increase thermotolerance (Zwack and Rashotte, 2015; Cortleven et al., 2019).

CKs also contribute to SAR and induced systemic resistance (ISR) through crosstalk with SA and JA during biotic stress (Cortleven et al., 2019). CK can enhance pathogen resistance by activating type-A ARR transcription factors, which regulate various defense genes. Some pathogens exploit CK pathways by producing CK-like molecules to manipulate host immunity and promote susceptibility. Plant pathogens including bacteria (e.g., *Rhodococcus fascians* and *Agrobacterium tumefaciens*), fungi (e.g., *Magnaporthe oryzae*, the causal agent of rice blast), and even some nematodes can synthesize genuine CKs or functional CK mimics. Genes for CK biosynthesis, such as those encoding iso-pentenyl transferase (IPT), are frequently found in pathogen genomes (Ma and Ma, 2016).

11.5.2 Hormonal crosstalk in plant defense

Plant hormones orchestrate intricate signaling networks that govern growth, development, and defense responses to various biotic and abiotic stresses. The activation of defensive mechanisms in plants is unforeseen not only on individual hormone activities but also on the characteristics of hormonal crosstalk, whether it is synergistic or antagonistic among various signaling pathways. This driving interaction allows plants to adapt environmental signals and adjust physiological responses accordingly.

The signaling pathways of ABA, SA, JA, and ET interact at multiple regulatory nodes, such as hormone-responsive transcription factors, to coordinate defense responses. Whole-plant adaptation and sustained development are crucial results of an effective defense system under stress conditions. The interplay between stress-responsive hormones (ABA, SA, JA, and ET) and growth-promoting hormones (auxins, GA, and CK) is important in regulating plant responses to stress.

Among phytohormones, ABA, SA, JA, and ET are reported as the primary mediators of plant defense mechanisms against both biotic and abiotic stressors. ABA is primarily associated with reactions to abiotic stressors, such as drought, salinity, thermal extremes, cold, and mechanical damage. Environmental stressors increase ABA levels, consequently activating stress-responsive genes and promoting physiological alterations such as stomatal closure and osmo-protectant accumulation. These responses include a generalist defense approach generally linked to systems that regulate drought tolerance (EL Sabagh et al., 2022).

Conversely, ABA, SA, JA, and ET are mostly involved in response to biotic stress, such as pathogen infection and herbivory, during which their endogenous concentrations markedly increase. Although recent reports indicate that plant defense is not solely regulated by these four hormones. Substantial intercommunication exists between stress-responsive and growth-regulatory hormones, including auxins, GAs, and CK. This hormonal interaction improves plant responses, ensuring maintenance of both defense and development processes under stressful conditions (Verma et al., 2016).

The intricate interactions and communication between several hormone pathways indicate the complexity of plant signaling systems. These interactions not only improve growth and developmental processes but also stimulate complex defense responses under stress settings.

Advancements in plant hormone research have illuminated the integration of hormonal signaling networks, particularly the functions of ABA, ET, SA, and JA in stress sensing and response. The initial stage of stress perception is described by variations in hormone level, which initiate downstream signaling pathways that affect metabolism and growth. These hormonal responses are promoted by a diverse range of signaling intermediates, including receptors, kinases, transcription factors, and additional regulatory components. The capacity of individual hormones to interact either synergistically or antagonistically across various signaling levels enables plants to prioritize specific responses based on the type and intensity of the stress experienced. This hormonal interaction serves as a vital regulatory mechanism that regulates the equilibrium between growth and defense. These insights are important for increasing our understanding of plant adaptive responses and have notable implications for agricultural biotechnology. Manipulating these hormonal pathways can facilitate crop resilience, enhance productivity, and affirm sustainable agricultural practices amid environmental problems.

11.6 Network integration of mobile signals

Network integration of mobile signals such as peptides, RNAs, and hormones in plants is a critical aspect of systemic communication, coordinating growth, development, and stress responses across the plant. These diverse mobile molecules travel in the plant via the vascular system and transmit information rapidly to orchestrate whole-plant adaptation.

11.6.1 Interplay and crosstalk between RNA, peptides, and hormones

The regulation of plant responses to internal and external signals is a multilayered process involving hormones, RNA molecules (especially noncoding RNAs), and SPs. These three components do not act in isolation but instead are intricately linked, forming an interconnected regulatory network. Hormone signaling pathways extensively crosstalk with peptide perception and RNA signaling networks, producing combined or synergistic outputs depending on environmental context (Wang et al., 2016; Kashtoh et al., 2025).

They act as long-distance signals central to systemic stress signaling where they regulate the expression of genes encoding RNAs and peptides and also modulate synthesis or processing of SPs (Zandalinas et al., 2020). On the other hand, these mobile peptides or RNAs can induce synthesis of hormones locally, closing signaling loops (Dunoyer et al., 2013; Xiao et al., 2025).

Mobile RNAs fine-tune hormonal responses or peptide-driven pathways by controlling receptor perception or hormone metabolism genes through RNA species such as miRNAs and siRNAs. miRNAs are broadly involve in regulation of mRNA stability and translation of hormone signaling components. Many hormone-responsive transcription factors such as MYC2 in JA signaling and EIN3 in ET signaling are often targets of miRNAs. Another miRNA, miR399, was found to coordinate phosphate uptake together with CK signaling to adjust root growth and phosphate acquisition (Li et al., 2024b).

SPs such as systemin, CLEs, CEPs, RALFs, and IDAs function as local as well as systemic signals and not only trigger signaling cascades (e.g., MAPKs and calcium fluxes) related to growth, development, and stress responses but also modulate hormone biosynthesis/signaling processes (Pandita et al., 2023). However, they often alter hormonal levels locally, which feed into systemic hormone signaling circuits. CLE25 peptides, which are induced under drought stress, move from roots to leaves and promote ABA biosynthesis for stomatal closure. CLE and CEP peptides influence CK or auxin synthesis, whereas systemin influences JA synthesis systemically in distant parts of the plant (Wang et al., 2016). Peptides like PSKs are modulated by auxin and JA and in turn influence cell proliferation and immunity. Peptides like Pip (pipecolic acid–derived signals) and lipid-based molecules induce SA biosynthesis and together propagate defense signals across the plant and induce SAR (Pandita et al., 2023).

11.6.2 Signaling networks and feedback loops

Plants rely on complex signaling networks to sense and respond to environmental and developmental cues. These networks include signal perception, transduction cascades, transcriptional reprogramming, and feedback regulation. A signaling network typically includes receptors (e.g., RLKs and hormone receptors like PYR/PYL for ABA), secondary messengers (Ca^{2+}, ROS, and IP_3), protein kinases and phosphatases (MAPKs and CDPKs), transcription factors (WRKY, MYB, NAC, and bZIP), and target gene activation or repression.

Multiple pathways often crosstalk, creating a web of interactions that allow plants to prioritize growth or defense responses. Like ABA and JA can act synergistically in stomatal closure, and SA and JA can be antagonistic in pathogen defense (Aerts et al., 2021). RNAs modulate the levels of key network components (e.g., miR398 regulates ROS scavengers under oxidative stress), and peptides relay signals between tissues, bridging local perception and systemic response (Li et al., 2022; Pandita et al., 2023).

Feedback mechanisms regulate signaling intensity and duration to prevent overreaction or maintain homeostasis. It can be either positive, which amplifies the signal (e.g., ABA promotes NCED3 expression, which leads to more ABA synthesis), or negative, which limits the signal (e.g., auxin-induced GH3 genes conjugate IAA, reducing free auxin). EIN3 and MYC2 in ET and JA signaling, respectively, are more examples of feedback loops. In JA signaling, JA-Ile binding to receptor COI1 triggers degradation of JAZ repressors and frees TFs (MYC2/3/4) that activate target genes, including those encoding enzymes for JA biosynthesis, constituting a positive feedback loop (Aerts et al., 2021). Hormones also regulate the expression and processing of peptides and RNA signals, completing feedback control loops. miRNAs also participate in feedback loops. For example, miRNAs suppress or enhance components participating in hormone pathways, which can regulate the expression or mobility of these RNAs in a feedback manner (Dong et al., 2022). Peptide signaling often induces or represses hormonal pathways, altering hormone levels systemically or locally. Peptide-receptor interactions themselves can be modulated by posttranslational modifications of peptides or receptors or by coreceptors (e.g., SERK/BAK1), forming regulatory loops (Song et al., 2025). Peptides like CEPs trigger systemic nitrogen signaling by inducing CK biosynthesis, which in turn regulates CEP expression, completing a feedback circuit (Zwack and Rashotte, 2015; Cortleven et al., 2019). Peptides also influence ROS signaling and other rapid systemic signals, integrating environmental stimuli into hormonal feedback networks. These feedback loops are dynamic and often involve epigenetic regulation, posttranslational modifications, and protein degradation pathways to ensure precise control.

11.7 Conclusion

The systemic stress response in plants represents a sophisticated and highly integrated communication network, essential for survival in a dynamic and often hostile environment. This chapter has detailed the major roles of three classes of mobile signaling molecules: sRNAs, peptides, and hormones. Rather than operating in isolation, these mobile signals function in a deeply interconnected web of crosstalk and feedback loops to coordinate response in the whole plant. sRNAs, particularly miRNAs, act as precise posttranscriptional regulators, moving systemically to tune the expression of stress-responsive genes and transcription factors, thereby coordinating metabolic and physiological adaptations to abiotic and biotic challenges. Various SPs serve as crucial short as well as long-distance messengers that often initiate localized cascades, which are then amplified into systemic defenses through interactions with hormonal pathways. Phytohormones such as ABA, SA, JA, and ET are the major runners of these responses, acting as master regulators that modulate growth and defense responses, activate antioxidant systems, and prime distant tissues for enhanced resilience.

The true sophistication of plant systemic signaling lies in the synergistic interplay between these components. Hormones regulate the biogenesis and movement of sRNAs, whereas peptides can trigger hormonal biosynthesis and sRNAs, in turn, modulate the receptors and components of both peptide and hormonal signaling pathways. This creates a resilient and self-regulating network capable of prioritizing responses, avoiding energy waste, and ensuring survival. By exploring the molecular basis of these mobile signals, we can unlock the potential to engineer crops with enhanced resilience to climate or nonclimate-induced stresses, reduce dependency on chemical pesticides, and ultimately preserve future food security.

References

Aerts, N., Pereira Mendes, M., Van Wees, S.C.M., 2021. Multiple levels of crosstalk in hormone networks regulating plant defense. Plant Journal 105 (2), 489–504. Available at https://doi.org/10.1111/tpj.15124, http://onlinelibrary.wiley.com/journal/10.1111/(ISSN)1365-313X.

Ahmad, I., Song, X., Hussein Ibrahim, M.E., Jamal, Y., Younas, M.U., Zhu, G., et al., 2023. The role of melatonin in plant growth and metabolism, and its interplay with nitric oxide and auxin in plants under different types of abiotic stress. Frontiers in Plant Science 14. Available at https://doi.org/10.3389/fpls.2023.1108507, https://www.frontiersin.org/journals/plant-science.

Arnao, M.B., Hernández-Ruiz, J., 2018. Melatonin and its relationship to plant hormones. Annals of Botany 121 (2), 195–207. Available at https://doi.org/10.1093/aob/mcx114.

Cao, Y., Yang, P.F., Li, M., 2025. Research progress of peptides discovery and function in resistance to abiotic stress in plant. Stress Biology 5 (1). Available at https://doi.org/10.1007/s44154-025-00220-1.

Chang, S., Xiao, F., 2025. Comprehensive review of plant small signaling peptides: from stress adaptation mechanisms to practical solutions for crop resilience. International Journal of Biological Macromolecules 299, 139971. Available at https://doi.org/10.1016/j.ijbiomac.2025.139971.

Chaudhary, S., Grover, A., Sharma, P.C., 2021. MicroRNAs: potential targets for developing stress-tolerant crops. Life (Chicago, Ill.: 1978) 11 (4). Available at https://doi.org/10.3390/life11040289, https://www.mdpi.com/2075-1729/11/4/289/pdf.

Cirillo, V., Molisso, D., Aprile, A.M., Maggio, A., Rao, R., 2022. Systemin peptide application improves tomato salt stress tolerance and reveals common adaptation mechanisms to biotic and abiotic stress in plants. Environmental and Experimental Botany 199, 104865. Available at https://doi.org/10.1016/j.envexpbot.2022.104865.

Cortleven, A., Leuendorf, J.E., Frank, M., Pezzetta, D., Bolt, S., Schmülling, T., 2019. Cytokinin action in response to abiotic and biotic stresses in plants. Plant, Cell & Environment 42 (3), 998–1018. Available at https://doi.org/10.1111/pce.13494.

Dong, Q., Hu, B., Zhang, C., 2022. microRNAs and their roles in plant development. Frontiers in Plant Science 13. Available at https://doi.org/10.3389/fpls.2022.824240.

Du, B., Haensch, R., Alfarraj, S., Rennenberg, H., 2024. Strategies of plants to overcome abiotic and biotic stresses. Biological Reviews 99 (4), 1524–1536. Available at https://doi.org/10.1111/brv.13079.

Dubois, M., Van den Broeck, L., Inzé, D., 2018. The pivotal role of ethylene in plant growth. Trends in Plant Science 23 (4), 311–323. Available at https://doi.org/10.1016/j.tplants.2018.01.003.

Dunoyer, P., Melnyk, C., Molnar, A., Slotkin, R.K., 2013. Plant mobile small RNAs. Cold Spring Harbor Perspectives in Biology 5 (7), a017897. Available at https://doi.org/10.1101/cshperspect.a017897.

EL Sabagh, A., Islam, M.S., Hossain, A., Iqbal, M.A., Mubeen, M., Waleed, M., et al., 2022. Phytohormones as growth regulators during abiotic stress tolerance in plants. Frontiers in Agronomy 4. Available at https://doi.org/10.3389/fagro.2022.765068, www.frontiersin.org/journals/agronomy#.

Fidler, J., Graska, J., Gietler, M., Nykiel, M., Prabucka, B., Rybarczyk-Płońska, A., et al., 2022. PYR/PYL/RCAR receptors play a vital role in the abscisic-acid-dependent responses of plants to external or internal stimuli. Cells 11 (8), 1352. Available at https://doi.org/10.3390/cells11081352.

Husain, T., Fatima, A., Suhel, M., Singh, S., Sharma, A., Prasad, S.M., et al., 2020. A brief appraisal of ethylene signaling under abiotic stress in plants. Plant Signaling & Behavior 15 (9), 1782051. Available at https://doi.org/10.1080/15592324.2020.1782051.

Jaillais, Y., Chory, J., 2010. Unraveling the paradoxes of plant hormone signaling integration. Nature Structural & Molecular Biology 17 (6), 642–645. Available at https://doi.org/10.1038/nsmb0610-642.

Jiang, C.H., Li, Z.J., Zheng, L.Y., Yu, Y.Y., Niu, D.D., 2023. Small RNAs: efficient and miraculous effectors that play key roles in plant–microbe interactions. Molecular Plant Pathology 24 (8), 999–1013. Available at https://doi.org/10.1111/mpp.13329, http://onlinelibrary.wiley.com/journal/10.1111/(ISSN)1364-3703.

Jones, J.D.G., Dangl, J.L., 2006. The plant immune system. Nature 444 (7117), 323–329. Available at https://doi.org/10.1038/nature05286.

Kashtoh, H., Rabbee, M.F., Baek, K.H., 2025. New insights into plant signaling mechanisms in biotic and abiotic stress. Plants 14 (13). Available at https://doi.org/10.3390/plants14131953, www.mdpi.com/journal/plants.

Khan, M.I.R., Khan, N.A., 2013. Salicylic acid and jasmonates: approaches in abiotic stress tolerance. Journal of Plant Biochemistry and Physiology 1 (4), e113.

Khan, M.I.R., Fatma, M., Per, T.S., Anjum, N.A., Khan, N.A., 2015. Salicylic acid-induced abiotic stress tolerance and underlying mechanisms in plants. Frontiers in Plant Science 6 JUNE. Available at https://doi.org/10.3389/fpls.2015.00462, http://journal.frontiersin.org/article/10.3389/fpls.2015.00462/pdf.

Khatoon, S., Kumari, S., Gandhi, M., Nagarwal, K., Sahoo, R.N., Iqbal, N., et al., 2025a. Appraising diverse metrics of nitric oxide in salt stress tolerance of high yielding wheat genotypes. India Nitric Oxide - Biology and Chemistry 156, 82–93. Available at https://doi.org/10.1016/j.niox.2025.03.005, https://www.sciencedirect.com/science/journal/10898603.

Khatoon, S., Mahajan, M., Khan, M.I.R., 2025b. Ethylene–potassium interactions safeguard ionic balance, chlorophyll biosynthesis, and source–sink metabolism to sustain wheat growth and yield responses under salt stress. Plant Stress 18. Available at https://doi.org/10.1016/j.stress.2025.101049, https://www.sciencedirect.com/science/journal/2667064X.

Khatoon, S., Mahajan, M., Kumari, S., Iqbal, N., Wahid, I., Khan, M.I.R., 2025c. Green-synthesized gold nanoparticles induce adaptation in photosynthetic responses, sugar and nitrogen metabolism, and seed yield of salt-stressed mustard plants. Clean Technologies and Environmental Policy 27 (10), 5683–5699. Available at https://doi.org/10.1007/s10098-024-02761-x, https://www.springer.com/journal/10098.

Khraiwesh, B., Zhu, J.K., Zhu, J., 2012. Role of miRNAs and siRNAs in biotic and abiotic stress responses of plants. Biochimica et Biophysica Acta - Gene Regulatory Mechanisms 1819 (2), 137–148. Available at https://doi.org/10.1016/j.bbagrm.2011.05.001.

Kim, J.S., Jeon, B.W., Kim, J., 2021. Signaling peptides regulating abiotic stress responses in plants. Frontiers in Plant Science 12. Available at https://doi.org/10.3389/fpls.2021.704490, https://www.frontiersin.org/journals/plant-science.

Kumari, S., Kaur, P., Mahajan, M., Nayak, S.R., Khanna, R.R., Rehman, M.T., et al., 2025. γ-aminobutyric acid (GABA) supplementation modulates phosphorus retention, production of carbon metabolites and defense metabolism under arsenic toxicity in wheat. Plant Science 356. Available at https://doi.org/10.1016/j.plantsci.2025.112504, https://www.sciencedirect.com/science/journal/01689452.

Kurepa, J., Smalle, J.A., 2023. Plant hormone modularity and the survival-reproduction trade-off. Biology 12 (8), 1143. Available at https://doi.org/10.3390/biology12081143.

Lacerda, A.F., Vasconcelos, E.A.R., Pelegrini, P.B., Grossi de Sa, M.F., 2014. Antifungal defensins and their role in plant defense. Frontiers in Microbiology 5. Available at https://doi.org/10.3389/fmicb.2014.00116, http://journal.frontiersin.org/Journal/10.3389/fmicb.2014.00116/full.

Lee, Z.H., Hirakawa, T., Yamaguchi, N., Ito, T., 2019. The roles of plant hormones and their interactions with regulatory genes in determining meristem activity. International Journal of Molecular Sciences 20 (16), 4065. Available at https://doi.org/10.3390/ijms20164065.

Li, J., Song, Q., Zuo, Z.-F., Liu, L., 2022. MicroRNA398: a master regulator of plant development and stress responses. International Journal of Molecular Sciences 23 (18), 10803. Available at https://doi.org/10.3390/ijms231810803.

Li, W., Ye, T., Ye, W., Liang, J., Wang, W., Han, D., et al., 2024a. S-acylation of a non-secreted peptide controls plant immunity via secreted-peptide signal activation. EMBO Reports 25 (2), 489–505. Available at https://doi.org/10.1038/s44319-023-00029-x.

Li, Z., Yang, J., Cai, X., Zeng, X., Zou, J.J., Xing, W., 2024b. A systematic review on the role of miRNAs in plant response to stresses under the changing climatic conditions. Plant Stress 14. Available at https://doi.org/10.1016/j.stress.2024.100674, https://www.sciencedirect.com/science/journal/2667064X.

Li, X., Zhu, L., Wang, H., Zhou, X., Wang, M., Li, L., et al., 2025. Peptide hormone-mediated regulation of plant development and environmental adaptability. Advanced Science 12 (34). Available at https://doi.org/10.1002/advs.202506590.

Liang, Y., Yang, X., Wang, C., Wang, Y., 2024. miRNAs: primary modulators of plant drought tolerance. Journal of Plant Physiology 301, 154313. Available at https://doi.org/10.1016/j.jplph.2024.154313.

Liu, G., Hu, Q., Zhang, X., Jiang, J., Zhang, Y., Zhang, Z., et al., 2022. Melatonin biosynthesis and signal transduction in plants in response to environmental conditions. Journal of Experimental Botany 73 (17), 5818–5827. Available at https://doi.org/10.1093/jxb/erac196.

Ma, K.W., Ma, W., 2016. Phytohormone pathways as targets of pathogens to facilitate infection. Plant Molecular Biology 91 (6), 713–725. Available at https://doi.org/10.1007/s11103-016-0452-0, www.wkap.nl/journalhome.htm/0167-4412.

Mahajan, M., Nazir, F., Jahan, B., Siddiqui, M.H., Iqbal, N., Khan, M.I.R., 2023. Salicylic acid mitigates arsenic stress in rice (*Oryza sativa*) via modulation of nitrogen–sulfur assimilation, ethylene biosynthesis, and defense systems. Agriculture 13 (7), 1293. Available at https://doi.org/10.3390/agriculture13071293.

Mahajan, M., Poor, P., Kaur, H., Aher, R.R., Palakolanu, S.R., Khan, M.I.R., 2025a. Salt stress tolerance and abscisic acid in plants: associating role of plant growth regulators and transcription factors. Plant Physiology and Biochemistry 228. Available at https://doi.org/10.1016/j.plaphy.2025.110303, https://www.sciencedirect.com/science/journal/09819428.

Mahajan, M., Khatoon, S., Prakash, S., Khan, M.I.R., 2025b. Plant stress hormones: fostering drought stress resilience. Optimizing Plant Health under Abiotic-Stress Environments. CRC Press, pp. 123–142.

Maida, Y., Masutomi, K., 2011. RNA-dependent RNA polymerases in RNA silencing. Biological Chemistry 392 (4), 299–304. Available at https://doi.org/10.1515/BC.2011.035.

Muhammad, M., Basit, A., Arooj, A., Dixit, G., Majeed, M., Sinha, D., et al., 2024. Roles of gibberellins in plant defense against biotic and abiotic stress. Plant Growth Regulators to Manage Biotic and Abiotic Stress in Agroecosystems. CRC Press, China, pp. 183–210. Available at http://doi.org/10.1201/9781003389507-9, https://www.taylorfrancis.com/books/edit/10.1201/9781003389507/plant-growth-regulators-manage-biotic-abiotic-stress-agroecosystems-kamel-abd-elsalam-heba-mohamed.

Nakagami, S., Kajiwara, T., Tsuda, K., Sawa, S., 2024. CLE peptide signaling in plant-microbe interactions. Frontiers in Plant Science 15. Available at https://doi.org/10.3389/fpls.2024.1481650.

Nazir, F., Mahajan, M., Khatoon, S., Albaqami, M., Ashfaque, F., Chhillar, H., et al., 2023. Sustaining nitrogen dynamics: a critical aspect for improving salt tolerance in plants. Frontiers in Plant Science 14. Available at https://doi.org/10.3389/fpls.2023.1087946, https://www.frontiersin.org/journals/plant-science.

Ng, L.M., Melcher, K., Teh, B.T., Xu, H.E., 2014. Abscisic acid perception and signaling: structural mechanisms and applications. Acta Pharmacologica Sinica 35 (5), 567–584. Available at https://doi.org/10.1038/aps.2014.7, http://www.nature.com.

Ozturk, M., Bhat, R.A., Ashraf, M., Tonelli, F.M.P., Unal, B.T., Dar, G.H., 2023. Phytohormones and Stress Responsive Secondary Metabolites. Elsevier, Turkey, pp. 1–291. Available at http://doi.org/10.1016/C2021-0-00172-3, https://www.sciencedirect.com/book/9780323918831.

Pandita, D., Bhat, J.A., Wani, S.H., ElSayed, A.I., Nawaz, G., Mukherjee, S., et al., 2023. Mobile signaling peptides: secret molecular messengers with a mighty role in plant life. Journal of Plant Growth Regulation 42 (10), 6801–6834. Available at https://doi.org/10.1007/s00344-023-11069-x, https://www.springer.com/journal/344.

Pattyn, J., Vaughan-Hirsch, J., Van de Poel, B., 2021. The regulation of ethylene biosynthesis: a complex multilevel control circuitry. New Phytologist 229 (2), 770–782. Available at https://doi.org/10.1111/nph.16873.

Qian, Z., He, L., Li, F., 2024. Understanding cold stress response mechanisms in plants: an overview. Frontiers in Plant Science 15. Available at https://doi.org/10.3389/fpls.2024.1443317.

Ren, C., Fan, P., Li, S., Liang, Z., 2023. Advances in understanding cold tolerance in grapevine. Plant Physiology 192 (3), 1733–1746. Available at https://doi.org/10.1093/plphys/kiad092.

Ren, G., Zhang, Y., Chen, Z., Xue, X., Fan, H., 2024. Research progress of small plant peptides on the regulation of plant growth, development, and abiotic stress. International Journal of Molecular Sciences 25 (7), 4114. Available at https://doi.org/10.3390/ijms25074114.

Samynathan, R., Venkidasamy, B., Shanmugam, A., Ramalingam, S., Thiruvengadam, M., 2023. Functional role of microRNA in the regulation of biotic and abiotic stress in agronomic plants. Frontiers in Genetics 14. Available at https://doi.org/10.3389/fgene.2023.1272446.

Seyfferth, C., Tsuda, K., 2014. Salicylic acid signal transduction: the initiation of biosynthesis, perception and transcriptional reprogramming. Frontiers in Plant Science 5. Available at https://doi.org/10.3389/fpls.2014.00697.

Singh, S., Husain, T., Singh, V.P., Tripathi, D.K., Prasad, S.M., Dubey, N.K., 2022. Role of ethylene in combating biotic stress. Ethylene in Plant Biology. Wiley, pp. 388–397. Available at http://doi.org/10.1002/9781119744719.ch18.

Skripnikov, A., 2023. Bioassays for identifying and characterizing plant regulatory peptides. Biomolecules 13 (12), 1795. Available at https://doi.org/10.3390/biom13121795.

Slavokhotova, A.A., Shelenkov, A.A., Andreev, Y.A., Odintsova, T.I., 2017. Hevein-like antimicrobial peptides of plants. Biochemistry (Moscow) 82 (13), 1659–1674. Available at https://doi.org/10.1134/s0006297917130065, https://www.springer.com/journal/10541.

Song, M., Zhou, Y., Li, G., Barashkova, A.S., Rogozhin, E.A., Chang, W., 2025. Peptides in plant–microbe interactions: functional diversity and pharmacological applications. Cell Surface 13. Available at https://doi.org/10.1016/j.tcsw.2025.100145, https://www.sciencedirect.com/science/journal/24682330.

Tiwari, R.K., Lal, M.K., Kumar, R., Mangal, V., Altaf, M.A., Sharma, S., et al., 2022. Insight into melatonin-mediated response and signaling in the regulation of plant defense under biotic stress. Plant Molecular Biology 109 (4-5), 385–399. Available at https://doi.org/10.1007/s11103-021-01202-3, https://link.springer.com/journal/11103.

Tomaž, Š., Gruden, K., Coll, A., 2022. TGA transcription factors—structural characteristics as basis for functional variability. Frontiers in Plant Science 13. Available at https://doi.org/10.3389/fpls.2022.935819.

Verma, V., Ravindran, P., Kumar, P.P., 2016. Plant hormone-mediated regulation of stress responses. BMC Plant Biology 16 (1). Available at https://doi.org/10.1186/s12870-016-0771-y, http://www.biomedcentral.com/bmcplantbiol/.

Wang, M., Wang, Q., Zhang, B., 2013. Response of miRNAs and their targets to salt and drought stresses in cotton (*Gossypium hirsutum* L). Gene 530 (1), 26–32. Available at https://doi.org/10.1016/j.gene.2013.08.009.

Wang, G., Zhang, G., Wu, M., 2016. CLE peptide signaling and crosstalk with phytohormones and environmental stimuli. Frontiers in Plant Science 6. Available at https://doi.org/10.3389/fpls.2015.01211 2016.

Wang, J., Song, L., Gong, X., Xu, J., Li, M., 2020. Functions of jasmonic acid in plant regulation and response to abiotic stress. International Journal of Molecular Sciences 21 (4), 1446. Available at https://doi.org/10.3390/ijms21041446.

Wang, Y., Mostafa, S., Zeng, W., Jin, B., 2021. Function and mechanism of jasmonic acid in plant responses to abiotic and biotic stresses. International Journal of Molecular Sciences 22 (16), 8568. Available at https://doi.org/10.3390/ijms22168568.

Wang, R., Fei, Y., Pan, Y., Zhou, P., Adegoke, J.O., Shen, R., et al., 2023. IMA peptides function in iron homeostasis and cadmium resistance. Plant Science 336, 111868. Available at https://doi.org/10.1016/j.plantsci.2023.111868.

Wang, Y., Wang, J., Sarwar, R., Zhang, W., Geng, R., Zhu, K.-M., et al., 2024. Research progress on the physiological response and molecular mechanism of cold response in plants. Frontiers in Plant Science 15. Available at https://doi.org/10.3389/fpls.2024.1334913.

Wang, S., Liang, Y., Jiang, Y., 2025. Plant CLE peptides: functions, challenges, and future prospects. Plant Hormones 1 (1), 0. Available at https://doi.org/10.48130/ph-0025-0006.

Xiao, F., Zhou, H., Lin, H., 2025. Decoding small peptides: regulators of plant growth and stress resilience. Journal of Integrative Plant Biology 67 (3), 596–631. Available at https://doi.org/10.1111/jipb.13873.

Xiong, L., Huang, Y., Liu, Z., Li, C., Yu, H., Shahid, M.Q., et al., 2022. Small EPIDERMAL PATTERNING FACTOR-LIKE2 peptides regulate awn development in rice. Plant Physiology 190 (1), 516–531. Available at https://doi.org/10.1093/plphys/kiac278, https://academic.oup.com/plphys/article/190/1/516/6605858.

Xu, Y., Yu, Z., Zhang, D., Huang, J., Wu, C., Yang, G., et al., 2018. CYSTM, a novel non-secreted cysteine-rich peptide family, involved in environmental stresses in *Arabidopsis thaliana*. Plant and Cell Physiology 59 (2), 423–438. Available at https://doi.org/10.1093/pcp/pcx202.

Yamada, M., Hsiao, Y.C., 2021. The roles of peptide hormones and their receptors during plant root development. Genes 12 (1), 1–13. Available at https://doi.org/10.3390/genes12010022, https://www.mdpi.com/2073-4425/12/1/22/pdf.

Yan, Y., Ham, B.K., 2022. The mobile small RNAs: important messengers for long-distance communication in plants. Frontiers in Plant Science 13. Available at https://doi.org/10.3389/fpls.2022.928729, https://www.frontiersin.org/journals/plant-science.

Yang, H., Fang, R., Luo, L., Yang, W., Huang, Q., Yang, C., et al., 2023. Uncovering the mechanisms of salicylic acid-mediated abiotic stress tolerance in horticultural crops. Frontiers in Plant Science 14. Available at https://doi.org/10.3389/fpls.2023.1226041.

Yuan, M., Ngou, B.P.M., Ding, P., Xin, X.-F., 2021. PTI–ETI crosstalk: an integrative view of plant immunity. Current Opinion in Plant Biology 62, 102030.

Yun, S.H., Khan, I.U., Noh, B., Noh, Y.S., 2024. Genomic overview of INA-induced NPR1 targeting and transcriptional cascades in *Arabidopsis*. Nucleic Acids Research 52 (7), 3572–3588. Available at https://doi.org/10.1093/nar/gkae019, https://academic.oup.com/nar/issue.

Zandalinas, S.I., Fichman, Y., Devireddy, A.R., Sengupta, S., Azad, R.K., Mittler, R., 2020. Systemic signaling during abiotic stress combination in plants. Proceedings of the National Academy of Sciences of the United States of

America 117 (24), 13810–13820. Available at https://doi.org/10.1073/pnas.2005077117, https://www.pnas.org/content/pnas/117/24/13810.full.pdf.

Zeier, J., 2021. Metabolic regulation of systemic acquired resistance. Current Opinion in Plant Biology 62, 102050. Available at https://doi.org/10.1016/j.pbi.2021.102050.

Zhang, H., Zhao, Y., Zhu, J.K., 2020. Thriving under stress: how plants balance growth and the stress response. Developmental Cell 55 (5), 529–543. Available at https://doi.org/10.1016/j.devcel.2020.10.012, https://www.cell.com/developmental-cell/home.

Zhang, F., Yang, J., Zhang, N., Wu, J., Si, H., 2022a. Roles of microRNAs in abiotic stress response and characteristics regulation of plant. Frontiers in Plant Science 13. Available at https://doi.org/10.3389/fpls.2022.919243.

Zhang, T., Li, X., Zhao, Q., Shi, Y., Hao, Y., You, C., 2022b. Genome-wide identification and functional characterization of the MdCLE peptide family in apple (*Malus × domestica*). Horticultural Plant Journal 8 (3), 279–288. Available at https://doi.org/10.1016/j.hpj.2021.12.003.

Zhao, X., He, Y., Liu, Y., Wang, Z., Zhao, J., 2024. JAZ proteins: key regulators of plant growth and stress response. The Crop Journal 12 (6), 1505–1516. Available at https://doi.org/10.1016/j.cj.2024.11.001.

Zhao, Y., Zheng, B., 2025. Local trafficking and long-distance transport of small RNAs in plants. Journal of Genetics and Genomics 52 (11), 1347–1358. Available at https://doi.org/10.1016/j.jgg.2025.03.011.

Zwack, P.J., Rashotte, A.M., 2015. Interactions between cytokinin signalling and abiotic stress responses. Journal of Experimental Botany 66 (16), 4863–4871. Available at https://doi.org/10.1093/jxb/erv172, http://jxb.oxfordjournals.org/.

CHAPTER

12

Heterotrimeric G-protein subunits in plant signal transduction: roles in plant development and stress responses

C. Vismaya, Jismon Jose, and Swarup Roy Choudhury

Department of Biology, Indian Institute of Science Education and Research (IISER), Tirupati, Andhra Pradesh, India

12.1 Introduction

Signal transduction is a fundamental biological process that enables cells to detect environmental cues and convert them into intracellular responses. Through regulatory mechanisms at the plasma membrane, external stimuli are recognized and relayed into the cell, leading to specific changes in gene expression and cellular behavior. A signaling cascade typically includes receptors that detect a ligand or external stimulus, transducers and secondary messengers that transmit the signal inward, amplifiers that increase its strength, and response regulators that trigger specific cellular actions or changes. In plants, for example, recognition of a pathogen-associated molecule (PAM) by a pattern-recognition receptor (PRR) can trigger a cascade in which calcium ions act as secondary messengers, protein kinases serve as amplifiers, and transcription factors function as response regulators, ultimately leading to the activation of defense-related gene expression. Transducers that convey signals from receptors to downstream effectors include heterotrimeric G-proteins (hereafter referred to as HGPs), small GTP-binding proteins such as Ras, which regulate kinase cascades, and protein kinases like mitogen-activated protein kinases (MAPKs) to propagate and amplify signals. HGP-mediated signaling plays a major role in both animals and plants by acting as a key mechanism for cells to perceive and respond to external stimuli. In animals, G-protein-coupled receptors (GPCRs) activate HGPs, which transmit signals inside the cell to regulate processes such as metabolism, sensory perception, immune responses, and development. This signaling pathway controls vital physiological functions,

Harnessing Plant-Stimulating Molecules
DOI: https://doi.org/10.1016/B978-0-443-44994-9.00007-9

including hormone action, neurotransmission, and cellular communication, making it essential for maintaining homeostasis and adapting to environmental changes (Pierce et al., 2002). In plants, although G-protein signaling mechanisms are somewhat distinct, they are equally important for perceiving environmental signals, such as hormones and stress factors, thereby influencing growth, development, and defense responses (Urano et al., 2016a).

The GPCR pathway employs HGPs as molecular switches that transduce signals from receptors to downstream effector proteins by cycling between active and inactive forms. HGPs consist of three subunits: Gα, Gβ, and Gγ. In the canonical animal model, HGPs act downstream of seven-transmembrane GPCRs, which serve as guanine nucleotide exchange factors (GEFs) by facilitating the exchange of GDP for GTP on Gα. In its inactive state, GDP-bound Gα forms a trimeric complex with the Gβγ dimer (Brandt and Ross, 1985; Gilman, 1987). Ligand binding to the GPCR triggers the exchange of GDP for GTP, activating Gα and causing its dissociation from Gβγ. The active GTP-bound Gα then interacts with specific effectors such as enzymes and ion channels to regulate their activity, while the liberated Gβγ dimer also activates distinct signaling pathways (Neves et al., 2002). Signal termination is mediated by GTPase-activating proteins (GAPs), such as regulators of G-protein signaling (RGS proteins), which accelerate GTP hydrolysis on Gα, returning it to the inactive GDP-bound form that reassociates with Gβγ to reform the heterotrimeric complex (Milligan and Kostenis, 2006; Maruta et al., 2021).

Unlike the canonical GPCR pathway in animals, plant G-protein signaling exhibits notable differences in receptor structure, activation mechanisms, and signaling dynamics, despite their shared role in transducing extracellular signals via G-proteins (Gilman, 1987; Urano et al., 2012). Although plant and animal HGPs are structurally conserved, they differ significantly in regulation and signaling behavior (Urano et al., 2013). Plants possess GPCR-like proteins, such as GCR1 and GCR2, but their functional association with HGPs remains unclear; GCR1 does interact with Gα subunits, and GCR2 was proposed as an abscisic acid (ABA) receptor, yet evidence does not fully support classical GPCR activity for these proteins (Perfus-Barbeoch et al., 2004; Liu et al., 2007; Chen, 2014). Unlike animals, plants do not have well-established seven-transmembrane-spanning (7TM) GPCRs that couple with G-proteins to initiate signaling. Instead, plant Gα proteins undergo spontaneous guanine nucleotide exchange and are constitutively active unless inhibited by negative regulatory factors (Trusov et al., 2012; Urano et al., 2013). This suggests that plant Gα subunits activate independently of classical GPCR-mediated facilitation. Plants have expanded their signaling repertoire by integrating inputs from diverse receptor families, in contrast to animals, where G-proteins predominantly associate with GPCRs.

Plant G-proteins interact with and are phosphorylated by receptor-like kinases (RLKs), suggesting that G-proteins link downstream signaling from RLK-mediated signal perception at the plasma membrane. The regulation of plant G-protein activity likely depends more on phosphorylation and dephosphorylation than on the GTP-GDP cycle typical of animal systems, reflecting the kinase activity of RLKs (Roy Choudhury and Pandey, 2015). In animals, signal termination typically involves cytoplasmic regulators of G-protein signaling (RGS proteins) that accelerate GTP hydrolysis on Gα. In plants, RGS proteins such as AtRGS1 combine a regulator of G-protein signaling domain with a predicted 7TM domain, localizing uniquely to the membrane and integrating signal modulation directly at the plasma membrane (Bender and Zipfel, 2018). The GTPase activity of RGS1, COLD1, and phospholipase Dα1 contributes to biochemical regulation and deactivation of plant Gα subunits (Urano and Jones, 2014; Roy Choudhury and Pandey, 2016).

A key distinction between plants and animals is the diversity of G-protein subunits encoded in their genomes and thus the potential for combinatorial signaling complexity. The human genome encodes 23 Gα, 5 Gβ, and 12 Gγ subunits (McCudden et al., 2005; Milligan and Kostenis, 2006; Birnbaumer, 2007). In contrast, the model plant *Arabidopsis thaliana* genome encodes one canonical Gα (GPA1), one Gβ (AGB1), and three Gγ subunits (AGG1, AGG2, and AGG3) (Ma et al., 1990; Weiss et al., 1994; Mason and Botella, 2000; Chakravorty et al., 2011). The rice genome encodes one Gα, one Gβ, and five Gγ subunits (Urano et al., 2016b; Pandey and Vijayakumar, 2018; Zhang et al., 2021a). Polyploid plants like soybean contain multiple gene copies reflecting genome duplication, with 4 Gα, 4 Gβ, and 10 Gγ subunits (Bisht et al., 2011; Roy Choudhury and Bisht, 2011). These multiple subunits enable plant G-proteins to orchestrate diverse physiological processes, including defense and development. Their functions encompass regulation of cell division and proliferation (Ullah et al., 2001; Chen et al., 2006), root and leaf growth (Lease et al., 2001; Ullah et al., 2003; Chen et al., 2006; Li et al., 2012), stomatal development (Wang et al., 2001; Coursol et al., 2003; Fan et al., 2008; Zhang et al., 2008; Nilson and Assmann, 2010; Yu and Assmann, 2016), and control of grain size (Fan et al., 2006; Huang et al., 2009; Mao et al., 2010; Li et al., 2012; Wendt et al., 2016; Vavilova et al., 2017; Kaur et al., 2018; Sun et al., 2018; Xu, Li and Li, 2019). Additionally, plant G-proteins are involved in innate immunity, sugar signaling, nutrient use efficiency, hormone signaling, and responses to abiotic stresses such as cold, drought, salinity, and alkalinity (Ullah et al., 2002; Trusov et al., 2006; Huang et al., 2006; Zhang et al., 2008; Colaneri et al., 2014; Chakravorty et al., 2015).

Plant HGP signaling comprises a distinctive and multifaceted network that differs fundamentally from the classical animal GPCR paradigm in both structure and regulatory mechanisms. This system plays a crucial role in incorporating diverse extracellular signals to regulate a wide array of physiological processes, including growth, development, stress responses, and immunity. Although this book chapter provides an overview of the functional and signaling roles of Gα and Gβ subunits, it places greater emphasis on summarizing the current understanding of Gγ subunits. We explore in detail the unique contributions of each G-protein subunit, their interactions within the heterotrimer, and their involvement in key processes, including cell division, hormone signaling, environmental stress adaptation, and defense mechanisms.

12.2 Functional role of the Gα subunit in plant signal transduction

12.2.1 Structural characteristics of the Gα subunit

Interest in the roles of HGPs in plant defense and development has grown significantly in recent years. Among the G-protein subunits, the Gα component has emerged as a crucial regulator in mediating plant responses to developmental signals and environmental cues. Structurally, the Gα subunit possesses inherent GTPase activity, cycling between a GTP-bound active state and a GDP-bound inactive form through the binding of guanine nucleotides. The canonical Gα protein is composed of two primary domains: a Ras-like GTPase domain responsible for GTP hydrolysis and a helical domain that influences nucleotide exchange and protects the nucleotide-binding site. This is followed by a switch region essential for the conformational shifts associated with Gα activation (Mohanasundaram and Pandey, 2023). In addition to these canonical Gα subunits, plants uniquely contain noncanonical extra-large Gα proteins (XLGs). These XLGs differ structurally from canonical Gαs and are exclusive to plants, although both forms share a common evolutionary origin (Urano et al., 2016b). XLGs are characterized by an extended N-terminal region containing an NLS (nuclear localization signal) and a larger Gα domain. Unlike canonical Gα subunits, XLGs lack typical myristoylation or palmitoylation motifs required for membrane anchoring. The presence of the NLS allows XLG proteins to localize to the nucleus, thereby expanding their functional roles and influence over signal transduction pathways beyond the cell surface to nuclear processes.

12.2.2 Functional roles of the Gα subunit in plant developmental signaling

In the model plant *Arabidopsis thaliana*, there is one canonical Gα subunit known as AtGPA1, along with three XLGs (AtXLG1, AtXLG2, and AtXLG3) (Chakravorty et al., 2011). At the morphological level, AtGPA1 plays an important role in determining leaf shape, rosette size, the angle of the hook, and hypocotyl elongation. It primarily promotes cell proliferation. Studies have shown that when the regulator of G signaling protein RGS1 is lost, AtGPA1 becomes more active, leading to changes in cell elongation and proliferation (Chen et al., 2003). Additionally, AtGPA1 controls sugar-related signaling pathways, influencing hypocotyl development and growth arrest through the RGS-mediated activity of AtGPA1 (Chen et al., 2006; Johnston et al., 2007; Urano et al., 2012). In other dicots, such as *Brassica juncea*, reducing the expression of *BjuGα1* through RNA interference causes notable decreases in plant height, flower and seed pod size, and seed weight. In contrast, plants that overexpress a constitutively active form of *BjuGα1*, which lacks GTPase activity, exhibit increased height, larger organs, and heavier seeds, without any negative effects on seed quality traits such as fatty acid, glucosinolate, oil, and protein content (Kumar and Bisht, 2020). Similarly, in cucumber, mutants of the Gα subunit protein (*CsGPA1*) exhibit shorter and broader fruits, produce more floral organs, and have smaller fruit cells. These mutants show altered expression of two key developmental genes: a reduction in *CRABS CLAW* (*CRC*), which promotes fruit length, and an increase in *WUSCHEL* (*WUS*), a regulator of stem cell activity. The CsGPA1 protein likely interacts with CsCLV1 to reduce floral number while promoting fruit elongation (Han et al., 2024) (Fig. 12.1 and Table 12.1).

Beyond the canonical Gα, the first XLG was found in *Arabidopsis* (Lee and Assmann, 1999). Analyses of XLG2 and XLG3 T-DNA knockout mutants, as well as an XLG1 knockdown mutant, have demonstrated the functions of XLGs in *Arabidopsis*, including reducing ABA-induced inhibition of seed germination (Ding et al., 2008), enhancing tolerance to salinity (Chakravorty et al., 2015), preventing stomatal proliferation (Chakravorty et al., 2015), and suppressing primary root growth and stimulating root waving (Pandey et al., 2008). Protein turnover-related PUB family E3 ubiquitin ligases are physically interacting with *Arabidopsis* XLGs (Fig. 12.1). Cytokinin hyposensitivity and

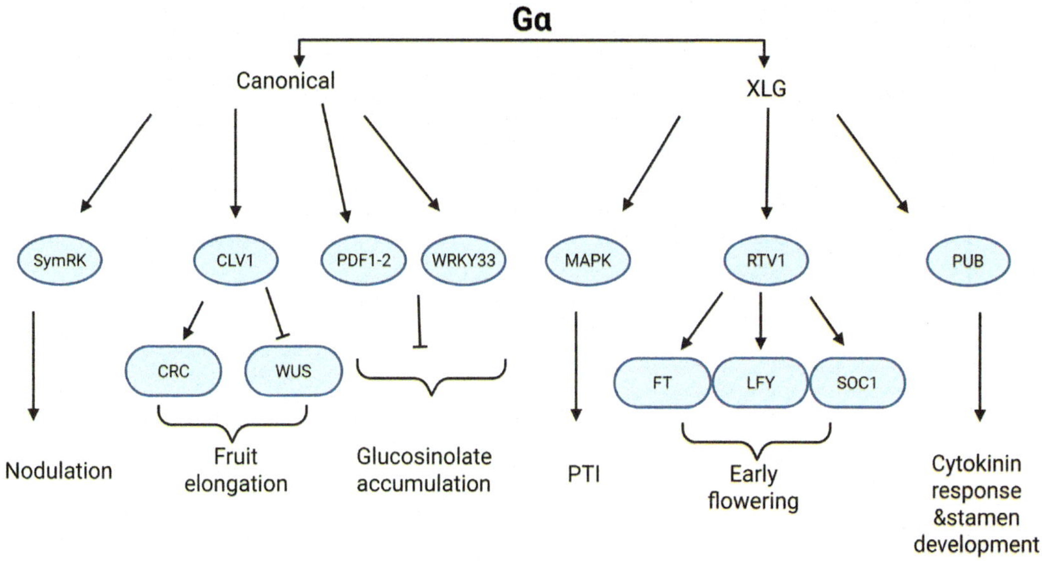

FIGURE 12.1 Pathway-specific activation of Gα subunits. The Gα interacts with the symbiosis receptor kinase (*SymRK*) to regulate nodulation. Canonical Gα can interact with CLV1, which positively regulates CRC and negatively regulates WUS, regulating the fruit elongation. By increasing the expression of defense markers PDF1.2 and WRKY33, Gα decreases glucosinolate accumulation. Noncanonical XLGs mediate pathogen-triggered immunity (*PTI*) through MAPK pathways. XLG2 interacts with the protein related to Vernalization 1 (*RTV1*), which, in turn, increases the expression of flowering-related genes such as flowering locus T (*FT*), leafy (*LFY*), and suppressor of overexpression of constans 1 (*SOC1*), leading to earlier flowering. Protein turnover-related PUB family E3 ubiquitin ligases physically interact with Arabidopsis XLGs and control stamen development and cytokinin response. *This presentation was created in BioRender.*

abnormalities in stamen development are shared by higher-order XLG mutants and mutants of the interacting PUBs (Wang et al., 2017). The immune response to bacterial and fungal pathogens is also significantly influenced by XLGs (Zhu et al., 2009; Maruta et al., 2015; Liang et al., 2016). They interact with MAPKKK3/5-MKK4/5-MPK3/6 to mediate pathogen-triggered immunity (PTI) (Wang et al., 2023) (Fig. 12.1). The noncanonical XLG2 subunit is involved in controlling the timing of flowering in *Arabidopsis*. XLG2 interacts with the protein related to Vernalization 1 (RTV1), and their interaction enhances RTV1's activity. This, in turn, increases the expression of flowering-related genes such as *flowering locus T* (*FT*), leafy (*LFY*), and suppressor of overexpression of constans *1* (*SOC1*), leading to earlier flowering (Heo et al., 2012) (Fig. 12.1). Additionally, XLG proteins negatively regulate root length: plants lacking all three XLGs develop longer primary roots when grown in the dark, a change not seen in plants lacking only one of the XLG genes (Ding et al., 2008; Urano et al., 2013).

12.2.3 Functional roles of the Gα subunit in biotic and abiotic stress responses

Gα subunits also play vital roles in plant responses to environmental stresses and pathogens. In *Arabidopsis*, mutation of the canonical Gα gene *GPA1* alters the plant's

TABLE 12.1 Functional characterization of heterotrimeric G-protein subunits across plant species.

Subunit	Organism	Response/Function	Interactor (physical or genetic)	References
AtGPA1	*Arabidopsis thaliana*	Controls leaf shape, rosette size, and the angle of the hook hypocotyl elongation	RGS1	Chen et al. (2003)
		Stomatal closure		Coursol et al. (2003)
		Nitrogen response	GCR1	Chakravorty et al. (2015), Chakraborty et al. (2019)
		Defense response	BAK1	Xue et al. (2020)
BjuGα1	*Brassica juncea*	Controls plant height, flower and seed pod size, and seed weight		Kumar and Bisht (2020)
CsGPA1	*Cucumis sativus* (Cucumber)	Promotes fruit elongation	CsCLV1	Han et al. (2024)
RGA1	*Oryza sativa (Rice)*	Thermotolerance		Feng et al. (2023)
		Nitrogen and salt stress response and gibberellic acid and brassinosteroids response		Ueguchi-Tanaka et al. (2000)
		Light stress response		Ferrero-Serrano and Assmann (2016)
MaGα1	*Morus alba L.* (Mulberry)	Drought and salt stress response		Liu et al. (2019)
LeGPA1	*Solanum lycopersicum (Tomato)*	Cold stress response		Guo et al. (2020)
		Controls fruit size and leaf shape		Wang et al. (2025)
GmGα1	*soybean*	Nodulation	SymRKα	Roy Choudhury and Pandey (2022)
BjuXLG3	*Brassica juncea*	Plant defenses		Tiwari, Kaur, and Bisht (2021)
XLG2	*Arabidopsis thaliana*	Cell death		Karimian et al. (2024)

AGB1	*Arabidopsis thaliana*	Controls plant height, leaf and fruit shape, floral bud length, and hypocotyl length		Ullah et al. (2003)
		Salt stress response	BBX21	Colaneri et al. (2014)
		Plant defense	AtRbohD and AtRbohF	Torres et al. (2002)
		Salicylic acid (SA) signaling	*SID2*	Torres et al. (2013)
		UPR-associated cell death	BiP/HSP70	Wang et al. (2007)
ZmGβ1	*Zea mays* (Maize)	Autoimmune response		Wu et al. (2020)
		Regulate kernel row number		
SlGB1	*Solanum lycopersicum* (Tomato)	Phenolamide biosynthesis	bHLH79	Wang et al. (2025)
RGB1	*Oryza sativa* (Rice)	Regulate cell number, seed development, and grain filling		Utsunomiya et al. (2011)
		Auxin response		Zhang et al. (2021b)
		Salt tolerance		Swain et al. (2017)
		Heat stress response		Biswas et al. (2019)
MtGbeta1	*Medicago truncatula*	Nodulation	PLC	Bovin et al. (2022)
PsGbeta1	*Pisum sativum*			
AGG1 and AGG2	*Arabidopsis thaliana*	Cell death and defense responses	*BIR1*	Liu et al. (2013)
AGG3		Cell proliferation. Regulates seed and organ size		Li et al. (2012)
RGG1 &RGG2	*Oryza sativa* (Rice)	Negatively regulates grain and panicle length by controlling cytokinin-mediated cell division		Tao et al. (2020)
				Miao et al. (2019)
DEP1&GGC2		Positively regulate grain length and filling		Zhang et al. (2019)
DEP1		Aerenchyma formation		Chen et al. (2021)

(*Continued*)

TABLE 12.1 (Continued)

Subunit	Organism	Response/Function	Interactor (physical or genetic)	References
DEP1		BR and auxin signaling	OsMYB86	Sun et al. (2015)
GS3		Negatively regulate grain size		Gao et al. (2015)
GS3		Alkaline stress response		Yu et al. (2023); Zhang et al. (2023)
RGG1		Salt stress response		Swain et al. (2017)
TT2		Heat stress response		Kan et al. (2022)
SlGGB1	*Solanum lycopersicum* (Tomato)	Controls seed size, fruit morphology, number of lateral roots		Subramaniam *et* al. (2016)
		Auxin and ABA signaling		
HvDEP1	*Hordeum vulgare* (Barley)	Regulates culm elongation and seed size		Wendt et al. (2016)
AT1	*Sorghum bicolor* (Sorghum)	Alkaline stress response	PIP2	Zhang et al. (2023)
ZmGS3	*Zea mays* (Maize)	Alkaline stress response		Yu et al. (2023); Zhang et al. (2023)
CsGG3.2	*Cucumis sativus* (Cucumber)	Cold stress response		Bai et al. (2018)
MaGγ1 & MaGγ2	*Morus alba L.* (Mulberry)	ABA signaling in response to drought	PP2Cs and SnRK2s	Liu et al. (2021)

Summary of functional studies performed on various plant species highlighting the physiological responses and associated genetic or physical interactors of different HGP subunits. *UPR*, Unfolded protein response; *ABA*, abscisic acid; *BR*, brassinosteroid.

response to the hormone ABA in guard cells that control stomatal movement. Although the *gpa1* mutant plants fail to inhibit inward potassium channels in response to ABA, their stomata still close normally due to alternative signaling pathways involving changes in cytosolic pH (Wang et al., 2001). Moreover, sphingosine-1-phosphate (S1P), a lipid signaling molecule produced in guard cells in response to ABA treatment, acts upstream of GPA1. The *gpa1* mutants show reduced sensitivity to S1P's effects on stomatal closure and ion channel regulation, although sphingosine kinase activity is not affected (Coursol et al., 2003).

Studies analyzing *Arabidopsis* mutants lacking both the GPCR-like GCR1 and the Gα subunit GPA1 revealed changes in the expression of nitrogen-responsive genes, suggesting that these proteins contribute to nutrient signaling (Chakravorty et al., 2015; Chakraborty et al., 2019). In defense responses, Arabidopsis Gα plays an important role in immune signaling triggered by bacterial elicitors. For example, Gα participates in the signaling network activated by the bacterial peptide flg22 by dissociating from receptor complexes after getting phosphorylated by BAK1 and enhancing reactive oxygen species (ROS) production through NADPH oxidase RbohD (Xue et al., 2020) (Table 12.1). Similarly, increased Gα transcripts are observed in peas infected with the fungus *Erysiphe pisi*, highlighting its role in pathogen response (Patel et al., 2016). In *Brassica juncea,* XLGs (*BjuXLG1*, *BjuXLG2*, and *BjuXLG3*) contribute to early defenses against *Sclerotinia sclerotiorum,* with *BjuXLG3* becoming more important as infection progresses. Knockdown of *BjuXLG* genes reduces the expression of defense markers PDF1.2 and WRKY33, and decreases glucosinolate accumulation, impairing resistance to this pathogen (Tiwari et al., 2021) (Fig. 12.1 and Table 12.1). Glucosinolates and their breakdown products are crucial for defense in Brassicaceae plants and appear to be modulated by XLGs (Sotelo et al., 2015; Abuyusuf et al., 2018; Tiwari et al., 2021).

In rice, the Gα subunit RGA1 negatively regulates thermotolerance by affecting energy and carbohydrate metabolism (Feng et al., 2023). It further regulates plant structure, grain size, seed germination, and quality characteristics (Yang et al., 2024). Mutations in *RGA1* result in reduced sensitivity to high nitrogen and salt stress and alter the expression of proteins involved in nitrogen use efficiency, suggesting that RGA1 modulates these responses. RGA1 also mediates responses to hormones such as gibberellic acid and brassinosteroids (Ueguchi-Tanaka et al., 2000). Changes in leaf temperature, photoprotection, and gene expression linked to photosynthesis in *rga1* mutants suggest its role in protecting rice plants from light stress (Ferrero-Serrano and Assmann, 2016). Other crop species also rely on Gα for stress tolerance. In mulberry, Gα negatively regulates drought and salt tolerance by controlling the detoxification of ROS (Liu et al., 2019). In tomato, overexpression of *LeGPA1* improves cold tolerance by regulating antioxidant systems and the ICE-CBF cold response pathway (Guo et al., 2020). Tomato Gα further influences fruit size and leaf shape, emphasizing its developmental importance (Wang et al., 2025; Xiang Wang et al., 2025). In soybean, Gα interacts with the symbiosis receptor kinase (SymRK) to regulate nodulation via phosphorylation (Roy Choudhury and Pandey, 2022) (Fig. 12.1 and Table 12.1). Together, these findings demonstrate that the Gα subunit serves as a central regulator of plant growth, development, and responses to both biotic and abiotic stresses across a range of plant species.

12.3 Role of Gβ in modulating G-protein signaling pathways

12.3.1 Structural features of the Gβ subunit

The Gβ subunit plays a critical and multifaceted role in cellular signaling that extends beyond its structural contribution to the HGP complex. Structurally, Gβ proteins are characterized by an N-terminal region and a series of seven WD-40 repeats, a conserved motif also observed in mammalian β subunits. The N-terminal segment is believed to interact directly with the Gα subunit through the formation of a coiled-coil domain, facilitating stable complex formation. Meanwhile, the WD-40 repeats serve as a platform for protein–protein interactions, enabling Gβ to engage with various effector proteins and regulatory molecules (Fujisawa et al., 2001). This unique architectural design enables the Gβ subunit to function as a central hub, precisely coordinating multiple signaling pathways within the cell. The Gβ subunit typically functions in close partnership with the Gγ subunit, forming the Gβγ dimer, which acts as a key signaling unit in plant cells. Although Gβ and Gγ work together, they can regulate distinct effectors, allowing the complex to modulate diverse signaling pathways and biological responses.

12.3.2 Functional roles of the Gβ subunit in plant developmental signaling

The Gβ subunit of HGPs plays a central role in regulating plant growth and development across diverse species. In *Arabidopsis thaliana*, mutants lacking the Gβ subunit (*agb1*) exhibit distinctive developmental abnormalities, including short, blunt fruits, rounded leaves, and reduced floral bud length (Ullah et al., 2003). A particularly notable feature of *agb1* mutants is the presence of siliques with blunt tips, in contrast to the elongated siliques of wild-type Col-0 plants. In addition, *agb1* mutants display a short hypocotyl phenotype. The zinc-finger transcription factor BBX21 functions as an integrator of light and ABA signaling in hypocotyl elongation. AGB1 directly interacts with BBX21, and genetic analyses have shown that the short hypocotyl phenotype of *agb1* is partially suppressed in *bbx21 agb1* double mutants. This suggests that AGB1 antagonizes BBX21 activity during hypocotyl development, with BBX21 acting downstream of AGB1. Overexpression of BBX21 in the *agb1* background further impairs hypocotyl shortening, underscoring the functional interplay between these proteins. Importantly, the C-terminal domain of BBX21 is essential for its interaction with AGB1 and for suppressing its transcriptional activation capacity (Xu et al., 2017) (Table 12.1).

While Arabidopsis *agb1* mutants remain viable and fertile, the situation in monocots such as maize (*Zea mays*) highlights species-specific differences in Gβ function. In maize, mutations in the Gβ subunit often result in lethality due to autoimmune responses (Wu et al., 2020) (Fig. 12.2). In the same study, another maize Gβ mutant was identified with attenuated autoimmune activity, which still maintains developmental phenotypes, thereby enabling detailed genetic screening. Functional analyses reveal that Gβ acts epistatically with Gα to regulate meristem size, suggesting that both subunits operate within the same signaling module. Furthermore, natural allelic variation in maize Gβ has been linked to kernel row number, an agronomic trait of major importance (Wu et al., 2020). By contrast, in *Arabidopsis*, loss of Gβ primarily accelerates leaf senescence and halts development under

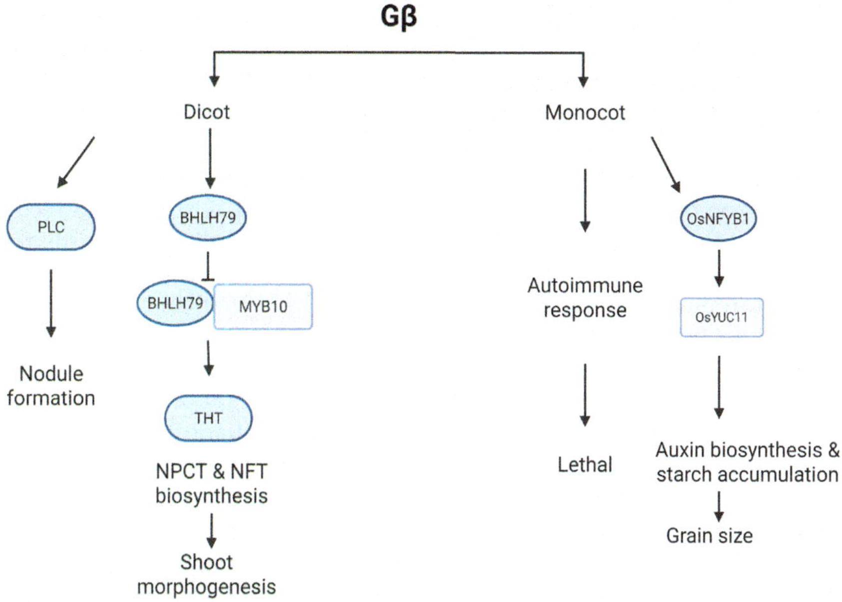

FIGURE 12.2 Functional roles of Gβ in signal transduction. Gβ regulates nodulation through the PLC-mediated pathway. Gβ interacts with the transcription factor bHLH79, preventing the formation of bHLH79-MYB10 complexes. These dimers are required for the activation of THT transcription. This is required for phenolamide (NPCT & NFT) biosynthesis, which contributes to the death of the shoot apex. Gβ interacts with OsNF-YB1, which directly activates OsYUC11, thereby regulating auxin biosynthesis, starch accumulation, and ultimately grain size. In monocots, the Gβ subunit often results in lethality due to autoimmune responses. *PLC*, Phospholipase C; *THT*, tyramine hydroxycinnamoyl transferase. *This presentation was created in BioRender.*

salt (Na^+) stress, emphasizing the divergence in developmental and stress-related roles between monocots and dicots (Colaneri et al., 2014).

A similar pattern of divergence is evident in tomato (*Solanum lycopersicum*), where a mutant of SlGB1 produces unique developmental consequences compared to both *Arabidopsis* and maize. The *slgb1* mutants exhibit aberrant shoot morphogenesis, resulting in the death of the shoot apex. This phenotype is linked to the overaccumulation of tyramine-derived phenolamides such as N-p-coumaroyltyramine (N-P-CT) and N-trans-feruloyltyramine (N-FT), synthesized by tyramine hydroxycinnamoyl transferases (THTs) (Wang et al., 2025). Mechanistic studies reveal that SlGB1 interacts with the transcription factor bHLH79 at the cytoplasm and cell membrane, preventing the formation of bHLH79-MYB10 complexes in the nucleus. Since these dimers are required for activation of THT transcription and phenolamide biosynthesis, disruption of their assembly contributes to shoot apex death in *slgb1* mutants (Fig. 12.2 and Table 12.1). Notably, such lethal phenotypes are characteristic of tomatoes and monocots, such as rice and maize, but are not observed in dicots, such as *Arabidopsis*.

In rice (*Oryza sativa*), suppression of the Gβ subunit (*RGB1*) produces yet another set of developmental outcomes. Knockdown plants exhibit browning at lamina joints and nodes, smaller seeds, reduced fertility, and impaired germination efficiency. RGB1 has been shown to regulate cell number, thereby positively influencing seed development and grain filling. Functional studies further suggest that RGB1 operates both independently and in coordination with the Gα subunit (RGA1) to control cell proliferation (Utsunomiya et al., 2011). During grain filling, the knockdown of RGB1 delays the expression of genes associated with starch biosynthesis and sucrose metabolism, resulting in reduced grain weight.

Additionally, grains from *RGB1*-deficient plants show lower auxin levels, a consequence of decreased expression of *OsNF-YB1* and *OsYUC11*. Importantly, OsNF-YB1 directly activates OsYUC11, thereby regulating auxin biosynthesis, starch accumulation, and ultimately grain size (Zhang et al., 2021b). These findings illustrate how Gβ signaling integrates hormonal regulation with carbohydrate metabolism to determine yield traits in cereals (Fig. 12.2).

In legumes, the functional importance of Gβ extends into symbiotic development. *Medicago truncatula* and *Pisum sativum* each possess a single Gβ-encoding gene (*MtGbeta1* and *PsGbeta1*, respectively). RNA interference-mediated suppression of these genes drastically reduces nodule formation, underscoring the essential role of Gβ in the establishment of symbiosis (Bovin et al., 2022). Reporter gene analyses demonstrated strong Gβ expression in developing nodule primordia and in young nodules, particularly within the meristem and invasion zones of mature nodules. Protein interaction studies further revealed that G-protein subunits interact with phospholipase C (PLC), suggesting crosstalk between HGP signaling and PLC-mediated pathways during nodulation (Bovin et al., 2022) (Fig. 12.2 and Table 12.1).

Taken together, these species-specific studies highlight the versatile yet divergent functions of the Gβ subunit across plant lineages. While *Arabidopsis* mutants illustrate developmental modulation in hypocotyl elongation and stress adaptation, monocot crops such as maize and rice exhibit lethal or yield-associated phenotypes, and tomato mutants reveal unique metabolic disruptions associated with phenolamide overproduction. In legumes, Gβ signaling integrates with symbiotic pathways to regulate nodule formation. Collectively, these findings underscore the central roles of Gβ in plant development, metabolism, and stress responses.

12.3.3 Functional roles of the Gβ subunit in stress responses

Plants must continually adjust to a plethora of environmental difficulties since they are rooted in place, unlike mobile organisms. Plant survival and productivity are seriously threatened by stresses, which range from biotic stresses, such as pathogen attacks, to abiotic pressures, including salinity, drought, and extreme temperatures. Plants require complex signaling networks that recognize stress and coordinate the appropriate physiological responses to survive in this harsh environment. HGPs have become important regulators inside these networks. Apart from the Gα subunit, the Gβγ dimer is also a crucial modulator of stress responses, as indicated by recent studies.

12.3.3.1 *Biotic stress response*

Plants employ microbe-associated molecular pattern (MAMP)–triggered immunity (MTI) as their primary defense mechanism to rapidly respond to biotic stress caused by bacteria, fungi, and oomycetes. PRRs situated at the plasma membrane recognize MAMP signals and initiate downstream defense signaling. One of the best-characterized examples is the perception of bacterial flagellin, where the PRR flagellin-insensitive 2 (FLS2), a RLK, specifically detects the conserved flg22 peptide (Gómez-Gómez and Boller, 2000; Chinchilla et al., 2007). Upon ligand recognition, FLS2 forms a heterodimer with the coreceptor BRI1-associated kinase 1 (BAK1), triggering a cascade of immune responses that includes transcriptional induction of defense genes, production of ROS, and activation of MAPK pathways. HGPs also play a crucial role in fine-tuning innate immunity in plants by regulating the formation of ROS, which are primarily generated by NADPH oxidases, and act as versatile signaling molecules during pathogen responses. In *Arabidopsis*, the heterotrimeric Gβ subunit AGB1 interacts with the NADPH oxidases AtRbohD and AtRbohF (Torres et al., 2002) (Table 12.1). Genetic studies support a functional link, as *agb1* null mutants are epistatic to NADPH oxidase null mutants. The combinatorial triple mutant *agb1, atrbohD,* and *atrbohF* displays *agb1*-like phenotypes, and *agb1* suppresses some phenotypes of the *atrbohD/atrbohF* double mutant (Torres et al., 2013; Liang et al., 2016). However, when challenged with the necrotrophic fungus *Plectosphaerella cucumerina*, the *agb1 atrbohD atrbohF* triple mutant exhibits enhanced susceptibility, compared to the single *agb1* and *atrbohD atrbohF* double mutants (Torres et al., 2013). These findings suggest that G-proteins and NADPH oxidases act within the same pathway during *Pseudomonas syringae* infection but participate in distinct pathways in defense against necrotrophic pathogens.

Further evidence indicates that AGB1-mediated defense is mainly independent of salicylic acid (SA) signaling. Isochorismate synthase 2 catalyzes the production of isochorismate, an intermediate in the biosynthesis of phyloquinone and SA. The *agb1 sid2* (isochorismate synthase 2) double mutant exhibits greater susceptibility to both *P. syringae* and *P. cucumerina* than either single mutant, suggesting that AGB1 contributes to defense independently of pathogen-induced SA biosynthesis or signaling (Torres et al., 2013) (Table 12.1). These observations highlight the complexity of signaling networks that coordinate defense activation in response to the nature of the pathogen. In addition to its role in PTI, Gβ also modulates plant sensitivity to stress caused by protein misfolding in the endoplasmic reticulum (ER) (Afrin et al., 2022). Treatment with the glycosylation inhibitor tunicamycin (Tm), a nucleoside antibiotic produced by *Streptomyces species* such as *S. lysosuperificus* and *S. clavuligerus*, induces the unfolded protein response (UPR). Compared to wild-type plants or *gpa1–4* (Gα-null) mutants, homozygous *agb1–2* (Gβ-null) seedlings display significantly higher resistance to Tm-induced growth suppression. In mature plants, Tm infiltration results in markedly reduced cell death in *agb1* mutants compared to wild-type plants. Consistently, *agb1* mutants exhibit weaker induction of BiP (binding immunoglobulin protein, an HSP70 chaperone) chaperone proteins and reduced transcriptional responses to Tm (Wang et al., 2007). Subcellular localization studies reveal that Arabidopsis Gβ protein predominantly resides in the ER, cofractionating with membrane-associated luminal BiP. Interestingly, the Gβ protein undergoes degradation during the UPR, whereas the Gα protein remains stable (Wang et al., 2007). Together, these results suggest that the Gβγ dimer functions as a signaling unit in the ER, promoting UPR-associated cell death in a manner distinct from its role in the canonical Gαβγ heterotrimer (Table 12.1).

Finally, susceptibility to necrotrophic fungi further distinguishes the defense roles of G-protein subunits. Compared to wild type, Arabidopsis *agb1* mutants exhibit increased susceptibility to *Alternaria brassicicola* and *Fusarium oxysporum* (Trusov et al., 2006).

Conversely, Gα mutants show reduced susceptibility compared to *agb1*, indicating that Gα and Gβ have diverse functions in this context. Notably, the *gpa1agb1* double mutant resembles the *agb1* single mutant, indicating that the Gβγ dimer is the primary signaling unit in defense against these necrotrophic pathogens (Trusov et al., 2006).

Taken together, these findings highlight the diverse roles of Gβ signaling in plant immunity. By interacting with NADPH oxidases, modulating ER stress responses, and functioning through the Gβγ dimer independently of Gα, the Gβ subunit integrates multiple pathways to shape plant resistance against biotrophic and necrotrophic pathogens.

12.3.3.2 Abiotic stress response

Traditionally regarded as a structural partner of the Gγ subunit, the Gβ subunit is now recognized as a dynamic regulator in its own right, particularly under abiotic stress conditions. Plant stress tolerance depends on the precise regulation of physiological functions, such as osmotic balance, detoxification of ROS, and stomatal movement, all of which are strongly influenced by Gβ signaling. Through its interactions with signaling proteins and transcriptional regulators, Gβ contributes to plant responses to drought, salinity, heat, and cold stress. Accumulating evidence shows that mutations or overexpression of *Gβ* genes significantly alter stress sensitivity across multiple species.

Expression studies in *Arabidopsis*, rice, peas, and rapeseed indicate that *Gβ* transcript levels are strongly induced under stress conditions. In Arabidopsis, AGB1 functions as a negative regulator of stomatal development; loss-of-function mutants display increased stomatal density, resulting in higher transpiration rates and greater drought sensitivity (Zhang et al., 2008). Beyond drought, AGB1 also plays an important role in ionic homeostasis under salinity stress (Yu and Assmann, 2015). The *agb1* mutants accumulate more Na^+ and less K^+ in both roots and shoots when grown in hydroponic conditions. They also exhibit larger stomatal apertures, higher stomatal conductance, and enhanced root-to-shoot transport of radiolabeled ^24Na^+ under transpiring conditions. Tracer assays further show that while initial Na^+ uptake rates are similar between wild-type and *agb1* roots, mutants display greater net Na^+ accumulation after 90 minutes, suggesting that AGB1 regulates Na^+ efflux mechanisms. In addition, AGB1 has been shown to suppress stress-induced cell death, indicating that a finely tuned balance of HGP activity is required to reconcile growth with survival under adverse conditions (Simoni et al., 2022).

Comparative studies highlight conservation of function across species. Arabidopsis AGB1 shares 76% sequence homology with rice Gβ (RGB1). Overexpression of *RGB1* in rice leads to a significant upregulation of stress-responsive antioxidant genes, including superoxide dismutase (SOD) and ascorbate peroxidase (APX), which play crucial roles in ROS detoxification (Biswas et al., 2019). Consistently, simultaneous overexpression of *RGB1* and *RGG1* in rice enhances antioxidant machinery and improves salt tolerance (Swain et al., 2017). Similar results have been observed in mulberry, where overexpression of *MaGβ1* in tobacco plants conferred enhanced salt tolerance. Under salinity stress, these transgenic plants accumulated more proline and exhibited higher activities of ROS-scavenging enzymes, including peroxidase (POD), SOD, and catalase (CAT) (Liu et al., 2019). Collectively, these findings suggest that Gβ enhances stress tolerance by boosting antioxidant defenses, thereby improving ROS detoxification capacity under salinity stress. Beyond salinity, Gβ also contributes to heat stress tolerance. In rice, *RGB1*-overexpressing transgenic plants exhibited a stronger induction of heat shock proteins (OsHSP1 and OsHSP2) in response to thermal stress. Since these proteins function as molecular chaperones to protect cellular structures, their increased expression is likely responsible for the improved thermotolerance observed in transgenic plants. Enhanced expression of ROS-scavenging enzymes in the same lines further supports the role of Gβ in integrating oxidative stress responses with heat adaptation (Biswas et al., 2019).

The Gβ has been associated with regulating programmed cell death through the UPR in the ER. *Arabidopsis* mutants lacking *Gβ* display increased resistance to tunicamycin-induced cell death, fail to form the large protein aggregates typically observed under ER stress, and show delayed and attenuated transcriptional activation of UPR genes compared to wild type (Wang et al., 2007). These findings underscore a distinct role for the Gβγ dimer in ER-localized stress signaling, independent of its canonical role within the HGP complex.

In summary, the Gβ subunit emerges as a versatile regulator of abiotic stress tolerance. By modulating stomatal development, ionic homeostasis, ROS detoxification, heat shock responses, hormone crosstalk, and ER stress signaling, Gβ integrates diverse physiological processes to optimize plant survival under challenging environmental conditions.

12.4 The multifaceted roles of the Gγ subunit in plant signaling

12.4.1 Structural features of Gγ subunits in plant signal transduction

In plants, Gγ subunits display notable structural diversity that underlies their varied functional roles within HGP signaling. These small polypeptides commonly feature a conserved C-terminal region that undergoes prenylation, a lipid modification essential for anchoring the Gβγ complex to the plasma membrane and determining subcellular localization. Based on differences in size, sequence, and domain organization, plant Gγ proteins are classified into distinct types (Trusov et al., 2007). Type A Gγ subunits resemble canonical animal counterparts and carry the classical CaaX prenylation motif, facilitating firm membrane attachment and stable dimerization with the Gβ subunit (Yalovsky, 2011). Furthermore, structural sequence analysis of all presently available Gγ subunits from land plants showed that tryptophan is present in the isoprenylation CaaX domain of "canonical" proteins that are primarily found in seed-bearing plants, compared to other eukaryotes, with very few exceptions (*Arabidopsis*). According to these findings, plant G-proteins are significantly less linked to their animal counterparts than previously believed (Trusov et al., 2012). Type B and type C Gγs lack this typical CaaX motif, but type C Gγs possess extended C-terminal domains highly enriched with cysteines, which contribute to membrane affinity and may participate in redox-sensitive regulatory mechanisms or additional protein–protein interactions (Maruta et al., 2021). The structural integrity and functional specificity of the Gβγ dimer depend on a tightly intertwined coiled-coil interaction between the Gβ and Gγ subunits, which is crucial for proper folding and assembly (Tennakoon et al., 2021). This interaction allows the complex to adopt conformations that facilitate selective engagement with a variety of downstream effectors and regulatory proteins. Beyond its structural role, the Gγ subunit, in conjunction with its partner Gβ, serves as a vital signaling unit that regulates diverse physiological processes. By anchoring the Gβγ complex to the plasma membrane, the Gγ subunit ensures accurate spatial and temporal control of signal transduction. Its contributions extend to enhancing the specificity, efficiency, and fidelity of intracellular signaling, underscoring the indispensable role of Gγ in coordinating plant cellular responses.

12.4.2 Gγ subunit–driven signaling in development

12.4.2.1 Gγ subunit–driven signaling in seed size regulation

Seed size is a fundamental trait in plant development, directly affecting reproductive fitness, agricultural productivity, and economic value. Although the genetic and hormonal controls of seed growth have been widely characterized, recent studies have highlighted the crucial contribution of HGPs, particularly the Gγ subunit, in this regulation (Sun et al., 2018; Wu et al., 2022). The Gγ subunit, functioning as part of the Gβγ dimer, is essential for mediating signaling events that coordinate cell proliferation and expansion during seed development, frequently interacting with hormonal pathways such as auxin and brassinosteroids (Jose and Roy Choudhury, 2020).

In *Arabidopsis thaliana*, there are three Gγ subunit genes (AGG1, AGG2, and AGG3); among them, AGG1 and AGG2 are canonical Gγs, and AGG3 is noncanonical, having a long cysteine-rich domain. Loss-of-function mutants for AGG3 (*agg3*) exhibit reduced seed and organ size, while *AGG3* overexpression lines display significantly increased seed and organ size, indicating that AGG3 is a positive regulator of both traits and controls the period of cell proliferation (Li et al., 2012). The introduction of the Arabidopsis *AGG3* gene into *Camelina sativa* under constitutive expression results in transgenic lines with higher seed mass, seed number per plant, and a 15%–40% increase in oil yield, further demonstrating the function of AGG3 in seed size and agricultural output (Roy Choudhury et al., 2014). In the case of the monocot model rice, there are five Gγ subunits: GS3 (grain size 3), DEP1 (dense and erect panicle 1), GGC2 (G-protein gamma subunit 2 of type C), RGG1, and RGG2 (Xiong et al., 2024). The functional analysis of RGG1 revealed that it negatively regulates grain and panicle length by controlling cytokinin-mediated cell division (Tao et al., 2020). RGG2, classified as a type B Gγ subunit, interacts with RGB1 to regulate cell division, thereby negatively affecting grain and organ size through modulation of gibberellic acid biosynthesis and signaling (Miao et al., 2019). GGC2, an atypical Gγ, acts together with DEP1 to positively regulate grain length and filling. Notably, the *dep1* mutant with a truncated C-terminus produces grains with decreased length and mass, but with an increased number of grains per panicle. DEP1/qPE9-1 enhances grain size by facilitating endosperm cell proliferation during grain development and by promoting starch accumulation during late filling stages (Zhang et al., 2019). The *GS3* gene is a negative regulator of grain size in rice. The *GS3* loss-of-function mutant (*gs3*) produces elongated grains as a result of a truncated protein lacking the essential GGL domain and C-terminal tail. The GS3 protein comprises an N-terminal organ size regulation (OSR) domain and a cysteine-rich C-terminal tail. Proper regulation of grain size depends on tail-mediated interactions between GS3, DEP1/GGC2, and RGB1. When GS3 proteins are present in greater amounts, they competitively bind to RGB1, which reduces the interaction of DEP1/GGC2 with RGB1, leading to the formation of shorter grains (Gao et al., 2015). Comparative studies indicate that, although Arabidopsis AGG3 and rice GS3 and DEP1 share structural similarities, their regulatory effects diverge: GS3 suppresses seed and organ growth, whereas AGG3 and DEP1 promote it (Sun et al., 2018). A recent study proved that in rice (*Oryza sativa* L.), the tails of three Gγ subunits, DEP1, GGC2, and GS3, determine their roles and define grain size and the signaling selectivity, which enables crops to transition from positive regulation to negative control, and it is fueled by repetitive elements and a link region present in plant-specific Gγ tails. The conserved Gγ head regions did not exhibit signaling specificity like the tails did, but the subsequent downstream signal transduction and grain size were affected by the altered contact between the mutant Gβ and Gγ. When considered as a whole, these results indicate that plant-specific Gγ tails influence signaling selectivity and are promising targets for manipulating G-proteins to maximize crop production. In *Solanum lycopersicum*, *SlGGB1*, the major type B Gγ subunit, forms complexes with Gβ and localizes to the cytoplasm, plasma membrane, and nucleus. Knockdown of *SlGGB1* leads to smaller seeds, altered fruit morphology, and an increased number of lateral roots. These SlGGB1-silenced lines are hypersensitive to exogenous auxin and exhibit impaired ABA responses during seed germination, as confirmed by transcriptome analysis (Subramaniam et al., 2016). In barley (*Hordeum vulgare*), HvDep1, a type 3 Gγ subunit, positively regulates culm elongation and seed size. Field studies have revealed that the agronomic impact of *hvdep1* loss-of-function mutants varies with environment, sometimes increasing and at other times decreasing harvestable yield (Wendt et al., 2016), suggesting that both genetic and environmental factors modulate the yield response to HvDep1 in barley.

Collectively, these studies highlight the pivotal role of plant Gγ subunits and their variants in the genetic and molecular regulation of seed and organ size, acting through diverse interactions with both hormonal pathways and other G-protein subunits. This complexity supports efforts to leverage Gγ signaling in breeding programs aimed at optimizing seed yield and quality.

12.4.2.2 *Gγ in aerenchyma formation*

Rice roots develop specialized tissue known as aerenchyma, which is crucial for survival under waterlogged or hypoxic conditions. Aerenchyma forms through the programmed breakdown of cortical cells, producing a spongy tissue with interconnected air spaces that facilitate internal gas exchange, particularly the movement of oxygen from root tips to shoots. This adaptation not only reduces respiratory energy costs by enabling efficient oxygen transport but also improves nitrogen use efficiency, enhances root activity under flooding conditions, alleviates mechanical stress, and contributes to overall resilience and yield. The HGP γ-subunit DEP1 has been implicated in hydrogen peroxide (H_2O_2)-mediated aerenchyma formation (Chen et al., 2021). In this process, ethylene induces the expression of respiratory burst oxidase homologs (RBOHs), which generate superoxide radicals ($O_2\cdot^-$) that are subsequently converted into H_2O_2 either spontaneously or enzymatically. Accumulated H_2O_2 diffuses into the cytoplasm, where it damages cellular structures and promotes cortical cell death, thereby initiating the formation of aerenchyma. While metallothioneins (MTs) in stelar and outer cell layers normally degrade H_2O_2, decreased MT expression in cortical cells

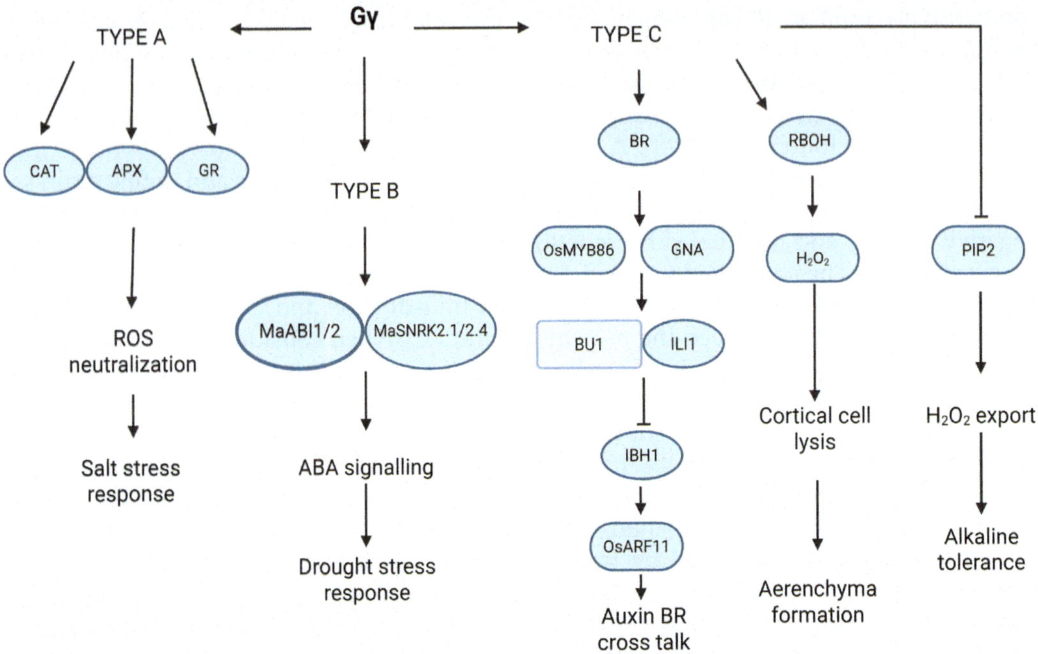

FIGURE 12.3 Functional role of plants Gγs. There are three classes of G gamma: type A, type B, and type C. Type A Gγ induces the production of CAT, APX, and GR antioxidant enzymes, which are crucial for neutralizing the excess ROS generated during salt exposure. Mulberry Gγ subunits MaGγ1 and MaGγ2 physically interact with ABA signaling proteins MaABI1/2 and MaSnRK2.1/2.4, thereby mediating signal crosstalk between G-proteins and ABA signaling. Type C Gγ, DEP1 interacts with the R2R3-MYB transcription factor OsMYB86 and directly promotes transcription of brassinosteroid upregulated 1 (BU1) (Table 12.1). BU1 and ILI1 cooperatively inhibit IBH1-mediated repression of auxin response factor 11 (OsAliuRF11), thereby linking BR signaling to auxin pathways. Type C Gγ, DEP1 enhances the RBOH expression, thus maintaining elevated endogenous H_2O_2 levels required for efficient aerenchyma formation. Type C Gγ, AT1, and GS3 negatively modulate the phosphorylation level of PIP2 aquaporins. Thus reduces the H_2O_2 export activity of PIP2s, leading to oxidative stress and alkaline sensitivity. *CAT*, Catalase; *APX*, ascorbate peroxidase; *ROS*, reactive oxygen species; *ABA*, abscisic acid; *RBOH*, respiratory burst oxidase homologs. *This presentation was created in BioRender.*

reduces H_2O_2 scavenging, allowing its buildup to trigger localized programmed cell death. Studies have shown that rice adventitious roots with upregulated *MT2b* and *RBOH* expression accumulate higher levels of H_2O_2, thereby promoting cortical cell lysis and aerenchyma development (Rajhi et al., 2011; Yamauchi et al., 2011). DEP1 enhances this process by increasing RBOH expression, thus maintaining elevated endogenous H_2O_2 levels necessary for efficient aerenchyma formation (Chen et al., 2021) (Fig. 12.3).

12.4.3 Gγ subunit–driven signaling in abiotic stress responses

12.4.3.1 Alkaline tolerance

Alkaline stress poses a significant challenge to plant growth, as high pH levels limit nutrient uptake and disrupt cellular homeostasis. Recent advances in signal transduction research have identified the Gγ subunit as a significant mediator of alkaline stress response. In sorghum, alkaline tolerance 1 (AT1) encodes a noncanonical Gγ subunit that is homologous to rice GS3 (Zhang et al., 2023). Genome-wide association analysis conducted on 352 sorghum accessions identified AT1 as a critical determinant of response to alkaline stress. Functional studies on near-isogenic lines of contrasting AT1 haplotypes revealed that the Sbat1 allele (Hap2), which encodes a truncated Gγ protein, increases sensitivity to alkaline conditions compared to the wild-type full-length SbAT1 (Hap1). This heightened sensitivity results from a frameshift mutation that truncates the C-terminus of SbAT1. Overexpression of full-length *AT1/GS3* in rice and sorghum decreases alkaline tolerance, and the effect is even more pronounced when the truncated form is overexpressed. Conversely, the knockout of AT1/GS3 in maize, rice, millet, and sorghum confers enhanced alkaline tolerance, indicating a conserved regulatory mechanism among monocot species. Mechanistically, AT1 and GS3 interact with aquaporin PIP2 proteins, influencing ROS homeostasis. Dysfunctional AT1/GS3 leads to diminished PIP2 phosphorylation, which restricts H_2O_2 export from cells, resulting in lower oxidative damage and increased tolerance to alkaline stress (Fig. 12.3 and Table 12.1). Thus Gγ subunit-mediated signaling modulates the plant's ability to adapt to high pH environments (Yu et al., 2023; Zhang et al., 2023).

12.4.3.2 *Salinity tolerance*

Salinity is an abiotic stress that negatively impacts crop yield. Several studies have drawn our attention to the role of Gγ subunits in regulating the saline stress response. In rice, overexpression of RGG1, which is a type A Gγ, resulted in increased tolerance to salt stress (Swain et al., 2017). These transgenic plants exhibited elevated transcript levels of antioxidant enzymes, such as CAT, APX, and GR, which help neutralize the excess ROS generated during salt exposure (Fig. 12.3). This enhanced antioxidant defense system helps in maintaining cellular redox balance under saline stress. The transcript levels of other G protein subunits, such as RGG2, RGB1, RGA1, DEP1, and GS3, also increase in RGG1-overexpressing plants, suggesting the coordinated regulation of the G protein complex during saline stress conditions (Swain et al., 2017). Additionally, it has been reported that RGG1 also interacts with stress-responsive proteins that mitigate protein aggregation and facilitate stress signaling. Studies on mulberry have shown that the Gβ and Gγ subunits simultaneously promote drought and salt tolerance by enhancing ROS detoxification, confirming the positive regulation of plant HGPs in stress adaptation (Liu et al., 2019).

12.4.3.3 *Cold stress tolerance*

Exposure to low temperatures imposes oxidative stress on sensitive crops such as cucumber, whose genome encodes six distinct Gγ proteins. The cucumber genome encodes six Gγ subunits, and cold treatment induces CsGG3.2, a type 3 Gγ subunit. Overexpression of CsGG3.2 resulted in enhanced tolerance to chilling injury. These transgenic lines exhibited increased activity of antioxidant enzymes, including glutathione reductase, POD, CAT, and SOD, in response to cold stress. These enzymes help in reducing the accumulation of malondialdehyde (MDA), ROS, oxygen-free radicals, and hydrogen peroxide. The reduction in ROS level and membrane lipid peroxidation shows the role of CsGG3.2 in cold signal transduction and adaptive defense (Bai et al., 2018). This suggests a future opportunity for developing cold stress–tolerant crop varieties by manipulating Gγ subunits (Bai et al., 2018).

12.4.3.4 *Thermotolerance*

The Gα, Gβ, and Gγ subunits that make up HGPs have become essential signal transducers among the molecular players coordinating plant responses to heat stress. Although the Gα and Gβ subunits have been thoroughly examined, newer studies have brought attention to the special and essential function of the Gγ subunit in regulating thermotolerance. The TT2 (THERMOTOLERANCE 2), a natural quantitative trait locus (QTL) that encodes a Gγ subunit and imparts thermotolerance in rice, has been found to have a unique molecular control mechanism that is separated from the removal of cytotoxic proteins, which is mediated by another QTL, TT14 (Kan et al., 2022). The *TT2* QTL showed thermotolerance without any yield penalty at both the vegetative and reproductive stages. A natural allele with loss of TT2 function displayed higher wax retention under high temperatures and improved heat tolerance. At the mechanistic level, the transcription factor SCT1 (sensing Ca^{2+} transcription factor 1) interprets Ca^{2+} signals through Ca^{2+}-dependent binding with calmodulin and serves as a repressor of downstream genes, such as wax synthesis regulatory 2 (OsWR2). When TT2 is disrupted, the heat-induced Ca^{2+} increase does not occur, which in turn weakens the interaction between calmodulin and SCT1, leading to the activation of OsWR2 and subsequent wax biosynthesis. To address food insecurity caused by global warming, it may be possible to develop highly thermotolerant cereal crop varieties without yield penalties by manipulating these Gγ pathway genes individually or in combination (Kan et al., 2022).

12.4.4 Gγ subunit–driven signaling in biotic stress responses

Plants are constantly challenged by diverse biotic stresses, including bacteria, fungi, viruses, and herbivorous insects. To defend themselves, they rely on intricate signaling networks that detect invading organisms and activate immune responses. The Gγ subunit, as part of the Gβγ dimer, plays an essential role in these defense mechanisms by transmitting signals from PRRs such as FLS2 and EFR to downstream immune pathways. Genetic studies of *agg1* and *agg2* mutants have demonstrated that both AGG1 and AGG2 are required for cell death and defense responses in bir1-1, as well as for pathogen-associated molecular pattern (PAMP)–triggered immunity, mediated by FLS2, EFR, and CERK1 (Liu et al., 2013). Notably, the *agg1agg2* double mutant shows markedly reduced PAMP-triggered defense responses and suppression of cell death in *bir1-1*, indicating that AGG1 and AGG2 are critical regulators of immunity and cell death (Lee et al., 2013; Liang et al., 2016) (Table 12.1). Further evidence links Gγ signaling to MAP kinase cascades. Both *agb1* and *agg1agg2* mutants exhibit reduced activation of MPK4, while MPK3 and MPK6 activation remains unaffected, suggesting that impaired MPK4 signaling contributes, at least in part, to the compromised PAMP-triggered immunity observed in these mutants (Zhang et al., 2021a). Interestingly, the flg22-induced oxidative burst remains intact in mutants of MPK4's upstream kinases MEKK1 and MKK1/MKK2, highlighting that MPK4 inhibition is a distinct regulatory checkpoint in Gβγ-mediated immunity (Liu et al., 2013; Bigeard et al., 2015).

Although the Gβγ dimer contributes to the promotion and/or maintenance of cell death, it is not the direct cause of this process. This observation aligns with previous studies that emphasize the role of HGPs in regulating, rather than initiating, defense responses (Li and Weigel, 2021). Importantly, the enhancement of cell death mediated by AGB1 and AGG1 requires the atypical Gα protein XLG2, but not GPA1 or XLG1, consistent with earlier findings that established a functional link between Gβγ signaling and XLG2 (Maruta et al., 2015; Liang et al., 2018). Moreover, the requirement of AGB1/AGG1 for the XLG2-dependent acceleration of cell death supports the notion that these signaling components might act in two distinct pathways (Karimian et al., 2024).

12.4.5 Role of Gγ in hormone signaling

The structural diversity of plant Gγ subunits suggests that this structural diversity of Gγ gives them functional specialization. Recent studies provide evidence for Gγ's role in regulating hormone responses. In rice, the Gγ subunit dense and erect panicle 1 (DEP1) positively regulates brassinosteroid (BR) signaling. BRs facilitate DEP1 nuclear translocation via grain number associated (GNA). Within the nucleus, DEP1 interacts with the R2R3-MYB transcription factor OsMYB86 and directly promotes transcription of brassinosteroid upregulated 1 (BU1), a downstream BR-responsive gene. This activation cascade is further reinforced by interactions among BU1, the bHLH protein ILI1 binding bHLH (IBH1), and the HLH protein increased leaf inclination 1 (ILI1) (Fig. 12.3). BU1 and ILI1 cooperatively inhibit IBH1-mediated repression of auxin response factor 11 (OsARF11), thereby linking BR signaling to auxin pathways and facilitating crosstalk between these two hormone systems (Li et al., 2025). Through this pathway, DEP1 serves as a central hub that integrates G-protein activity with canonical BR and auxin signaling in rice (Sun et al., 2015).

In the case of ABA, evidence from woody plants suggests that Gγ is involved in drought tolerance. Mulberry expresses multiple ABA signaling components, including *PYLs*, *PP2Cs*, *SnRK2s*, and *ABFs*, as well as ABA-responsive genes such as *MaRD29B*. These genes are strongly induced under ABA treatment and drought stress. Functional assays revealed that MaGγ1 and MaGγ2 interact with ABA signaling proteins MaABI1/2 and MaSnRK2.1/2.4, thereby mediating the signaling crosstalk between HGPs and ABA signaling (Fig. 12.3 and Table 12.1). Knockdown of MaGγ1/2 in mulberry, coupled with a transient activation assay in tobacco, confirmed that these subunits enhance ABA signaling to strengthen drought stress response (Liu et al., 2021). This provides the first direct evidence of Gγ subunits integrating with ABA signaling in a woody plant context.

In tomato, silencing of type B Gγ subunit SlGGB1 revealed its role in auxin and ABA signaling. SlGGB1-silenced plants exhibit hypersensitivity to exogenous auxin, despite showing endogenous auxin levels lower or comparable to wild-type plants. Interestingly, the silenced lines did not display strong ABA hyposensitivity under normal conditions, but they did exhibit ABA-related defects during seed germination. Transcriptome profiling revealed the aberrant expression of genes involved in ABA sensing, signaling, and response, supporting the conclusion that SlGGB1 mediates the crosstalk between auxin and ABA signaling in tomato (Subramaniam et al., 2016).

These studies demonstrate that the plant Gγ subunits are active regulators of hormone signaling. By playing an active role in BR, auxin, and ABA pathways, Gγ subunits act as a signaling hub that fine-tunes growth, development, and stress responses. Their role in coordinating crosstalk between different hormones highlights their importance in balancing developmental plasticity with adaptive stress tolerance.

12.5 Conclusion

HGPs play a vital part as molecular switches that coordinate a variety of physiological responses in plants. HGPs, composed of the Gα, Gβ, and Gγ subunits, control important functions such as hormonal signaling, stress adaptation, growth, and development. The distinctive activation methods of plant G-proteins, compared to their metazoan counterparts, set them apart and demonstrate evolutionary divergence. Recent findings highlight the interaction between G-proteins and RLKs, pointing to a regulatory paradigm based on phosphorylation that enhances our understanding of signal perception and transduction in plants. The Gα subunit is a dynamic modulator of HGP signaling, as its activation and inactivation regulate the signaling process. Functionally, the Gα subunit regulates several physiological functions, from responses to biotic and abiotic stressors to stomatal control, root architecture, and seed germination. Its interactions with downstream effectors and crosstalk with hormonal pathways, such as auxin, ABA, and gibberellins, signify its role in preserving cellular homeostasis and promoting adaptive growth. Current perspectives on plant Gβ subunits reveal that they are active controllers of cellular communication rather than just passive structural subunits. The Gβ subunit's role in growth control and environmental adaptability is underscored by its involvement in pathways regulated by ABA, brassinosteroids, and ROS. Its significance extends beyond traditional G protein pathways due to its capacity to interact with RLKs and other noncanonical signaling components (Yu et al., 2018). The Gγ subunit, once thought to be a silent partner, is now emerging as a significant factor in the orchestration of plant life, as it helps the HGP complex maintain its structural integrity and functional specificity. Cell division, organ patterning, hormone signaling, and responses to environmental stress are just a few of the many physiological processes that it affects. Recent research has demonstrated that variations in Gγ subunit isoforms produce distinct developmental consequences, suggesting that Gγ serves as a dynamic regulator with context-dependent functions, rather than merely a passive partner. The precise manipulation of Gγ subunits via gene-editing technologies has led to improvements in crop resilience, productivity, and adaptability (Sun et al., 2018; Kan et al., 2022; Yu et al., 2023; Zhang et al., 2023). Future G-protein research will probably focus more on novel interactions with noncanonical signaling pathways. Taken together, the HGPs are a promising area where basic research findings may lead to revolutionary agricultural advancements. In addition to improving our knowledge of plant signaling, the research paves the way for long-term fixes for world food security.

Acknowledgments

The authors are grateful for the funding from the Indian Institute of Science Education and Research (IISER), Tirupati; VC and JJ acknowledge the Council of Scientific and Industrial Research (CSIR), Government of India, for a PhD fellowship.

Declaration

During the preparation of this work, the author(s) used generative AI to improve the readability and language of the writing. After using, the author(s) carefully reviewed, revised, and edited the content to ensure accuracy and originality and took full responsibility for the final content of the publication.

References

Abuyusuf, M., Robin, A.H.K., Lee, J.H., Jung, H.J., Kim, H.T., Park, J.I., et al., 2018. Glucosinolate profiling and expression analysis of glucosinolate biosynthesis genes differentiate white mold resistant and susceptible cabbage lines. International Journal of Molecular Sciences 19 (12). Available at https://doi.org/10.3390/ijms19124037, https://www.mdpi.com/1422-0067/19/12/4037/pdf.

Afrin, T., Costello, C.N., Monella, A.N., Kørner, C.J., Pajerowska-Mukhtar, K.M., 2022. The interplay of GTP-binding protein AGB1 with ER stress sensors IRE1a and IRE1b modulates *Arabidopsis* unfolded protein response and bacterial immunity. Plant Signaling & Behavior 17 (1). Available at https://doi.org/10.1080/15592324.2021.2018857.

Bai, L., Liu, Y., Mu, Y., Anwar, A., He, C., Yan, Y., et al., 2018. Heterotrimeric G-protein γ subunit CsGG3.2 positively regulates the expression of CBF genes and chilling tolerance in cucumber. Frontiers in Plant Science 9. Available at https://doi.org/10.3389/fpls.2018.00488, http://journal.frontiersin.org/article/10.3389/fpls.2018.00488/full.

Bender, K.W., Zipfel, C., 2018. Plant G-protein activation: connecting to plant receptor kinases. Cell Research 28 (7), 697–698. Available at https://doi.org/10.1038/s41422-018-0046-2, http://www.nature.com/cr/index.html.

Bigeard, J., Colcombet, J., Hirt, H., 2015. Signaling mechanisms in pattern-triggered immunity (PTI). Molecular Plant 8 (4), 521–539. Available at https://doi.org/10.1016/j.molp.2014.12.022, https://www.cell.com/molecular-plant/home.

Birnbaumer, L., 2007. Expansion of signal transduction by G proteins. The second 15 years or so: from 3 to 16 α subunits plus βγ dimers. Biochimica et Biophysica Acta - Biomembranes 1768 (4), 772–793. Available at https://doi.org/10.1016/j.bbamem.2006.12.002, www.elsevier.com/locate/bbamem.

Bisht, N.C., Jez, J.M., Pandey, S., 2011. An elaborate heterotrimeric G-protein family from soybean expands the diversity of plant G-protein networks. New Phytologist 190 (1), 35–48. Available at https://doi.org/10.1111/j.1469-8137.2010.03581.x.

Biswas, S., Islam, M.N., Sarker, S., Tuteja, N., Seraj, Z.I., 2019. Overexpression of heterotrimeric G protein beta subunit gene (OsRGB1) confers both heat and salinity stress tolerance in rice. Plant Physiology and Biochemistry 144, 334–344. Available at https://doi.org/10.1016/j.plaphy.2019.10.005, http://www.journals.elsevier.com/plant-physiology-and-biochemistry/.

Bovin, A.D., Pavlova, O.A., Dolgikh, A.V., Leppyanen, I.V., Dolgikh, E.A., 2022. The role of heterotrimeric G-protein beta subunits during nodulation in medicago truncatula gaertn and *Pisum sativum* L. Frontiers in Plant Science 12. Available at https://doi.org/10.3389/fpls.2021.808573, https://www.frontiersin.org/journals/plant-science.

Brandt, D.R., Ross, E.M., 1985. GTPase activity of the stimulatory GTP-binding regulatory of adenylate cyclase, G(s). Accumulation and turnover of enzyme-nucleotide intermediates. Journal of Biological Chemistry 260 (1), 266–272.

Chakraborty, N., Kanyuka, K., Jaiswal, D.K., Kumar, A., Arora, V., Malik, A., et al., 2019. GCR1 and GPA1 coupling regulates nitrate, cell wall, immunity and light responses in *Arabidopsis*. Scientific Reports 9 (1). Available at https://doi.org/10.1038/s41598-019-42084-2, www.nature.com/srep/index.html.

Chakravorty, D., Gookin, T.E., Milner, M.J., Yu, Y., Assmann, S.M., 2015. Extra-large G proteins expand the repertoire of subunits in *Arabidopsis* heterotrimeric G protein signaling. Plant Physiology 169 (1), 512–529. Available at https://doi.org/10.1104/pp.15.00251, http://www.plantphysiol.org/content/169/1/512.full.pdf.

Chakravorty, D., Trusov, Y., Zhang, W., Acharya, B.R., Sheahan, M.B., McCurdy, D.W., et al., 2011. An atypical heterotrimeric G-protein β-subunit is involved in guard cell K+-channel regulation and morphological development in *Arabidopsis thaliana*. Plant Journal 67 (5), 840–851. Available at https://doi.org/10.1111/j.1365-313X.2011.04638.x.

Chen, Y., Chen, Y., Zhang, Y., Zhang, D., Li, G., Wei, J., et al., 2021. Heterotrimeric G protein γ subunit DEP1 is involved in hydrogen peroxide signaling and promotes aerenchyma formation in rice roots. Plant Signaling & Behavior 16 (5), 1889251. Available at https://doi.org/10.1080/15592324.2021.1889251.

Chen, J.G., Gao, Y., Jones, A.M., 2006. Differential roles of *Arabidopsis* heterotrimeric G-protein subunits in modulating cell division in roots. Plant Physiology 141 (3), 887–897. Available at https://doi.org/10.1104/pp.106.079202, http://www.plantphysiol.org/.

Chen, Y., Ji, F., Xie, H., Liang, J., Zhang, J., 2006. The regulator of G-protein signaling proteins involved in sugar and abscisic acid signaling in Arabidopsis seed germination. Plant Physiology 140 (1), 302–310. Available at https://doi.org/10.1104/pp.105.069872.

Chen, J.G., Willard, F.S., Huang, J., Liang, J., Chasse, S.A., Jones, A.M., et al., 2003. A seven-transmembrane RGS protein that modulates plant cell proliferation. Science 301 (5640), 1728–1731. Available at https://doi.org/10.1126/science.1087790.

Chen, J.-G., 2014. Heterotrimeric G-protein signaling in *Arabidopsis*. Plant Signaling & Behavior 3 (12), 1042–1045. Available at https://doi.org/10.4161/psb.3.12.6064.

Chinchilla, D., Zipfel, C., Robatzek, S., Kemmerling, B., Nürnberger, T., Jones, J.D.G., et al., 2007. A flagellin-induced complex of the receptor FLS2 and BAK1 initiates plant defence. Nature 448 (7152), 497–500. Available at https://doi.org/10.1038/nature05999, http://www.nature.com/nature/index.html.

Colaneri, A.C., Tunc-Ozdemir, M., Huang, J.P., Jones, A.M., 2014. Growth attenuation under saline stress is mediated by the heterotrimeric G protein complex. BMC Plant Biology 14 (1). Available at https://doi.org/10.1186/1471-2229-14-129, http://www.biomedcentral.com/1471-2229/14/129.

Coursol, S., Fan, L.M., Stunff, H.L., Splegel, S., Gilroy, S., Assman, S.M., 2003. Sphingolipid signalling in *Arabidopsis* guard cells involves heterotrimeric G proteins. Nature 423 (6940), 651–654. Available at https://doi.org/10.1038/nature01643.

Ding, L., Pandey, S., Assmann, S.M., 2008. *Arabidopsis* extra-large G proteins (XLGs) regulate root morphogenesis. Plant Journal 53 (2), 248–263. Available at https://doi.org/10.1111/j.1365-313X.2007.03335.x.

Fan, C., Xing, Y., Mao, H., Lu, T., Han, B., Xu, C., et al., 2006. GS3, a major QTL for grain length and weight and minor QTL for grain width and thickness in rice, encodes a putative transmembrane protein. Theoretical and Applied Genetics 112 (6), 1164–1171. Available at https://doi.org/10.1007/s00122-006-0218-1.

Fan, L.-M., Zhang, W., Chen, J.-G., Taylor, J.P., Jones, A.M., Assmann, S.M., 2008. Abscisic acid regulation of guard-cell K + and anion channels in Gβ- and RGS-deficient *Arabidopsis* lines. Proceedings of the National Academy of Sciences 105 (24), 8476–8481. Available at https://doi.org/10.1073/pnas.0800980105.

Feng, B., Xu, Y., Fu, W., Li, H., Li, G., Li, J., et al., 2023. RGA1 negatively regulates thermo-tolerance by affecting carbohydrate metabolism and the energy supply in rice. Rice 16 (1). Available at https://doi.org/10.1186/s12284-023-00649-w, https://thericejournal.springeropen.com.

Ferrero-Serrano, Á., Assmann, S.M., 2016. The α-subunit of the rice heterotrimeric G protein, RGA1, regulates drought tolerance during the vegetative phase in the dwarf rice mutant d1. Journal of Experimental Botany 67 (11), 3433–3443. Available at https://doi.org/10.1093/jxb/erw183, http://jxb.oxfordjournals.org/.

Fujisawa, Y., Kato, H., Iwasaki, Y., 2001. Structure and function of heterotrimeric G proteins in plants. Plant and Cell Physiology 42 (8), 789–794. Available at https://doi.org/10.1093/pcp/pce111, http://pcp.oxfordjournals.org/.

Gao, X., Zhang, X., Lan, H., Huang, J., Wang, J., Zhang, H., 2015. The additive effects of GS3 and qGL3 on rice grain length regulation revealed by genetic and transcriptome comparisons. BMC Plant Biology 15. Available at https://doi.org/10.1186/s12870-015-0515-4.

Gilman, A.G., 1987. G proteins: transducers of receptor-generated signals. Annual Review of Biochemistry 56, 615–649. Available at https://doi.org/10.1146/annurev.bi.56.070187.003151.

Gómez-Gómez, L., Boller, T., 2000. FLS2: An LRR receptor-like kinase involved in the perception of the bacterial elicitor flagellin in *Arabidopsis*. Molecular Cell 5 (6), 1003–1011. Available at https://doi.org/10.1016/s1097-2765(00)80265-8, www.molecule.org.

Guo, X., Li, J., Zhang, L., Zhang, Z., He, P., Wang, W., et al., 2020. Heterotrimeric G-protein α subunit (LeGPA1) confers cold stress tolerance to processing tomato plants (Lycopersicon esculentum Mill). BMC Plant Biology 20 (1). Available at https://doi.org/10.1186/s12870-020-02615-w, http://www.biomedcentral.com/bmcplantbiol/.

Han, L., Huang, Y., Li, C., Tian, D., She, D., Li, M., et al., 2024. Heterotrimeric Gα-subunit regulates flower and fruit development in CLAVATA signaling pathway in cucumber. Horticulture Research 11 (6). Available at https://doi.org/10.1093/hr/uhae110, https://academic.oup.com/hr.

Heo, J.B., Sung, S., Assmann, S.M., 2012. Ca 2+-dependent GTPase, extra-large G protein 2 (XLG2), promotes activation of DNA-binding protein related to vernalization 1 (RTV1), leading to activation of floral integrator genes and early flowering in *Arabidopsis*. Journal of Biological Chemistry 287 (11), 8242–8253. Available at https://doi.org/10.1074/jbc.M111.317412, http://www.jbc.org/content/287/11/8242.full.pdf+html.

Huang, X., Qian, Q., Liu, Z., Sun, H., He, S., Luo, D., et al., 2009. Natural variation at the DEP1 locus enhances grain yield in rice. Nature Genetics 41 (4), 494–497. Available at https://doi.org/10.1038/ng.352.

Huang, J., Taylor, J.P., Chen, J.G., Uhrig, J.F., Schnell, D.J., Nakagawa, T., et al., 2006. The plastid protein THYLAKOID FORMATION1 and the plasma membrane G-protein GPA1 interact in a novel sugar-signaling mechanism in *Arabidopsis*. The Plant Cell 18 (5), 1226–1238. Available at https://doi.org/10.1105/tpc.105.037259, http://www.plantcell.org/.

Johnston, C.A., Taylor, J.P., Gao, Y., Kimple, A.J., Grigston, J.C., Chen, J.-G., et al., 2007. GTPase acceleration as the rate-limiting step in *Arabidopsis* G protein-coupled sugar signaling. Proceedings of the National Academy of Sciences 104 (44), 17317–17322. Available at https://doi.org/10.1073/pnas.0704751104.

Jose, J., Roy Choudhury, S., 2020. Heterotrimeric G-proteins mediated hormonal responses in plants. Cellular Signalling 76, 109799. Available at https://doi.org/10.1016/j.cellsig.2020.109799.

Kan, Y., Mu, X.R., Zhang, H., Gao, J., Shan, J.X., Ye, W.W., et al., 2022. TT2 controls rice thermotolerance through SCT1-dependent alteration of wax biosynthesis. Nature Plants 8 (1), 53–67. Available at https://doi.org/10.1038/s41477-021-01039-0, http://www.nature.com/nplants/.

Karimian, P., Trusov, Y., Botella, J.R., 2024. Conserved role of heterotrimeric g proteins in plant defense and cell death progression. Genes 15 (1). Available at https://doi.org/10.3390/genes15010115, http://www.mdpi.com/journal/genes/.

Kaur, J., Choudhury, S.R., Vijayakumar, A., Hovis, L., Rhodes, Z., Polzin, R., et al., 2018. *Arabidopsis* type III Gγ protein AGG3 is a positive regulator of yield and stress responses in the model Monocot Setaria viridis. Frontiers in Plant Science 9. Available at https://doi.org/10.3389/fpls.2018.00109, http://journal.frontiersin.org/article/10.3389/fpls.2018.00109/full.

Kumar, R., Bisht, N.C., 2020. Heterotrimeric Gα subunit regulates plant architecture, organ size and seed weight in the oilseed Brassica juncea. Plant Molecular Biology 104 (4-5), 549–560. Available at https://doi.org/10.1007/s11103-020-01060-5, https://link.springer.com/journal/11103.

Lease, K.A., Wen, J., Li, J., Doke, J.T., Liscum, E., Walker, J.C., 2001. A mutant *Arabidopsis* heterotrimeric G-protein β subunit affects leaf, flower, and fruit development. The Plant Cell 13 (12), 2631–2641. Available at https://doi.org/10.1105/tpc.13.12.2631, http://www.plantcell.org/.

Lee, Y.R.J., Assmann, S.M., 1999. *Arabidopsis thaliana* 'extra-large GTP-binding protein' (AtXLG1): a new class of G-protein. Plant Molecular Biology 40 (1), 55–64. Available at https://doi.org/10.1023/A:1026483823176, www.wkap.nl/journalhome.htm/0167-4412.

Lee, S., Rojas, C.M., Ishiga, Y., Pandey, S., Mysore, K.S., 2013. *Arabidopsis* heterotrimeric G-proteins play a critical role in host and nonhost resistance against Pseudomonas syringae pathogens. PLoS ONE 8 (12). Available at https://doi.org/10.1371/journal.pone.0082445, http://www.plosone.org/article/fetchObject.action?uri=info%3Adoi%2F10.1371%2Fjournal.pone.0082445&representation=PDF.

Li, S., Liu, Y., Zheng, L., Chen, L., Li, N., Corke, F., et al., 2012. The plant-specific G protein γ subunit AGG3 influences organ size and shape in *Arabidopsis thaliana*. New Phytologist 194 (3), 690–703. Available at https://doi.org/10.1111/j.1469-8137.2012.04083.x.

Li, L., Weigel, D., 2021. One hundred years of hybrid necrosis: hybrid autoimmunity as a window into the mechanisms and evolution of plant-pathogen interactions. Annual Review of Phytopathology 59, 213–237. Available at https://doi.org/10.1146/annurev-phyto-020620-114826, http://www.annualreviews.org/journal/phyto.

Li, S., Zhao, Z., Liu, T., Zhang, J., Xing, X., Feng, M., et al., 2025. The G-protein γ subunit DEP1 facilitates brassinosteroid signaling in rice via a MYB–bHLH–ARF module. The Plant Cell 37 (5). Available at https://doi.org/10.1093/plcell/koaf122.

Liang, X., Ding, P., Lian, K., Wang, J., Ma, M., Li, L., et al., 2016. *Arabidopsis* heterotrimeric G proteins regulate immunity by directly coupling to the FLS_2 receptor. eLife 5 (2016). Available at https://doi.org/10.7554/eLife.13568, https://elife-publishing-cdn.s3.amazonaws.com/13568/elife-13568-v2-download.pdf.

Liang, X., Ma, M., Zhou, Z., Wang, J., Yang, X., Rao, S., et al., 2018. Ligand-triggered de-repression of *Arabidopsis* heterotrimeric G proteins coupled to immune receptor kinases. Cell Research 28 (5), 529–543. Available at https://doi.org/10.1038/s41422-018-0027-5, http://www.nature.com/cr/index.html.

Liu, J., Ding, P., Sun, T., Nitta, Y., Dong, O., Huang, X., et al., 2013. Heterotrimeric G proteins serve as a converging point in plant defense signaling activated by multiple receptor-like kinases. Plant Physiology 161 (4), 2146–2158. Available at https://doi.org/10.1104/pp.112.212431, http://www.plantphysiol.org/content/161/4/2146.full.pdf+html.

Liu, C., Hu, J., Fan, W., Zhu, P., Cao, B., Zheng, S., et al., 2021. Heterotrimeric G-protein γ subunits regulate ABA signaling in response to drought through interacting with PP2Cs and SnRK2s in mulberry (*Morus alba* L.). Plant Physiology and Biochemistry 161, 210–221. Available at https://doi.org/10.1016/j.plaphy.2021.02.017, http://www.journals.elsevier.com/plant-physiology-and-biochemistry/.

Liu, C., Xu, Y., Feng, Y., Long, D., Cao, B., Xiang, Z., et al., 2019. Ectopic expression of mulberry g-proteins alters drought and salt stress tolerance in tobacco. International Journal of Molecular Sciences 20 (1), 89. Available at https://doi.org/10.3390/ijms20010089.

Liu, X., Yue, Y., Li, B., Nie, Y., Li, W., Wu, W.H., et al., 2007. A G protein-coupled receptor is a plasma membrane receptor for the plant hormone abscisic acid. Science 315 (5819), 1712–1716. Available at https://doi.org/10.1126/science.1135882.

Ma, H., Yanofsky, M.F., Meyerowitz, E.M., 1990. Molecular cloning and characterization of GPA1, a G protein alpha subunit gene from *Arabidopsis thaliana*. Proceedings of the National Academy of Sciences 87 (10), 3821–3825. Available at https://doi.org/10.1073/pnas.87.10.3821.

Mao, H., Sun, S., Yao, J., Wang, C., Yu, S., Xu, C., et al., 2010. Linking differential domain functions of the GS3 protein to natural variation of grain size in rice. Proceedings of the National Academy of Sciences 107 (45), 19579–19584. Available at https://doi.org/10.1073/pnas.1014419107.

Maruta, N., Trusov, Y., Brenya, E., Parekh, U., Botella, J.R., 2015. Membrane-localized extra-large G proteins and Gβγ of the heterotrimeric G proteins form functional complexes engaged in plant immunity in *Arabidopsis*. Plant Physiology 167 (3), 1004–1016. Available at https://doi.org/10.1104/pp.114.255703, http://www.plantphysiol.org/content/167/3/1004.full.pdf.

Maruta, N., Trusov, Y., Jones, A.M., Botella, J.R., 2021. Heterotrimeric G proteins in plants: canonical and atypical Gα subunits. International Journal of Molecular Sciences 22 (21). Available at https://doi.org/10.3390/ijms222111841, https://www.mdpi.com/1422-0067/22/21/11841/pdf.

Maruta, N., Trusov, Y., Urano, D., Chakravorty, D., Assmann, S.M., Jones, A.M., et al., 2021. GTP binding by *Arabidopsis* extra-large G protein 2 is not essential for its functions. Plant Physiology 186 (2), 1240–1253. Available at https://doi.org/10.1093/PLPHYS/KIAB119, https://academic.oup.com/plphys/article/186/2/1240/6174699.

Mason, M.G., Botella, J.R., 2000. Completing the heterotrimer: isolation and characterization of an *Arabidopsis thaliana* G protein γ-subunit cDNA. Proceedings of the National Academy of Sciences 97 (26), 14784–14788. Available at https://doi.org/10.1073/pnas.97.26.14784.

McCudden, C.R., Hains, M.D., Kimple, R.J., Siderovski, D.P., Willard, F.S., 2005. G-protein signaling: back to the future. Cellular and Molecular Life Sciences 62 (5), 551–577. Available at https://doi.org/10.1007/s00018-004-4462-3.

Miao, J., Yang, Z., Zhang, D., Wang, Y., Xu, M., Zhou, L., et al., 2019. Mutation of RGG2, which encodes a type B heterotrimeric G protein γ subunit, increases grain size and yield production in rice. Plant biotechnology journal 17 (3), 650–664. Available at https://doi.org/10.1111/pbi.13005, http://onlinelibrary.wiley.com/journal/10.1111/.

Milligan, G., Kostenis, E., 2006. Heterotrimeric G-proteins: a short history. British Journal of Pharmacology 147 (1), S46. Available at https://doi.org/10.1038/sj.bjp.0706405.

Mohanasundaram, B., Pandey, S., 2023. Moving beyond the *Arabidopsis*-centric view of G-protein signaling in plants. Trends in Plant Science 28 (12), 1406–1421. Available at https://doi.org/10.1016/j.tplants.2023.07.014, www.elsevier.com/inca/publications/store/3/0/9/6/0/index.htt.

Neves, S.R., Ram, P.T., Iyengar, R., 2002. G protein pathways. Science 296 (5573), 1636–1639. Available at https://doi.org/10.1126/science.1071550.

Nilson, S.E., Assmann, S.M., 2010. The α-subunit of the *Arabidopsis* heterotrimeric G protein, GPA1, is a regulator of transpiration efficiency. Plant Physiology 152 (4), 2067–2077. Available at https://doi.org/10.1104/pp.109.148262, http://www.plantphysiol.org/cgi/reprint/152/4/2067.

Pandey, S., Monshausen, G.B., Ding, L., Assmann, S.M., 2008. Regulation of root-wave response by extra large and conventional G proteins in *Arabidopsis thaliana*. Plant Journal 55 (2), 311–322. Available at https://doi.org/10.1111/j.1365-313X.2008.03506.x.

Pandey, S., Vijayakumar, A., 2018. Emerging themes in heterotrimeric G-protein signaling in plants. Plant Science 270, 292–300. Available at https://doi.org/10.1016/j.plantsci.2018.03.001, http://www.journals.elsevier.com/plant-science/.

Patel, J.S., Sarma, B.K., Singh, H.B., Upadhyay, R.S., Kharwar, R.N., Ahmed, M., 2016. Pseudomonas fluorescens and trichoderma asperellum enhance expression of Gα subunits of the pea heterotrimeric G-protein during erysiphe pisi infection. Frontiers in Plant Science 6 (2016). Available at https://doi.org/10.3389/fpls.2015.01206, http://journal.frontiersin.org/article/10.3389/fpls.2015.01206/pdf.

Perfus-Barbeoch, L., Jones, A.M., Assmann, S.M., 2004. Plant heterotrimeric G protein function: insights from *Arabidopsis* and rice mutants. Current Opinion in Plant Biology 7 (6), 719–731. Available at https://doi.org/10.1016/j.pbi.2004.09.013.

Pierce, K.L., Premont, R.T., Lefkowitz, R.J., 2002. Seven-transmembrane receptors. Nature Reviews Molecular Cell Biology 3 (9), 639–650. Available at https://doi.org/10.1038/nrm908.

Rajhi, I., Yamauchi, T., Takahashi, H., Nishiuchi, S., Shiono, K., Watanabe, R., et al., 2011. Identification of genes expressed in maize root cortical cells during lysigenous aerenchyma formation using laser microdissection and microarray analyses. New Phytologist 190 (2), 351–368. Available at https://doi.org/10.1111/j.1469-8137.2010.03535.x.

Roy Choudhury, S., Bisht, N.C., Thompson, R., Todorov, O., Pandey, S., 2011. Conventional and novel gγ protein families constitute the heterotrimeric g-protein signaling network in soybean. PLoS ONE 6 (8). Available at https://doi.org/10.1371/journal.pone.0023361, http://www.plosone.org/article/fetchObjectAttachment.action?uri=info%3Adoi%2F10.1371%2Fjournal.pone.0023361&representation=PDF.

Roy Choudhury, S., Pandey, S., 2015. Phosphorylation-dependent regulation of G-protein cycle during nodule formation in soybean. The Plant Cell 27 (11), 3260–3276. Available at https://doi.org/10.1105/tpc.15.00517.

Roy Choudhury, S., Pandey, S., 2016. Interaction of heterotrimeric G-protein components with receptor-like kinases in plants: an alternative to the established signaling paradigm? Molecular Plant 9 (8), 1093–1095. Available at https://doi.org/10.1016/j.molp.2016.05.012.

Roy Choudhury, S., Pandey, S., 2022. SymRK -dependent phosphorylation of Gα protein and its role in signaling during soybean (*Glycine max*) nodulation. The Plant Journal 110 (1), 277–291. Available at https://doi.org/10.1111/tpj.15672.

Roy Choudhury, S., Riesselman, A.J., Pandey, S., 2014. Constitutive or seed-specific overexpression of *Arabidopsis* G-protein γ subunit 3 (AGG3) results in increased seed and oil production and improved stress tolerance in *Camelina sativa*. Plant biotechnology journal 12 (1), 49–59. Available at https://doi.org/10.1111/pbi.12115.

Simoni, E.B., Oliveira, C.C., Fraga, O.T., Reis, P.A.B., Fontes, E.P.B., 2022. Cell death signaling from endoplasmic reticulum stress: plant-specific and conserved features. Frontiers in Plant Science 13. Available at https://doi.org/10.3389/fpls.2022.835738, https://www.frontiersin.org/journals/plant-science.

Smrcka, A.V., 2008. G protein βγ subunits: central mediators of G protein-coupled receptor signaling. Cellular and Molecular Life Sciences 65 (14), 2191–2214. Available at https://doi.org/10.1007/s00018-008-8006-5.

Sotelo, T., Lema, M., Soengas, P., Cartea, M.E., Velasco, P., Drake, H.L., 2015. In vitro activity of glucosinolates and their degradation products against brassica-pathogenic bacteria and fungi. Applied and Environmental Microbiology 81 (1), 432–440. Available at https://doi.org/10.1128/aem.03142-14.

Subramaniam, G., Trusov, Y., Lopez-Encina, C., Hayashi, S., Batley, J., Botella, J.R., 2016. Type B heterotrimeric G protein γ-subunit regulates Auxin and ABA signaling in tomato. Plant Physiology 170 (2), 1117–1134. Available at https://doi.org/10.1104/pp.15.01675, http://www.plantphysiol.org/content/170/2/1117.full.pdf.

Sun, S., Chen, D., Li, X., Qiao, S., Shi, C., Li, C., et al., 2015. Brassinosteroid signaling regulates leaf erectness in oryza sativa via the control of a specific U-type cyclin and cell proliferation. Developmental Cell 34 (2), 220–228. Available at https://doi.org/10.1016/j.devcel.2015.05.019, https://www.cell.com/developmental-cell/home.

Sun, S., Wang, L., Mao, H., Shao, L., Li, X., Xiao, J., et al., 2018. A G-protein pathway determines grain size in rice. Nature Communications 9 (1). Available at https://doi.org/10.1038/s41467-018-03141-y, http://www.nature.com/ncomms/index.html.

Swain, D.M., Sahoo, R.K., Srivastava, V.K., Tripathy, B.C., Tuteja, R., Tuteja, N., 2017. Function of heterotrimeric G-protein γ subunit RGG1 in providing salinity stress tolerance in rice by elevating detoxification of ROS. Planta

245 (2), 367–383. Available at https://doi.org/10.1007/s00425-016-2614-3, link.springer.de/link/service/journals/00425/index.htm.

Tao, Y., Miao, J., Wang, J., Li, W., Xu, Y., Wang, F., et al., 2020. RGG1, involved in the cytokinin regulatory pathway, controls grain size in rice. Rice 13 (1). Available at https://doi.org/10.1186/s12284-020-00436-x, http://www.springerlink.com/content/1939-8425.

Tennakoon, M., Senarath, K., Kankanamge, D., Ratnayake, K., Wijayaratna, D., Olupothage, K., et al., 2021. Subtype-dependent regulation of Gβγ signalling. Cellular Signalling 82, 109947. Available at https://doi.org/10.1016/j.cellsig.2021.109947.

Tiwari, R., Kaur, J., Bisht, N.C., 2021. Extra-large G-proteins influence plant response to *Sclerotinia sclerotiorum* by regulating glucosinolate metabolism in *Brassica juncea*. Molecular Plant Pathology 22 (10), 1180–1194. Available at https://doi.org/10.1111/mpp.13096, http://onlinelibrary.wiley.com/journal/10.1111/.

Torres, M.A., Dangl, J.L., Jones, J.D.G., 2002. *Arabidopsis* gp91 phox homologues AtrbohD and AtrbohF are required for accumulation of reactive oxygen intermediates in the plant defense response. Proceedings of the National Academy of Sciences 99 (1), 517–522. Available at https://doi.org/10.1073/pnas.012452499.

Torres, M.A., Morales, J., Sánchez-Rodríguez, C., Molina, A., Dangl, J.L., 2013. Functional interplay between *Arabidopsis* NADPH oxidases and heterotrimeric G protein. Molecular Plant-Microbe Interactions 26 (6), 686–694. Available at https://doi.org/10.1094/MPMI-10-12-0236-RSpain, http://apsjournals.apsnet.org/doi/pdf/10.1094/MPMI-10-12-0236-R.

Trusov, Y., Chakravorty, D., Botella, J.R., 2012. Diversity of heterotrimeric G-protein subunits in plants. BMC Research Notes 5. Available at https://doi.org/10.1186/1756-0500-5-608.

Trusov, Y., Rookes, J.E., Chakravorty, D., Armour, D., Schenk, P.M., Botella, J.R., 2006. Heterotrimeric G proteins facilitate *Arabidopsis* resistance to necrotrophic pathogens and are involved in jasmonate signaling. Plant Physiology 140 (1), 210–220. Available at https://doi.org/10.1104/pp.105.069625, http://www.plantphysiol.org/.

Trusov, Y., Rookes, J.E., Tilbrook, K., Chakravorty, D., Mason, M.G., Anderson, D., et al., 2007. Heterotrimeric G protein γ subunits provide functional selectivity in Gβγ dimer signaling in *Arabidopsis*. The Plant Cell 19 (4), 1235–1250. Available at https://doi.org/10.1105/tpc.107.050096, http://www.plantcell.org/cgi/reprint/19/4/1235.

Ueguchi-Tanaka, M., Fujisawa, Y., Kobayashi, M., Ashikari, M., Iwasaki, Y., Kitano, H., et al., 2000. Rice dwarf mutant d1, which is defective in the α subunit of the heterotrimeric G protein, affects gibberellin signal transduction. Proceedings of the National Academy of Sciences 97 (21), 11638–11643. Available at https://doi.org/10.1073/pnas.97.21.11638.

Ullah, H., Chen, J.G., Temple, B., Boyes, D.C., Alonso, J.M., Davis, K.R., et al., 2003. The β-subunit of the *Arabidopsis* G protein negatively regulates auxin-induced cell division and affects multiple developmental processes. The Plant Cell 15 (2), 393–409. Available at https://doi.org/10.1105/tpc.006148, http://www.plantcell.org/.

Ullah, H., Chen, J.G., Wang, S., Jones, A.M., 2002. Role of a heterotrimeric G protein in regulation of *Arabidopsis* seed germination. Plant Physiology 129 (2), 897–907. Available at https://doi.org/10.1104/pp.005017, http://www.plantphysiol.org/.

Ullah, H., Chen, J.G., Young, J.C., Im, K.H., Sussman, M.R., Jones, A.M., 2001. Modulation of cell proliferation by heterotrimeric G protein in *Arabidopsis*. Science 292 (5524), 2066–2069. Available at https://doi.org/10.1126/science.1059040.

Urano, D., Chen, J.G., Botella, J.R., Jones, A.M., 2013. Heterotrimeric G protein signalling in the plant kingdom. Open Biology 3. Available at https://doi.org/10.1098/rsob.120186, http://www.rsob.royalsocietypublishing.org/site/misc/about.xhtml.

Urano, D., Jones, J.C., Wang, H., Matthews, M., Bradford, W., Bennetzen, J.L., et al., 2012. G protein activation without a GEF in the plant kingdom. PLoS genetics 8 (6). Available at https://doi.org/10.1371/journal.pgen.1002756, http://www.plosgenetics.org/article/fetchObjectAttachment.action?uri=info%3Adoi%2F10.1371%2Fjournal.pgen.1002756&representation=PDF.

Urano, D., Jones, A.M., 2014. Heterotrimeric G protein-coupled signaling in plants. Annual review of plant biology 65, 365–384. Available at https://doi.org/10.1146/annurev-arplant-050213-040133.

Urano, D., Miura, K., Wu, Q., Iwasaki, Y., Jackson, D., Jones, A.M., 2016a. Plant morphology of Heterotrimeric G protein mutants. Plant and Cell Physiology 57 (3), 437–445. Available at https://doi.org/10.1093/pcp/pcw002, http://pcp.oxfordjournals.org/.

Urano, D., Maruta, N., Trusov, Y., Stoian, R., Wu, Q., Liang, Y., et al., 2016b. Saltational evolution of the heterotrimeric G protein signaling mechanisms in the plant kingdom. Science signaling 9 (446). Available at https://doi.org/10.1126/scisignal.aaf9558, http://stke.sciencemag.org/content/sigtrans/9/446/ra93.full.pdf.

Urano, D., Phan, N., Jones, J.C., Yang, J., Huang, J., Grigston, J., et al., 2012. Endocytosis of the seven-transmembrane RGS1 protein activates G-protein-coupled signalling in *Arabidopsis*. Nature Cell Biology 14 (10), 1079–1088. Available at https://doi.org/10.1038/ncb2568.

Utsunomiya, Y., Samejima, C., Takayanagi, Y., Izawa, Y., Yoshida, T., Sawada, Y., et al., 2011. Suppression of the rice heterotrimeric G protein β-subunit gene, RGB1, causes dwarfism and browning of internodes and lamina joint regions. The Plant Journal 67 (5), 907–916. Available at https://doi.org/10.1111/j.1365-313x.2011.04643.x.

Vavilova, V., Konopatskaia, I., Kuznetsova, A.E., Blinov, A., Goncharov, N.P., 2017. DEP1 gene in wheat species with normal, compactoid and compact spikes. Federation BMC Genetics 18. Available at https://doi.org/10.1186/s12863-017-0583-6, http://www.biomedcentral.com/bmcgenet/.

Wang, J., Luo, Q., Deng, J., Liang, X., Li, Y., Wang, A., et al., 2025. The G-protein β subunit SlGB1 regulates tyramine-derived phenolamide metabolism for shoot apex growth and development in tomato. The Plant Cell 37 (4). Available at https://doi.org/10.1093/plcell/koaf070, https://academic.oup.com/plcell/issue.

Wang, S., Narendra, S., Fedoroff, N., 2007. Heterotrimeric G protein signaling in the *Arabidopsis* unfolded protein response. Proceedings of the National Academy of Sciences 104 (10), 3817–3822. Available at https://doi.org/10.1073/pnas.0611735104.

Wang, X.Q., Ullah, H., Jones, A.M., Assmann, S.M., 2001. G protein regulation of ion channels and abscisic acid signaling in *Arabidopsis* guard cells. Science 292 (5524), 2070–2072. Available at https://doi.org/10.1126/science.1059046.

Wang, X., Wang, Y., Zheng, Z., Cui, Y., 2025. GPA1 is a determinant of leaf width and fruit size in tomato. Plant Science 352, 112336. Available at https://doi.org/10.1016/j.plantsci.2024.112336.

Wang, Y., Wu, Y., Yu, B., Yin, Z., Xia, Y., 2017. EXTRA-LARGE G proteins interact with E3 ligases PUB4 and PUB2 and function in cytokinin and developmental processes. Plant Physiology 173 (2), 1235–1246. Available at https://doi.org/10.1104/pp.16.00816, http://www.plantphysiol.org/content/173/2/1235.full.pdf.

Wang, Y., Zhang, H., Wang, P., Zhong, H., Liu, W., Zhang, S., et al., 2023. *Arabidopsis* EXTRA-LARGE G PROTEIN 1 (XLG1) functions together with XLG2 and XLG3 in PAMP-triggered MAPK activation and immunity. Journal of Integrative Plant Biology 65 (3), 825–837. Available at https://doi.org/10.1111/jipb.13391, http://onlinelibrary.wiley.com/journal/10.1111/.

Weiss, C.A., Garnaat, C.W., Mukai, K., Hu, Y., Ma, H., 1994. Isolation of cDNAs encoding guanine nucleotide-binding protein beta-subunit homologues from maize (ZGB1) and *Arabidopsis* (AGB1). Proceedings of the National Academy of Sciences 91 (20), 9554–9558. Available at https://doi.org/10.1073/pnas.91.20.9554.

Wendt, T., Holme, I., Dockter, C., Preuß, A., Thomas, W., Druka, A., et al., 2016. HvDep1 Is a positive regulator of culm elongation and grain size in barley and impacts yield in an environment-dependent manner. PLoS ONE 11 (12), e0168924. Available at https://doi.org/10.1371/journal.pone.0168924.

Wu, L., Wang, X., Yu, Z., Cui, X., Xu, Q., 2022. Simultaneous improvement of grain yield and quality through manipulating two type C G protein gamma subunits in rice. International Journal of Molecular Sciences 23 (3), 1463. Available at https://doi.org/10.3390/ijms23031463.

Wu, Q., Xu, F., Liu, L., Char, S.N., Ding, Y., Je, B.I., et al., 2020. The maize heterotrimeric G protein β subunit controls shoot meristem development and immune responses. Proceedings of the National Academy of Sciences 117 (3), 1799–1805. Available at https://doi.org/10.1073/pnas.1917577116.

Xiong, M., Zhang, H., Huang, Y., Li, Y., 2024. G protein regulation in rice seed biology. Seed Biology 3, 1–11. Available at https://doi.org/10.48130/seedbio-0024-0008.

Xu, Db, Gao, Sq, Ma, Yn, Wang, Xt, Feng, L., et al., 2017. The G-Protein β Subunit AGB1 promotes hypocotyl elongation through inhibiting transcription activation function of BBX21 in *Arabidopsis*. Molecular Plant 10 (9), 1206–1223. Available at https://doi.org/10.1016/j.molp.2017.08.004.

Xu, R., Li, N., Li, Y., 2019. Control of grain size by G protein signaling in rice. Journal of Integrative Plant Biology 61 (5), 533–540. Available at https://doi.org/10.1111/jipb.12769, http://onlinelibrary.wiley.com/journal/10.1111/.

Xue, J., Gong, B.Q., Yao, X., Huang, X., Li, J.F., 2020. BAK1-mediated phosphorylation of canonical G protein alpha during flagellin signaling in *Arabidopsis*. Journal of Integrative Plant Biology 62 (5), 690–701. Available at https://doi.org/10.1111/jipb.12824, http://onlinelibrary.wiley.com/journal/10.1111/.

Yalovsky, S., 2011. 9 - Protein prenylation caax processing in plants. In: Tamanoi, F., Hrycyna, C.A., Bergo, M.O. (Eds.), The Enzymes. Academic Press, pp. 163–182.

Yamauchi, T., Rajhi, I., Nakazono, M., 2011. Lysigenous aerenchyma formation in maize root is confined to cortical cells by regulation of genes related to generation and scavenging of reactive oxygen species. Plant Signaling & Behavior 6 (5), 759–761. Available at https://doi.org/10.4161/psb.6.5.15417.

Yang, X., Lu, J., Shi, W.J., Chen, Y.H., Yu, J.W., Chen, S.H., et al., 2024. RGA1 regulates grain size, rice quality and seed germination in the small and round grain mutant srg5. BMC Plant Biology 24 (1). Available at https://doi.org/10.1186/s12870-024-04864-5, https://bmcplantbiol.biomedcentral.com/.

Yu, Y., Assmann, S.M., 2016. The effect of NaCl on stomatal opening in *Arabidopsis* wild type and agb1 heterotrimeric G-protein mutant plants. Plant Signaling and Behavior 11 (2). Available at https://doi.org/10.1080/15592324.2015.1085275, http://www.tandfonline.com/loi/kpsb20.

Yu, Y., Assmann, S.M., 2015. The heterotrimeric G-protein β subunit, AGB1, plays multiple roles in the *Arabidopsis* salinity response. Plant, Cell and Environment 38 (10), 2143–2156. Available at https://doi.org/10.1111/pce.12542, http://onlinelibrary.wiley.com/journal/10.1111/.

Yu, Y., Chakravorty, D., Assmann, S.M., 2018. The G protein β-subunit, AGB1, interacts with FERONIA in RALF1-regulated stomatal movement. Plant Physiology 176 (3), 2426–2440. Available at https://doi.org/10.1104/pp.17.01277, http://www.plantphysiol.org/content/176/3/2426.full.pdf.

Yu, F., Liu, C., Zhang, H., Xie, Q., 2023. A G protein γ subunit regulates crop alkaline sensitivity by modulating H_2O_2 transporter PIP2s. Journal of molecular cell biology 15 (3). Available at https://doi.org/10.1093/jmcb/mjad020.

Zhang, L., Hu, G., Cheng, Y., Huang, J., 2008. Heterotrimeric G protein α and β subunits antagonistically modulate stomatal density in *Arabidopsis thaliana*. Developmental Biology 324 (1), 68–75. Available at https://doi.org/10.1016/j.ydbio.2008.09.008, http://www.elsevier.com/inca/publications/store/6/2/2/8/1/6/index.htt.

Zhang, H., Xie, P., Xu, X., Xie, Q., Yu, F., 2021a. Heterotrimeric G protein signalling in plant biotic and abiotic stress response. Plant Biology 23, 20–30.

Zhang, D., Zhang, M., Liang, J., 2021b. RGB1 regulates grain development and starch accumulation through its effect on OsYUC11-mediated auxin biosynthesis in rice endosperm cells. Frontiers in Plant Science 12. Available at https://doi.org/10.3389/fpls.2021.585174, https://www.frontiersin.org/journals/plant-science.

Zhang, H., Yu, F., Xie, P., Sun, S., Qiao, X., Tang, S., et al., 2023. A Gγ protein regulates alkaline sensitivity in crops. Science 379 (6638). Available at https://doi.org/10.1126/science.ade8416.

Zhang, D., Zhang, M., Zhou, Y., Wang, Y., Shen, J., Chen, H., et al., 2019. The rice g protein γ subunit DEP1/qPE9–1 positively regulates grain-filling process by increasing auxin and cytokinin content in rice grains. Rice 12 (1). Available at https://doi.org/10.1186/s12284-019-0344-4, http://www.springerlink.com/content/1939-8425.

Zhu, H., Li, G.J., Ding, L., Cui, X., Berg, H., Assmann, S.M., et al., 2009. *Arabidopsis* extra large G-protein 2 (XLG2) interacts with the Gβ subunit of heterotrimeric G protein and functions in disease resistance. Molecular Plant 2 (3), 513–525. Available at https://doi.org/10.1093/mp/ssp001.

CHAPTER

13

CRISPR–Cas-enabled accelerated breeding pipelines for engineering climate-resilient crops under abiotic stress conditions

Aditi Dutt[1], *Arnika Keithellakpam*[1], *Adity Laskar*[1], *Anita Kachari*[1], *Channakeshavaiah Chikkaputtaiah*[1,2], *and Banashree Saikia*[1,2,*]

[1]Center for Biotechnology, Biological Sciences and Technology Division, CSIR-North East Institute of Science and Technology (CSIR-NEIST), Jorhat, Assam, India [2]Academy of Scientific and Innovative Research (AcSIR), Ghaziabad, Uttar Pradesh, India

13.1 Introduction

Abiotic stresses, including drought, salinity, heat, cold, and flooding, are significant environmental constraints restricting global crop productivity and food security. They cause a >50% reduction in yields of the key cereals and legumes in particular areas with poor infrastructure and high climate vulnerability (Ahmar et al., 2020; Pandey et al., 2022). The United Nations Food and Agriculture Organization (FAO) estimates that food production will need to increase by approximately 50% by 2050 to feed a growing world population, which is even more challenging considering the decline in arable land and increasing stress on water resources (FAO, 2017). Enhancing crop productivity via modern breeding methodologies is more essential than ever. Although traditional breed-based agriculture has served as the foundation for modern agricultural development, it is inherently slow. It can be time-bound, due to the time taken for generation cycles to be completed and a very restricted number of recombination events (Moose and Mumm, 2008; Dwivedi et al., 2008). Classically, it can take 10–15 years on average to develop a new crop variety, especially for

[*]Corresponding author. email address: banasreesaikia39@gmail.com

Harnessing Plant-Stimulating Molecules
DOI: https://doi.org/10.1016/B978-0-443-44994-9.00011-0

complex traits such as abiotic stress tolerance, which are controlled by multiple quantitative trait loci (QTLs) and susceptibility to genotype-by-environment interactions (Collard and Mackill, 2008; Varshney et al., 2021). Additionally, the polygenic nature and low heritability of such traits in stress conditions also limit the accuracy of their selection, which is also challenging for the breeding process (Mickelbart et al., 2015).

Speed breeding exploits environmental manipulation, i.e., lengthening the photoperiod, temperature control, and growth condition optimization, to speed up plants' growth and reproductive phases (Watson et al., 2018). Using this technique has enabled a threefold increase in the annual growth rate of crops grown under natural field conditions, reducing the time needed for genetic improvement of crops and releasing new variants (Ahmad et al., 2021). Advanced molecular biology and genomics have revolutionized crop improvement methods in recent years. One prominent example is genome editing using CRISPR/Cas technology. In 2013 CRISPR (Clustered Regularly Interspaced Short Palindromic Repeats) was first adapted for plant genome editing and has since proved to be a powerful tool for precise and targeted manipulation of plant genomes (Jinek et al., 2012; Belhaj et al., 2013). One of the possible applications of genome editing is to either introduce functional regulatory genes into superior cultivars or to eliminate harmful regulatory genes associated with susceptibility to abiotic stress. The CRISPR-speed breeding pipeline is an efficient, high-throughput breeding platform that can generate crops designed to tolerate abiotic stress conditions, with the potential to combine multiple traits. Additionally, speed breeding can cycle the edited lines quickly to be homozygous and generate varieties suitable for the field (Chimmili et al., 2022; Pandey et al., 2022).

Recent studies have demonstrated that this integrated approach is effective in editing genes such as *DREB*, *NHX1*, *HSP70*, and *HKT1–5*, which are linked to drought and salinity tolerance (Biswas et al., 2021; Chimmili et al., 2022; Li et al., 2022; Rai, 2022). Combining CRISPR with speed breeding reduces breeding time by over 50% and enables stacking multiple traits through multiplex genome editing. Beyond laboratory success, this strategy is being adopted by major crop improvement programs globally. Technologies like omics technology, genome-wide association studies (GWAS), and artificial intelligence (AI)-machine learning (ML) are being introduced to breeding to refine its trait targets and selection strategies. Since gene redundancy and multiple homeologs complicate specificity and efficiency, genome editing in polyploid crops, such as potatoes, wheat, and sugarcane, remains challenging (Chen et al., 2019). Similarly, a major challenge in many CRISPR-enabled breeding programs is achieving high-efficiency transformation and regeneration in resistant species, particularly in cereals and legumes (Gao, 2021).

The regulations regarding the classification of CRISPR-edited crops, particularly those developed through transgene-free methods such as ribonucleoprotein (RNP) delivery or SDN-1/SDN-2 edits, vary across different parts of the world. This makes it harder to sell them and get them used to being around the globe (Raza et al., 2021). For example, the United States, Brazil, and Argentina do not classify some genome-edited crops as genetically modified organisms (GMOs), whereas the European Union does, due to the precautionary principle. Ethical issues and public perceptions also need to be managed carefully and in advance. Many CRISPR-based changes are not transgenic, but most people are unaware of the distinction between genome editing and traditional genetic modification (GM). This lack of understanding can make people less likely to purchase something artificially edited, which in turn affects policy decisions made regarding genome-edited crops (Waltz, 2018). Therefore it is crucial to support transparent science communication, conduct thorough biosafety assessments, and collaborate effectively with all stakeholders, including farmers, consumers, regulators, and civil society. This method will help ensure that CRISPR-based breeding technologies are used fairly and responsibly.

In this chapter, we emphasize the use of CRISPR/Cas systems as a tool for speed breeding to engineer crops that are resilient to abiotic stresses. By summarizing significant target genes, experimental strategies, technical constraints, and regulatory controversies, we emphasize the pivotal role this technology is likely to play in the next generation of climate-resilient agriculture.

13.2 Speed breeding: accelerating crop breeding cycles

Climate change, including increased temperatures, more frequent floods, and droughts, is expected to cause new diseases and pest outbreaks more frequently, prompting a rapid response from plant breeders. Breeding results, including cross-generation mapping and assessment of desirable agronomic traits, help accelerate growth through the use of speed breeding techniques (Samantara et al., 2022). Speed breeding is a next-generation breeding approach that enables the growth of plants in controlled, manipulable conditions, thereby shortening the time and resources required to complete a breeding cycle by encompassing multiple generations within a single year (Ćeran et al., 2024).

13.2.1 Principle of speed breeding

Speed breeding is achieved through regulated photoperiod extension, which typically maintains 22 hours of light per day and night temperatures of around 22°C and 17°C, respectively. The optimal temperature varies by crop type, and optimal light quality is also applied. Ultimately, these shorten the growth period when a crop should grow and develop by speeding up photosynthesis and the time to move to the flowering, seed maturity, and harvesting stage. Speed breeding and single-seed descent are commonly employed to produce efficient breeding lines (Shyamala and Sasipriya, 2023).

13.2.2 Techniques used in speed breeding (e.g., controlled environment agriculture, long-period light exposure, and tissue culture)

Speed breeding is an innovative form of plant breeding that enables the creation of superior crops in a shorter time frame under carefully controlled environmental conditions. To achieve targeted breeding objectives and enhance crop quality, various strategies, including plant tissue culture, genetic engineering, molecular tools, modern breeding methods, the use of light-emitting diodes (LEDs), the application of growth regulators, immature seed germination, and micronutrient supplementation, are employed (Sharma et al., 2024). Recycled wastewater and agricultural waste are used in the controlled-environment plant growth system to produce plants. By reducing the carbon footprint and ensuring the stability of carbon sequestration, the controlled-environment plant growth chamber effectively offsets greenhouse gas emissions and mitigates any yield losses. It maintains a steady climate, allowing plants to develop, and enhances photosynthetic assimilation and metabolic efficiency (Blom et al., 2022).

Speed breeding primarily works by altering light intensity and duration, which, when perceived by photoreceptors, causes accelerated reproductive development in plants. These photoreceptors detect light and control the natural circadian cycle, which is the first and fastest responder to changing environmental conditions. High light intensity, increased CO_2, and proper temperature control can significantly increase photosynthetic rate, reducing the days to flower in plants (Rai, 2022).

In wheat systems, the modified speed breeding (mSB) method utilizes 22-hour photoperiods, which combine nearly 10 hours of natural sunlight with 12 hours of LED lighting. The Photosynthetic Photon Flux Density (PPFD) is approximately 200 $\mu mol \cdot m^{-2} \cdot s^{-1}$ during the daytime and around 65 $\mu mol \cdot m^{-2} \cdot s$ at night. Compared to an entirely artificial SB chamber, this regimen utilizes 80% less energy while producing similar flowering and grain production (Cha et al., 2023). Using a blue-enriched spectrum with reduced far-red components and a 10-hour photoperiod promoted the development of soybean plants with a short and small phenotype, flowering approximately 23 days after sowing, and reaching maturity in 77 days, which allowed for the production of over five generations annually. In contrast, flowering in amaranth occurred about 35 days after sowing, and in rice, it occurred about 60 days after sowing. However, in some amaranth accessions, the addition of far-red light shortened the time to flowering by over 10 days, and upto 20 days in some rice genotypes, further demonstrating the genotype-dependent effect of spectral composition on floral initiation in short-day species (Jähne et al., 2020).

Apart from light, several major crops, including soybeans, canola, wheat, barley, and maize, exhibit different temperature requirements across various growth stages. For instance, seed germination generally occurs within the range of 12°C–30°C, while optimal growth and flowering are achieved at 25°C–30°C (Wang et al., 2021). Additionally, winter wheat requires vernalization, or exposure to low temperatures, to transition from the vegetative phase to the reproductive stage (Zheng et al., 2023), whereas temperatures above 33°C can diminish pollen viability and increase male sterility in some crops. In SB protocols, each crop should have its maximum and minimum temperature ranges considered. Fan-and-pad cooling systems, foggers, and solar-powered air batteries are examples of devices used to control temperature (Ćeran et al., 2024).

Spice crops are frequently regenerated without the use of viruses or diseases by employing plant tissue culture techniques like cell culture, suspension culture, Bergmann plating technique, raft nurse tissue paper, seed-based culture, embryo-based culture, endosperm culture, ovary culture, ovule culture, anther culture, protoplast culture/somatic hybridization, and somaclonal variation. These methods are often used in conjunction with direct or indirect gene alteration technologies, including the CRISPR/Cas technology (Das et al., 2023), to create crops that can better withstand biotic and abiotic stressors. These crop variations in spice include biofortified population lines, resistant population lines, and much more. Additionally, to cultivate regenerated plant populations in controlled environments, technologies such as plant tissue culture and gene transfer techniques, including CRISPR/Cas, are integrated with soilless culture systems (Pandey et al., 2025).

13.2.3 Advantages of speed breeding in accelerating crop development

1. Speed breeding addresses issues such as low seed germination and inadequate vigor, making it more advantageous than the DH (doubled haploid) protocol. For example, it takes only 77 days to grow a full generation of durum wheat because its generation turnover is too fast and does not require specific genotypes or facilities, which makes it easier to adapt and modify crops (Potts et al., 2023).
2. Methods such as "SpeedGS," which integrate speed breeding and genomic selection, greatly enhance genetic gain. This technique speeds up selection and inbreeding in spring wheat by promoting greater recombination and early phenotypic selection (Gudi et al., 2022).
3. Compared to the usual 1–2 annual cycles, speed breeding allows for 4–6 generations each year, which significantly exceeds the typical 1–2 annual cycles. This speeds up selection, allowing breeders to analyze and develop numerous breeding cycles more quickly, and lowering the overall time needed to generate new kinds within a few years instead of decades (Pandey et al., 2022) (Table 13.1; Fig. 13.1).

13.3 CRISPR–Cas technology in crop improvement

13.3.1 Discovery and mechanism

Bacteria and archaea have an immunity mechanism called the CRISPR/Cas system, activated to combat plasmids and viruses that invade their bodies. Its foundation is the idea that nucleolytic cleavage occurs after accurate DNA-RNA recognition. CRISPR was adapted for gene editing in eukaryotic systems starting in 2012–13 (Koonin and Makarova, 2019). CRISPR/Cas9 is one of the most widely used systems in plants, derived from the bacterium *Streptococcus pyogenes*. It comprises two essential components: a nuclease enzyme (e.g., Cas9, Cas12a, or Cas13a) for cutting DNA and a single guide RNA (sgRNA) for guiding the Cas nuclease enzyme to the target site. A scaffold region and a ~20-nucleotide sequence complementary to the target DNA are present in the sgRNA. The Cas protein searches the genome for sequences that follow a protospacer adjacent motif (PAM) and then binds if the spacer sequence matches. A short nucleotide sequence known as PAM may appear upstream or downstream of the target site. Usually, it has two to six base pairs. A double-strand break (DSB) occurs at the target site when the Cas enzyme cleaves both DNA strands a few base pairs upstream of the PAM site after binding. Two primary cellular repair pathways then repair this DSB: nonhomologous end joining (NHEJ) is a repair mechanism for DNA that fixes DSBs without a homologous template, often resulting in insertions or deletions (indels). Homology-directed repair (HDR) is a more accurate process that uses a homologous donor DNA template to insert or correct sequences at the break site. This duality enables gene knockouts and precise allele edits, allowing for a broad spectrum of applications in stress tolerance, yield improvement, and disease resistance (Viviani et al., 2022).

TABLE 13.1 An overview of speed breeding protocols and applications for improving crop traits and breeding efficiency.

Protocol used	Crop	Photoperiod and temperature	Techniques	Tools and markers used	Traits enhanced	Application	References
Optimized LED Photoperiod + vernalization	Wheat	22 h light/2 h dark, 20°C– 22°C	Long photoperiod and regulated vernalization	Genomic selection, MAS, vernalization units, and LED growth chambers	Resistance to disease and stability of yield	Used for the rapid advancement of elite lines before multilocation field trials	Cha et al. (2023)
H_2O_2-Assisted Immature Seed Harvest	Wheat	22 h light; 15°C–20°C	Immature seed harvest	Growth chambers, H_2O_2 treatment setup	Faster germination, reduced cycle time	Applied to shorten the seed-to-seed interval in early generation selection	Kumar and Walia (2024)
Pod/Seed Chip Genotyping	Pigeon pea	13 h–8 h–13 h triphasic	SSD + chip-based MAS	LED lighting, SNP chip reader	Yield traits, disease resistance	Screens large hybrid populations for target alleles prematurity	Gangashetty et al. (2024)
Modified Short-Day + CO_2 Enrichment	Vegetable soybean (*Glycine max*)	10 h light; 30°C/25°C; 550 ppm CO_2	Short-day + CO_2, ovendry seed reset	CO_2-controlled chambers, seed dryers	Faster maturity, stable yield	Shortens maturity while preserving market-quality traits	Taku et al. (2024)
Groundnut Speed Breeding Review	Groundnut (*Arachis hypogaea*)	Controlled env. regimes	Light/thermal manipulation + SB	Growth chambers, MAS, GS, CRISPR	Yield, disease, drought, nutrition	Strategy review for rapid groundnut improvement	Pawar et al. (2023)
Single-Seed-Speed Bulks (SSSB)	Oat (*Avena sativa*)	22 h light/2 h dark; controlled	High-density planting in sand + prolonged photoperiod	High-density planting setups, sand media, and the single-seed descent method	Shorter flowering time, single-seed yield	Facilitates rapid development of breeding lines with minimal labor and greenhouse space	Kigoni et al. (2023)

LED, Light-emitting diode; *CRISPR*, Clustered Regularly Interspaced Short Palindromic Repeats.

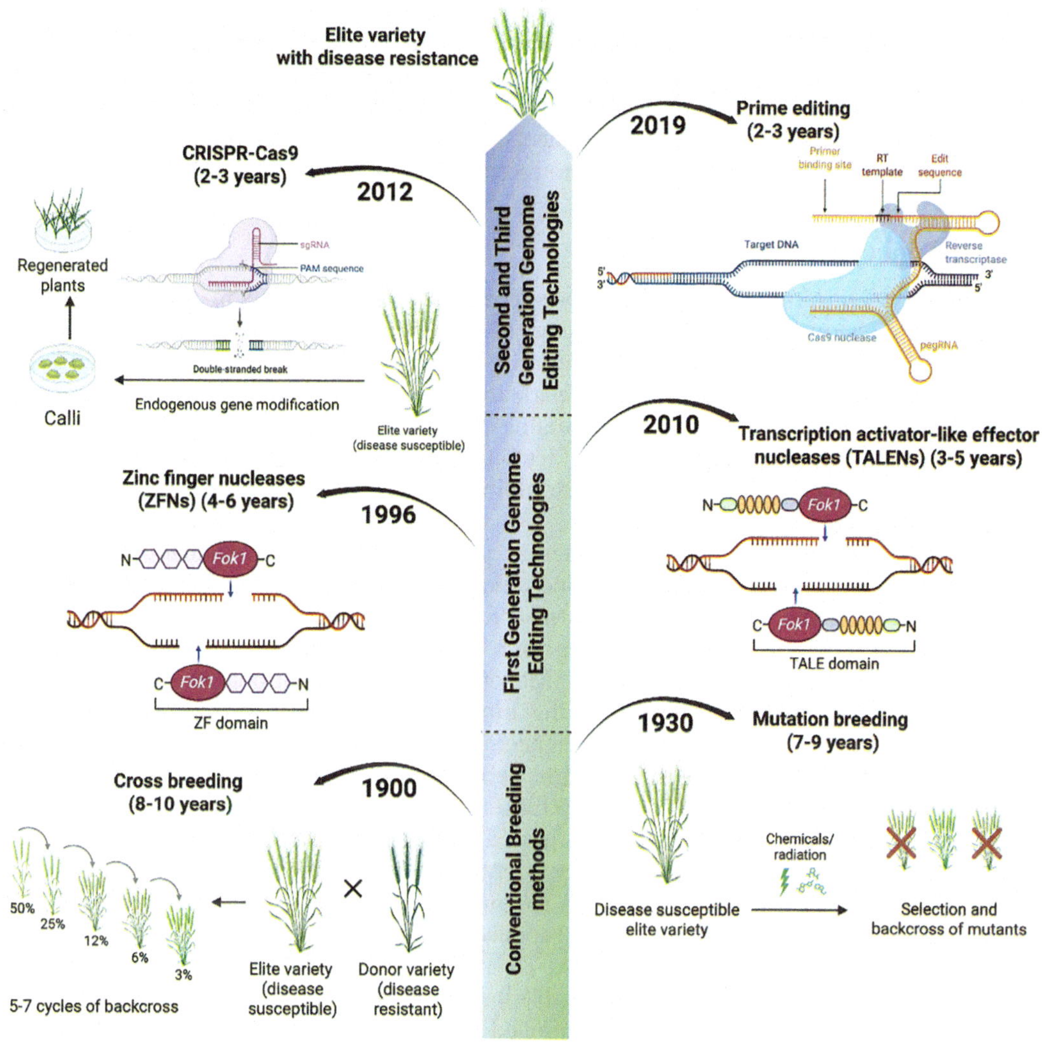

FIGURE 13.1 A timeline for the evolutionary history of plant breeding techniques.

13.3.2 Advantages over earlier genome editing technologies

Before the development of CRISPR, targeted genome modification was achieved using transcription activator-like effector nucleases (TALENs) and zinc-finger nucleases (ZFNs). TALENs and ZFNs are first-generation genome editing technologies well known before the CRISPR era. The drawback of using these platforms is that they are limited by laborious protein design, reduced efficiency, and constrained sequence flexibility. In contrast, CRISPR/Cas revolutionized gene editing by replacing complex protein engineering with RNA-guided targeting. Designing new edits became as simple as synthesizing a new sgRNA, significantly accelerating research and applied breeding timelines (Ali et al., 2023).

Multiplex editing was another key technology enabled by CRISPR, in which researchers could simultaneously knock out or alter various genes in a single transformation event using multiple sgRNAs. This is especially important for abiotic stress tolerance, as polygenic networks control traits resistant to salinity or drought. Compared to ZFNs and TALENs, CRISPR systems showed greater host compatibility across dicots and monocots, better on-target editing, and fewer off-target effects (especially with more recent variants) (Tariq et al., 2023).

13.3.3 Applications in crop plants

CRISPR/Cas was initially reported to be successfully used in plants in 2013, representing a historical milestone in plant biotechnology. Early experiments were initially proven in plants such as *Arabidopsis thaliana* and certain economically valuable crops, including rice (*Oryza sativa*), tobacco (*Nicotiana benthamiana*), and wheat (*Triticum aestivum*) (Viviani et al., 2022). Genes associated with hormone signaling, pathogen resistance, and plant architecture were among the primary targets of early tests to validate efficiency. Phytoene desaturase (PDS) knockouts, for instance, yielded clear albino phenotypes, which readily validated the success of editing.

By 2015, crop traits such as yield increases, quality improvements, nutritional enhancements, and enhanced resistance to biotic and abiotic stresses were being introduced through CRISPR/Cas. Establishing the required infrastructure and tools was another priority, which was addressed early on. For instance, specific vector systems were developed to target multiple genes in monocots and dicots (Ma et al., 2015). Additionally, highly effective multiplex genome editing toolkits utilizing binary vectors have been established. Even though the ultimate purpose was trait enhancement, these initial uses (2013–15) focused on establishing the necessary infrastructure and verifying the tool's basic functionalities across various plant crops. The extension of CRISPR/Cas use to more complex areas, such as abiotic stress tolerance, was primarily facilitated by this pioneering work.

Capitalizing on these initial achievements, scientists soon broadened CRISPR/Cas uses from simple gene knockouts to more sophisticated genome engineering in various crop systems. By late 2014 and 2015, the CRISPR system was being applied to engineer traits, such as disease resistance, nutrient enhancement, and drought tolerance. To engineer abiotic stress tolerance, scientists have developed multiplex editing systems that enable the simultaneous editing of multiple loci, which is crucial for addressing complex traits controlled by multiple genes (Ma et al., 2015; Gudeta, 2019). These developments were revolutionary for staple crops like wheat and rice, which frequently display polygenic stress responses. The deployment of CRISPR/Cas in model systems, such as *Arabidopsis* and *Nicotiana benthamiana*, helped validate sgRNA design rules, editing efficiency, and delivery methods, which were then applied to elite crop varieties. Successful edits in the biosynthesis of hormones, genes controlling plant height, and disease response have been reported, confirming CRISPR as an effective tool for manipulating agronomic traits (Chen et al., 2019; Tang and Tang, 2017). CRISPR/Cas9 has also been applied to tobacco to interfere with viral genes and induce resistance against RNA viruses, demonstrating the potential for engineered viral immunity in high-value crops (Scheben et al., 2017).

Between 2013 and 2015, foundations on genome editing platforms were established, which were crucial not only for demonstrating the efficacy of the system but also for creating the essential tools—multiplexing platforms, binary vectors, tissue-specific promoters, and transformation protocols—that facilitated the scale-up of genome editing from model systems in the lab to field crops (Harwood et al., 2017; Zhu et al., 2020). This research laid the foundation for applying CRISPR to mainstream breeding pipelines, particularly in response to the increasing need for climate-resilient agricultural solutions.

13.3.4 Next-generation CRISPR systems

Next-generation CRISPR tools—Cas12, Cas13, base editors, and prime editors—extend the versatility and precision of genome editing beyond what Cas9 offers. Cas12a (Cpf1) is most effective at recognizing T-rich PAM sequences and making staggered DNA cuts, enabling efficient multiplex editing, particularly in polyploid crops such as rice and banana (Molla et al., 2021). On the other hand, Cas13 targets RNA instead of DNA, enabling posttranscriptional gene silencing, transient gene regulation, and viral resistance, as demonstrated successfully in *Nicotiana* species (Kavuri et al., 2022). Single-nucleotide modifications are possible with base editors without resulting in DSBs. Cytosine base editors (CBEs) convert C•G to T•A, whereas adenine base editors (ABEs) convert A•T to G•C. These systems help introduce SNPs that are linked to trait enhancement, such as increased nutrient content and herbicide tolerance (Li et al., 2023). Apart from donor templates and DSBs, prime editing is a very versatile platform that can be used for insertions, deletions, and base substitutions. For traits linked to stress signaling and grain quality, it has been successfully implemented in wheat and rice (Hillary and Ceasar, 2022).

When combined, these cutting-edge CRISPR systems are revolutionizing plant breeding processes and accelerating the development of crops that can withstand climate change. Furthermore, precision agriculture pipelines are currently being driven by combinations of these tools with ML-guided trait prediction, omics technologies, and speed breeding. CRISPR-enabled breeding is being incorporated into international programs, such as USAID's Feed the Future projects and the CGIAR Genebank Platform, to accelerate the development of climate-resilient cultivars (Pankaj and Kumar, 2023). The full potential of CRISPR is expected to reach both underutilized orphan crops and major global staples, improving food security as transformation and regeneration protocols for resistant species like legumes, millets, and tubers continue to advance.

13.4 CRISPR–Cas for enhancing abiotic stress tolerance

13.4.1 Targeting key genes related to abiotic stress tolerance

Plant responses to abiotic stress are governed by distinct genetic factors, commonly grouped into "tolerance genes" (T genes) and "sensitivity genes" (S genes), reflecting their positive or negative regulatory roles. T genes positively enhance stress tolerance by activating protective physiological and molecular pathways, whereas S genes negatively regulate stress responses and contribute to susceptibility. A plethora of structural and regulatory stress-responsive genes, involved in stress-related regulatory networks, signaling pathways, and metabolite biosynthesis, can be precisely targeted using CRISPR/Cas technology to engineer stress-tolerant crop varieties (He et al., 2017; Jain, 2015; Razzaq et al., 2022; Zafar et al., 2020). The CRISPR-Cas system offers key capabilities for performing gene knockouts to validate and enhance the function of traits. Knocking out genes from the S-group of transcription factors (TFs) may reduce management stress sensitivity. Overexpressing genes from the T-group, or altering the promoter to upregulate T-group genes, can increase abiotic stress resilience (Razzaq et al., 2022; Zafar et al., 2020).

Structural genes, a subset of stress-related genes, form the basis of how plants perceive and respond to environmental stress. Structural genes are responsible for producing proteins that provide direct stress protection to cellular components, such as proteins that are late embryogenesis abundant (LEA), heat shock proteins (HSPs), and antioxidant enzymes (Trono and Pecchioni, 2022). Excessive accumulation of reactive oxygen species (ROS) disrupts cellular homeostasis under abiotic stress conditions, and its harmful effects can be male sterility, inhibition of photosynthesis, cell death, and inhibited yield (Panda et al., 2024; Petrov and Gechev, 2023). Multiple defense mechanisms, including regulation of gene expression, pathogen resistance, and ROS-mediated signaling for nitrogen fixation, have also been identified (Ribeiro et al., 2017).

Due to the bioinformatic and molecular genetic work by breeders and specialists in molecular biology, many T genes for antioxidant enzymes that mediate survival and tolerance to stress have been identified and incorporated into crops to improve their tolerances (Abu-Shahba et al., 2022; Zafar et al., 2020). Several examples include papain-like cysteine proteases (PLCPs). The *SPCP2* in sweet potato and *TaCP* in wheat were upregulated in response to abiotic stress, and when overexpressed in *Arabidopsis*, they enhanced drought tolerance (Liu et al., 2018; Zang et al., 2010; Chen et al., 2010). In parallel, oxidative stress mediated via S genes promotes programmed cell death and declines in antioxidant status by compromising homeostasis within hormonal signaling pathways and enhancing plant sensitivity; and knocking out S genes corresponding to this effect can improve tolerance by disrupting harmful signaling pathways (Zafar et al., 2020; Zhao et al., 2017). For example, *Oryza sativa* stress-related RING finger protein 1 (*OsSRFP1*) is an E3 ubiquitin ligase that acts as a negative regulator of abiotic stress. Accumulating H_2O_2 inhibits the activity of antioxidant enzymes in plant tissues. Knockdown of *OsSRFP1* leads to increased stress tolerance by reducing H_2O_2 accumulation and activating antioxidant defense pathways, which enhance antioxidant enzyme activity (Fang et al., 2015).

13.4.2 CRISPR for enhancing drought tolerance in crop plants

Drought stress represents a significant challenge to agricultural productivity, adversely affecting plant growth, development, and overall yield (Wahab et al., 2022). It triggers anatomical alterations that result in a range of physiological and biochemical reactions, including decreased photosynthesis, altered osmotic regulation, increased levels of drought-responsive proteins, and changes in the activity of antioxidant enzymes (Choudry et al., 2024; Yang et al., 2021). Staple grain crops such as rice, wheat, and maize are particularly vulnerable to water scarcity, which is expected to intensify with climate change. And by 2050, nearly half of the world's cultivated land is projected to experience severe drought conditions (Kaur et al., 2022; Kamanga et al., 2018; Shelake et al., 2022).

For example, regulating downregulatory genes, including *DERF1*, *PMS3*, *MSH1*, *MYB5*, and *SPP*, by CRISPR/Cas9 improved drought tolerance in rice (Zhang et al., 2014). The stress-activated protein kinase (*OsSAPK2*) gene knockout, achieved using CRISPR/Cas9 in rice, also resulted in impressive drought recovery and maintained biomass, with evidence of enhanced cellular protection and a more rapid recovery rate afterwards (Rai et al., 2024). Among oilseed crops, CRISPR/Cas9 has been used to edit the *A6 gene*. Repressor of GA 1-3 (RGA) resulting in increased drought tolerance in rapeseed (Wu et al., 2020). In tomato, CRISPR/Cas9 editing increased the drought resilience of several genes, including the gibberellin signaling gene *GID1* and the TF gene *LBD40, which is* related to stress adaptation (Liu et al., 2020). In *Solanum lycopersicum* (tomato), CRISPR mutants of the pipecolic acid (Pip) biosynthetic gene *SlALD1* showed enhanced drought resistance compared to the wild type. This phenotype was associated with enhanced CO_2 assimilation, increased photosystem activity, and increased activity of antioxidant enzymes (Wan et al., 2021).

13.4.3 CRISPR for enhancing salt tolerance in crop plants

Salinity is the second largest abiotic constraint to agriculture (in terms of area of fertile land and crop productivity) and currently affects more than 400 million hectares globally (Gajardo et al., 2023; Gong et al., 2020). Salinization and associated salt stress can occur through irrigation with saline water, irrigation in excess (Shrivastava and Kumar, 2015), overfertilization (Litalien and Zeeb, 2020), change of natural habitat to agricultural land (Chen and Mueller, 2018; Qadir et al., 2014), and climate change scenarios related to sea level rise due to global warming, flooding, and tsunamis (Casanova et al., 2016; Kumar et al., 2013). Salinity is projected to adversely impact over 50% of the global land area by 2050 (Jamil et al., 2011; Jha et al., 2019; Kaashyap et al., 2017). Salt stress leads to various intracellular plant responses such as accumulation of ROS, increased abscisic acid (ABA), changes in calcium homeostasis, activation of Ca^{2+}/calmodulin-dependent kinases, and the initiation of salt overly sensitive (SOS) homeostatic signaling pathways (Julkowska and Testerink, 2015).

Using CRISPR/Cas genome editing, researchers have mapped and edited genes linked to the salt stress response and have enhanced plant tolerance by tuning molecular pathways responsible for controlling ion balance and osmotic stability (Erdoğan and Cevher-Keskin, 2023; Shaheen et al., 2023; Kumar et al., 2023; Hualpa-Ramirez et al., 2024). For example, the knockout of *OsRR22*, a cytokinin signaling regulator, enhanced the survival of rice plants under saline conditions and had no negative impacts on their agronomic performance (Zhang et al., 2019). Additionally, targeting *OsNAC45* enhanced stress tolerance by modifying ABA-responsive transcriptional networks, suggesting that hormone crosstalk is crucial for stress adaptation (Zhang et al., 2020). In tomato, CRISPR mutations of the *SlMAPK3* gene have been shown to enhance salinity tolerance by reducing the excessive production of ROS and inducing the expression of genes related to ethylene signaling pathways (Shu et al., 2022).

CRISPR technology is increasingly being employed to specifically target (TFs involved in salt stress pathways, thereby providing a deeper understanding of the regulatory mechanisms at play (Angon et al., 2023). In fact, genomic editing on important TFs such as *DOF15* in rice (Qin et al., 2019), *NAC041* in rice (Bo et al., 2019), *ITPK1* in barley (Vlčko and Ohnoutková, 2020), *GTγ-2* in rice (Liu et al., 2020), and *PIL14* in rice (Mo et al., 2020) has all been shown to be involved in salt stress sensitivity and improve salt tolerance via additional stress-responsive phenotypes, leading to more consolidated tolerance.

13.4.4 CRISPR for mitigating the impact of heat stress in crop plants

Heat stress, exacerbated by global warming, is now one of the primary abiotic constraints to agricultural production in many parts of the world (Tubiello et al., 2007). Heat stress affects plant growth and productivity by impairing the efficiency of photosynthesis, destabilizing proteins, inhibiting pollen viability, and leading to excessive accumulation of ROS, which induces oxidative damage. Since plants thrive within a defined optimal temperature range, exposure to elevated temperatures significantly alters physiological processes, accelerating development and ultimately leading to substantial reductions in crop yield (Djanaguiraman et al., 2020; Singh et al., 2024). For instance, in wheat, exposure to high temperatures during anthesis (32°C/22°C) resulted in a ~25% reduction in photosynthetic rate and up to a 44% loss in grain yield compared to plants grown under optimal temperatures (22°C/14°C) (Djanaguiraman et al., 2020). Similar patterns have been documented in maize and broader crop studies, where yields begin to decline rapidly beyond thermal thresholds, underscoring the sensitivity of reproductive stages to heat stress (Hatfield and Prueger, 2015).

CRISPR/Cas enables the manipulation of genes involved in heat shock responses, antioxidant defense systems, and metabolic pathways affected by high temperatures (Pandey et al., 2024; Sagwal et al., 2025). For example, in tomato, CRISPR genome editing of the mitogen-activated protein kinase *SlMAPK3* gene revealed that it negatively regulates tolerance to heat stress because *SlMAPK3* mutants showed greater thermotolerance than wild-type plants. *SlMAPK3* was also found to regulate the expression of HSPs and heat shock factors (HSFs) (Yu et al., 2019). In maize, CRISPR/Cas9-targeted editing of the *TGMS5* gene has enabled the creation of thermosensitive male-sterile lines, a valuable resource for breeding heat-resilient hybrids (Li et al., 2017).

To address the impacts of global warming, acknowledging and utilizing the genetic determinants of heat stress tolerance is essential for developing crop varieties with greater acclimation to global warming. Phytochromes are appealing targets and have been established as key thermo-sensors (Jung et al., 2016; Legris et al., 2016). Knockout of *PYL1*, *PYL4*, and *PYL6*, using CRISPR/Cas9 technology, which encodes ABA receptors, has conferred enhanced high-temperature adaptability in rice (Miao et al., 2018). In lettuce, knockout of *NCED4*, a key enzyme in ABA biosynthesis, using CRISPR/Cas9, enabled seeds to germinate under elevated temperatures successfully. *LsNCED4* mutants may thus hold commercial value for cultivation in high-temperature regions (Bertier et al., 2018).

13.4.5 CRISPR for mitigating the impact of cold stress in crop plants

Low-temperature stress, including chilling (0°C–18°C) and freezing (<0°C), is a major abiotic factor that limits plant growth and causes substantial yield losses in many crops (Song et al., 2015). Exposure to low temperatures restricts the regular metabolic activity of plants, resulting in osmotic imbalance, excessive production of ROS, and a cascade of physiological disruptions. Mechanical injury and metabolic dysfunction caused by severe cold ultimately halt overall plant growth performances (Yadav, 2010). One of the main effects of cold stress on plants is the rigidification of cellular membranes, which restricts membrane fluidity and affects multiple physiological and biochemical processes. This stress further destabilizes protein structure, changes gene expression, decreases the redox activity of various essential enzymes, including antioxidant defense enzymes for ROS detoxification, and reduces the photosynthetic reaction efficiencies of plants (Ruelland et al., 2009; Muller et al., 2014; Adams et al., 2016).

A study utilized CRISPR/Cas9 to edit OsPIN5b, GS3, and OsMYB30 in rice simultaneously. This approach not only improved cold tolerance but also significantly enhanced yield-related traits. The triple mutants exhibited longer panicles, larger grain size, and improved survival under chilling conditions compared to wild-type controls (Zeng et al., 2018). In tomato, CRISPR/Cas9-mediated knockout of the *SlCBF1* gene resulted in small insertions and deletions, identifying *SlCBF1* as a regulator of chilling injury. A comparison of mutants revealed enhanced electrolyte leakage, a decrease in proline content, severe chilling symptoms, and elevated levels of indole acetic acid, confirming that the *SlCBF1* gene is a crucial regulator of the chilling stress response (Li et al., 2018). Annexin (OsANN3), a calcium-dependent phospholipid-binding protein, has also been identified as a key modulator in rice adaptation to cold stress. Targeted, CRISPR/Cas9-mediated modifications of OsANN3 increased relative electric conductivity, illustrating an important component of cold stress signaling networks involved in defense response (Shen et al., 2017). TFs play a crucial role in regulating gene expression in response to cold stress in plants (Mohamed et al., 2024; Shaheen et al., 2023; Yadav et al., 2023). For example, the OsWRKY63 TF gene was edited using CRISPR/Cas9, and the resulting rice lines had a greater level of cold stress tolerance (Li et al., 2021a,b).

NPR1 (*Nonexpresser of Pathogenesis-Related Genes 1*) is a cold-induced TF essential for cold tolerance. CRISPR/Cas9-mediated knockout of *NPR1* in tomato resulted in increased cold sensitivity, characterized by elevated levels of malondialdehyde (MDA), hydrogen peroxide (H_2O_2), and superoxide radicals (O_2^-), alongside reduced proline levels and diminished antioxidant enzyme activities compared to the wild type (Shu et al., 2023) (Fig. 13.2).

13.5 Accelerating genetic modification through CRISPR–Cas

Speed breeding is a way of shortening the breeding cycle, therefore allowing rapid generation progress and hence accelerating crop research. Speed breeding can be implemented using different techniques, such as increasing plants' daily light exposure and harvesting seeds early to minimize the generational period for day-neutral crops. Speed Breeding 3.0 uses advanced photobiological techniques to precisely manage environmental parameters, including light spectrum, intensity, photoperiod, temperature, and hormone regulation. This precise management accelerates plant growth and development, facilitating the production of as many as six generations each year in important crops such as wheat, rice, and barley (Kumar and Walia, 2024). In plants, CRISPR-Cas9 systems can be introduced through multiple delivery approaches, including transient and stable transformation methods such as protoplast transfection, agroinfiltration, and particle bombardment of callus tissues (Williams et al., 2024). The use of CRISPR/Cas offers considerable benefits for analyzing gene functions and establishing crop germplasm resources (Ahmad et al., 2023). It enables gene manipulation through knockout, knock-in, replacement, single-base editing, regulation, and epigenome

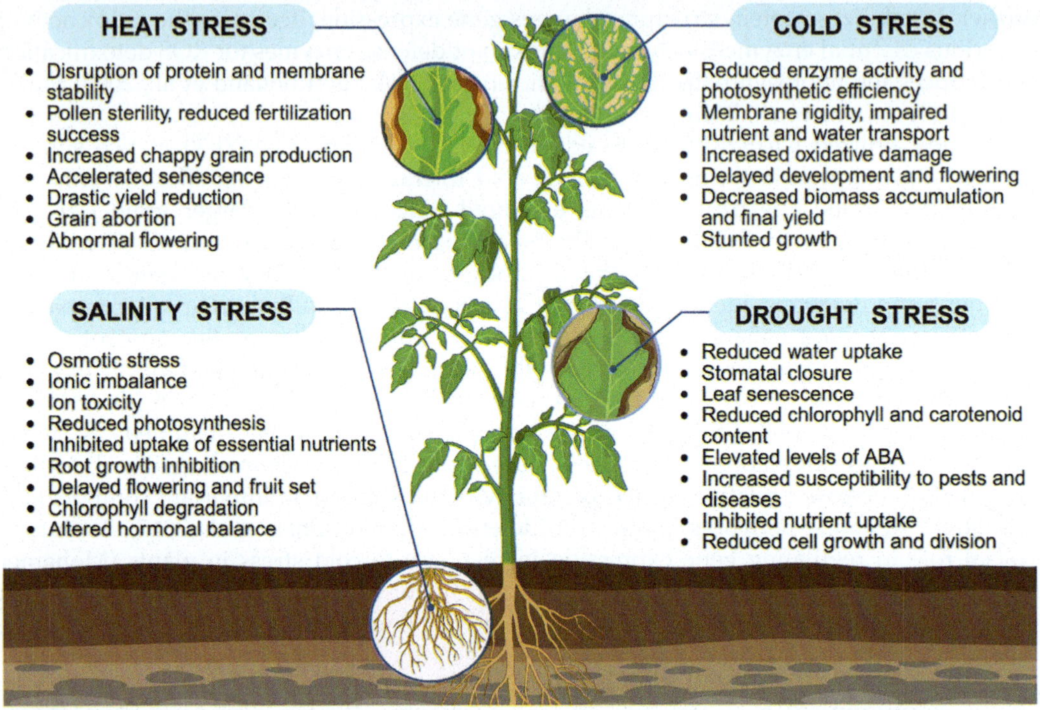

FIGURE 13.2 Various types of abiotic stresses (drought, salinity, heat, cold, etc.) and their impacts on crop growth and yield.

editing (Zhang et al., 2021a,b). For example, using CRISPR/Cas, KRN2 was identified as a convergently selected gene that regulates grain number in maize and rice. This deletion considerably increased grain yield without impacting other features, indicating a feasible technique for germplasm creation and de novo domestication. CRISPR/Cas is currently widely used to enhance features such as stress resistance, plant development, yield, and grain quality. In the United States, CRISPR-waxy corn hybrids with increased yield and agronomic performance have been produced and precommercialized (Tang et al., 2023).

Speed breeding optimizes environmental factors for faster crop improvement and generation turnover. To promote plant growth and reproduction, various parameters, including photoperiod, temperature, light intensity, and nutrient supply, are also adjusted (Patel and Gurjar, 2024). Combining CRISPR editing with speed breeding reduces the time necessary to stabilize altered alleles. Edited seeds can immediately begin accelerated cycling to achieve homozygosity faster. When paired with multiplex editing procedures, this approach enhances genetic gain per unit time and increases the production of climate-resilient, high-yielding cultivars (Haroon et al., 2022).

Several CRISPR/Cas-edited crop varieties have now progressed from experimental validation to official release, marking a significant milestone in precision breeding (Table 13.2). Table 13.3 summarizes globally released varieties developed specifically for abiotic stress tolerance, detailing their target genes, edited traits, and outcomes.

TABLE 13.2 Globally released CRISPR/Cas-edited crop varieties targeting abiotic stress tolerance.

Year of release/ approval	Country	Crop	Target gene(s)	Target trait	Method of modification	Source
2016	United States	Nonbrowning White button mushroom (*Agaricus bisporus*)	PPO gene silencing	Slower browning after harvesting, increased shelf life	DNA-free, ribonucleoprotein delivery	Evanega et al. (2024)
2021	Japan	*Tomato (Solanum lycopersicum)*	*SlMAPK3*	Heat and oxidative stress tolerance	*Agrobacterium*-mediated transformation	Zhang et al. (2018)
2023	Japan	*Waxy corn (Zea mays)*	*Wx1*	100% amylopectin starch for food industry	*Agrobacterium*-mediated transformation	Polidoros et al. (2024), Intrexon Corp. commercial trials
2024	United States	Romaine lettuce (Lactuca sativa)	PPO (polyphenol oxidase) gene family	Reduced browning, improved shelf life (~2 weeks longer)	*Agrobacterium*-mediated transformation	
2024	China	Wheat (*Triticum aestivum*)	*TaALS*	ALS herbicide resistance	*Agrobacterium*-mediated transformation	Dong, H., Huang, Y. and Wang, K., Genes, 12(6), p. 912.
2025	India	Rice (*Oryza sativa*) – *DRR Dhan 100 (Kamala)*	*SDN-1* (CRISPR-Cas9 edits on *Samba Mahsuri* background)	Drought and salinity tolerance; reduced methane emission	*Agrobacterium*-mediated transformation	ICAR, 2025, Official Release – Union Agriculture Minister announces genome-edited rice varieties Union Agriculture Minister Shri Shivraj Singh Chouhan Announces Two Genome-Edited Rice Varieties Developed in India | ICAR (2025)
2025	India	Rice (*Oryza sativa*) – *Pusa DST Rice 1*	*DST* gene edited using CRISPR-Cas9	Drought and salinity tolerance; yield stability under stress	*Agrobacterium*-mediated transformation	

CRISPR, Clustered Regularly Interspaced Short Palindromic Repeats.

TABLE 13.3 Overview of abiotic stress–related genes modified and analyzed through CRISPR/Cas technology.

Abiotic stress	Plant species	Target gene	Gene function	Modification	Method of transformation	Target trait	Results	References
Drought stress (DS)	Rice	*OsPUB7*	Negative regulator of drought and salinity stress	Gene knockout	*Agrobacterium tumefaciens*–mediated transformation	Drought stress	Enhanced drought stress	Kim et al. (2023)
	Rice	*OsPYL9*	Transcription factor	Gene knockout	*Agrobacterium tumefaciens*–mediated transformation	Drought stress	Improved yield under normal and DS conditions (in plant growth chamber)	Usman et al. (2020)
	Rice	*OsDST*	Transcription factor	Small deletions	*Agrobacterium tumefaciens*–mediated transformation	Drought stress; salt stress	Improved tolerance to NaCl, moderate tolerance to osmotic stress at seedling stage of the plant	Santosh Kumar et al. (2020)
	Rice	*OsERA1*	Transcriptional regulator/ signalling	Frameshift mutation	*Agrobacterium tumefaciens*–mediated transformation	Drought stress	Enhanced response to DS	Ogata et al. (2020)
	Rice	*OsNAC006*	Transcription factor	Gene knockout	*Agrobacterium tumefaciens*–mediated transformation	Drought stress	Enhanced plant growth under DS	Wang et al. (2020)
	Wheat	*TaDREB2*	Transcription factor	Small insertions and deletions	Protoplast transfection	Drought stress	Improved DS tolerance	Kim et al. (2018)
	Wheat	*TaGW2*	E3 ubiquitin ligase	Gene knockout	*Agrobacterium tumefaciens*–mediated transformation	Drought stress	Improved DS and grain yield than WT plants	Li et al. (2024a)
	Tomato	*SlNPR1*	Transcriptional coactivator	Small insertions and deletions	*Agrobacterium tumefaciens*–mediated transformation	Drought stress	Reduced DS tolerance in the mutants	Li et al. (2019)

	Tomato	*SlLBD40*	Transcription factor	Gene knockout	*Agrobacterium tumefaciens*–mediated transformation	Drought stress	DS tolerance in the knockout lines	Liu et al. (2020)
	Tomato	*SlMAPK3*	Signal transduction	Gene knockout	*Agrobacterium tumefaciens*–mediated cotyledon transformation	Drought stress	Reduced DS tolerance in the mutants in greenhouse conditions	Wang et al. (2017)
	Soybean	*Drb2a, Drb2b*	Transcription factor	In-frame, frame shift, and double homozygous mutant	*Agrobacterium tumefaciens*–mediated transformation	Drought stress	Increase in DS tolerance	Curtin et al. (2018)
	Soybean	*GmHdz4*	Transcription factor	Gene knockout	*Agrobacterium rhizogenes*–mediated transformation	Drought stress	Higher DS tolerance	Zhong et al. (2022)
	Soybean	*GmMYB118*	Base substitution	Base substitution	*Agrobacterium rhizogenes*–mediated transformation	Drought stress	Higher tolerance to DS and salt stress	Du et al. (2018)
	Potato	*StDRO2*	Control root system architecture (RSA)	Small deletion	*Agrobacterium tumefaciens*–mediated transformation	Drought stress	Improved DS resistance	Zhao et al. (2025)
	Maize	*ZmPL1*	Negatively regulates drought tolerance	Gene knockout	*Agrobacterium tumefaciens*–mediated transformation	Drought stress	Tolerance to DS	Wang et al. (2025)
	Maize	*ARGOS8*	Transcription factor	Promoter insertion and replacement	Biolistic-mediated transformation of embryo (DNA)	Drought stress	Enhanced DS	Shi et al. (2017)
Salt stress (SS)	Wheat	*TaHAG1*	Transcription factor	Gene knockout	*Agrobacterium tumefaciens*–mediated transformation	Salt stress	Enhanced SS tolerance	Zheng et al. (2021)
	Rice	*OsRR22*	Transcriptional regulator/ signalling	Gene knockout	*Agrobacterium tumefaciens*–mediated transformation	Salt stress	SS tolerance at the seedling stage	Zhang et al. (2019)

(*Continued*)

TABLE 13.3 (Continued)

Abiotic stress	Plant species	Target gene	Gene function	Modification	Method of transformation	Target trait	Results	References
	Tomato	*SlARF4*	Transcription factor	Deletion	*Agrobacterium tumefaciens*–mediated transformation	Salt stress	Enhanced SS tolerance	Bouzroud et al. (2020)
	Tomato	*HyPRP1*	Key role in plant developmental process and stress amelioration	Gene knockout	*Agrobacterium tumefaciens*–mediated transformation	Salt stress	High SS tolerance	Tran et al. (2023)
	Soybean	*GmAITR*	Transcriptional repressor	Gene knockout	*Agrobacterium tumefaciens*–mediated transformation	Salt stress	Enhanced SS tolerance	Wan et al. (2021)
	Barley	*HvGSK1.1*	Negative regulator of brassinosteroid (BR) signaling	Gene knockout	*Agrobacterium tumefaciens*–mediated transformation	Salt stress	Improved plant growth performance under SS conditions	Kloc et al. (2024)
	Barley	*HvITPK1*	Phosphate storage and abiotic stress signaling.	Gene knockout	*Agrobacterium tumefaciens*–mediated transformation	Salt stress	Enhanced salinity tolerance and reduced sensitivity	Vlčko and Ohnoutková (2020)
	Maize	*ZmG6PDH1*	Catalyzing the first step of the pentose phosphate pathway (PPP)	Gene knockout	Polyethylene glycol (PEG)-mediated protoplast transformation	Salt stress	Enhanced CS sensitivity	Li et al. (2023)
	Maize	*ZmHKT1*	Membrane transporter	Gene knockout	*Agrobacterium tumefaciens*–mediated transformation	Salt stress	Tolerance to SS	Zhang et al. (2018)

Heat stress (HS)	Rice	*OsGER4*	Rice root development	Gene knockout	*Agrobacterium tumefaciens*–mediated transformation	Heat stress	Improved root development and enhanced HS resilience in edited lines	Nguyen et al. (2023)
	Maize	*ZmTMS5*	Transcription factor	Small insertions and deletions	Particle bombardment	Heat stress	Creation of thermosensitive maize lines for hybrid development	Li et al. (2017)
	Maize	*ZmHsf20*	Transcription factor	Gene knockout	*Agrobacterium tumefaciens–mediated* transformation	Heat stress	Improved HS of maize seedlings	Li et al. (2024b)
	Tomato	*SlAGL6*	Transcription factor	Knockout of the second exon	*Agrobacterium tumefaciens mediated* transformation	Heat stress	Ensure fruit production under HS	Klap et al. (2017)
	Tomato	*BZR1*	Transcription factor	Loss-of-function mutation	*Agrobacterium tumefaciens mediated* transformation	Heat stress	Enhanced HS tolerance	Yin et al. (2020)
Cold stress (CS)	Potato	*VInv*	Starch-sucrose regulation in hexogenic pathway	Gene knockdown	*Agrobacterium tumefaciens–mediated* transformation	Cold stress	Enhanced CS	Yasmeen et al. (2022)
	Tomato	*SlCBF1*	Transcription factor	Small insertions and deletions	*Agrobacterium tumefaciens–mediated* transformation	Cold stress	Increase the ability to tolerate to CS	Li et al. (2018)
	Rice	*OsMYB30*	Transcription factor	InDel mutation	*Agrobacterium tumefaciens–mediated* transformation	Cold stress	Improved CS tolerance	Zeng et al. (2018)
	Rice	*OsAnn3*	Ca^{2+}-dependent phospholipid-binding proteins involved in plant development stress responses	Knockout of promoter	*Agrobacterium tumefaciens–mediated* transformation	Cold stress	Decreased CS tolerance	Shen et al. (2017)

CRISPR, Clustered Regularly Interspaced Short Palindromic Repeats.

13.5.1 Rapid advancement strategies for producing homozygous cultivars showing abiotic resilience

With the combination of CRISPR/Cas9 genome editing and haploid-inducer-mediated genome editing (HI-Edit/IMGE), it is now possible to create homozygous crop lines with abiotic stress tolerance features in just one to two generations. This significantly shortens breeding time frames (Sheng et al., 2025). Examples of CRISPR-mediated alteration of the genes underlying drought, heat, cold, and salt tolerance in staple crops are described, including the modification of *TaRPK1* in wheat or *ZmHDT103* in maize to increase drought resistance, followed by a fast accelerated cycle to homozygosity (Chen et al., 2024). With the CRISPR-Cas9 technology, the DST gene (drought and salt tolerance) was knocked out in rice to enhance its dual tolerance to salt and drought. Under drought stress, the resulting mutant plants retained more water in their leaves due to their wider leaves and lower stomatal density. At the seedling stage, these plants demonstrated enhanced tolerance to salt stress and a higher degree of tolerance to moderate osmotic stress (Rengasamy et al., 2024). The physiology of plants is significantly impacted by drought stress, which prevents the uptake of nutrients and other vital processes, such as transpiration, cell growth and elongation, turgor dynamics, and the regulation of gene expression. Using CRISPR/Cas9, several rice genes were identified and altered to create plants that can withstand abiotic stress (Table 13.2). For the development of salt-tolerant T2 homozygous mutant rice, the *OsRR22* locus was targeted and disrupted using Cas9 (Kaur et al., 2025). Disruption of *OsRR22* improved rice performance under elevated salinity (0.75%), without negatively affecting grain dimensions, yield, or biomass. A comparable strategy was applied to generate new rice germplasm at the seedling stage (He et al., 2023).

When haploids are produced and manipulated, chromosomes in haploid plants can be quickly doubled to create homozygous inbred lines. For instance, cereals can undergo haploid induction when the MATRILINEAL (MTL) and several other important genes are knocked out. For example, the ZmLG1 genome was edited using the IMGE system with a CAU5 HI line that carried the Cas9 cassette to produce haploids. The editing efficiency of the haploids was 4.1%. Using the DH approach in conjunction with CRISPR/Cas9 genome editing allowed for faster yields and more durable crop types (Ahmar et al., 2024).

13.5.2 CRISPR/Cas for targeted mutation of multiple genes involved in stress response

CRISPR/Cas genetic editing is an effective method for modifying multiple genes in plant stress response pathways, thereby accelerating efforts to enhance traits such as drought, heat, salinity, and heavy metal tolerance. Three rice ABA receptor genes (*OsPYL1, OsPYL4,* and *OsPYL6*) were deleted using CRISPR/Cas9, resulting in taller seedlings, increased panicle production, and enhanced heat tolerance in field conditions. This suggests that numerous signaling interruptions may be involved (Chakraborty and Wylie, 2025). According to a relevant study from 2024, precise in-locus activation was achieved by targeted knock-in of short transcriptional enhancers (STEs) into rice promoter regions. This allowed for the corresponding upregulation of up to four endogenous genes with transcript increases to 869-fold and stable inheritance through the T_3 generation (Yao et al., 2024).

Furthermore, rice's capacity to tolerate salt stress was substantially enhanced by CRISPR/Cas9-mediated deletion of TFs like *OsDST* and *OsbHLH024*, proving that targeting multiple stress-responsive regulators can enhance salinity tolerance (e.g., *OsDST* and *OsbHLH024* edits each improved salt stress outcomes) (Hossain et al., 2022). The success rate of targeting insertion is increased by novel methods like Cas9-Rap fusion system, which binds donor DNA in vivo to enhance HDR-based knock-ins. This results in a 72% in-frame insertion rate in rice, with faithful inheritance and minimal off-target effects (Zhou et al., 2025).

13.5.3 The potential for multigenerational crop development in a short time frame

Apart from effectively targeting many stress-responsive genes in a single generation, CRISPR/Cas genome editing accelerates the development of stable, heritable agricultural features through reduced breeding cycles and multigenerational advances, in addition to allowing precise targeting of several stress-response genes within a single generation. In rice, homozygous T_0 plants are commonly produced through CRISPR/Cas9-targeted mutations, which significantly reduce the breeding time frame by allowing agricultural features to be evaluated and passed on without further segregation (Park et al., 2022).

13.6 Challenges and limitations

CRISPR technology offers a groundbreaking opportunity to modify crops with precise GMs, enabling the development of crop varieties that resist abiotic stress (Angon et al., 2023). However, considerable technological, regulatory, and ecological challenges still impede an effective and practical application of these technologies to sustainable crop improvement (Hamdan et al., 2022).

13.6.1 Off-target effects

The unintended introduction of off-target mutations remains a significant limitation in CRISPR/Cas-mediated genome editing. In food crops, such intended mutations can disrupt key genes, potentially reduce yield, or introduce unwanted agronomic traits (Chen et al., 2024). They occur when CRISPR/Cas induces edits at unintended and nonspecific genomic sites, either due to gRNA mismatch tolerance or through gRNA-independent pathways (Jin et al., 2019; Robertson et al., 2022; Tsai and Joung, 2016). Most off-targets occur due to homologous sequences, but additional unwanted edits can occur in nearby target regions that have partially or unrelated sequences (Zhang et al., 2015). Several studies using CRISPR/Cas9 have found that more than half of RGEN-induced mutations occur at unintended sites, and this off-target activity defines an important limitation in plant genome editing (Adhikari and Poudel, 2020; Rasheed et al., 2021). Investigations have attempted to limit off-target effects within plant genome editing techniques as described directly using unbiased detection approaches for RGENs, cytosine and adenine (base) editors, prime editing, improving guide RNA designs, and creating higher fidelity Cas variants (Guo et al., 2023; He et al., 2024). Additionally, Cas12b has been reported to have minimal sgRNA-dependent off-target effects due to the reliability of strictly adhered to PAM requirements and minimal mismatch tolerance, indicating its potential for more accurate genome editing in rice studies (Gurel et al., 2023).

13.6.2 Delivery of CRISPR/Cas components into plant cells

Successful delivery of CRISPR/Cas DNA or RNP constructs into plant cells, especially in recalcitrant staple crops, such as rice, wheat, and maize, that have complex genetic architecture and poor tissue culture responses for regeneration of viable edited plants (Chen et al., 2024; Nakamura et al., 2023). There is also genotype variability, such as the numerous different varieties of wheat and maize, which exhibit varying plant responses to editing strategies. Consequently, establishing a universal editing procedure is challenging (Niraula et al., 2024). While conventional plant transformation systems for the delivery of CRISPR/Cas components, such as tissue culture, Agrobacterium-mediated transformation, and particle bombardment, are the most favorable methods of delivery, they are lengthy, labor-intensive, and introduce random somatic mutations or transgene integration (Zhang et al., 2021a,b; Gao, 2018; Kumar et al., 2023). Therefore there is a clear urgent need for new delivery systems to ensure efficient plant genome editing methods. A more promising means of delivery would involve the use of regeneration boosters to make tissue culture an option in recalcitrant species when using CRISPR and site-specific delivery into either the apical meristem or pollen grains to eliminate the need for tissue culture. Emerging delivery methods such as RNP delivery, viral vectors, nanoparticle carriers, and de novo meristem induction are being explored (Chen et al., 2024; Kumar et al., 2023; Yau et al., 2024). Nanoparticle-mediated delivery represents a promising strategy; however, its broader application is constrained by key challenges, including efficient intracellular release, stabilization of CRISPR components within plant cells, and the need to minimize cytotoxicity (Antony Ceasar and Ignacimuthu, 2023; Wang et al., 2019).

13.6.3 Genetic complexity of stress tolerance traits

Tolerance to abiotic stress is a polygenic agronomic trait governed by multiple genes. Achieving effective phenotypic changes for abiotic stress tolerance often requires mutagenesis of one or two gene copies, particularly in polyploid crop species (e.g., wheat, cotton, and potato) where gene redundancy is common (Angon et al., 2023; Ahmad et al., 2023). This is exemplified by studies targeting gene homologs in *Brassica napus*, which identified three gene copies in the reference genome, but only two in a specific pure line. Similarly, editing of BnSBE2 gene homologs revealed six genes in the database, while only three copies of BnSBE2 and one of BnSBE2.w were present in a particular cultivar. This high degree of gene redundancy complicates genome editing, especially for recessive traits, which require the elimination of all alleles to express the desired phenotype (AHMAD, 2023).

13.6.4 Ethical and regulatory issues and public perception

Although CRISPR crops are being developed and grown globally, their adoption is hindered by regulatory uncertainty of plant genome editing, divergent legal frameworks, ethical concerns, and public perception. These factors continue to challenge the integration of CRISPR into mainstream crop improvement initiatives (Chen et al., 2024; Gajardo et al., 2023).

The emergence of gene-edited crops has raised important regulatory and legislative questions worldwide. In many countries, GMOs containing foreign DNA are prohibited from commercial cultivation. However, CRISPR-edited plants, especially those developed without introducing transgenes, remain under debate. A central question is whether these should fall under existing GMO regulations or be treated separately, given their precision and the nature of the edits involved. Since different genome editing techniques can introduce a range of genetic changes from small deletions to targeted base substitutions, there is an urgent need to revisit and refine the current definitions of GM crops (Eckerstorfer et al., 2021; Hamdan et al., 2022; Kumar et al., 2023; Shaheen et al., 2023; Saeglitz and Bartsch, 2003). In this regard, several countries have completed or are close to finalizing their legislative frameworks, having already determined the regulatory pathway for genetically engineered (GE) crops.

In India, crop plants developed through genome-editing methods, CRISPR/Cas9, fall under the current regulatory framework that was originally designed for GE organisms. The primary piece of legislation is the Rules for the Manufacture, Use, Import, Export, and Storage of Hazardous Microorganisms/Genetically Engineered Organisms or Cells, 1989 (also referred to as "Rules 1989"), made under the Environment (Protection) Act, 1986 (Chimata and Bharti, 2019). Under the existing definitions of *Gene Technology* and *Genetic Engineering* in the Rules, 1989, India is obligated to regulate any products developed through the application of genome-editing technologies (Prasad and Chimata, 2023). In March 2022, the Ministry of Environment, Forest and Climate Change (MoEF&CC), Government of India, introduced revised Guidelines for the Safety Assessment of Genome-Edited Plants, distinguishing between SDN-1 and SDN-2 edits (which do not incorporate foreign DNA) and SDN-3 edits (which do). Under these guidelines, SDN-1 and SDN-2 plants are exempted from GMO regulations; however, the products derived through SDN-3 are still placed in the category of GMOs and will be regulated under "Rules 1989" (Sharma et al., 2025).

The EU currently adopts a process-based approach, classifying most gene-edited organisms, including those modified by CRISPR, as GMOs if the modification could not have occurred naturally (Menz et al., 2020). In contrast, the United States, under the USDA's Animal and Plant Health Inspection Service (APHIS), generally employs a product-based approach. Crops developed using gene-editing techniques that do not introduce foreign DNA (e.g., SDN-1 and SDN-2 categories of site-directed nucleases that involve small edits or deletions) may be exempt from the regulations applied to traditional GMOs, provided they do not pose a plant pest risk (Tachikawa and Matsuo, 2024). Additionally, the United Kingdom has recently joined countries such as Argentina, Brazil, Japan, and the United States to allow gene-editing tools for crop improvement. It has removed CRISPR-edited plants from GMO regulations, because the edits are no different from changes that would occur through natural selection and conventional breeding (Ghouri et al., 2023; Kumawat et al., 2024; Caccamo, 2023; Schmidt et al., 2021).

From an ethical point of view, two major concerns are front and center. First, there is the potential for unintended off-target mutations presenting human and environmental safety concerns, potentially leading to allergens and issues related to ecosystem integrity

(Távora et al., 2022). Second, the issue of equity and access captures the potential to widen the gap between smallholder farmers and large-scale farming operations, as the CRISPR patents primarily rest with multinational corporations (Munawar et al., 2023). The acceptance and success of gene-edited foods fundamentally rely on their acceptance by the consumer base. However, disinformation regarding GMOs has conditioned the consumer base, which remains a significant barrier to the acceptance of GMOs. Public concerns surrounding GMO products typically center on long-term health effects, potential environmental impacts related to, for example, cross-contamination and herbicide resistance, and ethical considerations related to the control of food systems and seed patents by corporate interests (Khokhar et al., 2023; Myskja and Myhr, 2020). Acceptance by consumers varies widely. For example, gene-edited plant products were supported by 45% of respondents (n=835) in China, compared to 36% for transgenic ones (Ortega et al., 2022). In Brazil, 84% of producers (n=37) were likely to plant transgenic beans, and 79% of consumers (n=100) were willing to eat them (da Silva Santos et al., 2022).

The socioeconomic landscape complicates CRISPR's integration into agriculture, especially as discussions have heightened the stakes surrounding intellectual property regimes and the increasing corporate control over seed availability (Chen et al., 2024; Wong and Chan, 2016). In specific contexts, such as India and much of Africa, the most significant challenge will be ensuring that CRISPR-based innovations, like drought-tolerant maize or disease-resistant cassava varieties, are available to both smallholder farmers and large agribusinesses (Munawar et al., 2023).

13.7 Future directions

The use of CRISPR/Cas genome editing as a speed breeding system represents a revolutionary development in crop science, and developments in next-generation genome editors, multi-omics integration, synthetic biology, and systems-level methodologies will likely influence future directions. There has never been a greater need for quicker, more accurate, and more adaptable breeding techniques as abiotic stress pressures increase due to climate change and global food security becomes more precarious.

13.7.1 Emerging CRISPR platforms

A key component of upcoming innovation is the expanding variety of CRISPR tools. Plant genomes and transcriptomes can now be altered with previously unheard-of precision and adaptability thanks to the development of Cas12, Cas13, base editors, and prime editors. Cas12 is very useful for editing AT-rich regulatory elements, which is advantageous for fine-tuning stress-responsive promoters, thanks to its staggered cuts and T-rich PAM recognition (Molla et al., 2021). Abiotic stress pathways can be manipulated reversibly using Cas13, which targets RNA rather than DNA, enabling dynamic, transgene-free control of gene expression (Kavuri et al., 2022).

Base editors, which allow for subtle edits in stress-associated alleles without DSBs, will become essential tools for single-nucleotide precision breeding. They have been used in early research to develop rice varieties resistant to salt and herbicides (Li et al., 2023). To modify the intricate regulatory networks that underlie stress resilience, prime editing—which can perform insertions, deletions, and all base conversions—offers the potential to rewrite DNA with surgical accuracy (Hillary & Caesar, 2022).

13.7.2 Multiplexing and trait pyramiding

Scientists have enabled a multiplex gene editing technique that simultaneously targets multiple genes and regulates tolerance against abiotic stresses, such as temperature fluctuations, salinity, and drought. Combining different traits is essential to producing plants that can withstand stressful environments. Thanks to new techniques, efficiency can be increased by delivering multiple guide RNAs in a single construct (Singha et al., 2022; Bharti et al., 2024). The existence of multiple copies of genes in crops like potatoes and wheat used to make plant breeding challenging; however, these new techniques simplify the process of overcoming these obstacles (Han et al., 2025).

13.7.3 Multi-omics integration for CRISPR-driven speed breeding

Multi-omics techniques, such as transcriptomics, proteomics, metabolomics, phenomics, and genomics, when integrated with CRISPR/Cas-driven speed breeding technology, have significantly enhanced the precision of genome editing. Multi-omics analysis can be utilized to create a blueprint for targeted genome editing by identifying crucial regulatory networks, putative genes, and biochemical pathways related to stress tolerance. Applying omics datasets to map molecular responses toward abiotic stresses has revealed potential target sites for CRISPR-mediated gene modification in root and tuber crops. These datasets help identify high-value targets, reduce off-target risks, and develop elite cultivars with improved traits in a shorter time frame when combined with ML and predictive breeding tools.

13.7.4 Artificial intelligence–guided CRISPR design and predictive breeding

AI and ML are the key technologies for future innovation in CRISPR-based breeding. With the help of AI in conjunction with GWAS, transcriptomics, and epigenomic datasets, we can create datasets based on chromatin accessibility and stress-induced gene expression profiles. These datasets help identify the best possible gene editing targets. To ensure increased accuracy and efficiency in gene editing, predictive models can also mitigate off-target risks and/or disruption of gene function before the editing is executed (Abdi et al., 2024; Roy et al., 2021). Every step of genome editing, from gRNA design to functional validation, may soon rely on these in silico design pipelines, which lessen the need for the empirical trial-and-error method.

Standalone CRISPR pipelines are already being adopted by national programs in China, India, and the United States, as well as international agricultural research organizations like the CGIAR and IRRI. To improve field performance, these platforms are transitioning from slow, multigenerational selection to precision-first strategies that incorporate multiplex editing, synthetic promoters, and even tissue-specific expression systems (Bharti et al., 2024).

The development of climate-resilient genotypes that can withstand drought, heat, salinity, and multiple concurrent stresses in a single genomic background depends on such innovations (Han et al., 2025).

13.8 Conclusion

CRISPR/Cas genome editing, when integrated with speed breeding, represents a paradigm shift in crop improvement, enabling the rapid and precise development of varieties that are tolerant to abiotic stress. Unlike traditional breeding, which may require a decade or more to fix complex traits, this combined approach can deliver stable, elite cultivars within a few generations by targeting multiple stress-responsive genes and accelerating generation turnover under controlled environments. Next-generation CRISPR/Cas platforms, such as Cas12, Cas13, base editors, and prime editors, expand the scope of edits from single-nucleotide changes to multigene trait pyramiding. Meanwhile, speed breeding ensures that these edits are quickly fixed and evaluated. This synergy has already been demonstrated in major crops for drought, salinity, heat, and cold tolerance and is poised to address the multistress challenges driven by climate change. However, technical barriers in transformation efficiency, genome editing in polyploids, off-target risk mitigation, and the establishment of enabling regulatory frameworks remain critical challenges. Addressing these requires advances in delivery systems, integration of multi-omics and AI-guided target prediction, and equitable access to CRISPR/Cas technology across diverse agricultural systems. In the coming decade, the CRISPR/Cas-based speed breeding pipeline offers a scalable and climate-smart breeding model with resource efficiency. By combining molecular precision with accelerated generation cycling, it can bridge the gap between laboratory innovation and field-ready solutions, delivering resilient, high-yielding crops at a pace aligned with the global food security needs.

Authors contribution

BS and CC: Conceptualized and designed the book chapter outline, revised and edited the original draft, acquired funding and resources, and provided overall supervision. AD: Prepared selected sections of the original draft, prepared tables and figures, and revised and edited the draft manuscript. AK, AL, and A Kachari: Prepared chosen sections of the original draft and revised and edited the draft manuscript. All authors proofread and approved the final version of the manuscript.

Ethics approval and consent to participate

Not applicable

Consent for publication

Not applicable

Availability of data and materials

Not applicable

Funding

This work was supported by the Council of Scientific and Industrial Research (CSIR), Government of India, as a Mission Project on Genome Editing for Crop Improvement (MMP-025301) and a CSIR In-house project (OLP-2503A) to CC. BS was supported by the Department of Science and Technology (DST), Govt. of India, through Postdoctoral Fellowship (DST/WISE-PDF/LS-112/2024(G).

Declaration of competing interest

The authors declare that they have no known competing financial interests or personal relationships that could have appeared to influence the work reported in this paper.

Acknowledgments

The authors are grateful to the Director, CSIR-North East Institute of Science and Technology (CSIR-NEIST), Jorhat, Assam, India, for providing encouragement and institutional support during the preparation of this book chapter. All figures were created using BioRender.com.

Data availability

No data is generated in this chapter.

References

Abdi, G., Patil, N., Mishra, P., Tariq, M., Dhariwal, R., Parveen, S., et al., 2024. Genomic Precision: Unveiling the Transformative Role of Genome Editing in Advancing Genomics Research and Applications. Springer Science and Business Media LLC, pp. 265–306. Available at http://doi.org/10.1007/978-981-97-3169-5_13.

Abu-Shahba, M.S., Mansour, M.M., Mohamed, H.I., Sofy, M.R., 2022. Effect of biosorptive removal of cadmium ions from hydroponic solution containing indigenous garlic peel and mercerized garlic peel on lettuce productivity. Scientia Horticulturae 293. Available at https://doi.org/10.1016/j.scienta.2021.110727, www.elsevier.com/inca/publications/store/5/0/3/3/1/6.

Adams, W.W., Stewart, J.J., Cohu, C.M., Muller, O., Demmig-Adams, B., 2016. Habitat temperature and precipitation of *Arabidopsis thaliana* ecotypes determine the response of foliar vasculature, photosynthesis, and transpiration to growth temperature. Frontiers in Plant Science 7 (2016). Available at https://doi.org/10.3389/fpls.2016.01026, http://journal.frontiersin.org/article/10.3389/fpls.2016.01026/full.

Adhikari, P., Poudel, M., 2020. CRISPR-Cas9 in agriculture: approaches, applications, future perspectives, and associated challenges. Malaysian Journal of Halal Research 3 (1), 6–16. Available at https://doi.org/10.2478/mjhr-2020-0002.

AHMAD, M., 2023. Plant breeding advancements with "CRISPR-Cas" genome editing technologies will assist future food security. Frontiers in Plant Science 14. Available at https://doi.org/10.3389/fpls.2023.1133036.

Ahmad, S., Makhmale, S., Bosamia, T.C., Sangh, C., 2021. Speed breeding: a potential tool for mitigating abiotic stresses. Abiotic Stress Management for Resilient Agriculture. Elsevier, pp. 237–257. Available at https://doi.org/10.1016/B978-0-12-822849-4.00012-7.

Ahmad, N., Fatima, S., Mehmood, M.A., Zaman, Q.U., Atif, R.M., Zhou, W., et al., 2023. Targeted genome editing in polyploids: lessons from Brassica. Frontiers in Plant Science 14. Available at https://doi.org/10.3389/fpls.2023.1152468, https://www.frontiersin.org/journals/plant-science.

Ahmar, S., Gill, R.A., Jung, K.H., Faheem, A., Qasim, M.U., Mubeen, M., et al., 2020. Conventional and molecular techniques from simple breeding to speed breeding in crop plants: recent advances and future outlook. International Journal of Molecular Sciences 21 (7). Available at https://doi.org/10.3390/ijms21072590, https://www.mdpi.com/1422-0067/21/7/2590/pdf.

Ahmar, S., Usman, B., Hensel, G., Jung, K.H., Gruszka, D., 2024. CRISPR enables sustainable cereal production for a greener future. Trends in Plant Science 29 (2), 179–195. Available at https://doi.org/10.1016/j.tplants.2023.10.016, https://www.sciencedirect.com/science/journal/13601385.

Ali, A., Zafar, M.M., Farooq, Z., Ahmed, S.R., Ijaz, A., Anwar, Z., et al., 2023. Breakthrough in CRISPR/Cas system: current and future directions and challenges. Biotechnology Journal 18 (8). Available at https://doi.org/10.1002/biot.202200642, http://onlinelibrary.wiley.com/journal/10.1002/(ISSN)1860-7314.

Angon, P.B., Mondal, S., Akter, S., Sakil, M.A., Jalil, M.A., 2023. Roles of CRISPR to mitigate drought and salinity stresses on plants. Plant Stress 8. Available at https://doi.org/10.1016/j.stress.2023.100169, www.journals.elsevier.com/plant-stress.

Antony Ceasar, S., Ignacimuthu, S., 2023. CRISPR/Cas genome editing in plants: dawn of Agrobacterium transformation for recalcitrant and transgene-free plants for future crop breeding. Plant Physiology and Biochemistry 196, 724–730. Available at https://doi.org/10.1016/j.plaphy.2023.02.030.

Belhaj, K., Chaparro-Garcia, A., Kamoun, S., Patron, N.J., Nekrasov, V., 2013. Editing plant genomes with CRISPR/Cas9. Trends in Plant Science 18 (11), 634–636. Available at https://doi.org/10.1016/j.copbio.2014.11.007.

Bertier, L.D., Ron, M., Huo, H., Bradford, K.J., Britt, A.B., Michelmore, R.W., 2018. High-resolution analysis of the efficiency, heritability, and editing outcomes of CRISPR/Cas9-induced modifications of NCED4 in lettuce (*Lactuca sativa*). G3: Genes, Genomes, Genetics 8 (5), 1513–1521. Available at https://doi.org/10.1534/g3.117.300396, http://www.g3journal.org/content/ggg/8/5/1513.full.pdf.

Bharti, D.R., Pramanik, B., Paul, A., Debnath, S., 2024. CRISPR-based precision breeding for crop improvement: current achievements and future directions. In: Barik, D., Mondal, A.K. (Eds.), Precision Breeding and Genome Editing. CRC Press, Boca Raton, FL, pp. 73–94. Available at https://doi.org/10.1201/9781003457237-6.

Biswas, D., Saha, S.C., Dey, A., 2021. CRISPR-Cas genome-editing tool in plant abiotic stress-tolerance. Plant Gene 26. Available at https://doi.org/10.1016/j.plgene.2021.100286, http://www.journals.elsevier.com/plant-gene/.

Blom, T., Jenkins, A., Pulselli, R.M., van den Dobbelsteen, A.A.J.F., 2022. The embodied carbon emissions of lettuce production in vertical farming, greenhouse horticulture, and open-field farming in the Netherlands. Journal of Cleaner Production 377, 134443. Available at https://doi.org/10.1016/j.jclepro.2022.134443.

Bo, W., Zhaohui, Z., Huanhuan, Z., Xia, W., Binglin, L., Lijia, Y., et al., 2019. Targeted mutagenesis of NAC transcription factor gene, OsNAC041, leading to salt sensitivity in rice. Rice Science 26 (2), 98–108. Available at https://doi.org/10.1016/j.rsci.2018.12.005, http://www.elsevier.com/wps/find/journaldescription.cws_home/713745/description#description.

Bouzroud, S., Gasparini, K., Hu, G., Barbosa, M.A.M., Rosa, B.L., Fahr, M., et al., 2020. Down regulation and loss of auxin response factor 4 function using CRISPR/Cas9 alters plant growth, stomatal function and improves tomato tolerance to salinity and osmotic stress. Genes 11 (3). Available at https://doi.org/10.3390/genes11030272, https://www.mdpi.com/2073-4425/11/3/272/pdf.

Caccamo, M., 2023. New precision-breeding law unlocks gene editing in England. Nature Biotechnology 41 (6), 752–753. Available at https://doi.org/10.1038/s41587-023-01795-8.

Casanova, M., Salazar, O., Oyarzún, I., Tapia, Y., Fajardo, M., 2016. Field monitoring of 2010-Tsunami impact on agricultural soils and irrigation waters: central Chile. Water, Air, and Soil Pollution 227 (11). Available at https://doi.org/10.1007/s11270-016-3113-7, http://www.kluweronline.com/issn/0049-6979/.

Ćeran, M., Miladinović, D., Đorđević, V., Trkulja, D., Radanović, A., Glogovac, S., et al., 2024. Genomics-assisted speed breeding for crop improvement: present and future. Frontiers in Sustainable Food Systems 8. Available at https://doi.org/10.3389/fsufs.2024.1383302.

Cha, J.K., Park, H., Choi, C., Kwon, Y., Lee, S.M., Oh, K.W., et al., 2023. Acceleration of wheat breeding: enhancing efficiency and practical application of the speed breeding system. Plant Methods 19 (1). Available at https://doi.org/10.1186/s13007-023-01083-1, https://plantmethods.biomedcentral.com/.

Chakraborty, A., Wylie, S.J., 2025. CRISPR/Cas9 for heat stress tolerance in rice: a review. Plant Molecular Biology Reporter 43 (3), 1047–1056. Available at https://doi.org/10.1007/s11105-025-01564-8, https://www.springer.com/journal/11105.

Chen, J., Mueller, V., 2018. Coastal climate change, soil salinity and human migration in Bangladesh. Nature Climate Change 8 (11), 981–985. Available at https://doi.org/10.1038/s41558-018-0313-8.

Chen, H.J., Su, C.T., Lin, C.H., Huang, G.J., Lin, Y.H., 2010. Expression of sweet potato cysteine protease SPCP2 altered developmental characteristics and stress responses in transgenic Arabidopsis plants. Journal of Plant Physiology 167 (10), 838–847. Available at https://doi.org/10.1016/j.jplph.2010.01.005.

Chen, K., Wang, Y., Zhang, R., Zhang, H., Gao, C., 2019. CRISPR/Cas genome editing and precision plant breeding in agriculture. Annual Review of Plant Biology 70 (1), 667–697. Available at https://doi.org/10.1146/annurev-arplant-050718–100049.

Chen, F., Chen, L., Yan, Z., Xu, J., Feng, L., He, N., et al., 2024. Recent advances of CRISPR-based genome editing for enhancing staple crops. Frontiers in Plant Science 15. Available at https://doi.org/10.3389/fpls.2024.1478398.

Chimata, M.K., Bharti, G., 2019. Regulation of genome edited technologies in India. Transgenic Research 28, 175–181. Available at https://doi.org/10.1007/s11248-019-00148-z, www.wkap.nl/journalhome.htm/0962-8819.

Chimmili, S.R., Kanneboina, S., Hanjagi, P.S., P.S., B., Sakhare, A.S., Senguttuvel, P., et al., 2022. Integrating advanced molecular, genomic, and speed breeding methods for genetic improvement of stress tolerance in rice next-generation. Plant Breeding Approaches for Stress Resilience in Cereal Crops. Springer Nature, India, pp. 263–283. Available at https://doi.org/10.1007/978-981-19-1445-4.

Choudry, M.W., Riaz, R., Nawaz, P., Ashraf, M., Ijaz, B., Bakhsh, A., 2024. CRISPR-Cas9 mediated understanding of plants' abiotic stress-responsive genes to combat changing climatic patterns. Functional and Integrative Genomics 24 (4). Available at https://doi.org/10.1007/s10142-024-01405-z, https://www.springer.com/journal/10142.

Collard, B.C.Y., Mackill, D.J., 2008. Marker-assisted selection: an approach for precision plant breeding in the twenty-first century. Philosophical Transactions of the Royal Society B: Biological Sciences 363 (1491), 557–572. Available at https://doi.org/10.1098/rstb.2007.2170, http://rstb.royalsocietypublishing.org/.

Curtin, S.J., Xiong, Y., Michno, J.M., Campbell, B.W., Stec, A.O., Čermák, T., et al., 2018. Crispr/cas9 and talens generate heritable mutations for genes involved in small rna processing of glycine max and medicago truncatula. Plant Biotechnology Journal 16 (6), 1125–1137. Available at https://doi.org/10.1111/pbi.12857.

Das, P., Chandra, T., Negi, A., Jaiswal, S., Iquebal, M.A., Rai, A., et al., 2023. A comprehensive review on genomic resources in medicinally and industrially important major spices for future breeding programs: status, utility and challenges. Current Research in Food Science 7, 100579. Available at https://doi.org/10.1016/j.crfs.2023.100579.

da Silva Santos, C.R., Teixeira, S.M., Cruz, J.E., Bron, P.C., 2022. Perception of producers and consumers on the adoption of genetically modified food: the case of the transgenic bean BRSFC401 RMD. Revista de Economia e Sociologia Rural 61 (2). Available at https://doi.org/10.1590/1806-9479.2022.250277, http://www.scielo.br/scielo.php?script=sci_serial&pid=0103-2003&lng=en&nrm=iso.

Djanaguiraman, M., Narayanan, S., Erdayani, E., Prasad, P.V.V., 2020. Effects of high temperature stress during anthesis and grain filling periods on photosynthesis, lipids and grain yield in wheat. BMC Plant Biology 20 (1). Available at https://doi.org/10.1186/s12870-020-02479-0.

Du, Y.T., Zhao, M.J., Wang, C.T., Gao, Y., Wang, Y.X., Liu, Y.W., et al., 2018. Identification and characterization of GmMYB118 responses to drought and salt stress. BMC Plant Biology 18 (1), 320. Available at https://doi.org/10.1186/s12870-018-1551-7.

Dwivedi, S., Perotti, E., Ortiz, R., 2008. Towards molecular breeding of reproductive traits in cereal crops. Plant Biotechnology Journal 6 (6), 529–559. Available at https://doi.org/10.1111/j.1467-7652.2008.00343.x.

Eckerstorfer, M.F., Grabowski, M., Lener, M., Engelhard, M., Simon, S., Dolezel, M., et al., 2021. Biosafety of genome editing applications in plant breeding: considerations for a focused case-specific risk assessment in the EU. BioTech 10 (3). Available at https://doi.org/10.3390/biotech10030010, https://www.mdpi.com/2673-6284/10/3/10/pdf.

Erdoğan, İ., Cevher-Keskin, B., Bilir, Ö., Hong, Y., Tör, M., 2023. Recent developments in CRISPR/Cas9 genome-editing technology related to plant disease resistance and abiotic stress tolerance. Biology 12 (7). Available at https://doi.org/10.3390/biology12071037, https://www.mdpi.com/journal/biology.

Evanega, S., Brown, Z., Bubeck, D., Falck-Zapeda, J., Chou, F.-L., Graham, N., et al., 2024. Applications, Benefits, and Challenges of Genome Edited Crops. Council for Agricultural Science and Technology. Available at http://doi.org/10.62300/hjvu4083.

Fang, H., Meng, Q., Xu, J., Tang, H., Tang, S., Zhang, H., et al., 2015. Knock-down of stress inducible OsSRFP1 encoding an E3 ubiquitin ligase with transcriptional activation activity confers abiotic stress tolerance through enhancing antioxidant protection in rice. Plant Molecular Biology 87 (4-5), 441–458. Available at https://doi.org/10.1007/s11103-015-0294-1, https://link.springer.com/journal/11103.

FAO, 2017. The Future of Food and Agriculture – Trends and Challenges. Food and Agriculture Organization of the United Nations. Available at https://www.fao.org/global-perspectives-studies/food-agriculture-projections-to-2050/en/.

Gajardo, H.A., Gómez-Espinoza, O., Boscariol Ferreira, P., Carrer, H., Bravo, L.A., 2023. The potential of CRISPR/Cas technology to enhance crop performance on adverse soil conditions. Plants 12 (9). Available at https://doi.org/10.3390/plants12091892, www.mdpi.com/journal/plants.

Gangashetty, P.I., Belliappa, S.H., Bomma, N., Kanuganahalli, V., Sajja, S.B., Choudhary, S., et al., 2024. Optimizing speed breeding and seed/pod chip based genotyping techniques in pigeonpea: a way forward for high throughput line development. Plant Methods 20 (1). Available at https://doi.org/10.1186/s13007-024-01155-w, https://plantmethods.biomedcentral.com/.

Gao, C., 2018. The future of CRISPR technologies in agriculture. Nature Reviews Molecular Cell Biology 19 (5), 275–276. Available at https://doi.org/10.1038/nrm.2018.2.

Gao, C., 2021. Genome engineering for crop improvement and future agriculture. Cell 184 (6), 1621–1635. Available at https://doi.org/10.1016/j.cell.2021.01.005.

Ghouri, M.Z., Munawar, N., Aftab, S.O., Ahmad, A., 2023. Regulation of CRISPR edited food and feed: legislation and future. GMOs and Political Stance. Academic Press, pp. 261–287. Available at https://doi.org/10.1016/B978-0-12-823903-2.00004-4.

Gong, Z., Xiong, L., Shi, H., Yang, S., Herrera-Estrella, L.R., Xu, G., et al., 2020. Plant abiotic stress response and nutrient use efficiency. Science China Life Sciences 63 (5), 635–674. Available at https://doi.org/10.1007/s11427-020-1683-x, http://www.springer.com/life+sciences/journal/11427.

Gudeta, D., 2019. Genome editing: tools and application in plants. Open Access Journal of Microbiology and Biotechnology 4 (1), 000135. Available at https://doi.org/10.23880/oajmb-16000135.

Gudi, S., Kumar, P., Singh, S., Tanin, M.J., Sharma, A., 2022. Strategies for accelerating genetic gains in crop plants: special focus on speed breeding. Physiology and Molecular Biology of Plants 28 (10), 1921–1938. Available at https://doi.org/10.1007/s12298-022-01247-8, https://www.springer.com/journal/12298.

Guo, C., Ma, X., Gao, F., Guo, Y., 2023. Off-target effects in CRISPR/Cas9 gene editing. Frontiers in Bioengineering and Biotechnology 11. Available at https://doi.org/10.3389/fbioe.2023.1143157, http://journal.frontiersin.org/journal/bioengineering-and-biotechnology#archive.

Gurel, F., Wu, Y., Pan, C., Cheng, Y., Li, G., Zhang, T., et al., 2023. On- and off-target analyses of CRISPR-Cas12b genome editing systems in rice. The CRISPR Journal 6 (1), 62–74. Available at https://doi.org/10.1089/crispr.2022.0072.

Hamdan, M.F., Karlson, C.K.S., Teoh, E.Y., Lau, S.E., Tan, B.C., 2022. Genome editing for sustainable crop improvement and mitigation of biotic and abiotic stresses. Plants 11 (19). Available at https://doi.org/10.3390/plants11192625, www.mdpi.com/journal/plants.

Han, X., Li, S., Zeng, Q., Sun, P., Wu, D., Wu, J., et al., 2025. Genetic engineering, including genome editing, for enhancing broad-spectrum disease resistance in crops. Plant Communications 6 (2), 101195. Available at https://doi.org/10.1016/j.xplc.2024.101195.

Haroon, M., Wang, X., Afzal, R., Zafar, M.M., Idrees, F., Batool, M., et al., 2022. Novel plant breeding techniques shake hands with cereals to increase production. Plants 11 (8), 1052. Available at https://doi.org/10.3390/plants11081052.

Harwood, W., Ricroch, A., Clairand, P., Harwood, W., 2017. Use of CRISPR systems in plant genome editing: toward new opportunities in agriculture. Emerging Topics in Life Sciences 1 (2), 169–182. Available at https://doi.org/10.1042/etls20170085.

Hatfield, J.L., Prueger, J.H., 2015. Temperature extremes: effect on plant growth and development. Weather and Climate Extremes 10, 4–10. Available at https://doi.org/10.1016/j.wace.2015.08.001, http://www.journals.elsevier.com/weather-and-climate-extremes/.

He, L., Shi, X., Wang, Y., Guo, Y., Yang, K., Wang, Y., 2017. Arabidopsis ANAC069 binds to C[A/G]CG[T/G] sequences to negatively regulate salt and osmotic stress tolerance. Plant Molecular Biology 93 (4-5), 369–387. Available at https://doi.org/10.1007/s11103-016-0567-3, www.wkap.nl/journalhome.htm/0167-4412.

He, F., Wang, C., Sun, H., Tian, S., Zhao, G., Liu, C., et al., 2023. Simultaneous editing of three homoeologues of TaCIPK14 confers broad-spectrum resistance to stripe rust in wheat. Plant Biotechnology Journal 21 (2), 354–368. Available at https://doi.org/10.1111/pbi.13956, http://onlinelibrary.wiley.com/journal/10.1111/(ISSN)1467-7652.

He, Y., Han, Y., Ma, Y., Liu, S., Fan, T., Liang, Y., et al., 2024. Expanding plant genome editing scope and profiles with CRISPR-FrCas9 systems targeting palindromic TA sites. Plant Biotechnology Journal 22 (9), 2488–2503. Available at https://doi.org/10.1111/pbi.14363, http://onlinelibrary.wiley.com/journal/10.1111/(ISSN)1467-7652.

Hillary, V.E., Ceasar, S.A., 2022. Prime editing in plants and mammalian cells: mechanism, achievements, limitations, and future prospects. BioEssays 44 (9). Available at https://doi.org/10.1002/bies.202200032, http://onlinelibrary.wiley.com/journal/10.1002/(ISSN)1521-1878.

Hossain, A., Rahman, M.M.E., Ali, S., Islam, T., Syed, M.A., Syed, T., et al., 2022. CRISPR-Cas9-Mediated Genome Editing Technology for Abiotic Stress Tolerance in Crop Plant. Elsevier BV, pp. 331–354. Available at http://doi.org/10.1016/b978-0-323-85665-2.00008-x.

Hualpa-Ramirez, E., Carrasco-Lozano, E.C., Madrid-Espinoza, J., Tejos, R., Ruiz-Lara, S., Stange, C., et al., 2024. Stress salinity in plants: new strategies to cope with in the foreseeable scenario. Plant Physiology and Biochemistry 208, 108507. Available at https://doi.org/10.1016/j.plaphy.2024.108507.

Jähne, F., Hahn, V., Würschum, T., Leiser, W.L., 2020. Speed breeding short-day crops by LED-controlled light schemes. Theoretical and Applied Genetics 133 (8), 2335–2342. Available at https://doi.org/10.1007/s00122-020-03601-4, link.springer.de/link/service/journals/00122/index.htm.

Jain, M., 2015. Function genomics of abiotic stress tolerance in plants: a CRISPR approach. Frontiers in Plant Science 6 (May). Available at https://doi.org/10.3389/fpls.2015.00375.

Jamil, A., Riaz, S., Ashraf, M., Foolad, M.R., 2011. Gene expression profiling of plants under salt stress. Critical Reviews in Plant Sciences 30 (5), 435–458. Available at https://doi.org/10.1080/07352689.2011.605739.

Jha, U.C., Bohra, A., Jha, R., Parida, S.K., 2019. Salinity stress response and 'omics' approaches for improving salinity stress tolerance in major grain legumes. Plant Cell Reports 38 (3), 255–277. Available at https://doi.org/10.1007/s00299-019-02374-5, link.springer.de/link/service/journals/00299/index.htm.

Jin, S., Zong, Y., Gao, Q., Zhu, Z., Wang, Y., Qin, P., et al., 2019. Cytosine, but not adenine, base editors induce genome-wide off-target mutations in rice. Science 364 (6437), 292–295. Available at https://doi.org/10.1126/science.aaw7166, https://science.sciencemag.org/content/364/6437/292/tab-pdf.

Jinek, M., Chylinski, K., Fonfara, I., Hauer, M., Doudna, J.A., Charpentier, E., 2012. A programmable dual-RNA-guided DNA endonuclease in adaptive bacterial immunity. Science 337 (6096), 816–821. Available at https://doi.org/10.1126/science.1225829, http://www.sciencemag.org/content/337/6096/816.full.pdf.

Julkowska, M.M., Testerink, C., 2015. Tuning plant signaling and growth to survive salt. Trends in Plant Science 20 (9), 586–594. Available at https://doi.org/10.1016/j.tplants.2015.06.008, www.elsevier.com/inca/publications/store/3/0/9/6/0/index.htt.

Jung, J.H., Domijan, M., Klose, C., Biswas, S., Ezer, D., Gao, M., et al., 2016. Phytochromes function as thermosensors in *Arabidopsis*. Science 354 (6314), 886–889. Available at https://doi.org/10.1126/science.aaf6005, http://science.sciencemag.org/content/sci/354/6314/886.full.pdf.

Kaashyap, M., Ford, R., Bohra, A., Kuvalekar, A., Mantri, N., 2017. Improving salt tolerance of chickpea using modern genomics tools and molecular breeding. Current Genomics 18 (6), 557–567. Available at https://doi.org/10.2174/1389202918666170705155252, http://www.benthamdirect.org/pages/all_b_bypublication.php.

Kamanga, R.M., Mbega, E., Ndakidemi, P., 2018. Drought tolerance mechanisms in plants: physiological responses associated with water deficit stress in *Solanum lycopersicum*. Advances in Crop Science and Technology 06 (03). Available at https://doi.org/10.4172/2329-8863.1000362.

Kaur, N., Sharma, S., Hasanuzzaman, M., Pati, P.K., 2022. Genome editing: a promising approach for achieving abiotic stress tolerance in plants. International Journal of Genomics 2022. Available at https://doi.org/10.1155/2022/5547231, http://www.hindawi.com/journals/ijg/.

Kaur, N., Qadir, M., Francis, D.V., Alok, A., Tiwari, S., Ahmed, Z.F.R., 2025. CRISPR/Cas9: a sustainable technology to enhance climate resilience in major Staple Crops. Frontiers in Genome Editing 7. Available at https://doi.org/10.3389/fgeed.2025.1533197, https://www.frontiersin.org/journals/genome-editing/about.

Kavuri, N.R., Ramasamy, M., Qi, Y., Mandadi, K., 2022. Applications of CRISPR/Cas13-based RNA editing in plants. Cells 11 (17), 2665. Available at https://doi.org/10.3390/cells11172665.

Khokhar, A., Shahbaz, M., Maqsood, M.F., Zulfiqar, U., Naz, N., Iqbal, U.Z., et al., 2023. Correction to: genetic modification strategies for enhancing plant resilience to abiotic stresses in the context of climate change. Functional & Integrative Genomics 23 (4). Available at https://doi.org/10.1007/s10142-023-01250-6.

Kigoni, M., Choi, M., Arbelaez, J.D., 2023. 'Single-Seed-SpeedBulks:' a protocol that combines 'speed breeding' with a cost-efficient modified single-seed descent method for rapid-generation-advancement in oat (*Avena sativa* L.). Plant Methods 19 (1). Available at https://doi.org/10.1186/s13007-023-01067-1, https://plantmethods.biomedcentral.com/.

Kim, D., Alptekin, B., Budak, H., 2018. CRISPR/Cas9 genome editing in wheat. Functional & Integrative Genomics 18 (1), 31–41. Available at https://doi.org/10.1007/s10142-017-0572-x.

Kim, M.S., Ko, S.R., Jung, Y.J., Kang, K.K., Lee, Y.J., Cho, Y.G., 2023. Knockout mutants of OsPUB7 generated using CRISPR/Cas9 revealed abiotic stress tolerance in rice. International Journal of Molecular Sciences 24 (6), 5338. Available at https://doi.org/10.3390/ijms24065338.

Klap, C., Yeshayahou, E., Bolger, A.M., Arazi, T., Gupta, S.K., Shabtai, S., et al., 2017. Tomato facultative parthenocarpy results from Sl AGAMOUS-LIKE 6 loss of function. Plant Biotechnology Journal 15 (5), 634–647. Available at https://doi.org/10.1111/pbi.12662.

Kloc, Y., Dmochowska-Boguta, M., Żebrowska-Różańska, P., Łaczmański, Ł., Nadolska-Orczyk, A., Orczyk, W., 2024. HvGSK1. 1 controls salt tolerance and yield through the brassinosteroid signaling pathway in barley. International Journal of Molecular Sciences 25 (2), 998. Available at https://doi.org/10.3390/ijms25020998.

Koonin, E.V., Makarova, K.S., 2019. Origins and evolution of CRISPR-Cas systems. Philosophical Transactions of the Royal Society B 374 (1772). Available at https://doi.org/10.1098/rstb.2018.0087.

Kumar, K., Kumar, M., Kim, S.R., Ryu, H., Cho, Y.G., 2013. Insights into genomics of salt stress response in rice. Rice 6 (1), 1–15. Available at https://doi.org/10.1186/1939-8433-6-27, http://www.springerlink.com/content/1939-8425.

Kumar, M., Prusty, M.R., Pandey, M.K., Singh, P.K., Bohra, A., Guo, B., et al., 2023. Application of CRISPR/Cas9-mediated gene editing for abiotic stress management in crop plants. Frontiers in Plant Science 14. Available at https://doi.org/10.3389/fpls.2023.1157678, https://www.frontiersin.org/journals/plant-science.

Kumar, K.P., Walia, P., 2024. Speed breeding in cereal crops: accelerating genetic improvement for rapid agricultural advancement. Journal of Experimental Agriculture International 46 (6), 465–477. Available at https://doi.org/10.9734/jeai/2024/v46i62498.

Kumawat, T., Agarwal, A., Saxena, S., Arora, S., 2024. An analysis of global policies and regulation on genome editing in plants. Gene Editing in Plants: CRISPR-Cas and Its Applications. Springer Science+Business Media, India, pp. 775–793. Available at https://link.springer.com/book/9789819985296.

Legris, M., Klose, C., Burgie, E.S., Rojas, C.C., Neme, M., Hiltbrunner, A., et al., 2016. Phytochrome B integrates light and temperature signals in *Arabidopsis*. Science 354 (6314), 897–900. Available at https://doi.org/10.1126/science.aaf5656, http://science.sciencemag.org/content/sci/354/6314/897.full.pdf.

Li, J., Zhang, H., Si, X., Tian, Y., Chen, K., Liu, J., et al., 2017. Generation of thermosensitive male-sterile maize by targeted knockout of the ZmTMS5 gene. Journal of genetics and genomics Yi chuan xue bao 44 (9), 465–468. Available at https://doi.org/10.1016/j.jgg.2017.02.002.

Li, R., Zhang, L., Wang, L., Chen, L., Zhao, R., Sheng, J., et al., 2018. Reduction of tomato-plant chilling tolerance by CRISPR-Cas9-mediated SlCBF1 mutagenesis. Journal of Agricultural and Food Chemistry 66 (34), 9042–9051. Available at https://doi.org/10.1021/acs.jafc.8b02177, http://pubs.acs.org/journal/jafcau.

Li, R., Liu, C., Zhao, R., Wang, L., Chen, L., Yu, W., et al., 2019. CRISPR/Cas9-Mediated SlNPR1 mutagenesis reduces tomato plant drought tolerance. BMC Plant Biology 19 (1), 38. Available at https://doi.org/10.1186/s12870-018-1627-4.

Li, Y., Liao, S., Mei, P., Pan, Y., Zhang, Y., Zheng, X., et al., 2021a. OsWRKY93 Dually functions between leaf senescence and in response to biotic stress in rice. Frontiers in Plant Science 12. Available at https://doi.org/10.3389/fpls.2021.643011, https://www.frontiersin.org/journals/plant-science.

Li, Z., Sun, P., Sun, P., Liang, Y.K., Ge, S., 2021b. OsBC1L1 and OsBC1L8 function in stomatal development in rice. Biochemical and Biophysical Research Communications 576, 40–47. Available at https://doi.org/10.1016/j.bbrc.2021.08.074, http://www.sciencedirect.com/science/journal/0006291X.

Li, Y., Wu, X., Zhang, Y., Zhang, Q., 2022. CRISPR/Cas genome editing improves abiotic and biotic stress tolerance of crops. Frontiers in Genome Editing 4. Available at https://doi.org/10.3389/fgeed.2022.987817.

Li, J., Zhang, C., He, Y., Li, S., Yan, L., Li, Y., 2023. Plant base and prime editing: current status and future perspectives. Journal of Integrative Plant Biology 65 (8), 1695–1712. Available at https://doi.org/10.1111/jipb.13425.

Li, S., Zhang, Y., Liu, Y., Zhang, P., Wang, X., Chen, B., et al., 2024a. The E3 ligase TaGW2 mediates transcription factor TaARR12 degradation to promote drought resistance in wheat. The Plant Cell 36 (3), 605–625. Available at https://doi.org/10.1093/plcell/koad307.

Li, Z., Li, Z., Ji, Y., Wang, C., Wang, S., Shi, Y., et al., 2024b. The heat shock factor 20-HSF4-cellulose synthase A2 module regulates heat stress tolerance in maize. The Plant Cell 36 (7), 2652–2667. Available at https://doi.org/10.1093/plcell/koae106.

Litalien, A., Zeeb, B., 2020. Curing the earth: a review of anthropogenic soil salinization and plant-based strategies for sustainable mitigation. Science of the Total Environment 698, 134235. Available at https://doi.org/10.1016/j.scitotenv.2019.134235.

Liu, H., Hu, M., Wang, Q., Cheng, L., Zhang, Z., 2018. Role of papain-like cysteine proteases in plant development. Frontiers in Plant Science 9. Available at https://doi.org/10.3389/fpls.2018.01717.

Liu, X., Wu, D., Shan, T., Xu, S., Qin, R., Li, H., et al., 2020. The trihelix transcription factor OsGTγ-2 is involved adaption to salt stress in rice. Plant Molecular Biology 103 (4-5), 545–560. Available at https://doi.org/10.1007/s11103-020-01010-1, https://link.springer.com/journal/11103.

Ma, X., Zhang, Q., Zhu, Q., Liu, W., Chen, Y., Li, H., et al., 2015. A robust CRISPR/Cas9 system for convenient, high-efficiency multiplex genome editing in plants. Molecular Plant 8 (8), 1228–1237. Available at https://doi.org/10.1016/j.molp.2015.04.007.

Menz, J., Modrzejewski, D., Hartung, F., Wilhelm, R., Sprink, T., 2020. Genome edited crops touch the market: a view on the global development and regulatory environment. Frontiers in Plant Science 11. Available at https://doi.org/10.3389/fpls.2020.586027, https://www.frontiersin.org/journals/plant-science.

Miao, C., Xiao, L., Hua, K., Zou, C., Zhao, Y., Bressan, R.A., et al., 2018. Mutations in a subfamily of abscisic acid receptor genes promote rice growth and productivity. Proceedings of the National Academy of Sciences 115 (23), 6058–6063. Available at https://doi.org/10.1073/pnas.1804774115.

Mickelbart, M.V., Hasegawa, P.M., Bailey-Serres, J., 2015. Genetic mechanisms of abiotic stress tolerance that translate to crop yield stability. Nature Reviews Genetics 16 (4), 237–251. Available at https://doi.org/10.1038/nrg3901, http://www.nature.com/reviews/genetics.

Mo, W., Tang, W., Du, Y., Jing, Y., Bu, Q., Lin, R., 2020. PHYTOCHROME-INTERACTING FACTOR-LIKE14 and SLENDER RICE1 interaction controls seedling growth under salt stress. Plant Physiology 184 (1), 506–517. Available at https://doi.org/10.1104/pp.20.00024.

Mohamed, H.I., Khan, A., Basit, A., 2024. CRISPR-Cas9 system mediated genome editing technology: an ultimate tool to enhance abiotic stress in crop plants. Journal of Soil Science and Plant Nutrition 24 (2), 1799–1822. Available at https://doi.org/10.1007/s42729-024-01778-x, https://www.springer.com/journal/42729.

Molla, K.A., Sretenovic, S., Bansal, K.C., Qi, Y., 2021. Precise plant genome editing using base and prime editors. Nature Plants 7, 636–649. Available at https://doi.org/10.1038/s41477-021-00991-1.

Moose, S.P., Mumm, R.H., 2008. Molecular plant breeding as the foundation for 21st century crop improvement. Plant Physiology 147 (3), 969–977. Available at https://doi.org/10.1104/pp.108.118232, http://www.plantphysiol.org/cgi/reprint/147/3/969.

Muller, O., Stewart, J.J., Cohu, C.M., Polutchko, S.K., Demmig-Adams, B., Adams, W.W., 2014. Leaf architectural, vascular and photosynthetic acclimation to temperature in two biennials. Physiologia Plantarum 152 (4), 763–772. Available at https://doi.org/10.1111/ppl.12226, http://onlinelibrary.wiley.com/journal/10.1111/(ISSN)1399-3054.

Munawar, N., Ahsan, K., Ahmad, A., 2023. CRISPR-edited plants' social, ethical, policy, and governance issues. Global Regulatory Outlook for CRISPRized Plants. Elsevier, United Arab Emirates, pp. 367–396. Available at https://www.sciencedirect.com/book/9780443184444.

Myskja, B.K., Myhr, A.I., 2020. Non-safety assessments of genome-edited organisms: should they be included in regulation? Science and Engineering Ethics 26 (5), 2601–2627. Available at https://doi.org/10.1007/s11948-020-00222-4, http://www.springerlink.com/content/1471-5546.

Nakamura, A., Yano, T., Mitsuda, N., Furubayashi, M., Ito, S., Sugano, S.S., et al., 2023. The sonication-assisted whisker method enables CRISPR-Cas9 ribonucleoprotein delivery to induce genome editing in rice. Scientific Reports 13 (1). Available at https://doi.org/10.1038/s41598-023-40433-w, https://www.nature.com/srep/.

Nguyen, T.T., Pham, D.T., Nguyen, N.H., Do, P.T., To, H.T.M., 2023. The Germin-like protein gene OsGER4 is involved in heat stress response in rice root development. Functional & Integrative Genomics 23 (3), 271.

Niraula, S., Khanal, R., Ghimire, N., 2024. CRISPR-Cas based precision genome editing: current advances and associated challenges in crop improvement and trait enhancement. Genetics & Applications 8 (2). Available at https://doi.org/10.31383/ga.vol8iss2ga05.

Ogata, T., Ishizaki, T., Fujita, M., Fujita, Y., 2020. CRISPR/Cas9-targeted mutagenesis of OsERA1 confers enhanced responses to abscisic acid and drought stress and increased primary root growth under non stressed conditions in rice. PLoS One 15 (12), e0243376. Available at https://doi.org/10.1371/journal.pone.0243376.

Ortega, D.L., Lin, W., Ward, P.S., 2022. Consumer acceptance of gene-edited food products in China. Food Quality and Preference 95. Available at https://doi.org/10.1016/j.foodqual.2021.104374, https://www.journals.elsevier.com/food-quality-and-preference.

Panda, S.K., Gupta, D., Patel, M., Vyver, C.V.D., Koyama, H., 2024. Functionality of reactive oxygen species (ROS) in plants: toxicity and control in poaceae crops exposed to abiotic stress. Plants 13 (15). Available at https://doi.org/10.3390/plants13152071, www.mdpi.com/journal/plants.

Pandey, S., Singh, A., Parida, S.K., Prasad, M., 2022. Combining speed breeding with traditional and genomics-assisted breeding for crop improvement. Plant Breeding 141 (3), 301–313. Available at https://doi.org/10.1111/pbr.13012, http://onlinelibrary.wiley.com/journal/10.1111/(ISSN)1439-0523.

Pandey, S., Divakar, S., Singh, A., 2024. Genome editing prospects for heat stress tolerance in cereal crops. Plant Physiology and Biochemistry 215, 108989. Available at https://doi.org/10.1016/j.plaphy.2024.108989.

Pandey, H., Misra, V., Sharma, A., Chatterjee, B., Sutradhar, M., Kumar, R., et al., 2025. Interventions of plant tissue culture techniques and genome editing in medicinally important spice crops. Medicinal Plant Biology 4. Available at https://doi.org/10.48130/mpb-0025-0019, https://www.maxapress.com/mpb.

Pankaj, Y.K., Kumar, V., 2023. CRISPR/Cas: the beginning of a new era in crop improvement. Genome Engineering for Crop Improvement. Available at https://doi.org/10.1007/978-3-031-28146-4_17.

Park, J.R., Kim, E.G., Jang, Y.H., Jan, R., Farooq, M., Ubaidillah, M., et al., 2022. Applications of CRISPR/Cas9 as new strategies for short breeding to drought gene in rice. Frontiers in Plant Science 13. Available at https://doi.org/10.3389/fpls.2022.850441, https://www.frontiersin.org/journals/plant-science.

Patel, S., Gurjar, D., 2024. Speed breeding: a transformative approach to accelerate crop improvement. Plant Cell Biotechnology and Molecular Biology 25 (7-8), 24–35. Available at https://doi.org/10.56557/pcbmb/2024/v25i7-88729.

Pawar, P., Talekar, N., Kumar, A., Salunkhe, H., 2023. Development and application of speed breeding technologies in groundnut (*Arachis hypogaea* L.): an advance approach. International Journal of Environment and Climate Change 13 (9), 1–13. Available at https://doi.org/10.9734/ijecc/2023/v13i92199.

Petrov, V., Gechev, T., 2023. ROS and abiotic stress in plants 2.0. International Journal of Molecular Sciences 24 (15), 11917. Available at https://doi.org/10.3390/ijms241511917.

Polidoros, A., Nianiou-Obeidat, I., Tsakirpaloglou, N., Petrou, N., Deligiannidou, E., Makri, N.-M., 2024. Genome-editing products line up for the market: will Europe harvest the benefits from science and innovation? Genes 15 (8), 1014. Available at https://doi.org/10.3390/genes15081014.

Potts, J., Jangra, S., Michael, V.N., Wu, X., 2023. Speed breeding for crop improvement and food security. Crops 3 (4), 276–291. Available at https://doi.org/10.3390/crops3040025, https://www.mdpi.com/journal/crops.

Prasad, A., Chimata, M.K., 2023. Regulation of genome-edited technologies: Indian policy perspective. Current Indian Science 01. Available at https://doi.org/10.2174/2210299x01666230223105053.

Qadir, M., Quillérou, E., Nangia, V., Murtaza, G., Singh, M., Thomas, R.J., et al., 2014. Economics of salt-induced land degradation and restoration. Natural Resources Forum 38 (4), 282–295. Available at https://doi.org/10.1111/1477-8947.12054Canada, http://onlinelibrary.wiley.com/journal/10.1111/(ISSN)1477-8947.

Qin, H., Wang, J., Chen, X., Wang, F., Peng, P., Zhou, Y., et al., 2019. Rice OsDOF15 contributes to ethylene-inhibited primary root elongation under salt stress. New Phytologist 223 (2), 798–813. Available at https://doi.org/10.1111/nph.15824, http://onlinelibrary.wiley.com/journal/10.1111/(ISSN)1469-8137.

Rai, K.K., 2022. Integrating speed breeding with artificial intelligence for developing climate-smart crops. Molecular Biology Reports 49 (12), 11385–11402. Available at https://doi.org/10.1007/s11033-022-07769-4, https://www.springer.com/journal/11033.

Rai, G.K., Khanday, D.M., Jamwal, G., Singh, M., 2024. CRISPR/Cas9 for enhancing resilience of crops to abiotic stresses. CRISPR and Plant Functional Genomics. CRC Press, India, pp. 133–143. Available at http://www.tandfebooks.com/doi/book/9781040015155.

Rasheed, A., Gill, R.A., Hassan, M.U., Mahmood, A., Qari, S., Zaman, Q.U., et al., 2021. A critical review: recent advancements in the use of CRISPR/Cas9 technology to enhance crops and alleviate global food crises. Current Issues in Molecular Biology 43 (3), 1950–1976. Available at https://doi.org/10.3390/cimb43030135, https://www.mdpi.com/1467-3045/43/3/135/pdf.

Raza, A., Razzaq, A., Mehmood, S.S., Hussain, M.A., Wei, S., He, H., et al., 2021. Omics: the way forward to enhance abiotic stress tolerance in *Brassica napus* L. GM Crops and Food 12 (1), 251–281. Available at https://doi.org/10.1080/21645698.2020.1859898, http://www.tandfonline.com/toc/kgmc20/current.

Razzaq, M.K., Akhter, M., Ahmad, R.M., Cheema, K.L., Hina, A., Karikari, B., et al., 2022. CRISPR-Cas9 based stress tolerance: new hope for abiotic stress tolerance in chickpea (*Cicer arietinum*). Molecular Biology Reports 49 (9), 8977–8985. Available at https://doi.org/10.1007/s11033-022-07391-4, https://www.springer.com/journal/11033.

Rengasamy, B., Manna, M., Thajuddin, N.B., Sathiyabama, M., Sinha, A.K., 2024. Breeding rice for yield improvement through CRISPR/Cas9 genome editing method: current technologies and examples. Physiology and Molecular Biology of Plants 30 (2), 185–198. Available at https://doi.org/10.1007/s12298-024-01423-y, https://www.springer.com/journal/12298.

Ribeiro, C.W., Korbes, A.P., Garighan, J.A., Jardim-Messeder, D., Carvalho, F.E.L., Sousa, R.H.V., et al., 2017. Rice peroxisomal ascorbate peroxidase knockdown affects ROS signaling and triggers early leaf senescence. Plant Science 263, 55–65. Available at https://doi.org/10.1016/j.plantsci.2017.07.009, http://www.journals.elsevier.com/plant-science/.

Robertson, G., Burger, J., Campa, M., 2022. CRISPR/Cas-based tools for the targeted control of plant viruses. Molecular Plant Pathology 23 (11), 1701–1718. Available at https://doi.org/10.1111/mpp.13252, http://onlinelibrary.wiley.com/journal/10.1111/(ISSN)1364-3703.

Roy, A., Purkaystha, S., Bhattacharyya, S., 2021. Advancement in molecular and fast breeding programs for climate-resilient agriculture practices. Genome Editing: A Tool from the Vault of Science for Engineering Climate-Resilient Cereals. Springer. Available at https://doi.org/10.1007/978-3-030-65912-7_4.

Ruelland, E., Vaultier, M.N., Zachowski, A., Hurry, V., 2009. Cold signalling and cold acclimation in plants. Advances in Botanical Research 49, 35–150. Available at https://doi.org/10.1016/S0065-2296(08)00602-2.

Saeglitz, C., Bartsch, D., 2003. Regulatory and associated political issues with respect to Bt transgenic maize in the European union. Journal of Invertebrate Pathology 83 (2), 107–109. Available at https://doi.org/10.1016/S0022-2011(03)00062-4, http://www.elsevier.com/inca/publications/store/6/2/2/8/8/3/index.htt.

Sagwal, V., Singla, A., Patwa, N., Gupta, O.P., 2025. CRISPR plants against heat stress: methods and applications. CRISPR for Climate-Smart and Sustainable Agriculture. Academic Press, pp. 165–175. Available at https://doi.org/10.1016/B978-0-443-24135-2.00008-6.

Samantara, K., Bohra, A., Mohapatra, S.R., Prihatini, R., Asibe, F., Singh, L., et al., 2022. Breeding more crops in less time: a perspective on speed breeding. Biology 11 (2). Available at https://doi.org/10.3390/biology11020275, https://www.mdpi.com/2079-7737/11/2/275/pdf.

Santosh Kumar, V.V., Verma, R.K., Yadav, S.K., Yadav, P., Watts, A., Rao, M.V., et al., 2020. CRISPR-Cas9 mediated genome editing of drought and salt tolerance (OsDST) gene in indica mega rice cultivar MTU1010. Physiology and Molecular Biology of Plants 26 (6), 1099–1110. Available at https://doi.org/10.1007/s12298-020-00819-w, http://www.springerlink.com/content/120962/.

Scheben, A., Wolter, F., Batley, J., Puchta, H., Edwards, D., 2017. Towards CRISPR /Cas crops – bringing together genomics and genome editing. New Phytologist 216 (3), 682–698. Available at https://doi.org/10.1111/nph.14702.

Schmidt, M.J., Gupta, A., Bednarski, C., Gehrig-Giannini, S., Richter, F., Pitzler, C., et al., 2021. Improved CRISPR genome editing using small highly active and specific engineered RNA-guided nucleases. Nature Communications 12 (1). Available at https://doi.org/10.1038/s41467-021-24454-5, http://www.nature.com/ncomms/index.html.

Shaheen, N., Ahmad, S., Alghamdi, S.S., Rehman, H.M., Javed, M.A., Tabassum, J., et al., 2023. CRISPR-Cas system, a possible "savior" of rice threatened by climate change: an updated review. Rice 16 (1). Available at https://doi.org/10.1186/s12284-023-00652-1, https://thericejournal.springeropen.com.

Sharma, A., Hazarika, M., Heisnam, P., Pandey, H., Devadas, V.A.S.N., Kesavan, A.K., et al., 2024. Controlled environment ecosystem: a cutting-edge technology in speed breeding. ACS Omega 9 (27), 29114–29138. Available at https://doi.org/10.1021/acsomega.3c09060, pubs.acs.org/journal/acsodf.

Sharma, M., Suma, R., Borgohain, T., Gogoi, D.J., Dutt, A., Chikkaputtaiah, C., et al., 2025. Multiomics approach to address alleviation of abiotic stress in root and tuber crops. Abiotic Stress in Underground Vegetables: Response and Mitigation. Elsevier, India, pp. 171–189. Available at https://www.sciencedirect.com/book/9780443239618.

Shelake, R.M., Kadam, U.S., Kumar, R., Pramanik, D., Singh, A.K., Kim, J.Y., 2022. Engineering drought and salinity tolerance traits in crops through CRISPR-mediated genome editing: targets, tools, challenges, and perspectives. Plant Communications 3 (6). Available at https://doi.org/10.1016/j.xplc.2022.100417, www.cell.com/plant-communications.

Shen, C., Que, Z., Xia, Y., Tang, N., Li, D., He, R., et al., 2017. Knock out of the annexin gene OsAnn3 via CRISPR/Cas9-mediated genome editing decreased cold tolerance in rice. Journal of Plant Biology 60 (6), 539–547. Available at https://doi.org/10.1007/s12374-016-0400-1, http://www.springer.com/life+sci/plant+sciences/journal/12374.

Sheng, H., Gao, P., Yang, C., Quilichini, T.D., Kochian, L.V., Datla, R., et al., 2025. Advances in genome editing through haploid induction systems. International Journal of Molecular Sciences 26 (10). Available at https://doi.org/10.3390/ijms26104779, http://www.mdpi.com/journal/ijms.

Shi, J., Gao, H., Wang, H., Lafitte, H.R., Archibald, R.L., Yang, M., et al., 2017. ARGOS 8 variants generated by CRISPR-Cas9 improve maize grain yield under field drought stress conditions. Plant Biotechnology Journal 15 (2), 207–216. Available at https://doi.org/10.1111/pbi.12603.

Shrivastava, P., Kumar, R., 2015. Soil salinity: a serious environmental issue and plant growth promoting bacteria as one of the tools for its alleviation. Saudi Journal of Biological Sciences 22 (2), 123–131. Available at https://doi.org/10.1016/j.sjbs.2014.12.001, http://www.ksu.edu.sa/sites/Colleges/CollegeofScinces/BotanyDepartment/SJBS/Default.aspx.

Shu, P., Li, Y., Li, Z., Sheng, J., Shen, L., 2022. SlMAPK3 enhances tolerance to salt stress in tomato plants by scavenging ROS accumulation and up-regulating the expression of ethylene signaling related genes. Environmental and Experimental Botany 193, 104698. Available at https://doi.org/10.1016/j.envexpbot.2021.104698.

Shu, P., Li, Y., Xiang, L., Sheng, J., Shen, L., 2023. SlNPR1 modulates chilling stress resistance in tomato plant by alleviating oxidative damage and affecting the synthesis of ferulic acid. Scientia Horticulturae 307, 111486. Available at https://doi.org/10.1016/j.scienta.2022.111486.

Shyamala, S.L., Sasipriya, S., 2023. Speed breeding- an emerging trend in crop improvement. International Journal of Plant & Soil Science 35 (22), 697–703. Available at https://doi.org/10.9734/ijpss/2023/v35i224180.

Singh, S., Mishra, V., Riyazuddin, R., Chugh, V., Upadhyay, S., 2024. CRISPR/Cas9-mediated genome editing for trait improvement and stress tolerance in Leguminosae (Legume Family). Plant Breeding. Available at https://doi.org/10.1111/pbr.13234, http://onlinelibrary.wiley.com/journal/10.1111/(ISSN)1439-0523.

Singha, D., Barik, D.P., Mondal, T.K., 2022. Strategies for multiplex genome editing in plants using CRISPR-based approaches. Plant Cell, Tissue and Organ Culture 151 (1), 1–14. Available at https://doi.org/10.3389/fpls.2021.72.

Song, Y., Liu, L., Feng, Y., Wei, Y., Yue, X., He, W., et al., 2015. Chilling- and freezing- induced alterations in cytosine methylation and its association with the cold tolerance of an alpine subnival plant, *Chorispora bungeana*. PLOS ONE 10 (8), e0135485. Available at https://doi.org/10.1371/journal.pone.0135485.

Tachikawa, M., Matsuo, M., 2024. Global regulatory trends of genome editing technology in agriculture and food. Japanese Society of Breeding, Japan Breeding Science 74 (1), 3–10. Available at https://doi.org/10.1270/jsbbs.23046, https://www.jstage.jst.go.jp/article/jsbbs/74/1/74_23046/_pdf.

Taku, M., Saini, M., Kumar, R., Debbarma, P., Rathod, N.K.K., Onteddu, R., et al., 2024. Modified speed breeding approach reduced breeding cycle to less than half in vegetable soybean [*Glycine max* (L.) Merr.]. Physiology and

Molecular Biology of Plants 30 (9), 1463–1473. Available at https://doi.org/10.1007/s12298-024-01503-z, https://www.springer.com/journal/12298.

Tang, W., Tang, A.Y., 2017. Applications and roles of the CRISPR system in genome editing of plants. Journal of Forestry Research 28 (1), 15–28. Available at https://doi.org/10.1007/s11676-016-0281-7, https://www.springer.com/journal/11676.

Tang, Q., Wang, X., Jin, X., Peng, J., Zhang, H., Wang, Y., 2023. CRISPR/Cas technology revolutionizes crop breeding. Plants 12 (17), 3119. Available at https://doi.org/10.3390/plants12173119.

Tariq, A., Mushtaq, M., Yaqoob, H., Bhat, B.A., 2023. CRISPR-Cas system in action: a golden window for efficient and precise genome editing. GM Crops & Food 14 (1), 45–62. Available at https://doi.org/10.1080/21645698.2023.2219111.

Távora, F.T.P.K., dos Santos Diniz, F.de A., de Moraes Rêgo-Machado, C., Chagas Freitas, N., Barbosa Monteiro Arraes, F., de Andrade, E.C., et al., 2022. CRISPR/Cas-and topical RNAi-based technologies for crop management and improvement: reviewing the risk assessment and challenges towards a more sustainable agriculture. Frontiers in Bioengineering and Biotechnology 10, 913728. Available at https://doi.org/10.3389/fbioe.2022.913728.

Tran, M.T., Son, G.H., Song, Y.J., Nguyen, N.T., Park, S., Thach, T.V., et al., 2023. CRISPR-Cas9-based precise engineering of SlHyPRP1 protein towards multi-stress tolerance in tomato. Frontiers in Plant Science 14. Available at https://doi.org/10.3389/fpls.2023.1186932, https://www.frontiersin.org/journals/plant-science.

Trono, D., Pecchioni, N., 2022. Candidate genes associated with abiotic stress response in plants as tools to engineer tolerance to drought, salinity and extreme temperatures in wheat: an overview. Plants 11 (23), 3358. Available at https://doi.org/10.3390/plants11233358.

Tsai, S.Q., Joung, J.K., 2016. Defining and improving the genome-wide specificities of CRISPR-Cas9 nucleases. Nature Reviews Genetics 17 (5), 300–312. Available at https://doi.org/10.1038/nrg.2016.28, http://www.nature.com/reviews/genetics.

Tubiello, F.N., Soussana, J.-F., Howden, S.M., 2007. Crop and pasture response to climate change. Proceedings of the National Academy of Sciences 104 (50), 19686–19690. Available at https://doi.org/10.1073/pnas.0701728104.

Union Agriculture Minister Shri Shivraj Singh Chouhan Announces Two Genome-Edited Rice Varieties Developed in India | ICAR, 2025. Available from: Icar.org.in. https://www.icar.org.in/en/union-agriculture-minister-shri-shivraj-singh-chouhan-announces-two-genome-edited-rice-varieties.

Usman, B., Nawaz, G., Zhao, N., Liao, S., Liu, Y., Li, R., 2020. Precise editing of the OsPYL9 gene by RNA-guided Cas9 nuclease confers enhanced drought tolerance and grain yield in rice (Oryza sativa L.) by regulating circadian rhythm and abiotic stress responsive proteins. International journal of molecular sciences 21 (21), 7854. Available at https://doi.org/10.3390/ijms21217854.

Varshney, R.K., Bohra, A., Yu, J., Graner, A., Zhang, Q., Sorrells, M.E., 2021. Designing future crops: genomics-assisted breeding comes of age. Trends in Plant Science 26 (6), 631–649. Available at https://doi.org/10.1016/j.tplants.2021.03.010, www.elsevier.com/inca/publications/store/3/0/9/6/0/index.htt.

Viviani, A., Spada, M., Giordani, T., Fambrini, M., Pugliesi, C., 2022. Origin of the genome editing systems: application for crop improvement. Biologia 77 (12), 3353–3383. Available at https://doi.org/10.1007/s11756-022-01142-3.

Vlčko, T., Ohnoutková, L., 2020. Allelic variants of CRISPR/Cas9 induced mutation in an inositol trisphosphate 5/6 kinase gene manifest different phenotypes in Barley. Plants 9 (2), 195. Available at https://doi.org/10.3390/plants9020195.

Wahab, A., Abdi, G., Saleem, M.H., Ali, B., Ullah, S., Shah, W., et al., 2022. Plants' physio-biochemical and phyto-hormonal responses to alleviate the adverse effects of drought stress: a comprehensive review. Plants 11 (13). Available at https://doi.org/10.3390/plants11131620, https://www.mdpi.com/2223-7747/11/13/1620/pdf?version=1655777519.

Waltz, E., 2018. With a free pass, CRISPR-edited plants reach market in record time. Nature Biotechnology 36 (1), 6–7. Available at https://doi.org/10.1038/nbt0118-6b.

Wan, L., Wang, Z., Tang, M., Hong, D., Sun, Y., Ren, J., et al., 2021. CRISPR-Cas9 gene editing for fruit and vegetable crops: strategies and prospects. Horticulturae 7 (7), 193. Available at https://doi.org/10.3390/horticulturae7070193.

Wang, J.W., Grandio, E.G., Newkirk, G.M., Demirer, G.S., Butrus, S., Giraldo, J.P., et al., 2019. Nanoparticle-mediated genetic engineering of plants. Molecular Plant 12 (8), 1037–1040. Available at https://doi.org/10.1016/j.molp.2019.06.010, https://www.cell.com/molecular-plant/home.

Wang, B., Zhong, Z., Wang, X., Han, X., Yu, D., Wang, C., et al., 2020. Knockout of the OsNAC006 transcription factor causes drought and heat sensitivity in rice. International Journal of Molecular Sciences 21 (7), 2288. Available at https://doi.org/10.3390/ijms21072288.

Wang, P., Luo, Q., Yang, W., Ahammed, G.J., Ding, S., Chen, X., et al., 2021. A novel role of pipecolic acid biosynthetic pathway in drought tolerance through the antioxidant system in tomato. Antioxidants 10 (12), 1923. Available at https://doi.org/10.3390/antiox10121923.

Wang, C., Zhou, Y., Wang, Y., Jiao, P., Liu, S., Guan, S., et al., 2025. CRISPR-Cas9-mediated editing of ZmPL1 gene improves tolerance to drought stress in maize. GM Crops & Food 16 (1), 1–16. Available at https://doi.org/10.1080/21645698.2024.2448869.

Wang, L., Chen, L., Li, R., Zhao, R., Yang, M., Sheng, J., et al., 2017. Reduced drought tolerance by CRISPR/Cas9-mediated SlMAPK3 mutagenesis in tomato plants. Journal of Agricultural and Food Chemistry 65 (39), 8674–8682. Available at https://doi.org/10.1021/acs.jafc.7b02745.

Watson, A., Ghosh, S., Williams, M.J., Cuddy, W.S., Simmonds, J., Rey, M.D., et al., 2018. Speed breeding is a powerful tool to accelerate crop research and breeding. Nature Plants 4 (1), 23–29. Available at https://doi.org/10.1038/s41477-017-0083-8, http://www.nature.com/nplants/.

Williams, K., Subramani, M., Lofton, L.W., Penney, M., Todd, A., Ozbay, G., 2024. Tools and techniques to accelerate crop breeding. Plants 13 (11), 1520. Available at https://doi.org/10.3390/plants13111520.

Wong, A.Y.T., Chan, A.W.K., 2016. Genetically modified foods in China and the United States: a primer of regulation and intellectual property protection. Food Science and Human Wellness 5 (3), 124–140. Available at https://doi.org/10.1016/j.fshw.2016.03.002, www.keaipublishing.com/en/journals/food-science-and-human-wellness/.

Wu, J., Yan, G., Duan, Z., Wang, Z., Kang, C., Guo, L., et al., 2020. Roles of the *Brassica napus* DELLA protein BnaA6.RGA, in modulating drought tolerance by interacting with the ABA signaling component BnaA10.ABF2. Frontiers in Plant Science 11. Available at https://doi.org/10.3389/fpls.2020.00577, https://www.frontiersin.org/journals/plant-science.

Yadav, S.K., 2010. Cold stress tolerance mechanisms in plants. A review. Agronomy for Sustainable Development 30 (3), 515–527. Available at https://doi.org/10.1051/agro/2009050.

Yadav, R.K., Tripathi, M.K., Tiwari, S., Tripathi, N., Asati, R., Chauhan, S., et al., 2023. Genome editing and improvement of abiotic stress tolerance in crop plants. Life 13 (7). Available at https://doi.org/10.3390/life13071456, www.mdpi.com/journal/life/.

Yang, X., Lu, M., Wang, Y., Wang, Y., Liu, Z., Chen, S., 2021. Response mechanism of plants to drought stress. Horticulturae 7 (3), 50. Available at https://doi.org/10.3390/horticulturae7030050.

Yao, Q., Shen, R., Shao, Y., Tian, Y., Han, P., Zhang, X., et al., 2024. Efficient and multiplex gene upregulation in plants through CRISPR-Cas-mediated knockin of enhancers. Molecular Plant 17 (9), 1472–1483. Available at https://doi.org/10.1016/j.molp.2024.07.009, https://www.sciencedirect.com/science/journal/16742052.

Yau, Y.Y., Easterling, M., Kumar, A., 2024. Advances in delivery of CRISPR–Cas reagents for precise genome editing in plants. Gene Editing in Plants: CRISPR-Cas and Its Applications. Springer Science+Business Media, United States, pp. 543–569. Available at https://link.springer.com/book/9789819985296.

Yasmeen, A., Shakoor, S., Azam, S., Bakhsh, A., Shahid, N., Latif, A., et al., 2022. CRISPR/Cas-mediated knockdown of vacuolar invertase gene expression lowers the cold-induced sweetening in potatoes. Planta 256 (6), 107. Available at https://doi.org/10.1007/s00425-022-04022-x.

Yin, W., Xiao, Y., Niu, M., Meng, W., Li, L., Zhang, X., et al., 2020. ARGONAUTE2 enhances grain length and salt tolerance by activating BIG GRAIN3 to modulate cytokinin distribution in rice. The Plant Cell 32 (7), 2292–2306. Available at https://doi.org/10.1105/tpc.19.00542.

Yu, W., Wang, L., Zhao, R., Sheng, J., Zhang, S., Li, R., et al., 2019. Knockout of SlMAPK3 enhances tolerance to heat stress involving ROS homeostasis in tomato plants. BMC Plant Biology 19 (1). Available at https://doi.org/10.1186/s12870-019-1939-z, http://www.biomedcentral.com/bmcplantbiol/.

Zafar, S.A., Zaidi, S.S.E.A., Gaba, Y., Singla-Pareek, S.L., Dhankher, O.P., Li, X., et al., 2020. Engineering abiotic stress tolerance via CRISPR/Cas-mediated genome editing. Journal of Experimental Botany 71 (2), 470–479. Available at https://doi.org/10.1093/jxb/erz476, http://jxb.oxfordjournals.org/.

Zang, Q.W., Wang, C.X., Li, X.Y., Guo, Z.A., Jing, R.L., Zhao, J., et al., 2010. Isolation and characterization of a gene encoding a polyethylene glycol-induced cysteine protease in common wheat. Journal of Biosciences 35 (3), 379–388. Available at https://doi.org/10.1007/s12038-010-0043-1.

Zeng, D.D., Yang, C.C., Qin, R., Alamin, M., Yue, E.K., Jin, X.L., et al., 2018. A guanine insert in OsBBS1 leads to early leaf senescence and salt stress sensitivity in rice (*Oryza sativa* L.). Plant Cell Reports 37 (6), 933–946. Available at https://doi.org/10.1007/s00299-018-2280-y, link.springer.de/link/service/journals/00299/index.htm.

Zhang, H., Zhang, J., Wei, P., Zhang, B., Gou, F., Feng, Z., et al., 2014. The CRISPR/Cas9 system produces specific and homozygous targeted gene editing in rice in one generation. Plant Biotechnology Journal 12 (6), 797–807. Available at https://doi.org/10.1111/pbi.12200, http://www.wiley.com/bw/journal.asp?ref=1467-7644.

Zhang, X.H., Tee, L.Y., Wang, X.G., Huang, Q.S., Yang, S.H., 2015. Off-target effects in CRISPR/Cas9-mediated genome engineering. Molecular Therapy Nucleic Acids 4 (11), e264. Available at https://doi.org/10.1038/mtna.2015.37, https://www.journals.elsevier.com/molecular-therapy-nucleic-acids.

Zhang, S., Wang, L., Zhao, R., Yu, W., Li, R., Li, Y., et al., 2018. Knockout of SlMAPK3 reduced disease resistance to *Botrytis cinerea* in tomato plants. Journal of Agricultural and Food Chemistry 66 (34), 8949–8956. Available at https://doi.org/10.1021/acs.jafc.8b02191.

Zhang, A., Liu, Y., Wang, F., Li, T., Chen, Z., Kong, D., et al., 2019. Enhanced rice salinity tolerance via CRISPR/Cas9-targeted mutagenesis of the OsRR22 gene. Molecular Breeding 39 (3). Available at https://doi.org/10.1007/s11032-019-0954-y, https://www.springer.com/journal/11032.

Zhang, X., Long, Y., Huang, J., Xia, J., 2020. OsNAC45 is involved in ABA response and salt tolerance in rice. Rice 13 (1). Available at https://doi.org/10.1186/s12284-020-00440-1, http://www.springerlink.com/content/1939-8425.

Zhang, D., Zhang, Z., Unver, T., Zhang, B., 2021a. CRISPR/Cas: a powerful tool for gene function study and crop improvement. Journal of Advanced Research 29, 207–221. Available at https://doi.org/10.1016/j.jare.2020.10.003, http://www.elsevier.com/wps/find/journaldescription.cws_home/722881/description#description.

Zhang, Y., Restall, J., Crisp, P., Godwin, I., Liu, G., 2021b. Current status and prospects of plant genome editing in Australia. In Vitro Cellular & Developmental Biology-Plant 57 (4), 574–583. Available at https://doi.org/10.1007/s11627-021-10188-y.

Zhao, J., Liu, P., Li, C., Wang, Y., Guo, L., Jiang, G., et al., 2017. LMM5. 1 and LMM5. 4, two eukaryotic translation elongation factor 1A-like gene family members, negatively affect cell death and disease resistance in rice. Journal of Genetics and Genomics 44 (2), 107–118. Available at https://doi.org/10.1016/j.jgg.2016.12.005.

Zhao, J., Yao, B., Peng, Z., Yang, X., Li, K., Zhang, X., et al., 2025. Splicing defect of StDRO2 intron 1 promotes potato root growth by disturbing auxin transport to adapt to drought stress. Horticultural Plant Journal 11 (2), 706–720. Available at https://doi.org/10.1016/j.hpj.2023.11.003.

Zheng, M., Lin, J., Liu, X., Chu, W., Li, J., Gao, Y., et al., 2021. Histone acetyltransferase TaHAG1 acts as a crucial regulator to strengthen salt tolerance of hexaploid wheat. Plant Physiology 186 (4), 1951–1969. Available at https://doi.org/10.1093/plphys/kiab187.

Zheng, Z., Gao, S., Wang, H., Liu, C., 2023. Shortening generation times for winter cereals by vernalizing seedlings from young embryos at 10 degree Celsius. Plant Breeding 142 (2), 202–210. Available at https://doi.org/10.1111/pbr.13074.

Zhong, X., Hong, W., Shu, Y., Li, J., Liu, L., Chen, X., et al., 2022. CRISPR/Cas9 mediated gene-editing of GmHdz4 transcription factor enhances drought tolerance in soybean (Glycine max [L.] Merr.). Frontiers in Plant Science 13, 988505. Available at https://doi.org/10.3389/fpls.2022.988505.

Zhou, Z., Xiao, J., Yin, S., Chen, Y., Yuan, Y., Zhang, J., et al., 2025. Cas9-Rep fusion tethers donor DNA in vivo and boosts the efficiency of HDR-mediated genome editing. Plant Biotechnology Journal 23 (6), 2006–2017. Available at https://doi.org/10.1111/pbi.70036, http://onlinelibrary.wiley.com/journal/10.1111/(ISSN)1467-7652.

Zhu, H., Li, C., Gao, C., 2020. Applications of CRISPR–Cas in agriculture and plant biotechnology. Nature Reviews Molecular Cell Biology 21 (11), 661–677. Available at https://doi.org/10.1038/s41580-020-00288-9.

Further reading

Jung, C., Seo, J.S., Han, S.W., Koo, Y.J., Kim, C.H., Song, S.I., et al., 2008. Overexpression of AtMYB44 enhances stomatal closure to confer abiotic stress tolerance in transgenic *Arabidopsis*. Plant Physiology 146 (2), 623. Available at https://doi.org/10.1104/pp.107.110981.

Wanga, M.A., Shimelis, H., Mashilo, J., Laing, M.D., 2021. Opportunities and challenges of speed breeding: a review. Plant Breeding 140 (2), 185–194. Available at https://doi.org/10.1111/pbr.12909.

Wu, J., Yang, R., Yang, Z., Yao, S., Zhao, S., Wang, Y., et al., 2017. ROS accumulation and antiviral defence control by microRNA528 in rice. Nature Plants 3 (1), 1–7. Available at https://doi.org/10.1038/nplants.2016.203.

Index

Note: Page numbers followed by "*f*" indicate figures and "*t*" indicate tables

Printed and bound by CPI Group (UK) Ltd, Croydon, CR0 4YY
14/00/2026
02216409-0011